Understanding Renewable Energy Systems

By the middle of this century, renewable energy must provide all of our energy supply if we are to phase out nuclear power and successfully stop climate change.

Now updated and expanded, the second edition of this textbook covers the full range of renewable energy systems and also includes such current trends as solar power storage, power-to-gas technologies, and the technology paths needed for a successful and complete energy transition. The topics are treated in a holistic manner, bringing together maths, engineering, climate studies, and economics, and enabling readers to gain a broad understanding of renewable energy technologies and their potential. Numerous examples are provided for calculations, and graphics help visualize the various technologies and mathematical methodologies. *Understanding Renewable Energy Systems* is an ideal companion for students of renewable energy at universities or technical colleges – on courses such as renewable energy, electrical engineering, engineering technology, physics, process engineering, building engineering, environment, applied mechanics, and mechanical engineering – as well as scientists and engineers in research and industry.

Volker Quaschning currently teaches and conducts research in the Renewable Energy Systems Department at Berlin's University of Applied Sciences (HTW), Germany. He has many years of experience in various fields of renewable energy, both within Germany and abroad.

Understanding Renewable Energy Systems

Second edition

Volker Quaschning

First edition published 2005
by Earthscan

This revised edition published 2016
by Routledge

2 Park Square, Milton Park, Abingdon, Oxon, OX14 4RN
and by Routledge
711 Third Avenue, New York, NY 10017

Routledge is an imprint of the Taylor & Francis Group, an informa business

This edition is an authorized translation of *Regenerative Energiesysteme 8.A.* published in German © 2013 Carl Hanser Verlag, Munich/FRG

British Library Cataloguing in Publication Data
A catalogue record for this book is available from the British Library

Library of Congress Cataloging-in-Publication Data
Quaschning, Volker, 1969– author.
[Regenerative Energiesysteme. English] Understanding renewable energy systems / Volker Quaschning. — Revised edition.
pages cm
Translation of: Regenerative Energiesysteme, 8.A. published in German, ©2013 Carl Hanser Verlag, Munich/FRG.
Includes bibliographical references and index.
1. Renewable energy sources. I. Title.
TJ808.Q3713 2016
333.79'4—dc23
2015027156

ISBN: 978-1-138-78194-8 (hbk)
ISBN: 978-1-138-78196-2 (pbk)
ISBN: 978-1-315-76943-1 (ebk)

Typeset in Bembo
by Apex CoVantage, LLC

Contents

Figures

Tables

1 Energy and climate protection

The term 'energy'

We hear the word 'energy' quite often without thinking about it much. It is used in a lot of different contexts. For instance, the term can be used in the sense of vitality, and we say that people have 'a lot of energy' to describe their temperament.

This book only deals with forms of – especially renewable – energy produced by technology, and they are described in terms of the laws of physics. Energy and power are two nearly inseparable terms. But because they are often confused, a clarification of the distinction between the two and other related terms is a good starting point.

In general, energy describes a system's ability to produce external effects, such as force along a path. A body's energy can change by taking up or exerting work. Here, energy occurs in a wide range of forms, including:

- mechanical energy
- potential energy
- kinetic energy
- heat or thermal energy
- magnetic energy
- mass
- electrical energy
- radiant energy
- and chemical energy.

In the definitions above, a litre of gasoline is a type of stored energy; when it is combusted, the force of an engine can move a car (having a certain mass) for a certain distance. The motion of the car is then a kind of work.

Heat is also a type of energy. We see its effects when the hot air rising from a candle moves the parts of a hanging mobile. A force is also needed for this motion.

The wind also contains energy, which can be used, say, to turn the blades of a wind turbine. The sun's rays can be used to generate heat. Radiation, especially solar radiation, is another form of energy.

Power is defined as the amount of work W done over time t.

$$P = \frac{\mathrm{d}W}{\mathrm{d}t} = \dot{W} \tag{1.1}$$

It therefore indicates how much time it took to perform a certain amount of work. For instance, work can be a person lifting a sack of cement up one metre. The work performed increases the sack's potential energy. If the sack is lifted twice as quickly, less time is needed, but the amount of power doubles although the same amount of work is performed.

Energy and work are expressed in **units** derived from the SI units of J (joules), Ws (watt-seconds), and Nm (newton metres). Power is measured in W (watts). Table 1.1 shows the conversion factors for the main units used today for energy equipment. In addition, there are a number of antiquated energy units, such as kilopond metres, kpm ($1\ \text{kpm} = 2.72 \cdot 10^{-6}$ kWh); erg ($1\ \text{erg} = 2.78 \cdot 10^{-14}$ kWh); electron volts (still used in physics), eV ($1\ \text{eV} = 4.45 \cdot 10^{-26}$ kWh); and (still common in the US) Btus (British Thermal Unit, 1 Btu = 1055.06 J = 0.000293071 kWh).

Because values can often be very small or very large and exponential indications are hard to follow, a number of prefixes are used as shown in Table 1.2.

Often, terms describing energy and power are confused along with the units defining them. When the wrong sizes are used, meanings change, leading to misunderstandings in the best case.

For example, let's take a newspaper article from the 1990s about a house with a solar roof. The article describes a photovoltaic system with a total output of 2.2 kW. The author then says that the price paid per kW of electricity sold to the grid was very low at 0.087 Deutschemarks, equivalent to roughly 0.05 cents today. But if the system were paid based on its power (kW represents the installed capacity), the entire system would receive 2.2 kW ×€0.087/kW = €0.19. The compensation for solar power was inadequate for a long time, to be sure, but probably no system owner had to make do with less than €0.20. Here,

Table 1.1 Conversion factors for various energy units

	kJ	*kcal*	*kWh*	*kg oe*	*m³ natural gas*
1 kilojoule (1 kJ = 1,000 W)	1	0.2388	0.000278	0.000024	0.000032
1 kilocalorie (kcal)	4.1868	1	0.001163	0.0001	0.00013
1 kilowatt-hour (kWh)	3,600	860	1	0.086	0.113
1 kg crude oil equivalent (oe)	41,868	10,000	11.63	1	1.319
1 m³ natural gas	31,736	7,580	8.816	0.758	1

Table 1.2 Prefixes and prefix abbreviations

Prefix	*Abbreviation*	*Value*	*Prefix*	*Abbreviation*	*Value*
Kilo	k	10^3 (thousand)	Milli	m	10^{-3} (thousandth)
Mega	M	10^6 (million)	Micro	μ	10^{-6} (millionth)
Giga	G	10^9 (billion)	Nano	n	10^{-9} (billionth)
Tera	T	10^{12} (trillion)	Piko	p	10^{-12} (trillionth)
Peta	P	10^{15} (quadrillion)	Femto	f	10^{-15} (quadrillionth)
Exa	E	10^{18} (quintillion)	Atto	a	10^{-18} (quintillionth)

the author was actually speaking of the electric energy sold to the grid and compensated at a rate of €0.087 per kilowatt-hour (kWh). If the system produced 1,650 kWh to sell to the grid, the owner would be paid €143.55 – 750 times more. This anecdote shows what a big difference a little 'h' can have if it is missing.

According to the laws of physics, energy cannot be generated, destroyed, or lost. Nonetheless, we speak of energy losses, energy generation, and energy production – despite the **law of conservation of energy,** which states:

Within an isolated system, total energy remains constant. Energy cannot be destroyed or created out of nothing; it can, however, be converted into various forms and exchanged between various parts of the system.

To see how energy can be converted from one form into another, let's go back to the example of gasoline and cars. Gasoline is a type of stored chemical energy. When it is combusted, thermal energy is created. The motor turns this thermal energy into kinetic energy, which is passed on to the car. When all of the gasoline has been consumed, the car stops. The energy has not, however, disappeared; rather, it was converted into potential energy (to the extent there is a difference in altitude), into the motor's waste heat, into friction (the tires), and given off as heat to the surroundings. We cannot, however, make use of this ambient heat in general. While driving, we converted most of the useful energy contained in the gasoline into ambient heat, which cannot be used. From our perspective, this energy is lost. When we speak of 'lost' or 'consumed' energy, we mean that we have converted high-quality energy into energy of lower, generally not usable quality.

The situation is different with photovoltaics. It converts sunlight directly into electrical energy. In common parlance, we say that the solar energy system 'generates electricity'. This description is incorrect within the laws of physics. It would be correct to say that the photovoltaic system converts a form of energy we cannot easily use (solar radiation) into higher-quality energy (electricity).

In this conversion, the energy can be used at different levels of efficiency. To illustrate how this works, let's take the example of boiling water.

The **thermal energy** Q needed to heat up a litre of water (m = 1 kg) from temperature ϑ_1 = 15 °C to ϑ_2 = 98 °C is relative to the heat capacity c of water c_{H_2O} = 4.187 kJ/(kg K), giving us

$$Q = c \cdot m \cdot (\vartheta_2 - \vartheta_1) \tag{1.2}$$

to produce Q = 348 kJ = 97 Wh.

A consumer magazine compared different water-heating appliances. The results are shown in Figure 1.1. In addition to various electrical appliances, gas stoves were also included. The comparison shows that gas stoves are the worst in terms of energy consumption even though its energy costs are the lowest. This difference comes about because we are comparing different types of energy here.

The electric stove uses electric energy to heat up the water. Electricity is quite rare in nature; natural occurrences include lightning and eels that use electrical charges to stun attackers. Electricity therefore first has to be made from another source of energy, such as coal in a power plant. Once again, an enormous amount of waste heat is created in the process and passed on to the ambient environment. In other words, only a small part of the

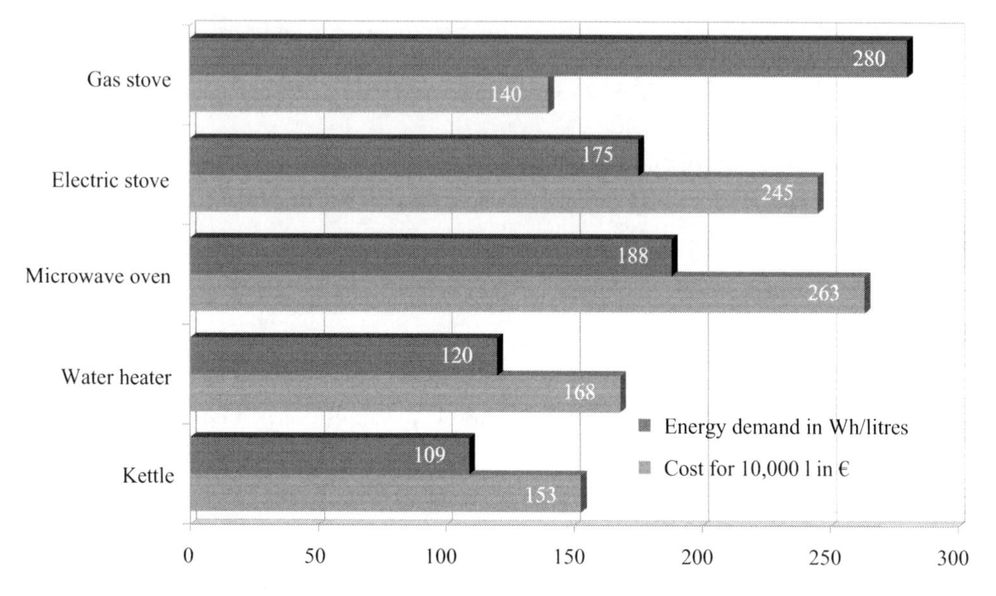

Figure 1.1 'What it costs to boil water' from 1994 [Sti94]

energy in the coal is converted into electrical energy, with the rest lost as waste heat. The quality of the conversion is referred to as **efficiency** η and is defined as follows:

$$\text{Efficiency} \quad \eta = \frac{\text{useful energy produced}}{\text{energy input}} \tag{1.3}$$

The average efficiency of a thermal power plant in Germany in the 1990s was around 34 % [Hof95]. The efficiency of modern power plants is slightly higher. Nonetheless, roughly 60 % of the energy consumed is nonetheless lost as waste heat when only 40 % comes out of the plant as electrical energy.

When energy technologies are used, a distinction therefore needs to be made between various **stages of energy conversion**: primary energy, final energy, and useful energy, as shown in Table 1.3.

The previously calculated amount of heat thus represents useful energy, while the numbers in Figure 1.1 are final energy. But when we compare the efficiency of gas and

Table 1.3 Primary energy, final energy, and useful energy

Term	*Definition*	*Energy type or source*
Primary energy	Energy in its original form, not technically processed	Such as crude oil, coal, uranium, solar radiation, and wind
Final energy	Energy as provided to consumers	Such as natural gas, heating oil, fuel, electricity, and district heat
Useful energy	Energy as it is consumed	Such as artificial lighting, heat, drive energy for machines and vehicles

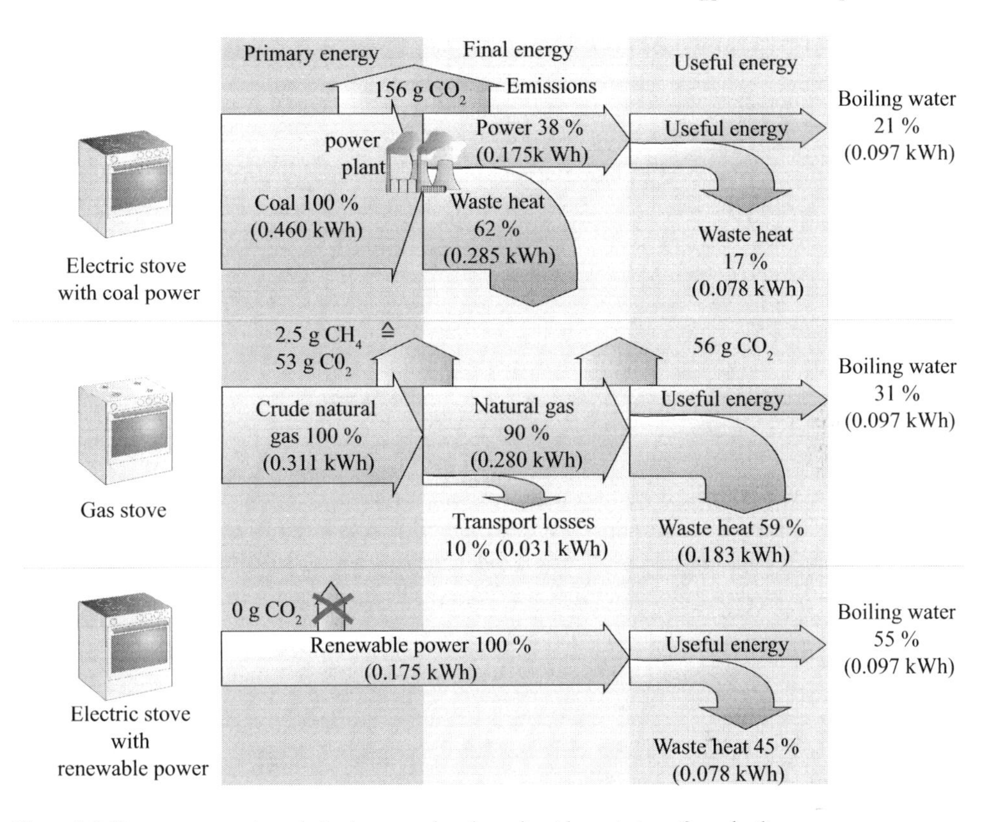

Figure 1.2 Energy conversion chain, losses and carbon dioxide emissions from boiling water

electricity, we should start with primary energy because both of them are quite different forms of final energy.

For electricity, we need to start with the source of energy fed into the plant, such as coal. Natural gas used to heat water is also final energy. When natural gas is transported to consumers, there are also losses, though they are minor in comparison to a power plant generating electricity. We then find that the electric stove's primary energy consumption is 460 Wh = 1,656 kJ, some 50 % greater than the gas stove's consumption although final energy consumption is more than 30 % lower. Figure 1.2 shows another comparison of the energy conversion chains based on the example of water heated by an electric stove and a gas stove.

Clearly, gas stoves are better for water heating than electrical stoves if the electricity is made with conventional fuels in terms of primary energy consumption, the decisive factor for environmental impacts. This example illustrates how important the distinction between primary energy, final energy, and useful energy is. Otherwise, we draw incorrect conclusions, as the comparison of gas and electric stoves in Figure 1.1 shows.

Trends in energy demand

Trends in global energy demand

At the end of the eighteenth century, crude oil and coal were still marginal sources of energy. Wood still covered most of the energy demand for heat. Progress had, however, been made in the use of water and wind. Both of them were used in mills and pump water.

In 1769, James Watt invented a useful steam engine, thereby opening the door to industrialization. Gradually, steam engines – and later, combustion engines – replaced wind and water mills. Coal became the biggest source of energy, followed by crude oil starting at the beginning of the twentieth century as automobiles became more common. As a source of energy, firewood increasingly became marginal in industrialized countries. And whereas traditional water mills fit well into landscapes, modern hydropower became giant technical facilities.

After the Great Depression of 1929, energy consumption rose dramatically. After the Second World War, consumption of natural gas increased, followed by nuclear power starting in the 1960s, though crude oil and coal remained dominant. The share of nuclear power in primary energy demand is still relatively small even today. Fossil energy – coal, crude oil, natural gas, etc. – currently cover some 85 % of global demand for primary energy.

Figure 1.3 shows the trend in global crude oil consumption; 1 million tons of crude oil is equivalent to 42 PJ = $42 \cdot 10^{15}$ J. The trend is indicative of overall energy consumption. After the Second World War, production volumes rose exponentially. As a result of the two oil price crises in 1973 and 1979, production volumes dropped noticeably, but only temporarily. Overall economic growth and the increase in energy consumption were set back by around four years as a result.

Table 1.4 shows **global primary energy consumption** by energy carrier for different years. We should keep in mind, however, that energy statistics for primary electricity – such

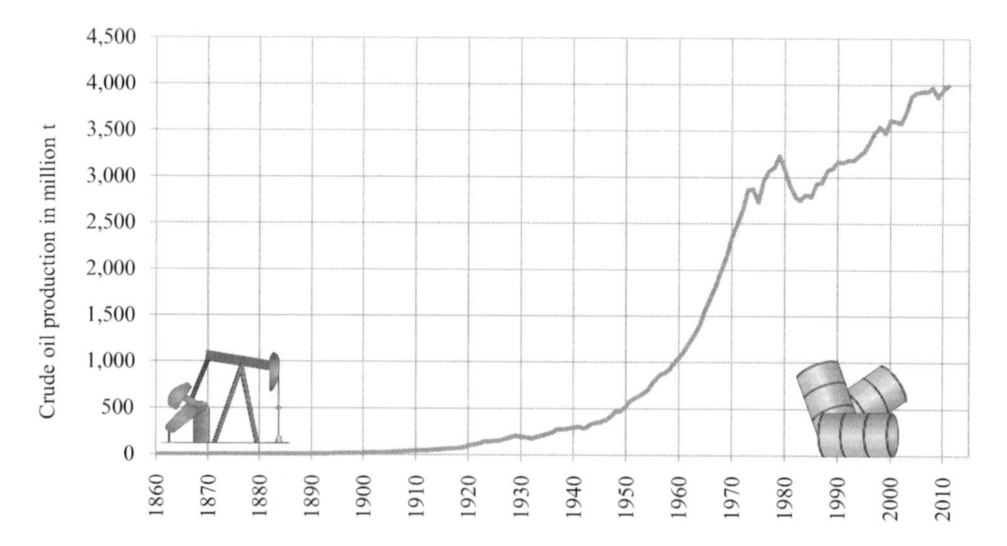

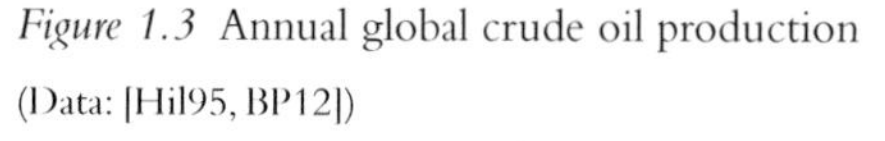

Figure 1.3 Annual global crude oil production

(Data: [Hil95, BP12])

Table 1.4 Global primary energy consumption without biomass and 'other' [Enq95, BP12]

PJ	1925	1938	1950	1960	1980	1995	2011
Solid fuels[1)]	36,039	37,856	46,675	58,541	77,118	94,973	155,856
Solid fuels[2)]	5,772	11,017	21,155	43,921	117,112	136,666	169,864
Natural gas	1,406	2,930	7,384	17,961	53,736	81,056	121,594
Hydropower[3)]	771	1,774	3,316	6,632	16,732	23,873	33,121
Nuclear power[3)]	0	0	0	0	6,741	22,027	25,081
Total	43,988	53,577	76,473	127,055	271,439	358,595	505,516

1) Lignite, hard coal, etc. 2) Petroleum products 3) Weighted with an efficiency of 38 %

as hydropower and nuclear power – are sometimes based on different calculations. Usually, nuclear plants are assumed to have an efficiency of 33 to 38 % in such statistics in line with the usual efficiencies for fossil power plants. If nuclear and hydropower are then compared without adjusting for this effect, the share of nuclear in global power demand seems to be much greater than the share of hydropower although the opposite is true.

Table 1.4 does not contain other energy sources, such as biomass (firewood and plant residue), wind power, and solar energy, which collectively covered some 50,000 PJ of primary energy consumption in 2007.

Over the next few years, global energy consumption will continue to grow strongly. In industrialized countries, energy consumption will not grow as quickly as it will in numerous emerging economies, which have a lot of catching up to do in terms of economic growth. Furthermore, the global population will continue to grow considerably over the next few decades. It is therefore realistic to expect energy demand to increase by a factor of 3 to 6 by the end of the century. The problems of our current energy supply will only worsen as a result, as will the consequences of the greenhouse effect; likewise, fossil fuel reserves will only be used up faster.

Figure 1.4 shows that energy demand varies greatly across the planet. Europe, Asia, and North America make up a large share of primary energy demand, though the population of Asia is six times greater than in Europe – and 10 times greater than in North America. Despite their large populations, the relatively underdeveloped countries of South America and Africa make up only a small share of global primary energy consumption today. Below, we will have an opportunity to discuss the unequal distribution of energy consumption in the context of per capita carbon emissions, which are closely related to energy consumption.

Trends in energy demand in Germany

Up to the end of the 1970s, energy demand in Germany rose constantly; it was assumed that economic growth and energy consumption are closely related. The oil crises of the 1970s and 1980s changed this thinking and people's behaviour. Energy conservation became more fashionable, and when cars were banned from the autobahns on Sundays, the empty roads were a clear sign of dependence on fossil fuels. People began to seriously investigate ways to increase the use of renewable energy. Lower oil prices calm down energy markets

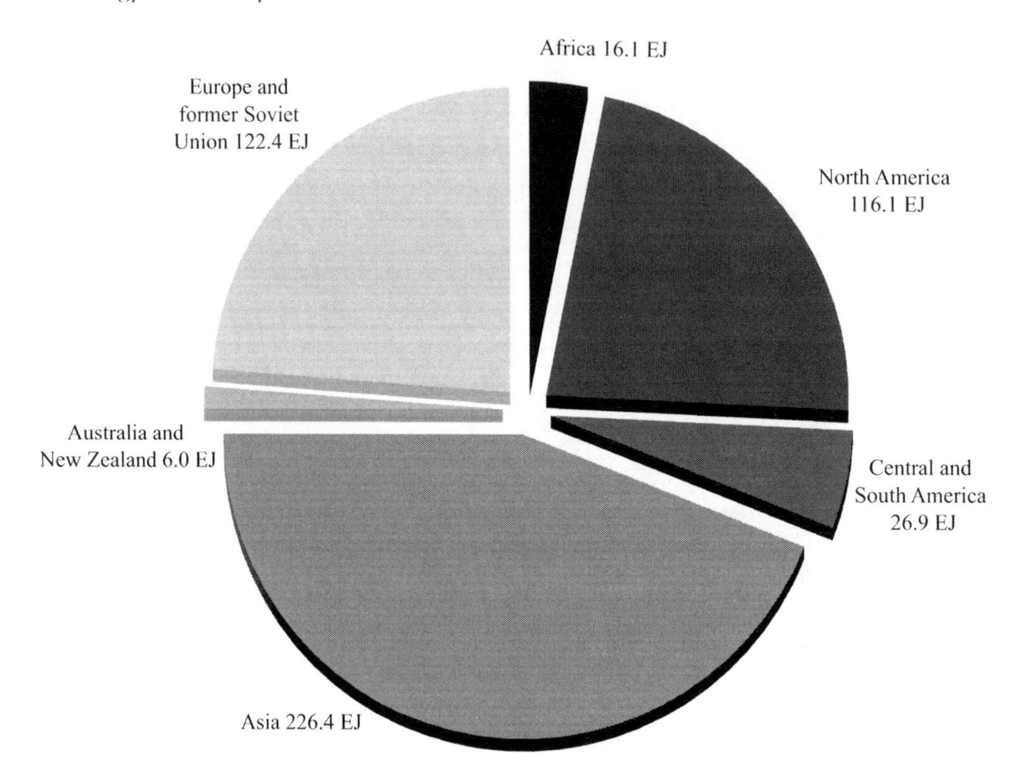

Figure 1.4 Global primary energy consumption in 2011 by region
(Data: [BP12])

eventually, so these new approaches once again faded into the background, and people went back to their old wasteful consumption of energy resources. In the new millennium, oil prices once again began to skyrocket, and growing public interest in climate protection opened up new perspectives for energy supply. Nonetheless, many obstacles – a lot of them unnecessary – remain on the path towards restructuring energy supply.

But since the beginning of the 1980s, a few details have changed fundamentally. Economic growth continues unabated, but energy consumption has stagnated at a high level; a growing number of people now realize that energy consumption and gross national product are not necessarily related – greater prosperity is possible even if energy demand stagnates or decreases.

Other events also significantly affected energy consumption in the 1980s and 1990s. Despite growing opposition, more nuclear plants continued to be built, eventually reducing excess power generation capacity as power consumption fell far short of expectations; the result was lower coal consumption. In 1986, the accident at the nuclear plant in Chernobyl, Ukraine, spelled doom for popular support for nuclear energy. New nuclear plants were off the table, and the share of nuclear power in primary energy demand remained flat in Germany for a long time at around 10 %. After the accident in Fukushima, Germany once again resolved to phase out nuclear by 2022, and the share of nuclear power has decreased since.

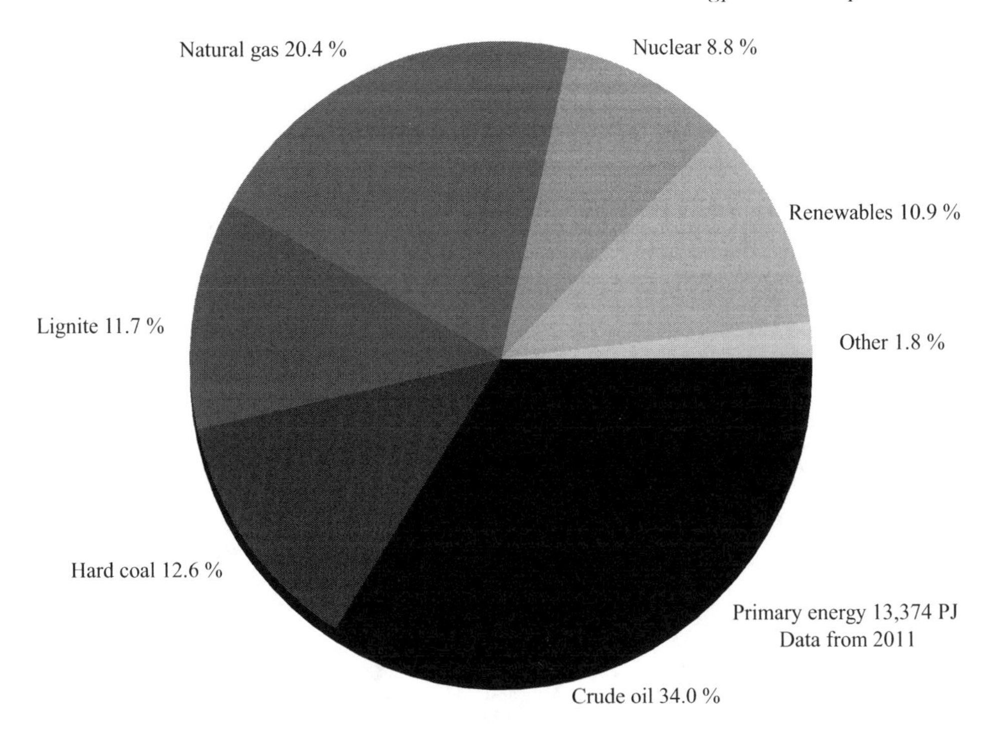

Figure 1.5 Shares of various energy sources in primary energy consumption in Germany in 2011 (Data: [AGEB12])

After the fall of the Berlin Wall and reunification, large sectors of industry in the former East Germany were shut down. The result was lower total energy consumption in Germany; without this one-off effect, consumption would have increased. Lignite usage fell considerably, but there was also great pressure on hard coal because of cost. Energy demand shifted to other energy sources: natural gas and renewables, such as wind power, solar energy, and biomass.

Nonetheless, the German energy sector still largely focuses on the use of fossil energy. In 2011, it still made up around 79 % of primary energy demand (Figure 1.5). In the meantime, however, renewables have grown considerably. For a long time, they were neglected. Aside from hydropower, they were often listed under 'other' if mentioned at all.

In terms of the **structure of energy consumption** in Germany, there have only been minor shifts up to now. Roughly, the sectors of industry, households, and transport each consume roughly the same amount of energy.

The growth of renewables is clear to see within the German power sector. Figure 1.7 shows the considerable growth. Over the past 20 years, production of renewable electricity has risen more than fivefold. If the boom in renewables continues unabated, Germany will have 90 to 100 % renewable electricity by 2040.

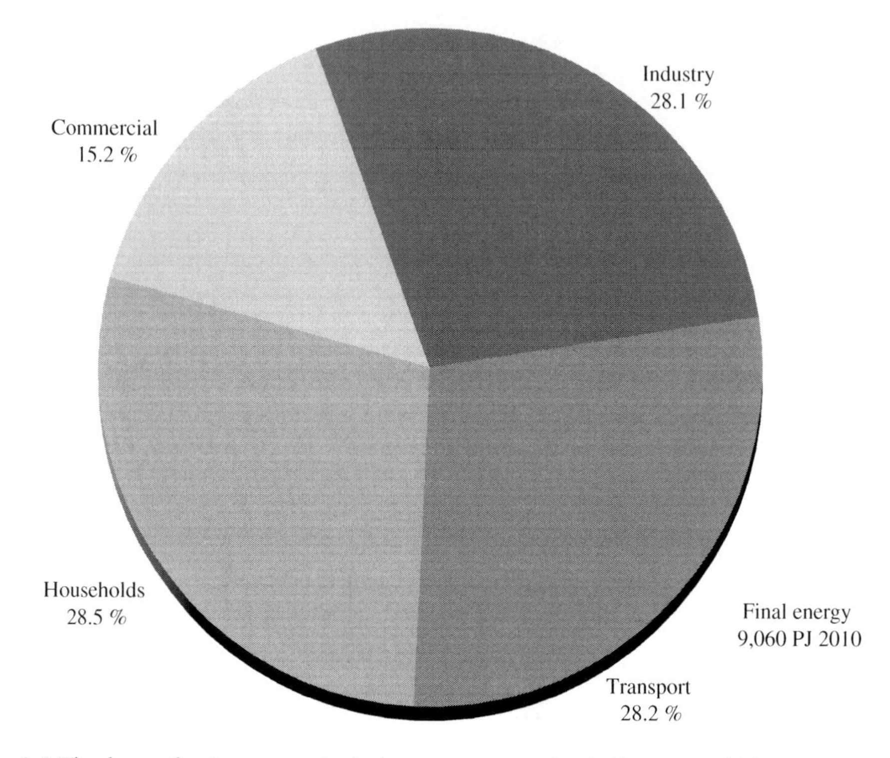

Figure 1.6 The shares of various sectors in final energy consumption in Germany in 2010 (Data: [BMWi])

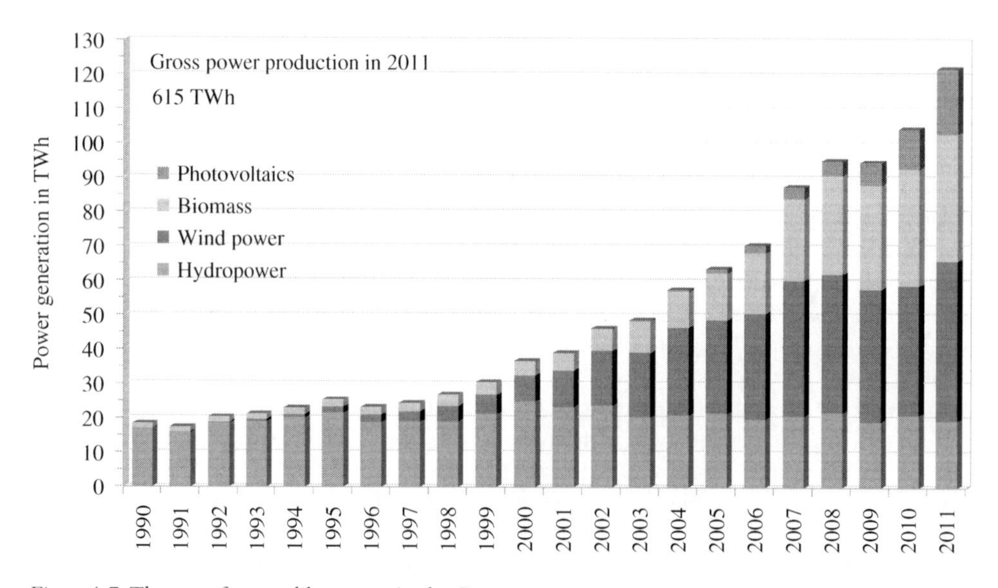

Figure 1.7 The use of renewable energy in the German power sector

The range of conventional energy sources

As explained above, our current energy supply still largely comes from fossil sources. Natural gas, crude oil, hard coal, and lignite took millennia to build up many millennia ago. They mainly consist of plant and animal matter and are thus stored biomass from long ago. Over the past 100 years, we have consumed a lot of these fossil resources. As more fossil reserves are tapped, exploration becomes more difficult, more technically complex, riskier, and as a result more expensive. If we continue to consume fossil energy at the current rate or even faster, all economically exploitable oil and gas reserves will have been used up by the end of the twenty-first century, with only coal reserves being available a bit longer (Table 1.5). In other words, the fossil resources that took millions of years to build up will have been completely consumed in only a few generations. Future generations will have to do without these energy sources.

It is hard to determine the exact fossil fuel reserves reported today as they are based on areas already explored. We can only roughly estimate how much will be discovered. But even if large new resources are found, fossil energy will remain finite. We can only extend the range by a few years.

When we talk about reserves, we mean that these raw materials whose exploitation has been demonstrated to be technically affordable through explorations, test drills, and measurements. In addition, there are resources, raw materials, that have not been proven to be recoverable at feasible prices; it is therefore risky to count on them. The statistical range is the product of an energy source's reserves divided by current annual consumption. If energy consumption increases, the range shortens; if additional resources become exploitable, the range is extended.

The planet's uranium reserves needed to fuel nuclear plants are also limited. Global estimated resources come in at around 15 million tons, 10 million tons of which have yet to be discovered – and are purely speculative (Table 1.6). Currently, nuclear power only meets around 5 % of primary energy demand worldwide. If nuclear were to cover all of primary energy demand, proven, affordable reserves would only last for three years. Breeder reactors would increase the range a bit, but nuclear fission would still be based on limited resources and is therefore not an alternative to fossil fuels.

In light of the limited reserves of those resources that conventional energy technologies require, little of the technology used today will survive the twenty-first century. We need to start making our energy sector ready for this transition today. There are a lot of reasons why

Table 1.5 Fossil energy reserves in 2010

	Crude oil	*Natural gas*	*Hard coal*
Recoverable reserves	216.9 billion t	192.1 trillion m^3	728.0 billion t
Production in 2010	4.044 billion t[1)]	3,282 trillion m^3	6.02 billion t
Range at current rates of production	54 years	59 years	121 years
Additional resources	298.1 billion t[1)]	530.5 trillion m^3 [1)]	17 204 billion t[2)]
Cumulative production	162.7 billion t	96.1 trillion m^3	n/a

(Data: [BMWi])

1) Conventional and non-conventional, including oil sands and shale gas
2) Total resources

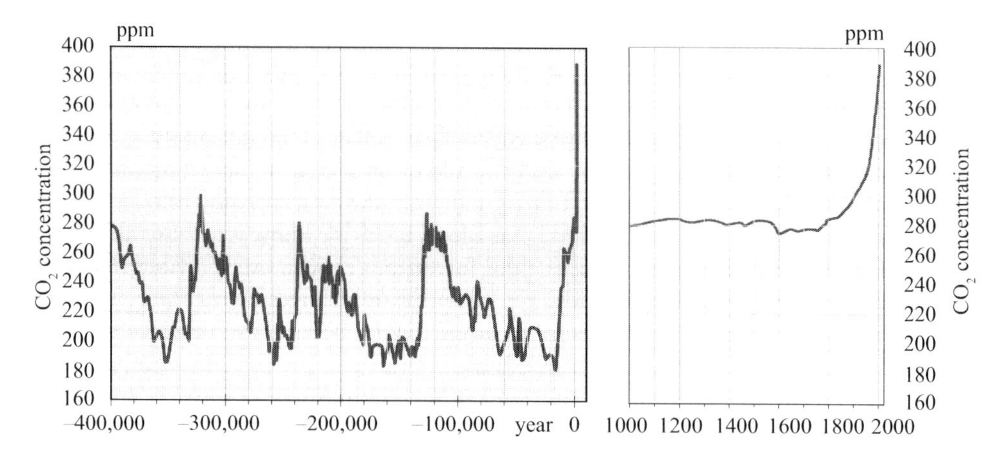

Figure 1.9 Development of carbon dioxide concentrations in the atmosphere over the last 400,000 years and in the last 1,000 years

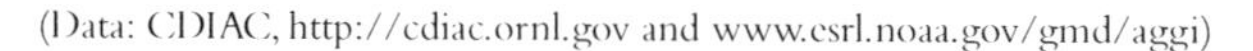
(Data: CDIAC, http://cdiac.ornl.gov and www.esrl.noaa.gov/gmd/aggi)

of the smaller amounts of methane are critical. In 2010, the average concentration of methane in the troposphere was 1.8 ppm, roughly 2½ times the pre-industrial level of 0.715 ppm.

Chlorofluorocarbons (CFCs), for instance, are used as a refrigerant in refrigerators and in large amounts as an aerosol in spray cans. CFCs are mainly known for their destructive impact on the ozone layer 10 to 50 kilometres up in the stratosphere. A global agreement has therefore been reached to reduce CFC production gradually, resulting in slower growth in atmospheric concentrations, which may even have been reduced. In this discussion, the greenhouse potential of CFCs (11 % of the greenhouse effect) has only played a minor role. While a lot of the alternatives to CFCs – such as CFC-23 and R134a – no longer damage the ozone layer, their greenhouse potential is similar.

Nitrous oxide (N_2O) is emitted when tropical rain forests are burned and, in particular, when fertilizer nitrates are used in agriculture. In 2010, the share of N_2O was only 19 % above the preindustrial level at 0.323 ppm, but N_2O is also persistent in the atmosphere for a long time.

Ozone (O_3) is also created near the surface of the earth more often when pollutants are emitted, such as from the combustion of fossil fuels in automobiles. **Stratospheric water vapour** (H_2O) as a result of civilization also plays a role in the greenhouse effect. A number of other gases may play a role in the greenhouse gas effect, though their impact is hard to assess.

The various greenhouse gases can be attributed to different groups of causes as follows:

The use of fossil fuels	50 %
Industry	19 %
Tropical rain forests (burning and rotting)	17 %
Agriculture	14 %

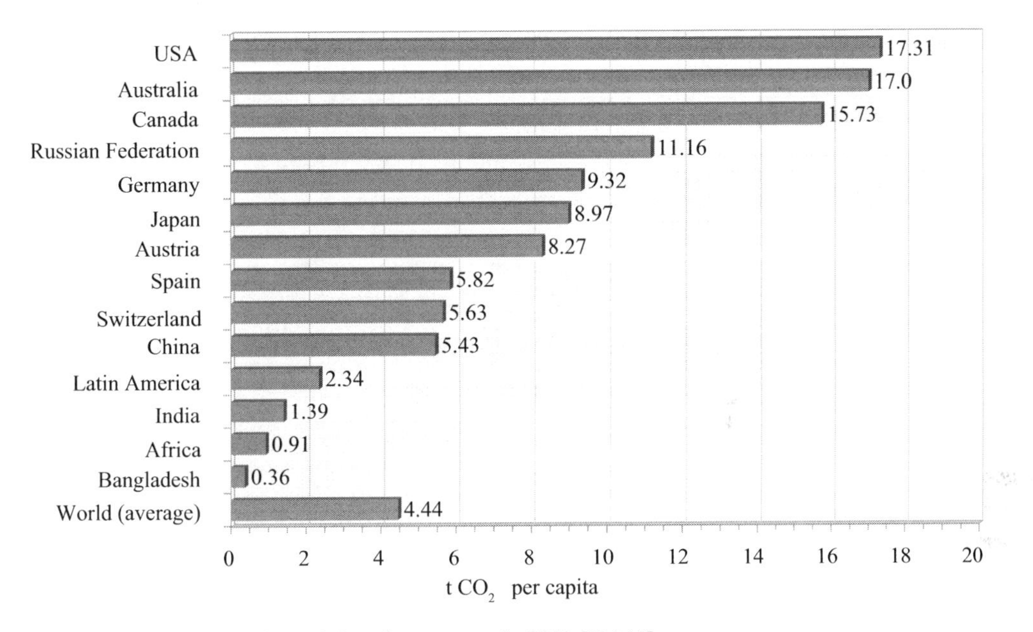

Figure 1.10 Per capita CO_2 emissions from energy in 2010 [IEA12]

The levels are quite different from one country/region to another. In developing countries, the main factor is the burning of tropical rain forests and agriculture, while the combustion of fossil fuels plays the biggest role in the greenhouse effect in industrialized countries. Energy consumption and resulting carbon emissions thus vary greatly from one part of the world to another, as Figure 1.10 shows.

Germans emit 10 times more carbon than Africans; North Americans, 19 times more. If everyone on the earth emitted as much carbon as North Americans, global carbon emissions would increase more than fourfold, and the anthropogenic greenhouse effect would more than double.

For a long time, the causes of climate change were very controversial. Even today, studies are published that cast doubt on the anthropogenic greenhouse effect. For instance, the increase in near-surface average temperatures by more than 0.7 °C over the past 100 years is passed off as a natural fluctuation. The authors of such studies usually come from groups that would be detrimentally affected by changes in the energy sector.

Today, numerous indicators reveal that climate change is slowly underway [IPC07, EEA10]:

- Fifteen of the hottest 16 8-year span since temperature measurements began in 1850 occurred from 1995 to 2010.
- The 2000s was the hottest decade since temperature measurements began.
- Global snow coverage has shrunk by more than 10 % since the late 1960s.
- In 1982, the Arctic was covered with 7.5 million km² of ice in the summer, but that area had decreased to 3.5 million km² by 2012.
- Glaciers in the Alps lost two thirds of their volume between 1850 and 2009.

- From 1961 to 2003, sea level rose by 1.8 mm per year, a figure that had even risen to 3.1 mm annually between 1993 and 2008.
- In the twentieth century, precipitation in northern latitudes increased by 0.5 to 1 % per decade.
- In Africa and Asia, the frequency and intensity of droughts increased.

The consequences of the anthropogenic greenhouse effect cannot be forecast in detail. We can only use various climate models to assess the possible effects of a greater concentration of greenhouse gases.

If the anthropogenic greenhouse effect and, in particular, consumption of fossil fuels does not slow down, carbon concentrations in the atmosphere will more than double relative to preindustrial levels over the next century. Global average temperatures will then increase by more than 2 °C above the current level by the end of the twenty-first century. In general, temperatures are expected to increase between 1.1 °C and 6.4 °C, depending on how much greenhouse gas is emitted. These temperature increases are similar to those between the Ice Age 18,000 years ago and the current warm phase, with the exception that the changes currently taking place will happen in only around 100 years, while the transition from the last Ice Age to the current warm phase took around 5,000 years.

A temperature increase of 2 °C in total – more than an increase of 0.1 °C per decade – is considered a level that could have catastrophic effects on civilization, food supply, and ecosystems. The greenhouse effect continues unabated and will probably have a devastating impact on global forests and agriculture. Food supply for humanity will worsen considerably as agricultural production decreases. The result will be famine and increasing migration – with all of the social problems that entails. Furthermore, we can expect global warming it to increase the intensity and frequency of storms both in the temperate zone and in the tropics, resulting in tremendous devastation. Sea level will rise by around a metre over the course of the century. Indeed, if temperatures increase by 2 °C, sea level is expected to increase by 2.7 metres by 2300 [Sch12].

If temperatures increase even more, we could even see the seas rise by as much as 30 metres or more. The ice on Greenland alone would raise sea level by seven metres if it all melted. As the catastrophic floods in recent years show, the result would be disastrous for coastal regions. In 1991, an estimated 139,000 people died in Bangladesh alone during floods. Numerous regions and islands close to sea level might disappear from the map altogether before the end of this century.

There is a general consensus that the greenhouse effect can no longer be stopped. Even the limit for global warming of no more than 2 °C is only realistic if considerable changes are made. For us to have a 67 % probability of staying within the **target of two degrees**, no more than 750,000 Mt of carbon dioxide can be emitted between 2010 and 2050 [WBG08]. Until recently, it was assumed that the world could stay within that carbon budget if carbon emissions were reduced by 80 % by 2050. Now we know that emissions will practically have to be completely offset between 2040 and 2050 because emissions have continued to increase so much in the past few years (Figure 1.11).

Essentially, we would need to stop using fossil fuels altogether by mid-century and switch to an energy supply based exclusively on renewables. Technologically and economically, this is easily feasible. We would, however, need to increase the rate at which the energy sector is transitioning considerably.

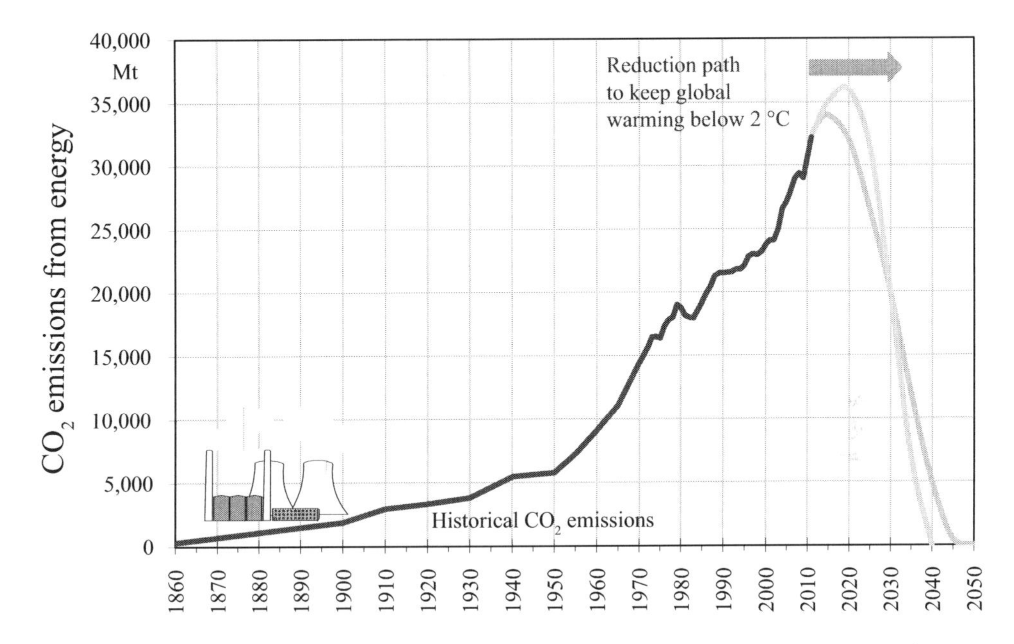

Figure 1.11 Reduction paths for global carbon dioxide emissions to limit global warming to less than 2 °C (Data: [WBG08])

In principle, climate protection targets can easily be met without reducing our level of industrial prosperity. The important thing is that everyone in society comes to understand the need for this reduction – and the possible consequences of not doing so. As this book shows, we already have enough options to cover our energy demand without fossil fuels. But this radical change that is needed will meet with resistance from the large number of parties that reap great financial benefits from our current energy supply. For the benefit of future generations, this resistance must be overcome as quickly as possible.

Today, representatives of companies from the conventional energy sector take every opportunity they can to cast doubt on the technical feasibility of a 100 % supply of renewable energy – when they are not calling into question climate change itself. Yet, these questions have already been clearly answered. We no longer have to decide whether doing without fossil energy is a good idea. In terms of the need to reduce carbon emissions considerably, we merely have to decide at what point our society is ready to do without it.

Nuclear power as a remedy for the greenhouse effect

Nuclear fission

Other sources of energy need to be found to cover our energy demand because we need to reduce consumption of fossil fuels considerably over the next few decades in order to limit the greenhouse effect. One option is nuclear power, and a distinction is made here between nuclear fission and nuclear fusion.

All nuclear plants currently in use are fission plants for the production of electricity. In this process, atoms of the uranium isotope ^{235}U are bombarded with neutrons, thereby splitting the uranium. Along with other fission products, two new atoms are produced: krypton ^{90}Kr and barium ^{143}Ba. Free neutrons ^{1}n are also created, which can also impact and break apart uranium nuclei. The mass of all of these products after fission is lower than before fission. This mass change releases energy ΔE as heat that can be used in a system. The entire process can be described in the following equation

$$^{235}_{92}\text{U} + ^{1}_{0}\text{n} \rightarrow ^{90}_{36}\text{Kr} + ^{143}_{56}\text{Ba} + 3\,^{1}_{0}\text{n} + \Delta E \quad (1.4)$$

for the nuclear reaction. Uranium does not occur in nature in a from that can be used in nuclear plants; nuclear fuel first has to be produced from uranium ore. Rock containing more than 0.1 % uranium oxide is considered worth processing as uranium ore. Uranium mining produces large amounts of overburden, which not only contains harmless rock, but also numerous radioactive residues that can cause cancer and other ailments. The uranium-235 needed only makes up 0.7 % of the uranium dioxide made from the uranium ore. Uranium-238, which cannot be directly used for nuclear fission, makes up a greater share. In large, energy-intensive processing facilities, the uranium has to be 'enriched', meaning that the share of fissile uranium-235 is increased to around 2 to 4 %.

At the end of 2011, 437 nuclear plants with a total generation capacity of 389,367 MW were in operation worldwide. The average nuclear plant thus had a capacity of 891 MW. The share of nuclear power in global primary energy is, however, relatively small at around 5 % at present. In Germany, nuclear made up 8.8 % of primary energy demand in 2011. That year, nuclear power made up around 17.6 % of electricity supply in the country. The share of nuclear power in electricity supply varies greatly from one country to another, as Figure 1.12 shows.

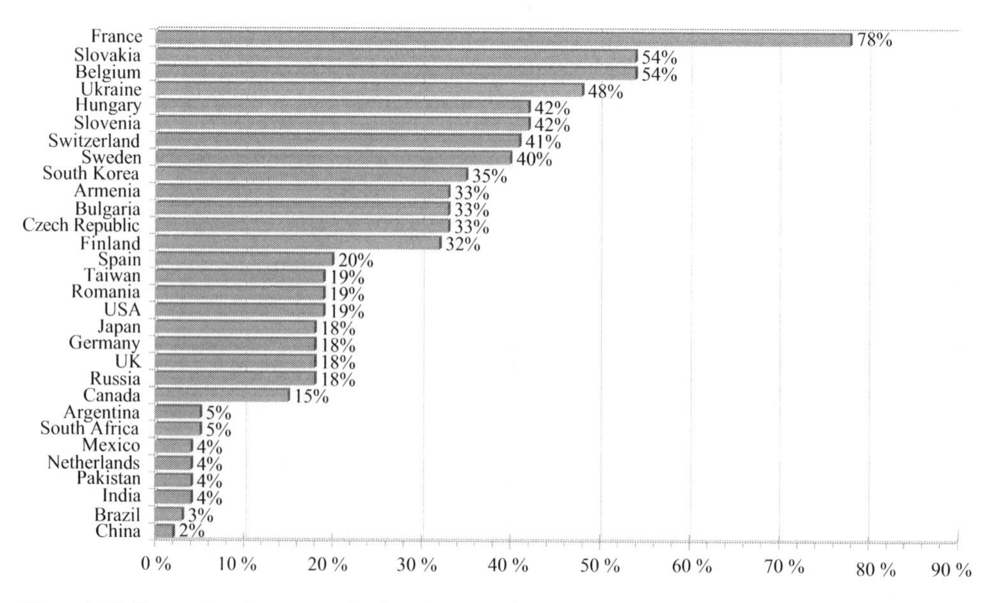

Figure 1.12 Share of nuclear power in electricity supply in 2011

(Data: [atw12])

In France, nuclear makes up the greater share of electricity, while other industrialized nations – such as Denmark and Austria – do without nuclear entirely. After the reactor meltdown in Chernobyl, Italy resolved to forgo nuclear power. Doing without nuclear power does not necessarily mean greater carbon emissions from fossil energy. Norway currently gets far more than 90 % of its electricity from hydropower, while Iceland has both hydropower and geothermal. The UK could get all of its electricity from wind power.

For nuclear power to cover all of the demand currently met by fossil fuels, at least 10,000 new nuclear plants would need to be built worldwide. Because a nuclear plant has a service life of around 30 years, all of the plants currently in operation would need to be renewed during this time, resulting in the need for a new nuclear reactor every day. In the process, nuclear plants would need to be built in politically unstable countries. There the risk of nuclear accidents from sabotage, war, negligence, and the military use of nuclear energy would be much greater.

As explained above, the world's uranium resources are limited. If nuclear power replaces some fossil energy, these resources will be used up faster. Fast breeders could extend this range, though they would not fundamentally change the limited nature of uranium reserves. For this reason alone, nuclear fission is not an alternative to fossil energy.

In 2010, 53,663 tons of uranium was produced worldwide. Given the low concentrations of uranium ore (see above) and the need to enrich this uranium, tremendous amounts of ore would need to be processed. The mining process itself has a tremendous impact on the environment, with the entire area being covered with radioactivity. Uranium is generally transported across long distances. During processing, large amounts of energy are consumed, and a lot of material and energy is needed to build the plants. Admittedly, no carbon is emitted during the nuclear fission process, but considerable amounts of carbon are emitted during the entire lifecycle – from power plant construction to uranium mining and plant/waste disposal. Though this amount of carbon is less than with a coal plant, the emissions are far above similar indirect carbon emissions from wind turbines.

The transport and storage of radioactive materials pose a very different type of risk. On the one hand, uranium and fuel rods need to be transported to various processing facilities and power plants; on the other, spent fuel rods and radioactive waste have to be taken to intermediate storage for further processing and eventually to final repositories. Under normal nuclear plant operation, dangerous, highly radioactive products are produced, and spent fuel rods pose great risks from radioactivity. One of the numerous radioactive materials is plutonium, a highly dangerous element making up nearly 1 % of nuclear waste. If a human inhales a microgram – a millionth of a gram – of plutonium, that person is likely to die from lung cancer. Theoretically, a single gram of plutonium could wipe out an entire city. There is no guarantee of absolute safety when transporting nuclear waste; an accident could always cause radioactive materials to be released. The final repository is also problematic because these materials remain deadly for millennia.

Even when a nuclear plant runs properly, there are risks. For instance, nuclear plants constantly release small amounts of radioactivity. Recently, cases of leukaemia in children living near nuclear plants in Germany have been found to have increased.

The greatest danger, however, stems from a maximum credible accident at a power plant. If such an accident occurred in Western Europe, millions of people would be affected. Large amounts of radioactivity would be released, making considerable areas unliveable for a long time; countless people and animals would die of radiation or eventually develop cancer as a result. The accidents in Harrisburg, Pennsylvania; Chernobyl, Ukraine;

and Fukushima, Japan, show that such accidents cannot be completely ruled out – and we have not even mentioned the horror scenario of terrorist attacks.

On 28 March 1979, an accident occurred in **Harrisburg**, the capital of Pennsylvania, where large amounts of radioactivity were released. Plant and animal life was impacted, and the number of stillbirths in the vicinity increased dramatically after the accident.

On 26 April 1986, a severe accident also occurred at the **Chernobyl** nuclear plant in Prypjat, Ukraine, a town with a population of 50,000 at the time. Even in Germany, the radioactivity released in Ukraine led to dramatically higher levels. A number of workers who tried to fix the problem on site paid for the mission with their lives. Various studies have found that the number of miscarriages and cases of cancer increased considerably as a result of this radiation.

On 11 March 2011, a severe earthquake and subsequent tsunami severely damaged the Japanese nuclear power plant at **Fukushima**. Engineers did manage to ramp down the reactors, but subsequent cooling could not be sufficiently maintained to accommodate the decay heat. The fuel rods overheated, and there were a number of explosions and fires. One of the reactor hulls blew off completely, releasing large amounts of radioactivity.

Nuclear energy facilities cannot only be used for civilian purposes, but also militarily. Not surprisingly, the military is behind the growth of nuclear power stations in a number of countries. The use of nuclear power in politically unstable countries can lead to international crises. Examples from recent history include Iran, Iraq, and North Korea. The more nuclear power is promoted, the greater the possibility of 'nuclear crises' and the risk of terrorists getting hold of radioactive material.

The use of nuclear power for peaceful purposes thus entails a lot of risks whose effects are hard to calculate. Because nuclear power is not the only way to ensure a low-carbon energy supply, the call for a nuclear phaseout is more than justified.

Nuclear fusion

At present, a lot of hopes rest on – and a lot of money is being invested in – a completely new nuclear power technology: nuclear fusion. The sun itself is a fusion reactor, in which the fusion of hydrogen releases energy. On earth, this process is to be reproduced by fusing deuterium ^{2}D and tritium ^{3}T to produce helium ^{4}He. A neutron ^{1}n and energy ΔE would be released in the process. The following formula describes the process:

$$ {}_{1}^{2}\mathrm{D} + {}_{1}^{3}\mathrm{T} \rightarrow {}_{2}^{4}\mathrm{He} + {}_{0}^{1}\mathrm{n} + \Delta E \tag{1.5} $$

But before the reaction can begin, the particles have to be heated up to temperatures exceeding a million °C. Because no material can withstand such temperatures, other technologies are being tested, such as enclosing the plasma of reacting materials in strong magnetic fields.

The raw materials needed for nuclear fusion are available on earth in great quantities, so that the range of resources is not an issue for the technology. On the other hand, it is unclear whether nuclear fusion will ever work. Critics point out that the technology is only 50 years away today – and will only be 50 years away tomorrow.

But even if this technology is ever ready for commercial use, there are a number of reasons why we should not pursue nuclear fusion. First, it is much more expensive than current nuclear fission. For economic reasons alone, renewable energy is a preferable alternative.

Second, nuclear fusion will also produce radioactive materials that can pose a risk. Third, tremendous amounts of funding that could have been used elsewhere have been invested in nuclear fusion. Finally, if the technology ever works, it will be too late. If we want to combat the greenhouse effect, we urgently need alternatives today. If we want to protect the climate, we cannot wait for a fusion reactor to be developed at some indeterminate time in the future.

The use of renewable energy

If we are to consume far less fossil energy and nuclear power is not an alternative, the question is what our future energy supply can look like. To begin with, energy productivity has to continue to increase considerably. In other words, the same demand for useful energy has to be met with far less primary energy; in line with lower primary energy consumption, carbon emissions can be reduced.

Nonetheless, the growing global population and the need for economic growth in developing countries will contribute to an increase in energy demand. Lovins and Weizsäcker describe this issue well in their book *Factor Four*. Over the next 50 years, we need to double prosperity even as we cut our consumption of energy and natural resources in half [Wei96]. Renewables will play a decisive role on the path towards this goal because only they can meet global energy demand in a climate-friendly manner.

Energy sources are considered renewable if they are infinite within a timeframe relevant for humanity. Renewable energy can be divided into three groups of solar, planetary, and geothermal energy (Figure 1.13). The potential annual energy supply from these three sources breaks down worldwide as follows:

- Solar energy 3,900,000,000 PJ/a
- Planetary energy (gravitation) 94,000 PJ/a
- Geothermal energy 996,000 PJ/a

Wind and precipitation are energy forms caused by natural energy conversion. They can be used to provide heat, electricity, and fuel.

The annual supply of renewable energy is magnitudes greater than global energy demand. Theoretically, renewables could easily cover global energy demand, but a transition will not necessarily be smooth. Rather, if we want to use renewables on a large scale, we need to set up a completely different energy sector from the one constructed in the past few decades.

The conventional energy sector is largely based on fossil energy. The goal is currently to extract, transport, and convert it into other forms of energy in microeconomically optimized central facilities as affordably as possible. The benefit of fossil energy is that it is immediately available, so consumers can use it whenever they wish.

In contrast, some forms of renewable energy, such as wind power, fluctuate considerably. If the energy sector runs completely on renewables, the focus cannot be merely on converting one form of energy into another (such as electricity), but also on ensuring the availability of energy. One option is large-scale energy storage; others are global energy transport and demand management – when energy consumption is adjusted to suit supply. The question is therefore not whether renewables will be able to provide all of our energy supply, but rather what share each specific source of renewable energy will have – and how quickly these shares can be reached in order to mitigate the greenhouse effect. Here, we see that

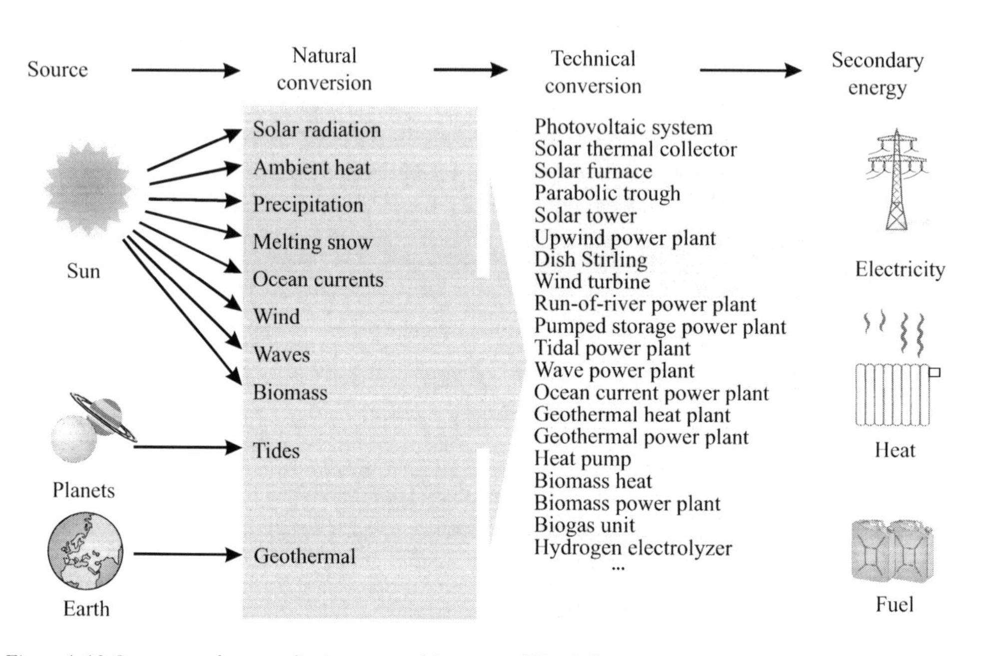

Figure 1.13 Sources and ways of using renewable energy [Qua09]

transitioning our energy sector is one of the biggest challenges of the twenty-first century. Below, we briefly present the wide range of renewable energy sources useful today.

Geothermal energy

Geothermal energy is energy from heat inside the earth. In the core of the earth, temperatures reach 4,600 °C. The energy released from radioactive decay is the main cause of these temperatures. Ninety-nine percent of the planet is hotter than 1,000 °C. Within the earth's crust, there is relatively little heat, but it quickly increases as we go deeper. Volcanic eruptions are an impressive display of the tremendous energy processes that take place within the earth. Because of the great temperature differences between the earth's core and its crust, heat constantly flows from the interior to the exterior. The energy contained in this heat is roughly equivalent to global primary energy demand. High temperatures are needed, however, for geothermal energy to be tapped. Such depths can be reached by means of drilling. Geothermal is only economically interesting in regions with geothermal anomalies. In such cases, temperatures are high enough for energy to be exploited at relatively shallow depths. The countries with considerable potential include the Philippines, Italy, Mexico, Japan, Iceland, New Zealand, and the United States. In 2010, some 10,717 MW of geothermal power plant capacity was installed worldwide, providing more than 60 TWh of electricity. In addition, numerous systems use geothermal energy as a heat source. Chapter 8 presents geothermal energy in greater detail.

Planetary energy

The planets exert gravitational forces on each other, but the greatest effects on earth are felt from the moon; these forces constantly change as the planets and the moon move relative to each other. The effects are best seen in the tides on coasts. Enormous amounts of energy are needed to move the tremendous masses of water during ebb and flow.

If the tides are especially high, tidal power plants can harvest this ocean energy. At high tide, water flows through turbines into a basin, and the water flows back through the turbines into the sea at low tide. In the process, the turbines produce electricity. Worldwide, only a few tidal power plants are in operation. The dams and basins needed are similar to large hydropower plants and change the natural landscapes considerably. Relatively little energy can come from tidal power plants in theory. Chapter 7 discusses this technology in greater detail.

Solar energy

By far the largest source of renewable energy is the sun. A large amount of solar energy – $3.9 \cdot 10^{24}$ J = $1.08 \cdot 10^{18}$ kWh – reaches the earth's service from the sun each year. This amount is nearly 10,000 times greater than global primary energy demand – and hence far more than all available fossil and nuclear energy reserves. If we manage to harness only 1/10,000 of the incident solar energy on earth, we could meet all of mankind's energy demand from the sun. Figure 1.14 visualizes the dimensions with circles of energy.

A distinction is made between direct and indirect solar energy when we discuss how to use it. When solar energy is used directly, technical equipment directly exploits incident insolation. Indirect solar energy is spoken of when solar heat is converted through natural processes into other forms of energy, such as wind, flowing water, and plant growth. These indirect forms of solar energy can also be exploited as sources of energy in technical systems.

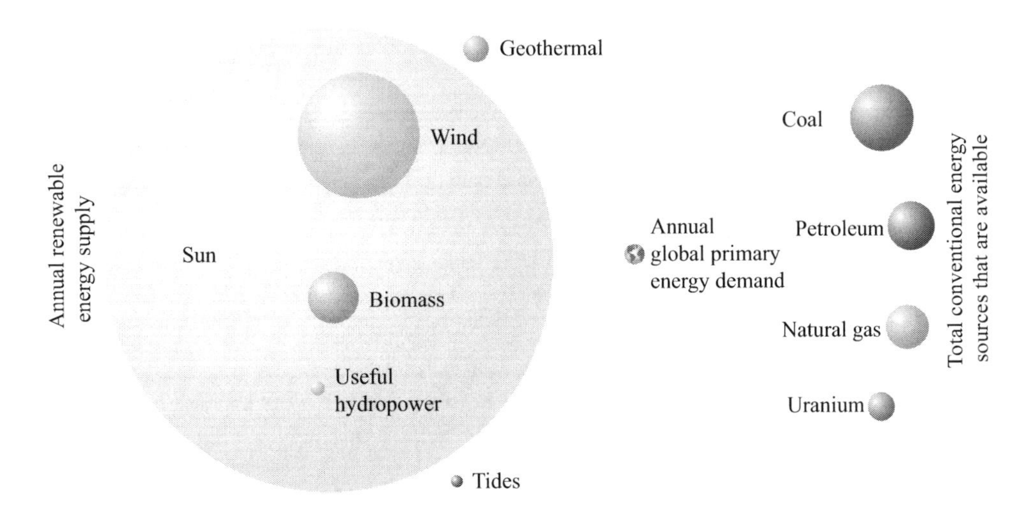

Figure 1.14 Energy circles. The amount of incident sunlight that reaches the earth each year exceeds both global annual energy consumption and total energy reserves many times over [Qua09]

Because solar energy is the biggest source of renewable energy and the theory of insolation applies for all systems that exploit solar energy directly, this topic is covered in detail in Chapter 2. Below, various technologies used to harvest solar energy directly and indirectly are briefly presented. They will be dealt with in greater detail in later chapters.

Direct use of solar energy

A wide range of technologies can be used to harvest solar energy directly, including the following:

- solar thermal power plants
- solar collectors for heat supply
- photovoltaics – solar cells that generate electricity
- photodissociation for fuel production.

SOLAR THERMAL POWER PLANTS

Solar thermal power plants convert solar heat into electricity. This technology also comes in different forms:

- parabolic troughs
- power towers
- dish Stirling
- solar updraft towers.

The global potential of solar thermal power plants is tremendous. Theoretically, they could easily cover all of civilization's energy demand. In the midterm, these power plants can become completely competitive with power plants running on fossil fuel or uranium. However, they are only economical in regions with a lot of direct sunlight. With the exception of updraft towers, plants exploiting solar heat to generate electricity concentrate sunlight; hence the umbrella term 'concentrated solar power'.

In 2012, global installed capacity of parabolic trough power plants exceeded 2,000 megawatts. Most of them had been built in the US and Spain. At present, several GW of additional power plants are currently planned or under construction. Chapter 4 deals with concentrated solar power in greater detail; Chapter 3, updraft towers.

SOLAR COLLECTORS FOR HEAT PRODUCTION

Solar thermal can not only be used to produce high-temperature heat and electricity, but also to cover demand for low-temperature heat, such as is needed for space heating and hot water. While collector systems for space heating are still quite rare, collectors are used in many parts of the world already to provide hot water. In 2011, nearly 1.3 million m^2 of collectors had been installed in Germany. That number amounts, however, to only 183 m^2 of collector area per 1,000 residents of the country. That year, Austria had nearly twice as much – 472 m^2 per 1,000 residents – while sunny Cyprus was able to post 700 m^2 per 1,000 residents. But the largest collector market in the world by far in terms of volume is currently China (Table 1.8). Chapter 3 talks about non-concentrated solar thermal collectors in greater detail.

Table 1.8 Newly installed glazed solar thermal collector area from 1990 to 2011

	Annual newly installed area in 1,000 m²										
	1990	*2000*	*2002*	*2004*	*2006*	*2008*	*2009*	*2010*	*2011*	*2012*	*2013*
China	720	5,563	7,700	13,500	18,000	31,000	42,000	49,000	*	63,900	*
Turkey	300	675	490	1,200	700	930	1,633	1,658	*	1,624	*
Germany	35	620	540	750	1,500	2,100	1,615	1,150	1,270	1,150	1,020
Italy	13	45	45	58	186	421	475	490	415	330	297
Brazil	n/a	n/a	48	n/a	434	555	391	473	*	626	*
Spain	9	40	66	90	175	434	391	337	267	226	229
France	15	24	27	52	220	388	265	256	251	250	190
Austria	40	153	153	183	293	348	356	280	230	206	179
Israel	250	390	380	70	222	278	292	316	*	330	*
Greece	204	181	152	215	240	298	206	214	230	243	227
USA	235	37	57	47	125	225	186	225	*	256	*
Japan	543	339	290	251	264	221	154	162	*	170	*

(Data: [EST03; EST12; IEA12c])

* No data were available for 2011 when this table went to press

In addition to collector systems that actively exploit solar energy, there are also ways of passively taking advantage of it. For instance, buildings can be optimally oriented with well-planned glass façades and transparent insulation. By combining the active and passive use of solar energy, buildings can become carbon-neutral; they then get all of the energy they need for space heating and hot water from the sun. Indeed, a number of homes in Germany even produce more energy than needed for these purposes; they are called Plus Energy Homes.

PHOTOVOLTAICS

Photovoltaics is a promising technology for the use of solar energy to generate electricity. Solar cells directly convert sunlight into electrical energy. In Germany alone, more than 200 GW of photovoltaics could be installed on roofs and elsewhere, and the figure increases to more than 1,000 GW if we include uncultivated fields [Qua00]. In total, far more than 1,000 TWh/a would be produced in the process, an amount far greater than the current domestic demand for electricity. It would not, however, make much sense to build that many photovoltaics because the result would be great excesses in production requiring large, expensive storage. Rather, it makes the most sense to combine photovoltaics with other sources of renewable energy, such as wind power, hydropower, and biomass, which complement each other well.

For a long time, a small number of countries were the main drivers in the global growth of photovoltaics. In 2011, more than 75 % of global installed photovoltaic capacity was still spread across only six countries (Table 1.9). New PV installations have grown annually between 20 and 80 %. Although photovoltaics only accounted for around 50 TWh of electricity generated worldwide in 2011, costs are falling quickly, and if growth rates continue the technology will quickly have the potential to become one of the main pillars of global energy supply. Chapter 5 deals with photovoltaics in detail.

Table 1.9 Global installed photovoltaic capacity in GW_p

	1992	*1995*	*2000*	*2002*	*2004*	*2006*	*2008*	*2009*	*2010*	*2011*	*2012*	*2013*
Germany	0.006	0.02	0.11	0.30	1.07	2.93	6.16	9.96	17.37	24.82	32.4	35.7
Italy	0.008	0.02	0.02	0.02	0.03	0.05	0.46	1.18	3.50	12.80	16.3	17.4
Japan	0.019	0.04	0.33	0.64	1.13	1.71	2.14	2.63	3.62	4.91	7.0	13.9
Spain	0.004	0.01	0.01	0.02	0.04	0.12	3.46	3.52	3.92	4.26	5.1	5.2
USA	0.044	0.07	0.14	0.21	0.38	0.62	1.17	1.62	2.53	3.97	7.2	12.0
China	0.000	0.00	0.00	0.00	0.00	0.08	0.14	0.30	0.80	3.00	7.0	18.3
Other	0.069	0.19	0.49	0.75	1.05	1.34	2.47	3.79	7.26	15.94	27.0	36.5
World	0.15	0.34	1.10	1.94	3.70	6.85	16.0	23.0	39.0	70.0	102	139

(Data: [IEA12b, Qua12])

Indirect use of solar energy

Solar energy is used indirectly when natural conversion processes turned insolation into another form of energy that can be used by equipment to provide useful energy. One example of indirect solar energy is hydropower. The sun's rays cause bodies of water to evaporate. Precipitation is then collected in streams and rivers, which flow back into the sea. Along the way, the water's kinetic energy and potential energy can be harnessed in power plants. The indirect forms of solar energy include:

- evaporation, precipitation, and flowing water
- melting snow
- waves
- ocean currents
- biomass growth
- the heating of the earth's surface and the atmosphere
- wind.

HYDROPOWER

Hydropower covers a wide range of indirect ways of using solar energy. In terms of global renewable power generation, hydropower is by far the biggest source.

Hydropower has been consistently growing for decades, whereas wind power and photovoltaics only began growing strongly in the 1990s. As a result, around 900 GW of hydropower had already been installed by 2009, producing 3,145 TWh net per year.

However, the further growth of hydropower has already reached its limits in some parts of the world. Large hydropower plants in particular generally have tremendous impacts on the local environment and are therefore not uncontroversial. Nonetheless, it seems it will still be possible to double global installed capacity. But because the potential of solar and wind power is much greater, hydropower can be expected to step down from its role as the largest source of renewable electricity in the next few decades. Table 1.10 shows which countries have the most installed hydropower capacity. Aside from Norway, the main hydropower countries in Europe are those in the Alps along with Spain and Sweden.

Table 1.10 Global installed hydropower capacity in GW

	1980	*1990*	*1995*	*2000*	*2002*	*2004*	*2006*	*2007*	*2008*	*2009*	*2010*	*2011*
China	20.3	36.0	52.1	79.4	86.1	105.2	128.6	145.3	171.5	196.8	219.0	231.0
Brazil	27.5	56.6	51.3	61.1	65.3	69.0	73.4	76.9	77.9	79.3	80.7	82.5
USA	81.7	73.9	78.6	79.4	79.4	77.6	77.8	77.9	77.9	78.5	78.8	78.7
Canada	47.9	59.2	64.6	67.2	69.0	70.7	72.7	73.3	74.2	74.9	74.9	74.9
Russia	1)	1)	44.0	43.9	44.8	45.5	46.1	46.8	46.8	47.0	47.4	47.3
India	12.2	18.8	21.0	25.1	26.9	32.6	36.6	38.1	39.3	39.6	40.6	42.4
Norway	19.4	25.8	27.4	26.8	26.3	26.1	27.4	27.6	28.1	28.2	28.4	28.4
Other	252.9 2)	305.4 2)	289.3	309.0	318.7	326.3	333.1	337.3	339.8	345.8	352.9	363.9
World	461.9	575.7	628.3	691.9	716.5	753.0	795.7	823.2	855.5	890.1	922.7	949.1

(Data: [EIA12])

1) Included in 'other'
2) Including Russia

Germany has relatively little potential. In 2009, only 4.76 GW was installed. Chapter 7 discusses the use of hydropower in detail.

WIND POWER

More than 100 years ago, when power played a major role. Technically advanced windmills were used to grind grain and pump water; in North America, thousands of 'western mills' for pumping water dotted the landscape. All wind mills are purely mechanical systems. Only in the past few decades have wind turbines for the generation of electricity become commonplace. In 2010, wind turbine manufacturers made 4.97 billion € in revenue in Germany, where 96,100 people were employed in the sector. The export ratio of German turbine manufacturers was 66 % in 2010. In mid-2012, Germany had 22,594 wind turbines with an installed capacity of 30,001 MW in operation [Dewi12]. At current growth rates, wind power will be able to make up more than 20 % of German power supply by 2020.

Germany has considerable wind power potential. Even with consideration of strict conditions for the siting of wind farms, a total of 189 GW can be installed on land, enough to produce 390 TWh/a – more than 60 % of Germany's power demand [BWE11]. The potential of offshore wind is even greater. Clearly, wind power can play a major role in reducing greenhouse gas emissions from the German power sector. The potential of wind power in other countries varies greatly. In the UK, it is much greater than in Germany, for instance. The British could get 1,760 TWh/a of electricity from wind power [Sel90], several times the country's power demand. Along with Germany, Spain, the US, Denmark, China, and India are the main growth markets for wind power (Table 1.11). But other countries plan to expand capacity considerably in the next few years. Chapter 6 talks about the use of wind energy for power generation in greater detail.

BIOMASS PRODUCTION

Biomass is by far the largest source of renewable energy in terms of global primary energy consumption. Biomass consists of organic materials that grow in nature, but also waste

Table 1.11 Global installed wind power capacity in GW

	1994	*1998*	*2000*	*2002*	*2004*	*2006*	*2008*	*2010*	*2011*	*2012*	*2013*
China	0.03	0.20	0.35	0.47	0.76	2.60	12.21	44.73	62.73	75.32	91.42
USA	1.54	2.14	2.61	4.67	6.74	11.60	25.17	40.30	46.92	60.01	61.09
Germany	0.60	2.88	6.11	12.00	16.63	20.62	23.90	27.19	29.06	31.27	34.25
Spain	0.08	0.88	2.84	5.04	8.26	11.62	16.75	20.62	21.67	22.78	22.96
India	0.20	0.99	1.22	1.70	3.00	6.27	9.65	13.07	16.08	18.42	20.5
Denmark	0.54	1.42	2.34	2.88	3.12	3.14	3.18	3.75	3.87	4.16	4.77
Other	0.52	1.64	2.99	5.31	9.11	18.37	23.93	47.98	58.01	71.14	83.46
World	3.51	10.15	18.46	32.07	47.62	74.22	120.8	197.6	238.4	283.1	318.1

(Data: [Qua12])

materials from living and dead organisms. Each year, the earth produces some $1.55 \cdot 10^{11}$ t/a of biomass, equivalent to $3 \cdot 10^{21}$ J of energy content. Roughly 1 % of that amount is used worldwide as a source of heat, covering some 11 % of global primary energy demand. In particular in developing countries, the share of biomass in energy supply is very high, exceeding 80 % in such countries as Ethiopia, Mozambique, and Nepal. But biomass is not always used sustainably. The use of biomass is only renewable – and hence, climate-neutral – when the amount consumed does not exceed the amount that grows back. If it is overused, the result is desertification, which steps up the greenhouse effect.

There are various ways to use biomass. In addition to combustion for the provision of heat and electricity, biomass can be liquefied, gasified, or fermented to produce alcohol in a number of conversion processes. The great benefit of biomass is that stored energy can be used to meet current energy demand, unlike solar power and wind power, which greatly fluctuate in supply. Biomass will therefore play a crucial role in making a largely renewable power supply more reliable.

In 2011, Germany had 5,479 MW of biomass capacity installed, which generated 31.9 TWh of electricity that year, equivalent to around 5 % of total power supply. In addition, organic waste is also used. It made up a much larger share of heat supply at around 124 TWh, equivalent to 9.5 %. The same year, biofuels in the amount of 34.2 TWh were consumed, equivalent to 5.5 % of German demand for fuel in transport [BMU]. Chapter 9 talks about the use of biomass in greater detail.

LOW-TEMPERATURE HEAT

Sunlight heats up both the surface of the earth and the atmosphere. Because different layers of the atmosphere and different parts of the earth's surface heat up differently, the air begins to flow to balance out these differences; in the process, this wind can be used to generate energy (see above). Solar heat is stored in soil for hours, days, and even months. Heat pumps take advantage of the low-temperature heat in the earth, groundwater, and air. It is hard to make a distinction between solar energy and geothermal when we talk about using low-temperature heat from the soil.

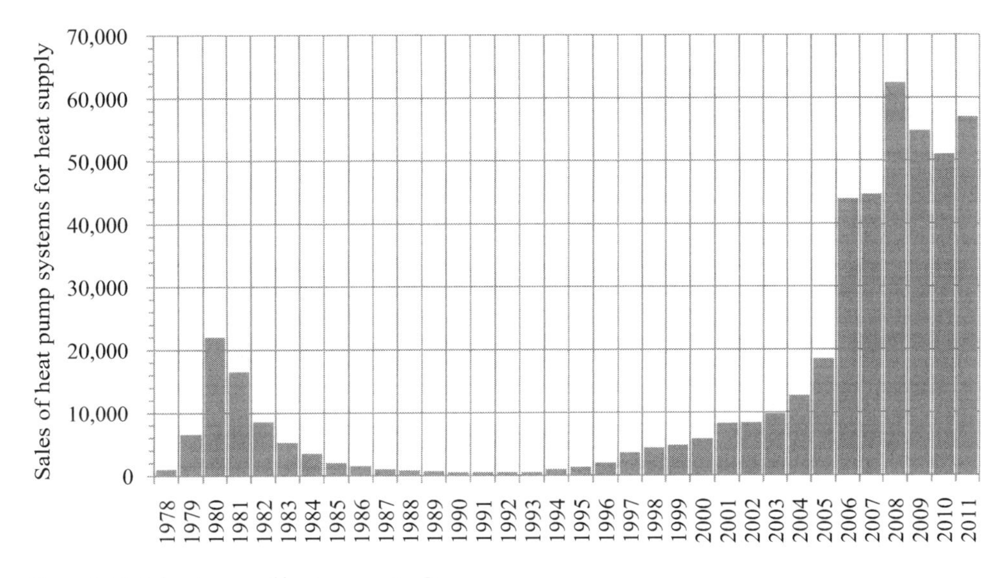

Figure 1.15 Sales figures of heat pumps in Germany

If the energy used to power heat pumps is renewable, the technology provides climate-neutral useful heat. For these reasons, heat pumps are considered a promising technology. At the beginning of the 1980s, the German market volume was quite significant, but it had nearly collapsed by the end of the 1990s. Since then, sales of heat pumps have successfully recovered (Figure 1.15). Chapter 8 goes into heat pumps in greater detail.

FUEL CELLS AND HYDROGEN PRODUCTION

Fuel cells and hydrogen are often spoken of within the context of renewables. Nonetheless, a distinction should be made. Hydrogen is not a renewable source of energy in and of itself; rather, it is an energy carrier that has to be created out of an energy source. Energy is therefore needed to produce hydrogen. This energy can come from renewables. At present, only small amounts of hydrogen are produced for the chemicals sector. Most of this hydrogen is made from fossil fuels. As long as hydrogen mainly comes from fossil sources, it makes more financial and ecological sense to consume the fossil energy directly.

Fuel cells can directly convert hydrogen and similar energy carriers – such as natural gas and methanol – into electricity. If fuel cells run on natural gas, the environmental payback is hardly impressive. Nonetheless, expectations are great for fuel cells, which are eventually to run on hydrogen. Especially in the transport sector, fuel cells are held to be an ecological alternative. If we ever have a renewable hydrogen economy, these expectations may be at least partly fulfilled. Chapter 10 talks about fuel cells and hydrogen production in greater detail.

Future energy demand, the energy transition, and climate protection

The increasing consumption of fossil fuels is the main cause of man-made climate change. As explained above, our consumption of fossil fuels needs to be reduced considerably in order to mitigate the negative consequences of global warming.

Unfortunately, civilization continues to march down the wrong path. If we do not change the direction of our energy policy fundamentally, fossil fuel consumption – and along with it, carbon emissions – will continue to increase for the time being. Yet, an energy transition is possible. By 2040, a carbon-free energy supply is possible.

Trends in global energy demand

The Intergovernmental Panel on Climate Change (IPCC) is a scientific institution founded in 1988 by the World Meteorological Organisation (WMO) and the United Nations Environmental Programme (UNEP). The IPCC is renowned for its reports, which serve as the basis for political consultations and decisions. In 2007, the IPCC received the Nobel Peace Prize for its work. The IPCC has investigated numerous scenarios for future trends in this century. Each of them takes account of various assumptions for economic growth, population growth, and consumption of fossil energy. The goal of these investigations is to show the impact our actions have on the climate.

Below, the results of six scenarios – A1F1, A1B, A1T, A2, B1, and B2 – are briefly presented [IPC00]:

- A1 assumes rapid economic growth, with population growth stagnating in mid-century. Three technological developments are then tracked. Scenario A1F1 assumes that fossil energy will be used intensively, while scenario A1B assumes there will be a balance between all energy sources; finally, scenario A1T assumes that carbon-free renewable energy will be used.
- Scenario A2 assumes a mixture of trends – population growth continues but outstrips technological developments.
- Scenario B1 assumes similar population growth as A1, but the trend towards an information and service sector society moves more quickly, leading to a reduction in material intensity and the rollout of cleaner technologies that go easier on resources.
- Scenario B2 assumes continuing population growth, but at a lower level than in A2. The focus is on economic development of social and ecological sustainability here. New technological trends in this scenario proceed more slowly, however, than in scenario B1.

Table 1.12 and Table 1.13 show the main assumptions for these scenarios in 2020, 2050, and 2100. In comparison, the columns for 1990 and 2010 show actual data, with 1990 being the reference year. Table 1.14 shows the outcome of carbon dioxide emissions in the various scenarios along with the effect on the climate, such as the increase in carbon dioxide concentrations in the atmosphere, the rise in average global annual temperature, and rising sea level. The margins of error for these calculations are relatively large because of the level of uncertainty in the underlying models. For instance, the indication of rising sea level does not include the massive melting of continental ice on Greenland. As a result, the sea level might rise much more than indicated in the table.

Table 1.12 Assumptions for population and economic growth up to 2100 in various IPCC scenarios

	Global population in billions					*Global gross domestic product in trillion USD$*				
	1990	*2010*	*2020*	*2050*	*2100*	*1990*	*2010*	*2020*	*2050*	*2100*
A1F1	5.3	6.8	7.6	8.7	7.1	21	34	53	164	525
A1B	5.3	6.8	7.4	8.7	7.1	21	34	56	181	529
A1T	5.3	6.8	7.6	8.7	7.0	21	34	57	187	550
A2	5.3	6.8	8.2	11.3	15.1	21	34	41	82	243
B1	5.3	6.8	7.6	8.7	7.0	21	34	53	136	328
B2	5.3	6.8	7.6	9.3	10.4	21	34	51	110	235

(Data: [IPC00; IEA12])

Table 1.13 Assumptions for the development of primary energy demand and the share of carbon-free energy by 2100 in various IPCC scenarios

	Primary energy demand in EJ					*Carbon-free primary energy demand in %*				
	1990	*2010*	*2020*	*2050*	*2100*	*1990*	*2010*	*2020*	*2050*	*2100*
A1F1	351	560	669	1,431	2,073	18	20	15	19	31
A1B	351	560	711	1,347	2,226	18	20	16	36	65
A1T	351	560	649	1,213	2,021	18	20	21	43	85
A2	351	560	595	971	1,717	18	20	8	18	28
B1	351	560	606	813	514	18	20	21	30	52
B2	351	560	566	869	1,357	18	20	18	30	49

(Data: [IPC00])

Table 1.14 Development of CO_2 emissions by 2100 along with the increase in CO_2 concentrations in the atmosphere, the increase in average annual temperatures, and rising sea level in various IPCC scenarios

	CO_2 emissions in Gt					*Increase by 2100 for CO_2 concentration*	*Temperature increase in °C*	*Sea level rise in cm*
	1990	*2009*	*2020*	*2050*	*2100*			
A1F1	26.0	37.0	46.5	87.6	103.4	280 %	2.4–6.4	26–59
A1B	26.0	37.0	46.1	60.1	49.5	210 %	1.7–4.4	21–48
A1T	26.0	37.0	37.7	45.1	15.8	165 %	1.4–3.8	26–59
A2	26.0	37.0	44.7	63.8	106.7	250 %	2.0–5.4	23–51
B1	26.0	37.0	38.9	41.4	15.4	150 %	1.1–2.9	18–38
B2	26.0	37.0	33.0	40.3	48.8	180 %	1.4–3.8	20–43

(Data: [IPC00, IPC01, IPC07])

The findings of these scientific investigations are sobering. Even in the scenario with the lowest impact on the climate, temperatures will rise by at least 1.4 °C this century. If we throw in the increase of around 0.7 °C that has already taken place, temperatures will increase by 2 to more than 6 °C, with the consequences being unforeseeable.

Table 1.15 Specific CO_2 emission factors of energy carriers [UNF98]

Energy carrier	*kg CO_2/kWh*	*kg CO_2/GJ*	*Energy carrier*	*kg CO_2/kWh*	*kg CO_2/GJ*
Wood[1)]	0.39	109.6	Crude oil	0.26	73.3
Peat	0.38	106.0	Kerosene	0.26	71.5
Brown coal	0.36	101.2	Gasoline	0.25	69.3
Hard coal	0.34	94.6	Refinery gas	0.24	66.7
Heating oil	0.28	77.4	LPG	0.23	63.1
Diesel	0.27	74.1	Natural gas	0.20	56.1

1) Non-sustainable use without reforestation

In all of the scenarios, even when fossil energy is consumed intensively, renewables are expected to grow strongly. The increase in carbon-free energy is between 400 and 2,500 %. Apparently, such growth still does not suffice to mitigate climate impacts. Not until the end of the twenty-first century is a completely carbon-free energy supply based on renewables expected to be possible.

Greater efficiency and the use of lower-carbon energy carriers are additional ways of reducing carbon emissions. For instance, Table 1.15 shows that emissions are roughly cut in half when natural gas replaces coal to provide the same amount of energy. However, global natural gas reserves are much smaller than coal reserves, so that lower-carbon fossil energy can only be a temporary solution.

None of these scenarios represents effective climate protection. For instance, a complete switch to renewable energy by mid-century would reduce carbon emissions far more than in the most optimistic scenarios, A1T and B1. Global temperatures would then not rise as much, nor would the sea level.

International climate protection

Internationally, politicians have also come to appreciate the need to reduce global carbon emissions in order to protect the earth's climate. In 1997, the Kyoto Protocol, signed after intensive negotiations, specified that emissions were to be reduced by 5.2 % below the level of 1990 no later than 2012. This compromise represents merely a moderate reduction in the growth of emissions, however, because a large number of developing and emerging countries were allowed to increase their emissions considerably outside of that reduction.

The signatories do not have to reduce their admissions at home, though. Flexible mechanisms allow countries to invest abroad in order to reduce emissions, and those specific investments will be attributed to the investing country. 'Carbon sinks' (forests, etc.) are also to be subtracted from total emissions. Because of such loopholes, international climate protection agreements have faced severe criticism. But without such mechanisms, the agreements would not have been signed at all in their current form.

Table 1.16 shows reduction obligations by country along with changes in greenhouse gas emissions over the past few years. Within the EU, the targets vary greatly from one country to another. While Germany and Denmark had a target reduction of 21 %, the UK's target was a 12.5 % reduction; Spain was allowed to increase its emissions by 15 %, compared with 25 and 27 % increases in Greece and Portugal.

Table 1.16 Reduction obligations in the Kyoto Protocol and actual trends among signatories [UNF12]

	Target	*Greenhouse gas emissions*[1)] *in Mt/y*			*Change 1990–2008*
		1990	*2000*	*2010*	
EU-15	–8 %	4,249	4,139	3,798	–10.6 %
Liechtenstein, Monaco, and Switzerland	–8 %	53	52	55	+2.2 %
Bulgaria, Estonia, Latvia, Lithuania, Romania, Slovakia, Slovenia, and the Czech Republic	–8 %	891	464	441	–46.2 %
USA	–7 %	6,151	7,072	6,602	+10.4 %
Japan	–6 %	1,267	1,342	1,258	–0.7 %
Canada	–6 %	589	718	692	+17.4 %
Poland and Hungary	–6 %	679	462	469	–31.0 %
Croatia	–5 %	31	26	29	–9.1 %
New Zealand	±0 %	60	69	72	+19.8 %
Belarus	±0 %	139	79	89	–35.7 %
Russia	±0 %	3,349	2,040	2,202	–34.2 %
Ukraine	±0 %	930	396	383	–58.8 %
Norway	+1 %	50	53	54	+8.2 %
Australia	+8 %	418	494	543	+29.8 %
Iceland	+10 %	4	4	5	+29.7 %
Total	–5.2 %	18,798	17,411	16,889	–10.2 %

1) CO_2, CH_4, N_2O, FKW, and SF_6 are converted into equivalent CO_2; agriculture and forestry are not taken into account.

The signatories have made most of their reductions in relation to the economic collapse of the East Block. In the foreseeable future, however, these countries are expected to once again have increasing emissions, so that other countries will increasingly have to reduce their emissions at home. But there are positive examples of western countries reducing emissions tremendously. In particular, the UK had reduced its emissions by 22.6 % as of 2010, largely because natural gas replaced coal. The UK thus reached its target for 2012 far in advance.

In contrast, the United States has increased its emissions and failed to comply with the original 7 % reduction of carbon emissions in the Kyoto Protocol, which the country never ratified.

International efforts to protect the climate have not led to tangible success yet. Perhaps the best thing about them is that they have drawn public attention to climate protection. The Kyoto Protocol expired in 2012, and since then negotiations have focused on a successor. It is unlikely that we will get binding targets for comprehensive, effective climate change. On the other hand, rising energy prices and rapid progress in developing renewables are reasons to hope that climate protection will soon be attractive worldwide even without international agreements.

Germany's Energiewende and climate protection

Carbon emission trends in Germany

In the past few years, Germany is one of the countries that has had its foot on the brakes in international climate protection negotiations. Thanks to the restructuring of the former East Germany, carbon emissions were reduced considerably at the beginning of the 1990s.

As German Chancellor, Helmut Kohl promised a 25 % reduction in carbon emissions by 2005 – a target that was not reached. This goal was even reduced in the Kyoto negotiations to 21 % by 2012. This target was at least met. Along with reunification, the main driver of this reduction was the economic crisis (Figure 1.16); efforts to protect the climate only played a marginal role. To date, climate protection policies and the growth of renewables have not sufficed to effectively protect the climate.

Events at the nuclear power plant near Fukushima showed that the use of nuclear power entails extreme risks even in high-tech countries. Germany's nuclear phaseout therefore now has broad support throughout society. But this stance on nuclear power poses an additional challenge for climate protection. After all, climate targets now have to be reached during the accelerated phaseout of nuclear power, which will only work if renewables are strongly supported and efficiency is consistently pursued. Up to now, German politicians have only half-heartedly taken the steps needed.

Renewable energy supply in Germany

Compared with other countries, Germany does not have great potential for renewables; no individual source can cover the country's energy supply alone. Germany therefore needs

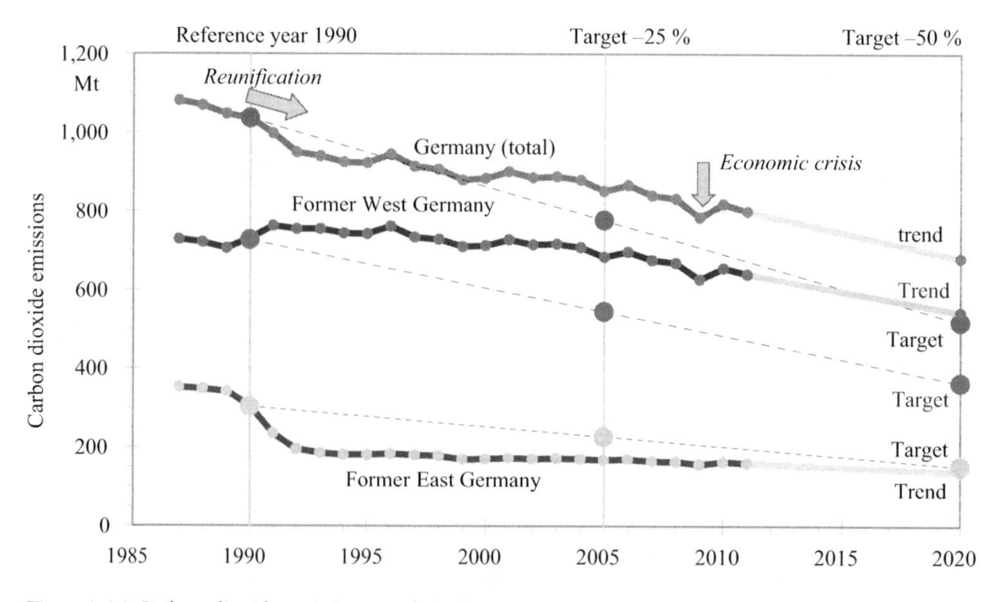

Figure 1.16 Carbon dioxide emission trends in Germany

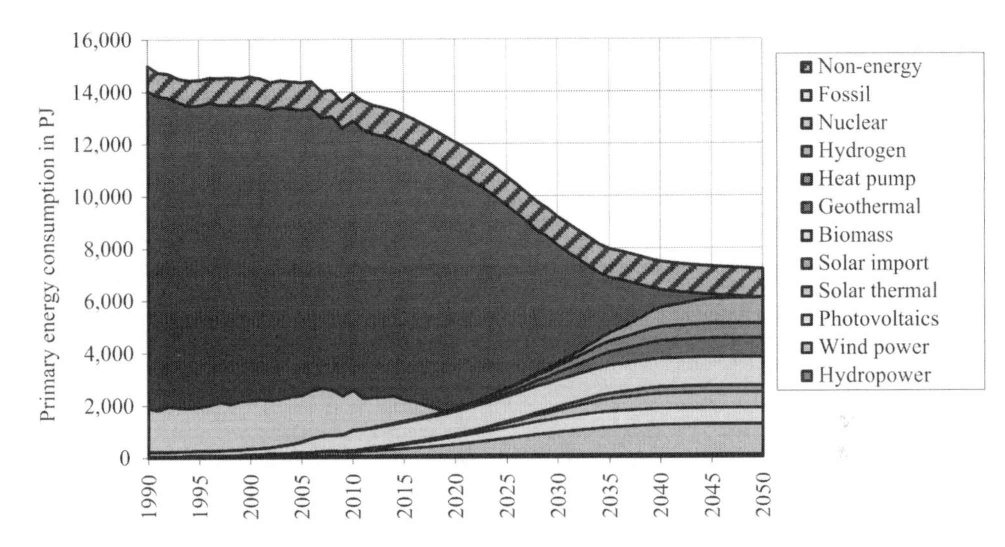

Figure 1.17 Previous development of primary energy demand in Germany from 1990 to 2011 along with a scenario for sustainable development up to 2050

a broad mix of different types of renewable sources. Figure 1.17 shows how **primary energy consumption** is likely to develop along with generation up to 2050. Consumption could be reduced in the process by around 50 %. Fossil and nuclear power plants lose 40 to 65 % of the primary energy put into them as waste heat through their cooling towers. Most power generators running on renewables do not have such losses. Statistically, great savings are therefore possible in terms of primary energy by switching generators from fossil and nuclear to renewables. The rest of the savings come from actual gains in efficiency.

POWER SUPPLY

Figure 1.18 shows an ambitious **scenario for power supply** that will reduce carbon emissions during an accelerated nuclear phaseout. A mixture of different renewable power generators will ensure power supply in the future. In Germany, wind power and photovoltaics have the greatest potential, followed at quite a distance by biomass use.

While the energy consumed for heat and transport are expected to decrease in terms of primary energy, no reduction is expected for electricity. New electrical applications – such as heat pumps and electric mobility – will more than make up for efficiency gains, leading to greater power consumption.

The scenario assumes that some renewable electricity will be imported. It can either come in the form of wind or solar power from other parts of Europe or northern Africa. Imports increased supply security and reduce the need for storage because fluctuations in the supply of wind and solar power compensate for each other over large areas. In principle, however, a 100 % supply of renewable energy within Germany would also be possible.

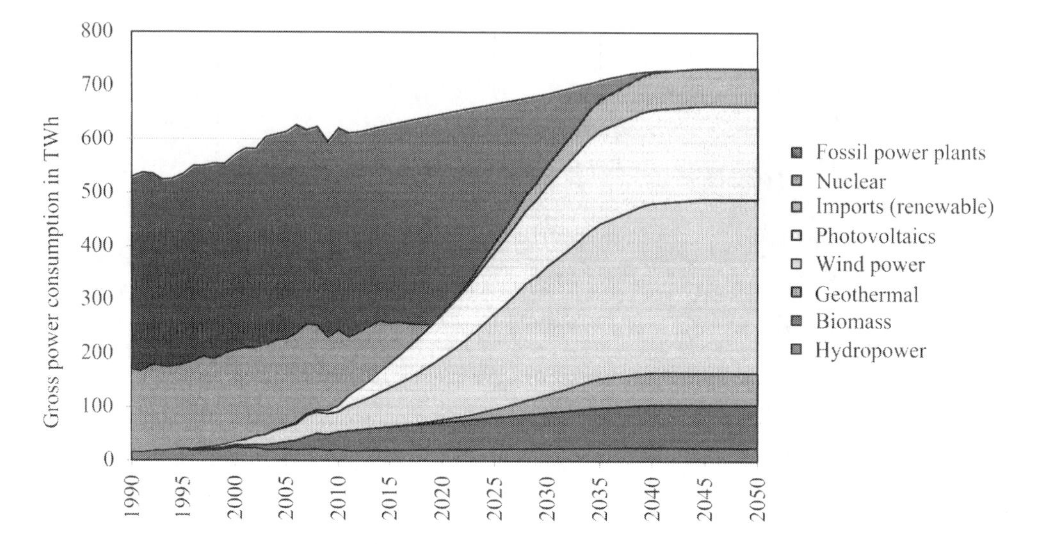

Figure 1.18 Previous development of gross power consumption in Germany from 1990 to 2011 along with a scenario for sustainable development up to 2050

The scenario described here clearly shows that Germany can sustainably get all of its energy from renewables alone. The potential exists. Renewable energy technologies are now quite sophisticated and largely competitive. Growth continues to clearly outstrip expectations from only a few years ago. Even faster growth is within the realms of possibility. If Germany manages to set an example for the world, global climate protection goals could be reached even sooner than currently expected. We can still keep the planet from warming up by more than 2 °C.

The steps that need to be taken for a successful energy transition in the power sector are well known and should be taken as soon as possible:

- Renewable power generation capacity needs to be installed quickly.
- Inflexible nuclear and lignite power plants need to be decommissioned soon.
- Grids need to be upgraded for the integration of renewable generation capacity.
- Smart grids need to be set up so that demand can better follow supply.
- New short-term storage needs to be built.
- Gas turbines need to be added as reserve and peak power plants.
- The power grid needs to be coupled with gas networks.

Transitioning energy supply entails enormous changes. The job is not done simply by setting up renewable power plants. Wind turbines and photovoltaics, which provide fluctuating energy supply, will dominate in the future. Because they do not ramp up and down as well, conventional nuclear and lignite power plants are unsuitable as bridge technologies; they will slow down the switch to a future-proof power supply. It is simply not possible, for instance, for nuclear and conventional lignite plants to be taken offline in the morning when the sun starts shining and bring them online quickly again in the evening. Security issues rule out such ramping with nuclear plants. Likewise, the extreme thermal loads would also wear down and possibly even damage lignite plants ramping up and down every day.

Because of the fluctuations in power supply, new storage capacity and concepts will be needed in order to ensure great reliability of supply (Figure 1.19). Today, Germany has around 40 GWh of pumped storage capacity, not even enough to cover German power demand for an hour; if Germany gets all of its electricity from renewables, enough storage to cover 1 to 4 weeks will be needed. Storage demand will not be greater than that because different sources of renewable energy, such as wind and photovoltaics, complement each other well, including seasonally.

The focus needs to be on storage solutions that are quickly available; after all, our power supply will need to be restructured before 2050 if we are to reach our climate protection targets. Because new pumped-storage facilities can only be built in a few places, battery storage is the primary solution for short-term storage to bridge a couple of hours. Lead batteries are relatively inexpensive and already available in great quantities. Other more powerful battery technologies will replace lead batteries over the midterm.

At present, storing power as a gas seems to be the most promising solution to bridge multiple days and weeks. The conversion of renewable electricity into gas as a storage medium is called **power-to-gas or P2G**. To underscore its renewable character, the term 'green gas' has also been coined. Germany has some 20 billion m³ of **natural gas storage** capacity, equivalent to 200 TWh of primary energy content. That volume is 5,000 times greater than current pumped-storage capacity. Even if we take into account the losses in the conversion of gas into electricity, Germany's existing gas storage capacity already largely suffices for a complete switch to renewable power.

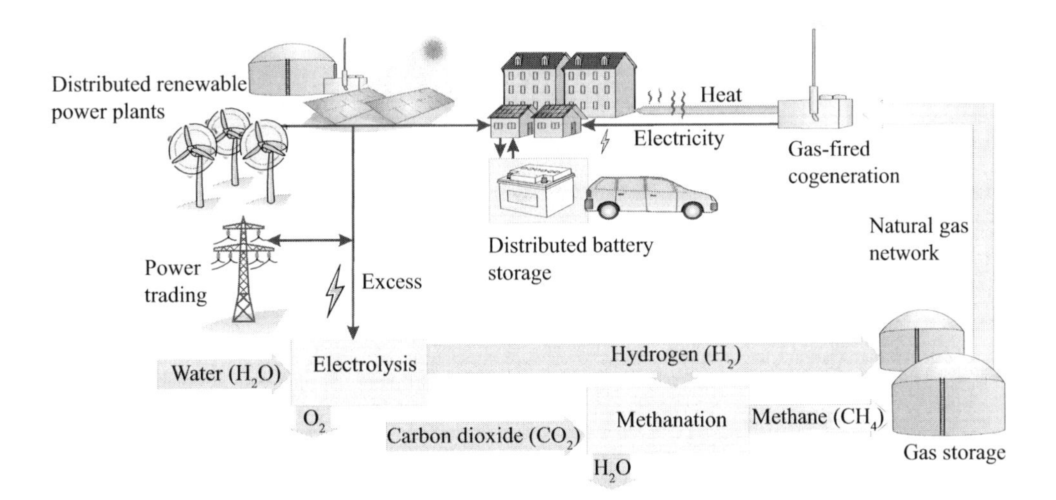

Figure 1.19 Storage concept for a renewable power supply

Electrolysis can be used to produce hydrogen from excess renewable electricity. Some of this hydrogen can be directly sold to natural gas networks. Larger amounts of green hydrogen can be used as a substitute for natural gas when carbon is added by means of a process called **methanation**. Natural gas networks can also be upgraded to accommodate a larger share of hydrogen (see Chapter 10).

As need be, gas turbines can turn the stored gas back into electricity, thereby ensuring power supply when there is not enough wind and solar power. Gas turbines are also much more flexible than coal and nuclear plants. It therefore makes sense to start replacing coal plant capacity with new gas turbines today. These turbines will initially run on natural gas but will be gradually transitioned to renewable gas to provide a reliable, carbon-free power supply.

HEAT SUPPLY

Unlike electricity demand, demand for heat can be reduced considerably. Roughly half of it can be offset through efficiency over the next 30 years, with the remainder being covered by renewables. To reach this goal, renovation rates will have to be stepped up considerably, however. In addition to optimal insulation, mechanical ventilation with heat recovery can reduce heat losses considerably.

A wide range of renewable technologies will also cover heat supply in the future. No single one of them has the potential to cover all of Germany's heat demand. Various technological and financial options are optimal depending on the specific location and building type. Figure 1.20 shows the various building blocks of renewable heat supply.

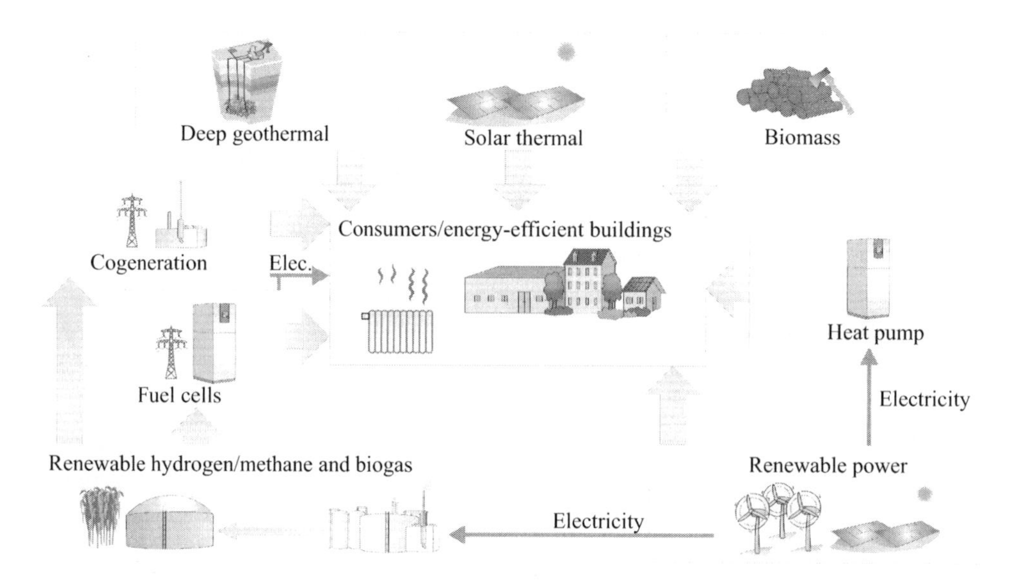

Figure 1.20 Building blocks of a renewable heat supply

TRANSPORT

For climate protection to be effective, considerable changes need to be made over the next few years in the transport sector. At present, transportation is largely based on fossil fuels, such as gasoline, diesel, and kerosene. More efficient motors will not help us reduce consumption of fossil fuels enough to reach our greenhouse gas targets. Likewise, replacing fossil fuels with biomass can only help reduce greenhouse gas emissions to a limited extent. The agricultural land available for energy crops will not suffice for the production of enough biofuels to replace fossil fuels.

We therefore urgently need alternatives. For road transport, there are great hopes for electric mobility. Electric motors are far more efficient than internal combustion engines and would therefore reduce energy demand in the transport sector considerably. Energy storage is a major challenge, but efficient batteries and hydrogen would be able to provide the storage capacity needed. If the electricity used to charge batteries or produce the hydrogen comes from renewables, the transport sector could become climate-neutral.

Transitioning energy supply

In conclusion, we already have the technologies and concepts for a sustainable energy supply. Transitioning our energy supply will, however, entail major changes in current supply structures. There will be winners and losers. Major utilities, for instance, refused to face the challenges of this transition for a long time. Their main involvement in renewables was limited to the further operation of existing hydropower plants. In 2011, the share of renewables in power production at Germany's four biggest energy providers was far below the national average (Table 1.17).

A number of researchers, politicians, and energy providers believe that our future energy supply should largely be based on **central renewable facilities**, such as offshore wind farms and photovoltaics/CSP in southern Europe and northern Africa. They believe such

Table 1.17 The composition of power generation at Germany's largest power providers in 2011 [AGEB12; EnB12; eon12; RWE12; Vat12]

Power provider	*RWE*	*e.on*[3]	*Vattenfall*	*EnBW*	*Germany*	*Climate protection*
Power generation	205.7 TWh	88.9 TWh	66.3 TWh	59.3 TWh	612 TWh	> 620 TWh
Renewables	4.3 %	3.5 %	[1] 3.9 %	10.8 %	19.9 %	>> 90 %
of which, hydropower	k.A.	3.0 %	3.9 %	>9 %	3.1 %	4 %
Nuclear power	16.7 %	51.1 %	0 %	47.7 %	17.7 %	0 %
Brown coal	36.0 %	[2]	80.7 %	[2]	24.6 %	0 %
Hard coal	23.2 %	45.4 %	9.2 %	[2] 38.1 %	18.7 %	0 %
Natural gas	18.7 %	[2]	5.6 %	[2]	13.7 %	<< 10%
Other	1.1 %	n/a	0.6 %	3.4 %	5.4 %	<< 10 %

1) Biomass included under Other
2) All fossil plants reported together
3) e.on provides insufficient figures for Germany and does not answer requests for more information. The table therefore contains several of my own estimates.

a supply would be the least expensive. Critics argue that a large number of high-voltage power lines would be needed, and they would meet with popular resistance. Not everyone is convinced that the cost of offshore wind and CSP can be reduced as much as expected, however. The investors will also have to have deep pockets because of the tremendous upfront cost of central power plants, which means large power providers will not face competition from small investors. And because these new investments will compete with existing fossil and nuclear plants, the transition will be unnecessarily slowed down. In the midterm, the high share of outdated nuclear and lignite plants will prevent large utilities from becoming a driving force in a transition to renewables. Indeed, these firms are using their influence on politicians to slow down this transition as much as they can in order to protect their high profits from their fossil and nuclear assets. Various environmental protection organizations therefore recommend that consumers switch to independent green power providers [Nat12].

Those who want a faster transition therefore call for **distributed renewables**. Such systems are smaller, can be built more quickly, and do not require tremendous upfront investments. One result would also be greater competition with established power suppliers. Another would be less need for high-voltage lines. In return, the distribution grid would have to be upgraded and local storage expanded. A lot of consumers would make their own power, such as from their own solar rooftops, which would democratize energy supply.

In practice, the most likely outcome is a mixture of the two concepts. But the longer large power providers wait, the greater the share of distributed renewables will be. It is hard to judge the extent to which competition between renewables and our current supply system leads to additional costs because the two systems are not optimized to lower costs. Judged against the high cost of future damage from climate change, the cost of accidents like the one in Fukushima, and problems caused by the rising price of fossil fuel, such questions about the cost of the transition seem to be a distraction. The main question is how we can make our energy supply sustainable in order to protect living standards for future generations.

2 Solar radiation

Introduction

The sun is by far the largest source of renewable energy. As explained in the previous chapter, geothermal and gravity are minor compared with solar energy. The sun's rays can be directly used in solar thermal or photovoltaic systems. Wind power and hydropower can be considered indirect forms of solar energy because the motion of wind and water are partly related to the sun. Because exact knowledge of insolation is crucial for calculations and simulations of a wide range of renewable energy systems, this chapter deals with the topic. It mainly focuses on photometry calculations. The main photometric parameters are shown in Table 2.1, with those pertaining to radiation physics being the most important for the use of solar energy. Those related to light merely concern the visible spectrum, while solar energy systems can also exploit invisible ultraviolet and infrared parts of the spectrum.

A number of the calculations below require physical constants, which are collected in the Appendix.

The sun: a fusion reactor

The sun is the centre of our solar system. It has been shining at its current intensity for around the past five billion years and will probably remain in existence for roughly the same timeframe. The sun is around 80 % hydrogen, 20 % helium, and only 0.1 % other elements. Table 2.2 shows the main data in a comparison of the sun with the earth.

The sun's radiation comes from **nuclear fusion processes**. In various intermediate reactions, four hydrogen nuclei (protons ^{1}p) fuse into a single helium nucleus (alpha particle $^{4}\alpha$)

Table 2.1 Major parameters in radiation physics and light [DIN5031]

Radiation-physics parameters			*Light parameters*		
Name	*Symbol*	*Unit*	*Name*	*Symbol*	*Unit*
Radiant energy	Q_e	Ws	Luminous energy	Q_v	lm s
Radiant flux	Φ_e	W	Luminous flux	Φ_v	lm
Radiant emittance	M_e	W/m²	Luminous emittance	M_v	lm/m²
Radiant intensity	I_e	W/sr	Luminous intensity	I_v	cd = lm/sr
Radiance	L_e	W/(m² sr)	Luminance	L_v	cd/m²
Irradiance	E_e	W/m²	Illuminance	E_v	lx = lm/m²
Irradiation	H_e	Ws/m²	Light exposure	H_v	lx s

Units: W = watt; m = metre; s = second; sr = steradiant; lm = lumen; lx = lux; cd = candela

Table 2.2 Data for the sun and earth

	Sun	*Earth*	*Ratio*
Diameter	1,391,320 km	12,756 km	1 : 109
Circumference	4,370,961 km	40,075 km	1 : 109
Surface	$6.081 \cdot 10^{12}$ km²	$5.101 \cdot 10^{8}$ km²	1 : 11,897
Volume	$1.410 \cdot 10^{18}$ km³	$1.0833 \cdot 10^{12}$ km³	1 : 1,297,590
Mass	$1.9891 \cdot 10^{30}$ kg	$5.9742 \cdot 10^{24}$ kg	1 : 332,946
Average density	1.409 g/cm³	5.516 g/cm³	1 : 0.26
Standard gravity (surface)	274.0 m/s²	9.81 m/s²	1 : 28
Surface temperature	5,777 K	288 K	1 : 367
Temperature at centre	15,000,000 K	6,700 K	1 : 2,200

consisting of two neutrons ^{1}n and two positively charged protons ^{1}p. In the process, two positrons e^+ and two neutrinos ν_e are created. The formula for the gross reaction, which is illustrated in Figure 2.1, is thus:

$$4\,{}^1_1\text{p} \rightarrow {}^4_2\alpha + 2e^+ + 2\nu_e + \Delta E \tag{2.1}$$

When we compared the mass of these nuclei before and after the reaction, we see that the total mass shrinks during the reaction. The particle masses are indicated in Table 2.3.

The differential mass Δm can be calculated as:

$$\Delta m = 4 \cdot m({}^1\text{p}) - m({}^4\alpha) - 2 \cdot m(e^+) \tag{2.2}$$

producing $\Delta m = 4 \cdot 1.00727647\text{ u} - 4.0015060883\text{ u} - 2 \cdot 0.00054858\text{ u} = 0.02650263\text{ u}$.

In this calculation, the mass of the neutrinos ν_e is not taken into account. The mass of a positron e^+ is equivalent to that of an electron e^-.

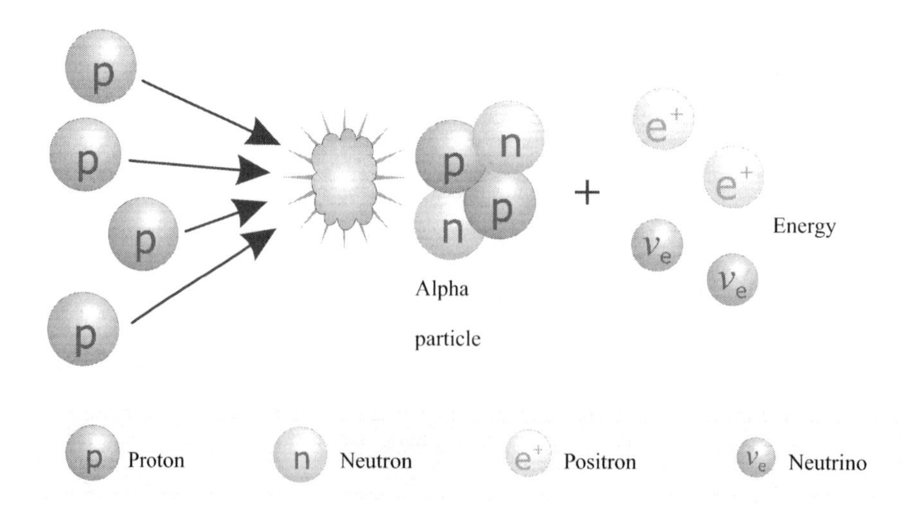

Figure 2.1 Fusion of four hydrogen nuclei, producing a helium nucleus (alpha particle)

Table 2.3 Various particle and nuclide masses (1 u = $1.660565 \cdot 10^{-27}$ kg)

Particle/nuclide	*Mass*	*Particle/nuclide*	*Mass*
Electron (e^-)	0.00054858 u	Hydrogen (1H)	1.007825032 u
Proton (^1p)	1.00727647 u	Helium (4He)	4.002603250 u
Neutron (^1n)	1.008664923 u	Alpha particle (4α)	4.0015060883 u

The total mass of all of the particles after fusion is thus lower than the sum of all of the particles before the fusion reaction. The mass defect Δm results from the conversion of Δm into the energy released ΔE, with ΔE being calculated as follows:

$$\Delta E = \Delta m \cdot c^2 \tag{2.3}$$

With $c = 2.99792458 \cdot 10^8$ m/s, the energy released during fusion is $\Delta E = 3.955 \cdot 10^{-12}$ J = 24.687 MeV. The various masses and the related energy difference can be explained based on the binding energy E_b of a nucleus ^{N+Z}K. A nucleus consists of N neutrons ^1n and Z protons ^1p. For a nucleus created from the meeting of protons and neutrons to be stable, its binding energy has to be released when protons and neutrons come together. The binding energy of a helium nucleus can be calculated from the mass difference between the alpha particle and two neutrons plus two protons.

Above, the electrons in the electron shell are not taken into account; only the atomic nuclei are. A hydrogen atom ^1H has one electron in its shell, compared with two electrons in the shell of ^4He. Two of the four electrons in the hydrogen atoms are found in the helium atom. The other two electrons are annihilated with the positrons, meaning that two electrons and two protons are converted into radiant energy. This radiant energy is thus four times greater than the mass of an electron at 2.044 MeV.

In total, this fusion reaction releases 26.731 MeV. While that may not sound like a lot of energy, the large number of fusion processes add up to release $3.8 \cdot 10^{26}$ Ws of energy each second.

Every second, the sun loses 4.3 million tons of mass ($\Delta \dot{m} = 4.3 \cdot 10^9$ kg/s). The radiant energy $\Phi_{e,S}$ of the sun is thus:

$$\Phi_{e,S} = \Delta \dot{m} \cdot c^2 = 3{,}845 \cdot 10^{26} \text{ W} \tag{2.4}$$

If this value is divided by the sun's surface A_{sun}, the product is the **radiant emittance of the sun**:

$$M_{e,S} = \frac{\Phi_{e,S}}{A_{sun}} = 63.3 \frac{\text{MW}}{\text{m}^2} \tag{2.5}$$

Every square metre of the sun's surface has a radiant emittance of 63.3 MW. A quarter square kilometre of the sun's surface radiates around 500 EJ each year, roughly equivalent to the entire primary energy demand worldwide. Of course, only a small part of this energy reaches the earth.

We can think of the sun as a black body. The **Stefan-Boltzmann law**

$$M_e(T) = \sigma \cdot T^4 \tag{2.6}$$

can then be used to determine the **sun's surface temperature** (T_{sun}). The Stefan-Boltzmann constant $\sigma = 5.67051 \cdot 10^{-8}$ W/(m² K⁴) produces:

$$T_{Sun} = \sqrt[4]{\frac{M_{e,S}}{\sigma}} = 5777 \text{ K} \tag{2.7}$$

If we imagine a sphere around the sun with a radius equivalent to the average distance between the centre of the earth and the centre of the sun ($r_{SE} = 1.496 \cdot 10^8$ km), the same amount of radiant energy that passes through the surface of the sun A_S also passes through the sphere A_{SE} (Figure 2.2). The sun's radiant emittance per square metre $M_{e,S}$ is, however, much greater than the sphere's irradiance E_e.

$M_{e,S} \cdot A_S = E_e \cdot A_{SE}$ and $A_{SE} = 4 \cdot \pi \cdot r_{SE}^2$ are added to produce the irradiance E_e:

$$E_e = M_{e,S} \cdot \frac{A_S}{A_{SE}} = M_{e,S} \cdot \frac{r_S^2}{r_{SE}^2} \tag{2.8}$$

It is equivalent to the extra-terrestrial irradiance of the earth on this sphere. The distance between the sun and the earth is not constant, but varies over the course of the year

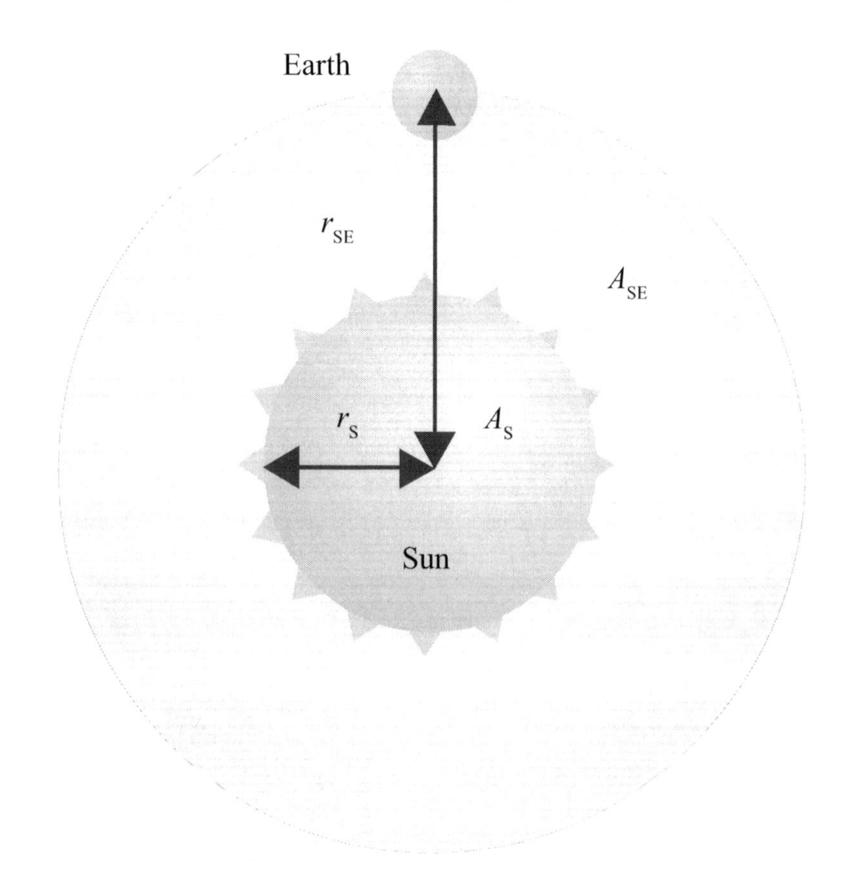

Figure 2.2 The same radiant energy passes through the surface of the sphere with the radius r_{SE} and through the sun

between $1.471 \cdot 10^8$ km and $1.521 \cdot 10^8$ km; as a result, the irradiance E_e varies between 1,321 W/m² and 1,412 W/m². The average is called the **solar constant** E_0 and is calculated as:

$$E_0 = 1360.8 \pm 0.5 \ \frac{\mathrm{W}}{\mathrm{m}^2} \tag{2.9}$$

Outside of the earth's atmosphere, this value can be measured with a surface directly facing the sun [Kop11]. The variations over the course of the year relative to the day of the year J can be calculated as follows:

$$E_0(J) = E_0 \cdot \left(1 + 0.0334 \cdot \cos\left(0.0172° \cdot J - 0.04747°\right)\right) \tag{2.10}$$

In addition to the total irradiance incident on the earth, the spectral composition of sunlight is very important for the use of solar energy. Sunlight comes in the form of photons at various wavelengths λ. Waves ranging from a length of 380 to 780 nm or 0.38 to 0.78 µm are visible to humans. Table 2.4 shows the colours of these wavelengths.

The sun can be roughly thought of as a black body, with a temperature corresponding to the sun's surface temperature of 5,777 K. A black body's **spectral radiance** $L_{e,\lambda}$ is relative to the wavelength λ for an absolute temperature T and can be calculated with the following formula according to Planck:

$$L_{e,\lambda} = \frac{c_1}{\lambda^5} \cdot \frac{1}{\exp\left(\dfrac{c_2}{\lambda \cdot T}\right) - 1} \cdot \frac{1}{\Omega_0} \tag{2.11}$$

Here,

$$c_1 = 2 \cdot h \cdot c^2 = 1{,}191 \cdot 10^{-16} \ \mathrm{Wm}^2 \tag{2.12}$$

and $$c_2 = \frac{h \cdot c}{k} = 1.439 \cdot 10^{-2} \ \mathrm{m\,K} \tag{2.13}$$

Table 2.4 Wavelengths of various hues

Hue	*Wavelength in nm*	*Hue*	*Wavelength in nm*
Ultraviolet	< 380	Yellow-green	560–570
Violet	380–450	Greenish yellow	570–575
Blue	450–482	Yellow	575–580
Greenish blue	482–487	Yellowish orange	580–585
Cyan	487–492	Orange	585–595
Bluish green	492–497	Reddish orange	595–620
Green	497–530	Red	620–780
Yellowish green	530–560	Infrared	> 780

1 µm = 1,000 nm, 1 nm = 0.001 µm

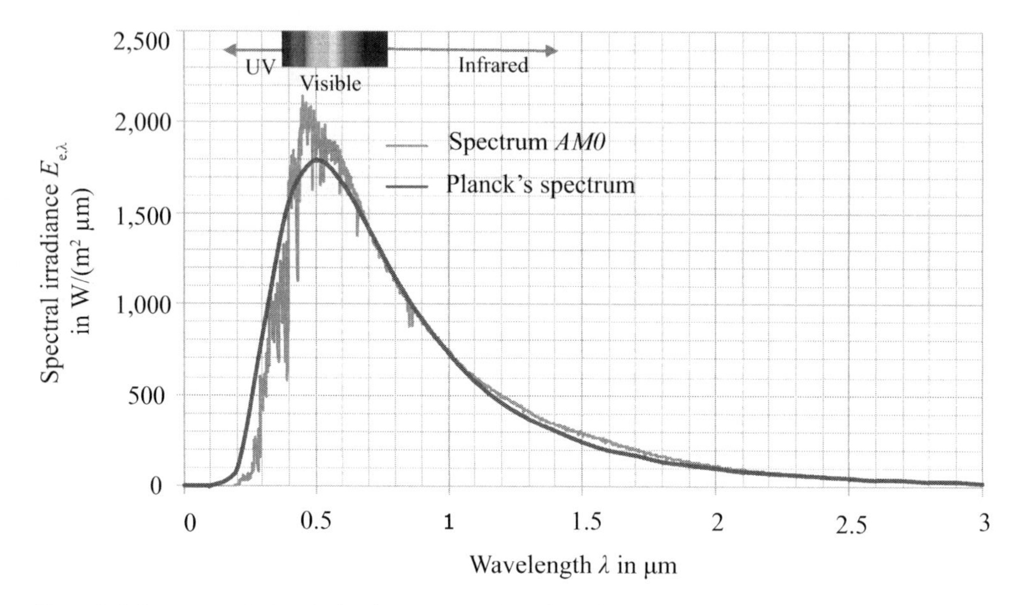

Figure 2.3 Spectrum *AM0* and Planck's Spectrum for a black body at a temperature of 5,777 K

$\Omega_0 = 1$ sr is only needed to correct for the unit balance; steradiant sr represents the unit of the solid angle. If a body radiates in all directions in space, the specific radiant emittance $M_{e,\lambda}$ (relative to the wavelength λ) and spectral irradiance is:

$$E_{e,\lambda} = \frac{r_S^2}{r_{SE}^2} \cdot M_{e,\lambda} = \frac{r_S^2}{r_{SE}^2} \cdot \pi \cdot L_{e,\lambda} \tag{2.14}$$

The previously determined irradiance E_e is the product of the integration of the spectral irradiance relative to the wavelength $E_{e,\lambda}$:

$$E_e = \int E_{e,\lambda} \mathrm{d}\lambda \tag{2.15}$$

In reality, the spectrum has a measurable, but minor deviation from the ideal of a black body (Figure 2.3). The real spectrum outside the earth's atmosphere is called spectrum *AM0*. In this extra-terrestrial spectrum, 7 % of the irradiance stems from the ultraviolet part of the spectrum, with 47 % coming from the visible part of the spectrum, and 46 % from the infrared section.

Irradiance on the earth

The irradiance measured on the earth is generally lower than that measured in space. Irradiance is reduced in the atmosphere. Here, distinctions are made between the following:

- reductions from atmospheric reflections
- reductions from atmospheric absorption (mainly O_3, H_2O, O_2, and CO_2)

- reduction as a result of Rayleigh scattering
- reduction as a result of Mie scattering.

Various gas particles in the atmosphere are the cause of reduction as a result of absorption. The absorption of various atmospheric components – such as water vapour, ozone, oxygen, and carbon dioxide – is highly selective, covering only a few parts of the solar spectrum.

Figure 2.4 shows spectrum *AM0* in space and *AM1.5*g on earth. Spectrum *AM1.5*g clearly shows the dips as a result of absorption by various gas particles. In a clear sky with the sun at 41.8° on a surface at an angle of 37° facing the sun, spectrum *AM1.5*g can be measured. Its total irradiance is 1,000 W/m². Spectrum *AM1.5*g serves as a reference in the classification of photovoltaic modules.

Reduction from **Rayleigh scattering** occurs when light hits molecular components of the atmosphere with a diameter much smaller than the wavelength of light. The impact of Rayleigh scattering increases as the wavelength of light decreases.

The reduction caused by **Mie scattering** is the result of particulate matter (pollution) in the air. These particles have a diameter much larger than wavelengths of light. Mie scattering differs from one location to another. It is weakest at high altitudes and strongest in industrial areas with heavily polluted air.

Table 2.5 shows various levels of reduction relative to the height of the sun γ_S. Other reductions can occur as a result of weather effects, such as heavy clouds, snow, and rain.

When the sun is low in the sky and its rays have a longer way to travel through the atmosphere, losses in the atmosphere increase. The relationship between the elevation of the sun γ_S and **air mass** (*AM*) is defined as follows:

$$AM = \frac{1}{\sin \gamma_S} \tag{2.16}$$

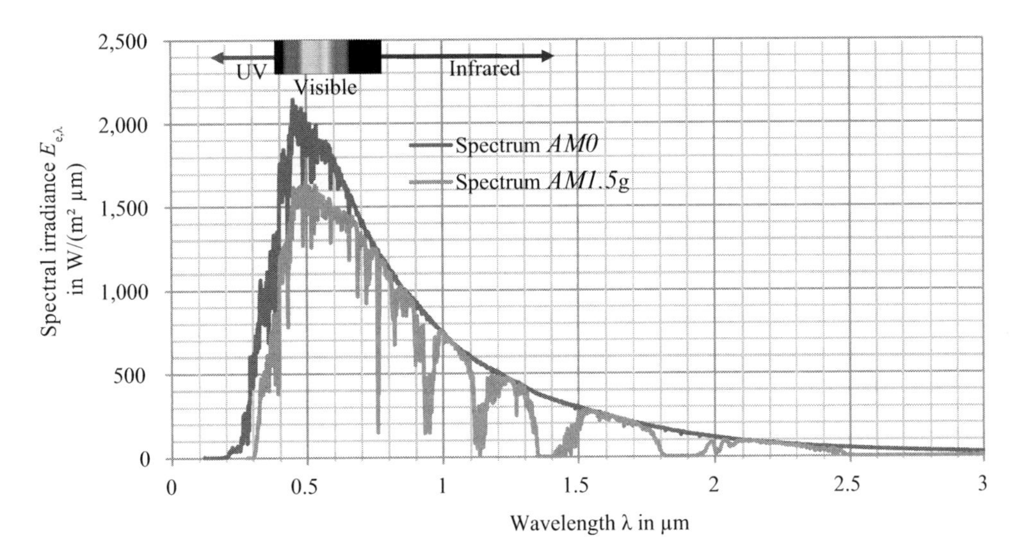

Figure 2.4 Spectrums of sunlight. *AM0*: the spectrum in space; *AM1.5*g: the spectrum on the earth with the sun at 41.8° and a surface facing the sun at an angle of 37°

The *AM* value indicates how often the path taken by sunlight was the shortest through the atmosphere. At a 90° angle, $AM = 1$; in space, it is zero.

Figure 2.5 shows the highest elevation of the sun and the resulting *AM* value for various days of the year in Berlin and Cairo.

The measurable irradiance E_0 is reduced as it passes through the atmosphere as a result of the aforementioned influences. Transmittance τ describes these effects. Reflections from

Table 2.5 Reduction influences relative to the height of the sun [Sch70]

γ_S	*AM*	*Absorption*	*Rayleigh scattering*	*Mie scattering*	*Total reduction*
90°	1.00	8.7 %	9.4 %	0–25.6 %	17.3–38.5 %
60°	1.15	9.2 %	10.5 %	0.7–29.5 %	19.4–42.8 %
30°	2.00	11.2 %	16.3 %	4.1–44.9 %	28.8–59.1 %
10°	5.76	16.2 %	31.9 %	15.4–74.3 %	51.8–85.4 %
5°	11.5	19.5 %	42.5 %	24.6–86.5 %	65.1–93.8 %

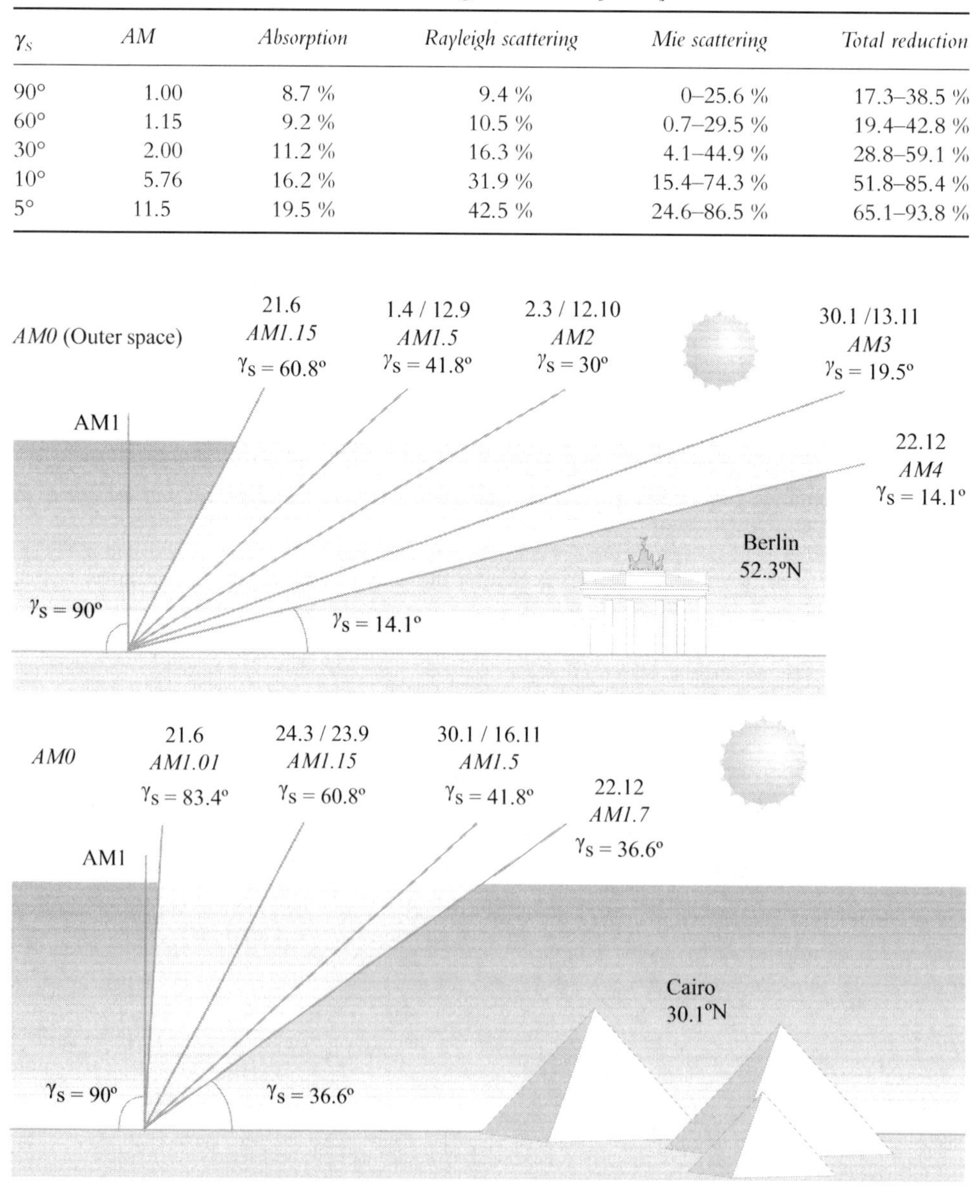

Figure 2.5 Sun paths and *AM* values for various days in Berlin and Cairo

bright clouds or snow-covered surfaces with degree of reflection ρ can also increase irradiance locally. The irradiance on a horizontal surface on the planet's surface – called 'global irradiance'– is calculated as follows:

$$E_{G,hor} = E_0 \cdot \tau_{Absorption} \cdot \tau_{Rayleigh} \cdot \tau_{Mie} \cdot \tau_{Clouds} \cdot (1+\rho) \tag{2.17}$$

Under optimal weather conditions and when the sun is high in the sky, irradiance can reach around 1,000 W/m². If there is a lot of reflection, the levels can even briefly exceed 1,200 W/m². In Germany, however, such levels rarely occur.

In the morning and evening, irradiance drops because the path light takes through the atmosphere is longer. Likewise, the sun is lower in the sky in the winter, so the path is longer – resulting in lower irradiance. In addition, clouds can reduce irradiance considerably. Figure 2.6 shows diurnal curves for global irradiance in Karlsruhe, Germany, on a cloudless summer day, a cloudless winter day, and a winter day with heavy cloud cover.

Irradiance over time $E_{G,hor}$ produces irradiation $H_{G,hor}$. Irradiance is usually measured in discrete intervals Δt, so irradiation is usually the sum of:

$$H_{G,hor} = \int E_{G,hor} dt = \sum E_{G,hor} \cdot \Delta t \tag{2.18}$$

Table 2.6 shows average monthly and annual global irradiation at four locations in Germany. The values for July are up to 10 times greater than those for December.

The **global annual irradiation** across all of Germany is just above 1,000 kWh/m². At latitudes north of Germany, annual global irradiation ranges from 700 kWh/m² to

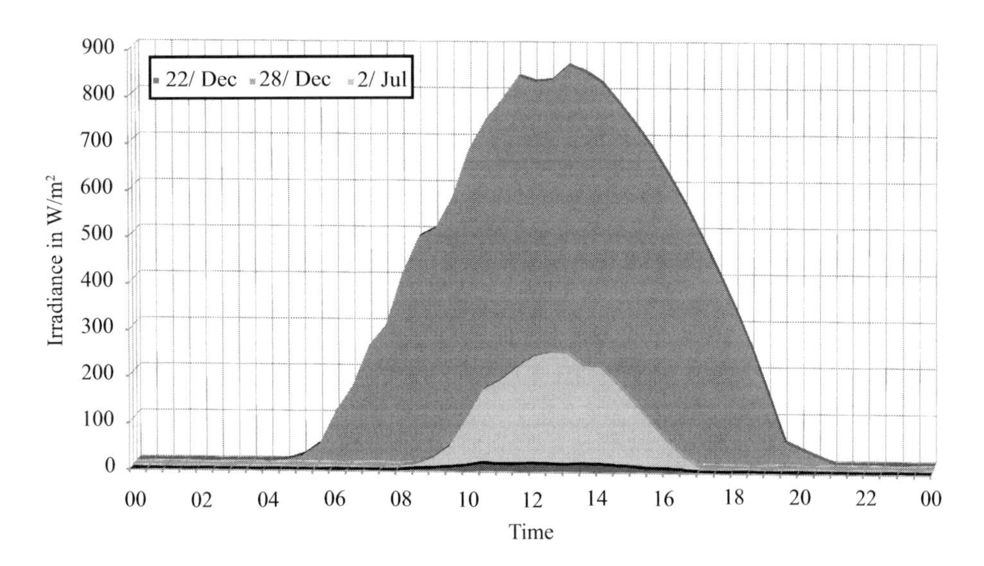

Figure 2.6 Diurnal curves of global irradiance in Karlsruhe on 2 July, 22 December, and 28 December 1991

Table 2.6 Long-term averages (1998–2010) of monthly global irradiation [JRC10]

kWh/m²	*Jan*	*Feb*	*Mar*	*Apr*	*May*	*Jun*	*Jul*	*Aug*	*Sep*	*Oct*	*Nov*	*Dec*	*Year*
Berlin	19	33	75	128	160	166	158	134	94	51	26	15	1059
Kassel	20	34	77	123	150	162	154	132	90	52	25	16	1037
Stuttgart	29	45	85	130	153	174	164	140	99	62	36	24	1139
Freiburg	29	45	84	129	153	172	166	141	104	63	38	24	1150

Table 2.7 Averages (1998–2010) of monthly global irradiation in kWh/m²

Latitude	*Bergen*	*Stockholm*	*Berlin*	*London*	*Vienna*	*Nice*	*Rome*	*Antalya*	*Almería*
	60°24'N	*59°21'N*	*52°28'N*	*51°31'N*	*48°15'N*	*43°39'N*	*41°48'N*	*36°53'N*	*36°50'N*
Jan	6	9	19	25	29	55	54	74	88
Feb	19	25	33	39	47	75	74	88	99
Mar	53	61	75	79	91	126	124	147	157
Apr	96	107	128	121	143	154	157	175	195
May	135	164	160	154	167	196	202	219	228
Jun	155	161	166	164	171	220	223	243	242
Jul	140	161	158	161	176	229	240	248	248
Aug	101	120	134	133	150	197	203	223	221
Sep	61	74	94	97	104	139	146	174	165
Oct	26	36	51	56	64	97	102	131	126
Nov	8	13	26	31	34	61	62	89	93
Dec	4	5	15	21	22	47	48	70	78
Year	803	938	1059	1070	1197	1595	1639	1883	1942

(Data: [JRC10])

1,000 kWh/m². It can exceed 1,800 kWh/m² in southern Europe. In deserts like the Sahara, local values can even exceed 2,500 kWh/m². But latitude alone does not suffice as a reliable indication of annual global irradiation. For instance, levels are roughly the same in Berlin and Stockholm, which is 7° north of the German capital, while the level in London is lower though the British capital is slightly further south (Table 2.7).

Within Europe, there are great differences in monthly irradiation. The main difference is between summer and winter. In Bergen, Norway, the disparity between June and December is 40:1, compared with only 3.3:1 in Lisbon, Portugal.

In the Sahara, average annual irradiation is around 2,350 kWh/m², twice as great as in Germany. Across the desert's total area of 8.7 million km², the Sahara has enough solar potential to cover global primary energy demand 200-fold. Only 48,500 km² of this area – roughly 1.5 times the size of Brandenburg – would suffice to cover all of mankind's energy demand worldwide with solar energy. Clearly, these figures show that it is possible to cover our entire energy demand with solar energy alone.

The solar energy incident within Germany is enough to cover German primary energy demand 100-fold. From one year to another, however, irradiation varies with the weather and changing climate conditions. Figure 2.8 shows the curve for irradiation over 70 years in Potsdam, Germany. There is a slight, but salient, increase in annual irradiation since the

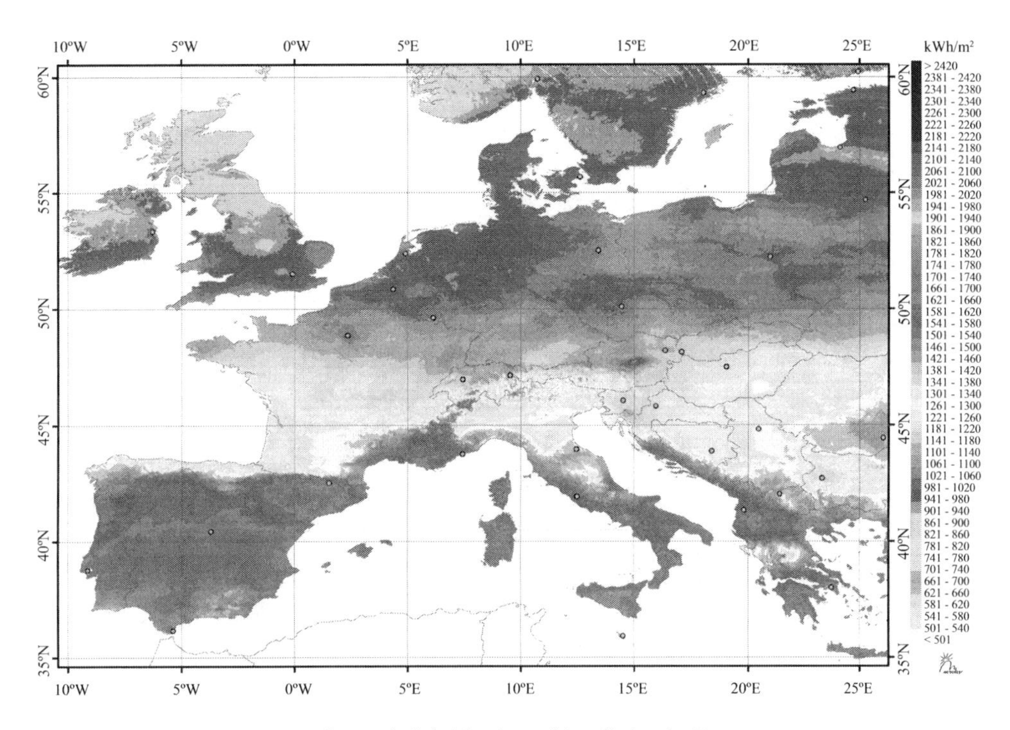

Figure 2.7 Long-term average of annual global horizontal irradiation in Europe

(Source: Meteotest; Meteonorm database (www.meteonorm.com))

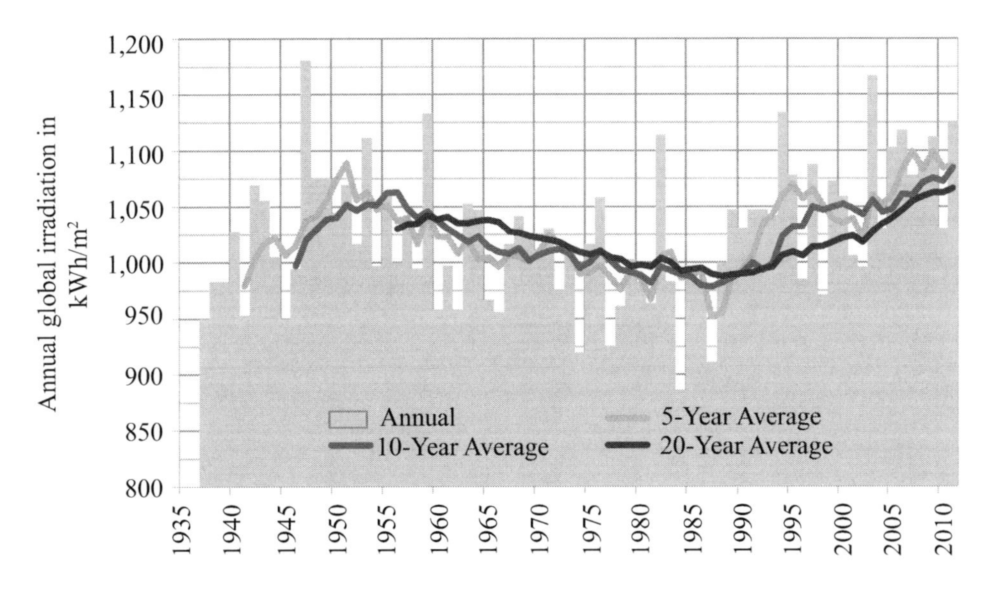

Figure 2.8 Annual irradiation in Potsdam from 1937 to 2011 along with the rolling average

(Data: DWD; Deutscher Wetterdienst – Germany's National Meteorological Service (www.dwd.de))

end of the 1980s. One reason may be that stricter pollution regulations have reduced the concentration of particles in the atmosphere that absorb sunlight.

Horizontal irradiance

As described above, sunlight is scattered and reflected by particles as it passes through the atmosphere. In space, almost all irradiance is direct, but the horizontal plane on the earth is exposed to both direct and diffuse light (Figure 2.9). Direct sunlight comes only in one direction – straight from the sun – and it is what causes shade behind objects with clear contours. In contrast, diffuse light does not come from any particular direction.

The composition of total/global irradiance $E_{G,hor}$ is calculated as follows from the direct irradiance $E_{dir,hor}$ and the diffuse irradiance $E_{diff,hor}$:

$$E_{G,hor} = E_{dir,hor} + E_{diff,hor} \quad (2.19)$$

Table 2.8 shows the shares of direct and diffuse light for an entire year in Berlin. The share of **diffuse light** is especially great on days with low overall irradiation and can reach 100 %. In contrast, the diffuse share drops below 20 % on days with a lot of global irradiation. Figure 2.10 shows typical values for daily direct and diffuse irradiation in Germany. The chart also shows that the share of diffuse light can exceed 50 % on average even in the summer.

Annual diffuse irradiation often only differs slightly even across great geographical deviations, as shown in Table 2.9. In contrast, there are tremendous differences in direct

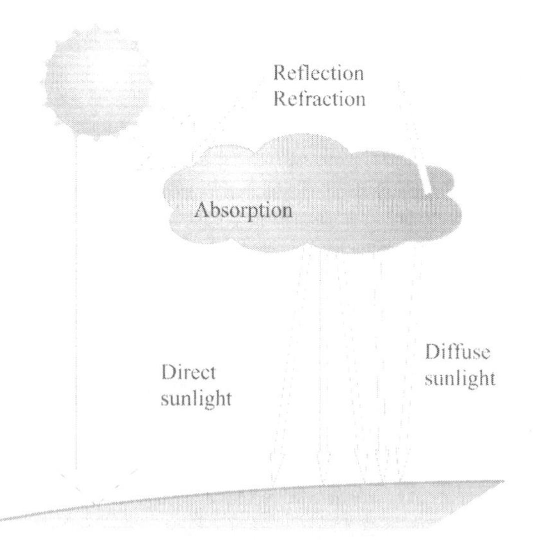

Figure 2.9 Sunlight passing through the atmosphere

Table 2.8 Averages (1998–2010) for direct and diffuse global irradiation in Berlin in kWh/m² [JRC10]

	Jan	*Feb*	*Mar*	*Apr*	*May*	*Jun*	*Jul*	*Aug*	*Sep*	*Oct*	*Nov*	*Dec*	*Year*
Direct	5	12	31	69	81	81	71	67	41	19	10	4	487
Diffuse	14	22	44	59	78	85	87	67	53	32	17	11	572

Table 2.9 Average (1998–2010) annual direct and diffuse global irradiation in kWh/m² [JRC10]

	Bergen	*Stockholm*	*Berlin*	*London*	*Vienna*	*Nice*	*Rome*	*Antalya*	*Almería*
Direct	305	450	487	486	599	1,037	1,065	1,337	1,418
Diffuse	498	488	572	594	599	588	574	546	524

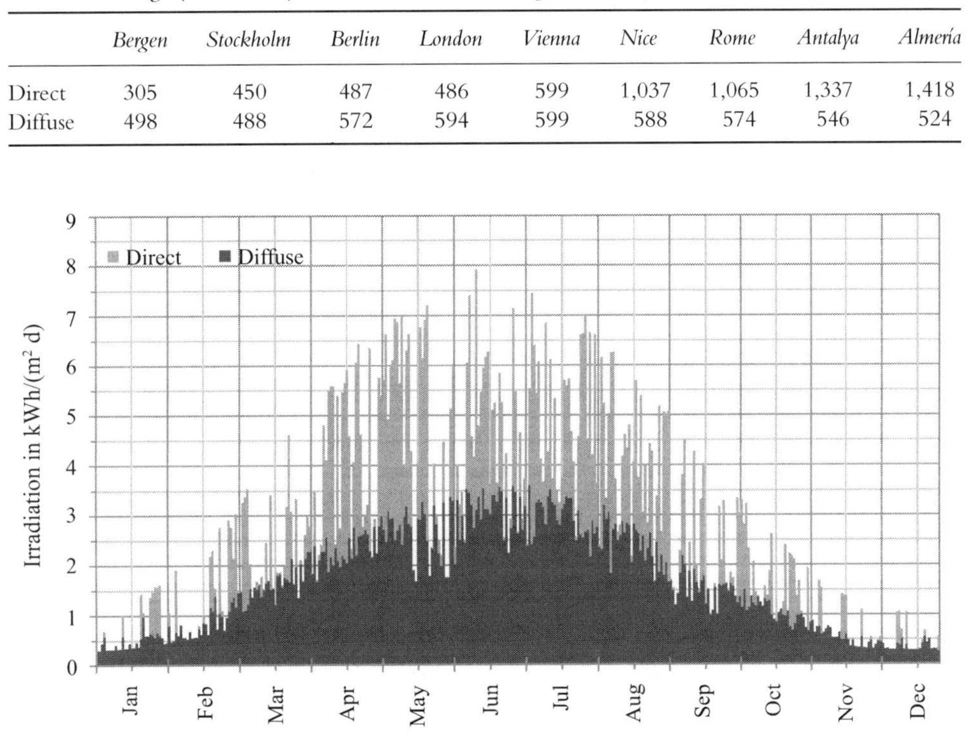

Figure 2.10 Daily sums of direct and diffuse irradiation in Berlin

irradiation. The share of direct sunlight is much greater in southern Europe than in northern Europe. In the summer, it can exceed 70 %, and it is clearly above 50 % even in the winter.

If a distinction needs to be made for global irradiance between direct and diffuse light but no separate measurements are available, statistics can be used to calculate the difference [Rei89]. Hourly values for global irradiance $E_{G,hor}$, the solar constant E_0, and the height of the sun γ_S produce the factor:

$$k_T = \frac{E_{G,hor}}{E_0 \cdot \sin \gamma_S} \tag{2.20}$$

which can be used to determine diffuse irradiance $E_{diff,hor}$ from global irradiance $E_{G,hor}$ and the elevation of the sun γ_S:

$$E_{diff,hor} = E_{G,hor} \cdot (1{,}020 - 0{,}254 \cdot k_T + 0{,}0123 \cdot \sin \gamma_S) \text{ for } k_T \leq 0.3$$

$$E_{diff,hor} = E_{G,hor} \cdot (1{,}400 - 1{,}749 \cdot k_T + 0{,}177 \cdot \sin \gamma_S) \text{ for } 0.3 < k_T < 0.78$$

$$E_{diff,hor} = E_{G,hor} \cdot (0{,}486 \cdot k_T - 0{,}182 \cdot \sin \gamma_S) \text{ for } k_T \geq 0.78 \tag{2.21}$$

Figure 2.12 shows a visualization of the formula. Here, we clearly see that there is a small share of diffuse light even on clear summer days of great total irradiance ($k_T \rightarrow 1$), but

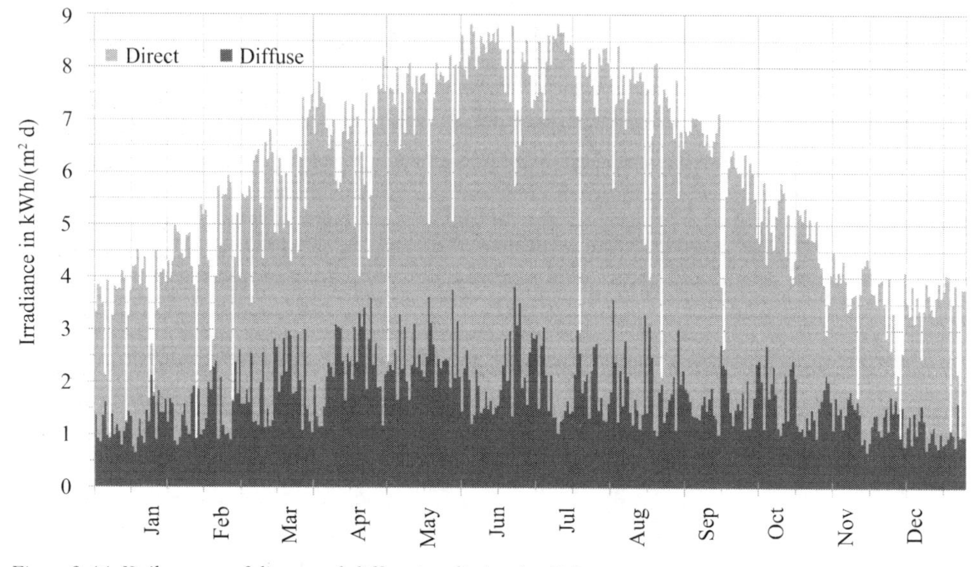

Figure 2.11 Daily sums of direct and diffuse irradiation in Cairo

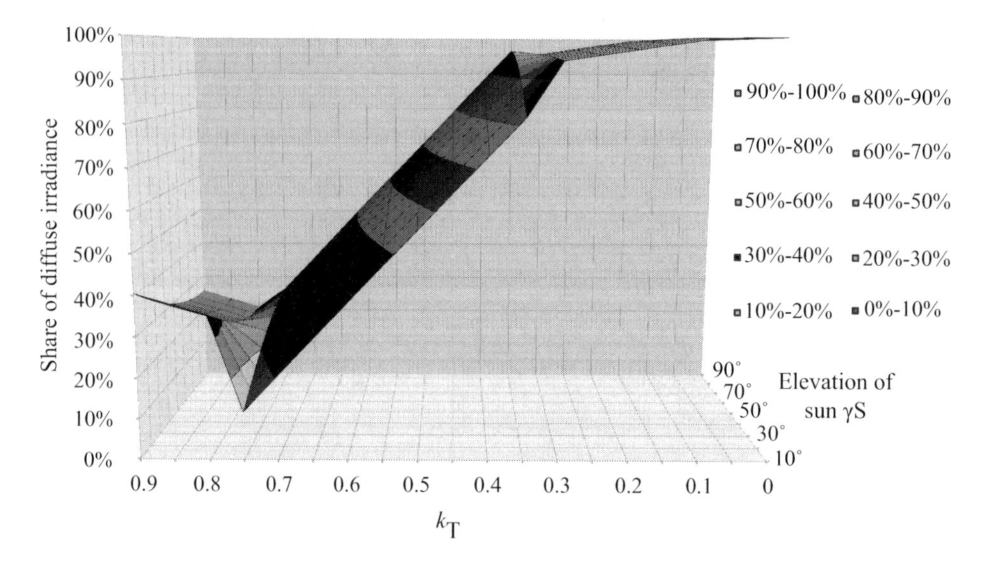

Figure 2.12 The share of diffuse irradiance relative to k_T and γ_S

the share increases to nearly 100 % on days of heavy cloud cover and low total irradiance ($k_T \rightarrow 0$). The next section describes ways to determine solar elevation γ_S.

Solar position and angle of incidence

A lot of calculations require an exact indication of solar elevation. The position of the sun can be unambiguously specified for every place on the planet based on two angles: solar elevation (or height) γ_S and **solar azimuth** α_S. DIN5034 defines a solar elevation as the angle between the centre of the sun and the horizon as seen from the viewer's position. The solar azimuth describes the angle between the geographic north and the vertical circle through the centre of the sun ($0° \mathrel{\hat{=}}$ N, $90° \mathrel{\hat{=}}$ O, $180° \mathrel{\hat{=}}$ S, $270° \mathrel{\hat{=}}$ W).

Unfortunately, the angles are not indicated and symbols not used in a unified fashion internationally. In addition to the definitions from [DIN5034] used in this book, definitions from EN ISO 9488 [DIN9488] and NREL are also common. In sources of information about solar elevation, the only thing that varies is the symbol, but definitions of solar azimuth vary greatly. Before making comparisons, you therefore first have to compare definitions.

In addition to the viewer's geographic location, solar elevation and solar azimuth are relative to the date and time of day. The angle between the centre of the sun and the celestial equator, the **solar declination** δ, which changes over the course of a year between $+23°26.5' \geq \delta \geq -23°26.5'$, plays the biggest role. In addition, there are seasonal fluctuations in the length of a solar day.

In the DIN algorithm, the parameter

$$J' = 360° \cdot \frac{\text{day of the year}}{\text{number of days in the year}}$$

is used to calculate the solar declination

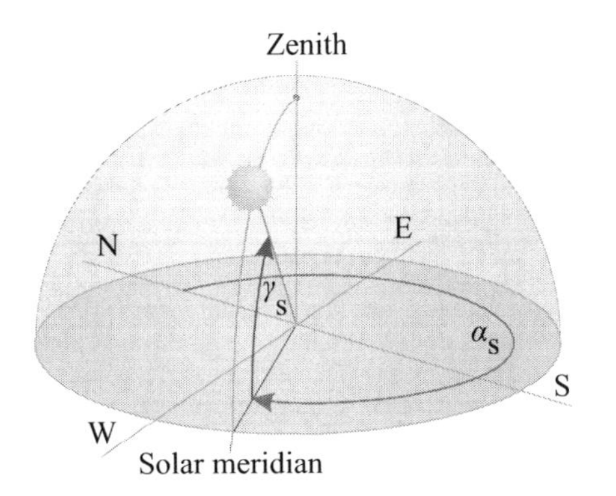

Figure 2.13 Designations of angles for the position of the sun in accordance with [DIN5034]

Table 2.10 Angle definitions and symbols for solar azimuth

Source	*Symbol*	*N*	*NE*	*E*	*SE*	*S*	*SW*	*W*	*NW*
NREL	γ	0°	45°	90°	135°	180°	225°	270°	315°
EN ISO 9488 for φ>0 [1]	γ	180°	225°	270°	315°	0°	45°	90°	135°
EN ISO 9488 for φ>0 [2]	γ	0°	270°	270°	225°	180°	135°	90°	45°
DIN 5034–2 (this book)	α_S	0°	45°	90°	135°	180°	225°	270°	315°

φ: Latitude, 1) north of the equator, 2) south of the equator

$$\delta(J') = \{0.3948 - 23.2559 \cdot \cos(J' + 9.1°) \\ -0.3915 \cdot \cos(2 \cdot J' + 5.4°) - 0.1764 \cdot \cos(3 \cdot J' + 26°)\}° \tag{2.22}$$

relative to time equation

$$Zgl(J') = \{0.0066 + 7.3525 \cdot \cos(J' + 85.9°) \\ +9.9359 \cdot \cos(2 \cdot J' + 108.9°) + 0.3387 \cdot \cos(3 \cdot J' + 105.2°)\} \ \text{min} \tag{2.23}$$

The local time LZ and the time zone (such as Central European Time CET = 1 h, Central European daylight-savings time CET+1 = 2 h) are relative to geographical length λ to determine the average local time

$$MOZ = LZ - timezone + 4 \cdot \lambda \cdot \text{min}/° \tag{2.24}$$

from which the time equation Zgl can be used to calculate solar time

$$WOZ = MOZ + Zgl \tag{2.25}$$

A location's geographical width φ and hour angle

$$\omega = (12.00 \ \text{h} - WOZ) \cdot 15°/\text{h} \tag{2.26}$$

can be used to calculate the elevation of the sun γ_S and its azimuth α_S:

$$\gamma_S = \arcsin(\cos\omega \cdot \cos\varphi \cdot \cos\delta + \sin\varphi \cdot \sin\delta) \tag{2.27}$$

$$\alpha_S = \begin{cases} 180° - \arccos\dfrac{\sin\gamma_S \cdot \sin\varphi - \sin\delta}{\cos\gamma_S \cdot \cos\varphi} & \text{for } WOZ \leq 12{:}00 \ \text{h} \\ 180° + \arccos\dfrac{\sin\gamma_S \cdot \sin\varphi - \sin\delta}{\cos\gamma_S \cdot \cos\varphi} & \text{for } WOZ > 12{:}00 \ \text{h} \end{cases} \tag{2.28}$$

The SUNAE algorithm promises somewhat greater accuracy than the DIN algorithm in calculating the sun's position [Wal78, Wil81, Kam90]; it also takes account of refracted sunlight in the atmosphere, which produces erroneous values in some locations. Other algorithms that are even less exact were developed by NREL; see SOLPOS and SPA on the Internet (www.nrel.gov/rredc/models_tools.html).

The sun's path across the sky is hard to detect based on these relatively complicated algorithms for a calculation of the ecliptic. Therefore, **ecliptic diagrams** are generally used; they calculate and visualize solar as a move and height for various days of the year. The values can be interpolated for data between the days calculated. Figure 2.14 shows the ecliptic diagram for Berlin; Figure 2.15, for Cairo. The time of day is given as a parameter in ecliptics.

The incident angle θ_{hor} of sunlight on a horizontal plane can be directly derived from the solar elevation γ_S. This angle is called the **solar zenith angle** θ_Z:

$$\theta_{hor} = \theta_Z = 90° - \gamma_S \tag{2.29}$$

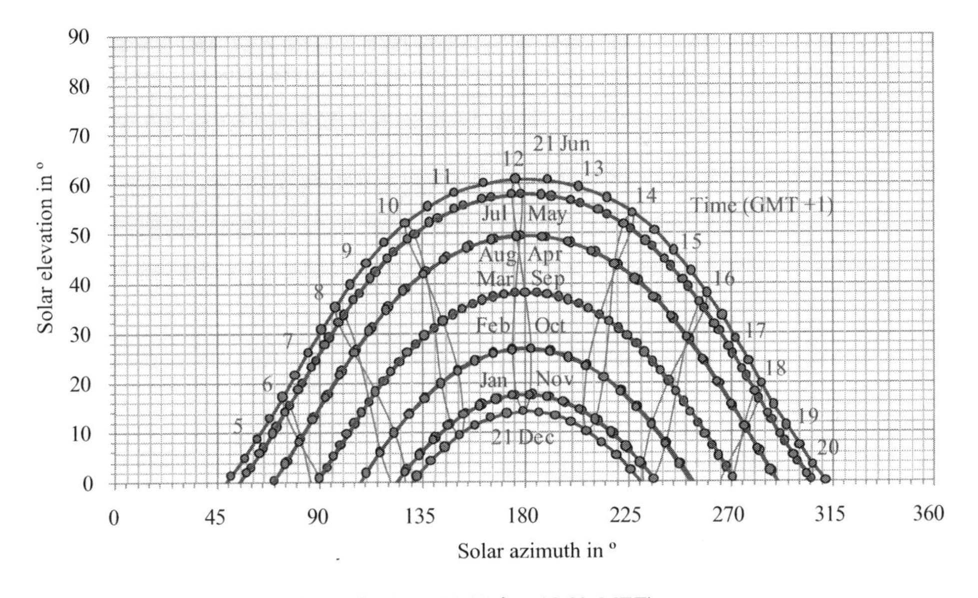

Figure 2.14 Ecliptic diagram for Berlin (φ = 52.3°, λ = 13.2°, MEZ)

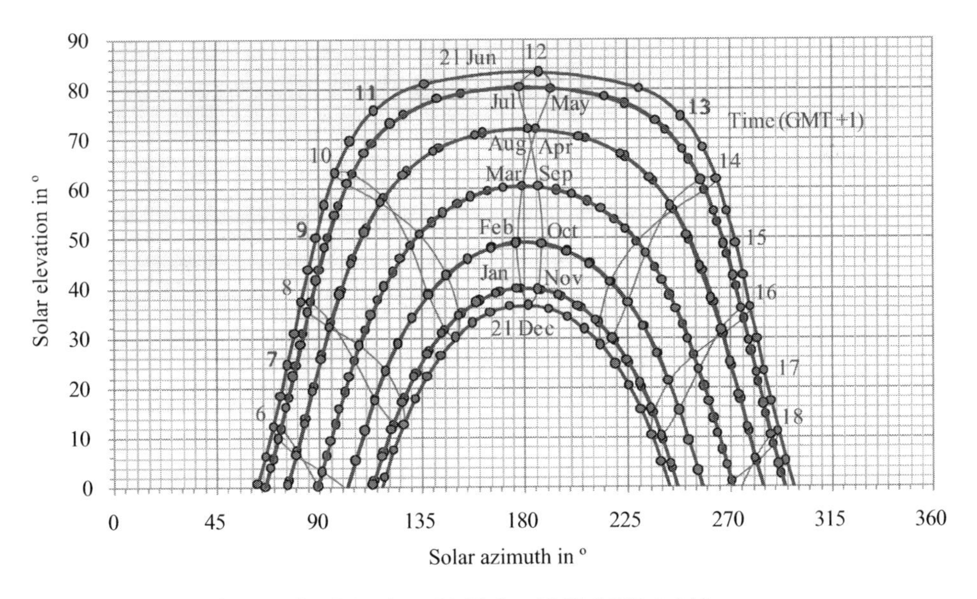

Figure 2.15 Ecliptic diagram for Cairo (φ = 30.1°, λ = 31.3°, MEZ + 1 h)

It is a bit trickier to calculate the incident angle θ_{gen} of a slanted plane relative to azimuth angle α_E and height angle γ_E. If the plane is turned clockwise (facing west), α is positive. Likewise, if it is turned counterclockwise (facing east), α_E is negative. Figure 2.16 shows how such angles are related to each other.

The incident angle θ_{gen} is the angle between a vector **s** towards the sun and the normal vector **n** of the plane. Because the sun's position has been determined in spherical

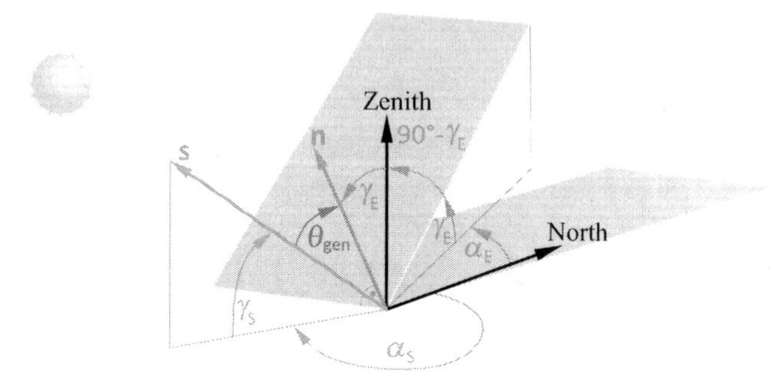

Figure 2.16 Calculating the angle of solar incidence on a slanted surface

coordinates up to now, it is converted into Cartesian coordinates with based vectors facing north, west, and the zenith. The following then applies for the vectors **s** and **n**:

$$s = (\cos\alpha_S \cdot \cos\gamma_S, -\sin\alpha_S \cdot \cos\gamma_S, \sin\gamma_S)^T \tag{2.30}$$

$$n = (-\cos\alpha_E \cdot \sin\gamma_E, \sin\alpha_E \cdot \sin\gamma_E, \cos\gamma_E)^T \tag{2.31}$$

Both vectors are standardized, so the **incident angle** θ_{gen} of the sunlight on the slanted plane can be calculated from the dot product of the two vectors:

$$\begin{aligned}\theta_{gen} &= \arccos(s \cdot n)\\ &= \arccos(-\cos\alpha_S \cdot \cos\gamma_S \cdot \cos\alpha_E \cdot \sin\gamma_E\\ &\quad -\sin\alpha_S \cdot \cos\gamma_S \cdot \sin\alpha_E \cdot \sin\gamma_E + \sin\gamma_S \cdot \cos\gamma_E)\\ &= \arccos(-\cos\gamma_S \cdot \sin\gamma_E \cdot \cos(\alpha_S - \alpha_E) + \sin\gamma_S \cdot \cos\gamma_E)\end{aligned} \tag{2.32}$$

Irradiance on a slanted plane

The global irradiance $E_{G,gen}$ on a slanted plane can be calculated from the direct and diffuse irradiance $E_{dir,gen}$ and $E_{diff,gen}$ along with the amount reflected by the ground $E_{refl,gen}$, which is not incident on the horizontal plane:

$$E_{G,gen} = E_{dir,gen} + E_{diff,gen} + E_{refl,gen} \tag{2.33}$$

Direct irradiance on a slanted plane

Figure 2.17 shows horizontal area A_{hor} with the same radiant flux Φ as the smaller area A_s perpendicular to the angle of incident sunlight:

$$\Phi_{dir,hor} = E_{dir,hor} \cdot A_{hor} = \Phi_{dir,s} = E_{dir,s} \cdot A_s \text{ and}$$
$$A_s = A_{hor} \cdot \cos\theta_{hor} = A_{hor} \cdot \sin\gamma_S \text{ produce}$$

$$E_{dir,s} = \frac{E_{dir,hor}}{\sin\gamma_S} \geq E_{dir,hor} \tag{2.34}$$

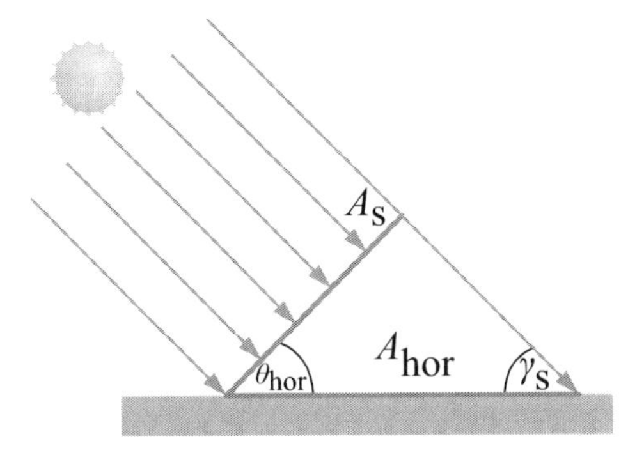

Figure 2.17 Irradiance incident on horizontal area A_{hor} and an area perpendicular A_s to the angle of incidence

This relation shows that direct irradiance $E_{dir,s}$ on a surface perpendicular to the sun is always greater than the direct irradiance $E_{dir,hor}$ on the horizontal plane. The difference is taking advantage of when photovoltaic and solar thermal systems face the sun at a perpendicular angle to increase energy yield.

Similarly, θ_{gen} in (2.32) produces the following for a slanted surface $E_{dir,s} = \dfrac{E_{dir,gen}}{\cos\theta_{gen}}$

It follows for direct irradiance on a slanted surface that

$$E_{dir,gen} = E_{dir,hor} \cdot \frac{\cos\theta_{gen}}{\sin\gamma_S} \tag{2.35}$$

If the irradiance on the horizontal is known, it can be converted for any slanted surface by using the formula above.

Diffuse irradiance on a slanted plane

In calculating diffuse irradiance $E_{diff,gen}$ on a slanted plane, a distinction is made between the isotropic and anisotropic approach. The former assumes that the composition of light is the same from all directions, meaning that the radiance is evenly distributed.

In contrast, diffuse irradiance is lower on a slanted surface than on a horizontal plane because the share of diffuse light behind the slanted plane does not exist in the latter case. For a plane at angle γ_E, the diffuse irradiance $E_{diff,gen}$ is determined from the diffuse irradiance intensity $E_{diff,hor}$ as follows in the **isotropic approach**:

$$E_{diff,gen} = E_{diff,hor} \cdot \tfrac{1}{2} \cdot (1 + \cos\gamma_E) \tag{2.36}$$

This formula, however, is only useful for rough estimates or cloudy conditions. In general, the anisotropic approach needs to be used to calculate diffuse irradiance because radiance varies greatly depending on the cardinal direction, especially under a cloudless sky. In this way, an increase in brightness on the horizon and near the sun can be observed. Below, two models are described to take account of this fact.

Klucher's model [Klu79] is a relatively straightforward calculation of diffuse irradiance $E_{diff,gen}$ on a slanted surface:

$$F = 1 - \left(\frac{E_{diff,hor}}{E_{G,hor}} \right)^2 \text{ produces}$$

$$E_{diff,gen} = E_{diff,hor} \cdot \tfrac{1}{2} \cdot (1 + \cos \gamma_E) \cdot \left(1 + F \cdot \sin^3 \frac{\gamma_E}{2} \right) \cdot \left(1 + F \cdot \cos^2 \theta_{gen} \cdot \cos^3 \gamma_S \right) \quad (2.37)$$

The Perez model [Per86, Per87, Per90] provides a more exact calculation of diffuse irradiance on a slanted surface, though it is more complicated.

This model defines atmospheric clearness index ε and an atmospheric brightness index Δ, which are calculated with the incident angle θ_{hor} of insolation on the horizontal plane (measured in rad), the constant $\kappa = 1.041$, the solar constant E_0, of direct and diffuse irradiance on the horizontal plane and air mass ($AM = 1 / \sin \gamma_S$):

$$\varepsilon = \frac{\dfrac{E_{diff,hor} + E_{dir,hor} \cdot \sin^{-1} \gamma_S}{E_{diff,hor}} + \kappa \cdot \theta_{hor}^3}{1 + \kappa \cdot \theta_{hor}^3} \quad (2.38)$$

$$\Delta = AM \cdot \frac{E_{diff,hor}}{E_0} \quad (2.39)$$

Now, the circumsolar brightening coefficient F_1 and the horizon brightening coefficeient F_2 can be calculated. Table 2.11 shows the constants F_{11} to F_{23} needed for this calculation. Eight different atmospheric clearness classes (ε-class$= 1 - 8$) correspond to the atmospheric clearness index ε.

With

$$F_1 = F_{11}(\varepsilon) + F_{12}(\varepsilon) \cdot \Delta + F_{13}(\varepsilon) \cdot \theta_{hor} \text{ and} \quad (2.40)$$

$$F_2 = F_{21}(\varepsilon) + F_{22}(\varepsilon) \cdot \Delta + F_{23}(\varepsilon) \cdot \theta_{hor} \quad (2.41)$$

and $a = \max(0; \cos \theta_{gen})$ and $b = \max(0.087; \sin \gamma_S)$, the diffuse irradiance $E_{diff,hor}$ on the horizontal plane can be used to calculate the diffuse irradiance $E_{diff,gen}$ on a slanted plane:

$$E_{diff,gen} = E_{diff,hor} \cdot \left[\frac{1}{2} \cdot (1 + \cos \gamma_E) \cdot (1 - F_1) + \frac{a}{b} \cdot F_1 + F_2 \cdot \sin \gamma_E \right] \quad (2.42)$$

Reflection from the ground

The isotropic approach suffices to calculate the reflection from the ground $E_{refl,gen}$. Anisotropic approaches only improve accuracy here to a negligible extent.

For a slanted plane with an angle γ_E, the irradiance reflected by the ground $E_{refl,gen}$ can be calculated with the **albedo value** A of total irradiance $E_{G,hor}$ on the horizontal plane:

$$E_{refl,gen} = E_{G,hor} \cdot A \cdot \tfrac{1}{2} \cdot (1 - \cos \gamma_E) \quad (2.43)$$

Table 2.11 Constants to determine F_1 and F_2 relative to ε [Per90]

ε-class	*1*	*2*	*3*	*4*	*5*	*6*	*7*	*8*
ε	*1,000–1,065*	*1,065–1,230*	*1,230–1,500*	*1,500–1,950*	*1,950–2,800*	*2,800–4,500*	*4,500–6,200*	*6,200–∞*
F_{11}	−0.008	0.130	0.330	0.568	0.873	1.132	1.060	0.678
F_{12}	0.588	0.683	0.487	0.187	−0.392	−1.237	−1.600	−0.327
F_{13}	−0.062	−0.151	−0.221	−0.295	−0.362	−0.412	−0.359	−0.250
F_{21}	−0.060	−0.019	0.055	0.109	0.226	0.288	0.264	0.156
F_{22}	0.072	0.066	−0.064	−0.152	−0.462	−0.823	−1.127	−1.377
F_{23}	−0.022	−0.029	−0.026	−0.014	0.001	0.056	0.131	0.251

Table 2.12 Albedo of various environments [Die57, TÜV84]

Ground	*Albedo A*	*Ground*	*Albedo A*
Grass (July and August)	0.25	Asphalt	0.15
Lawn	0.18–0.23	Forest	0.05–0.18
Dry grass	0.28–0.32	Heather and sandy areas	0.10–0.25
Disused fields	0.26	Water surface ($\gamma_S > 45°$)	0.05
Bare ground	0.17	Water surface ($\gamma_S > 30°$)	0.08
Gravel	0.18	Water surface ($\gamma_S > 20°$)	0.12
Concrete, weathered	0.20	Water surface ($\gamma_S > 10°$)	0.22
Concrete, clean	0.30	Fresh snow	0.80–0.90
Cement, clean	0.55	Old snow	0.45–0.70

The albedo value A has the greatest impact on the calculation's accuracy. If it cannot be measured, the values in Table 2.12 can be used as rough estimates. If the surroundings are not known, $A = 0.2$ is generally used.

Radiation gains from angle and tracking

If a surface tracks the sun so that the angle of incidence is practically zero, energy yield increases. The share of direct irradiation is much greater on a surface that optimally faces the sun, and this factor is almost entirely responsible for the improvement. On days with a lot of irradiation and a large share of direct sunlight, tracking can provide relatively great irradiation gains. In the summer, tracking increases the gains by around 50 % when the weather is good; in the summer, by 300 % or more (Figure 2.18).

On cloudy days with a large share of diffuse light, the energy gains from a slanted surface can be lower because the share of diffuse light is reduced by the share behind the slanted surface. But most of the energy gains from tracking come in the summer. On the one hand, the absolute energy gains are greater than in the winter; on the other, the share of cloudy days is much greater in the winter.

A distinction is made between single and dual-axis tracking. A system can only optimally face the sun at all times with dual-axis tracking, as shown in Figure 2.18. Because dual-axis tracking is more technically complicated, single-axis tracking is often preferred. The system can either track the sun on a daily basis or by season. It is relatively easy to track the sun over the year. You simply need to change the angle of the system in relatively long intervals (weeks or months). Theoretically, these changes can even be made manually.

At German latitudes, **dual-axis tracking** can produce energy gains exceeding 30 % from photovoltaic systems. Single-axis tracking increases energy gains by around 20 %. In areas with a greater irradiation, the energy gains are a bit greater. Tracking requires more equipment, however, which also entails higher costs. The stands have to be movable and able to withstand tremendous loads during storms, for instance. Maintenance of the tracking system has to be ensured over the system's entire service life. An electric motor or thermo-hydraulic system can be used for the drive. If an electric motor is used, electricity is needed, which reduces the energy gains from the tracking system. If the tracking equipment fails,

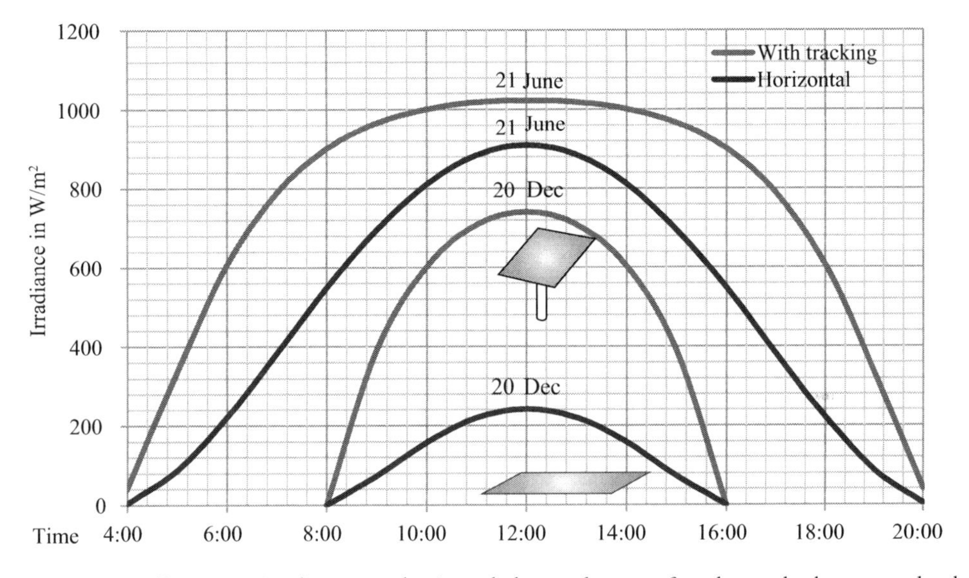

Figure 2.18 Differences in irradiance on a horizontal plane and on a surface that tracks the sun on cloudless days at 50° latitude

(Data: [DIN4710])

the system can be stuck in an unfavourable position, thereby reducing energy yield until the matter has been remedied.

The greater energy gains only make up for the drawbacks in some cases for photovoltaic systems, so that it is often done without tracking. Because non-concentrating solar collector systems often have immovable tubing, tracking is not used in such systems.

The situation is different, however, for parabolic trough systems and solar towers; these systems must track the sun. Concentrators that use mirrors can only make use of direct sunlight. Tracking is then needed to optimally concentrate sunlight.

Systems that do not track the sun can have their angle changed in order to increase energy yield. Photovoltaic systems used all year round have an optimal angle of around 30° facing south at German latitudes. If the system is only used in the summer, more acute angles of 10° to 20° provide greater energy gains. In the winter, however, the angle should be much steeper. The optimum is then around 60° to 70° (Table 2.13). Solar thermal systems used all year round are generally best used at steeper angles of around 45°.

Often, solar energy systems are installed on slanted roofs that are not always optimally oriented to the south. The energy yield can then be lower as a result. But the energy losses are relatively small over a wide range of angles (Figure 2.19).

In areas with great overall irradiation, adjusting the orientation of the surface can change the energy yield slightly. In Lisbon, 14 % more energy is produced if the system faces 30° to the south relative to a horizontal plane, compared with a 12 % gain in Berlin. If the surface faces 90° to the northeast, global irradiation drops to 40 % of the value for the horizontal plane, compared with 46 % in Berlin. In the southern hemisphere, the solar energy systems

Table 2.13 Averages (1998–2010) of monthly and annual irradiation on surfaces with various orientations in kWh/m²

	Jan	*Feb*	*Mar*	*Apr*	*May*	*Jun*	*July*	*Aug*	*Sep*	*Oct*	*Nov*	*Dec*	*Year*
Horizontal	19	33	75	128	160	166	158	134	94	51	26	15	1059
10° South	23	39	84	138	165	170	162	142	103	58	33	19	1140
30° South	30	48	96	150	167	167	161	148	116	70	44	25	1220
37° South	32	50	98	151	165	163	157	148	118	72	47	27	1230
45° South	34	52	100	150	160	156	152	145	118	74	49	28	1220
60° South	36	53	98	142	145	139	136	134	115	75	52	30	1150
90° South	33	47	81	104	95	87	87	94	91	65	49	29	861
45° SE/SW	28	44	89	140	157	156	150	139	108	65	40	23	1140
45° E/W	17	30	68	115	141	146	139	120	86	46	24	14	944
90° E/W	11	20	45	76	90	92	87	78	57	31	16	9	611

(Data: [JRC10])

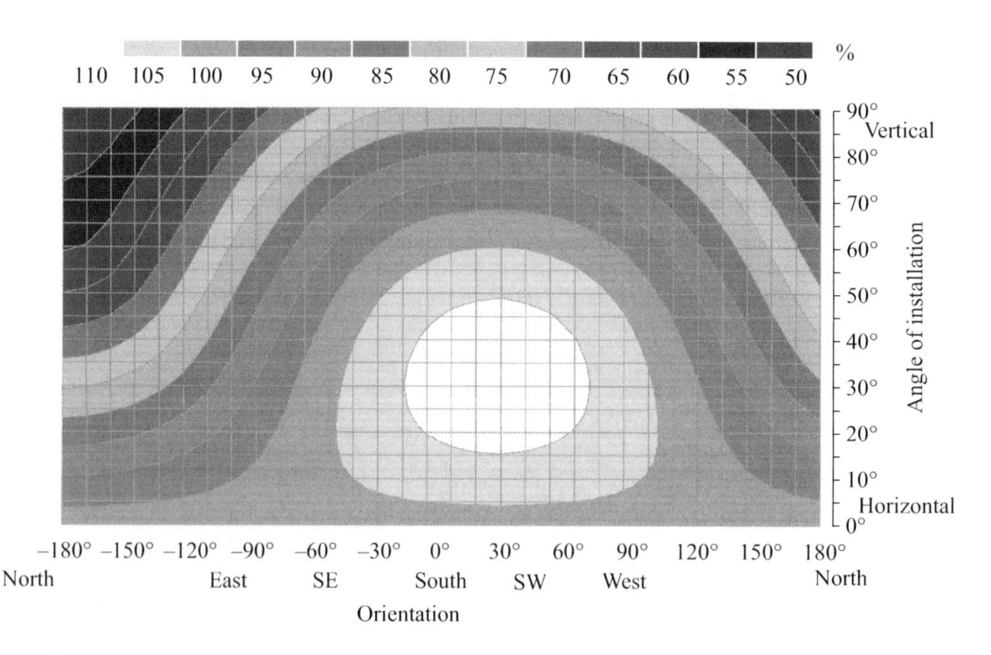

Figure 2.19 Changes in annual solar irradiation in Berlin relative to orientation and angle compared with a horizontal plane

should face the north to increase energy yield. Near the equator, angles should be as small as possible because the sun is close to the zenith for a large part of the year.

If losses or gains from an angle are to be calculated for other time frames or locations, hourly irradiance values on the horizontal plane are needed. They then need to be converted on an hourly basis using the formulae above for a slanted plane. Figure 2.20 summarizes the necessary calculations.

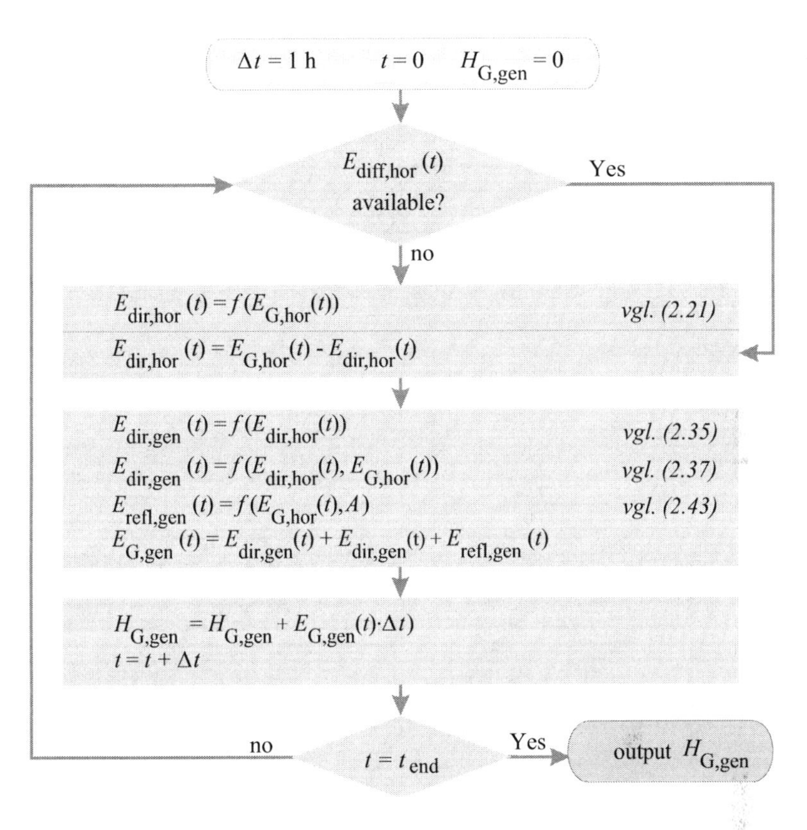

Figure 2.20 Calculating irradiance on a slanted plane from known data for horizontal insolation

Calculating losses from shading

All of the calculations described above assume that there is no shading from surrounding objects to reduce irradiance. In reality, however, completely unshaded surfaces are hard to find, and the losses from shading can be significant, especially for photovoltaic systems. Below, we therefore discuss how shading can be taken into consideration when calculating irradiance.

Including the surrounding area

Let us assume that a photovoltaic system is to be installed on a single-family home. There are numerous trees and shrubs in the surrounding area, along with a neighbouring building. First, a point of reference must be specified. The lowest point of the solar energy system where the greatest probability of shading is expected is a good starting point. All azimuth and height angles in the analyses below stem from this reference point.

The height and azimuth angles need to be determined for the chosen reference point relative to each object in the surroundings. Azimuth angle α can be determined using a compass, while height angle γ can be calculated using simple geometry, as shown in the example of a tree in Figure 2.21.

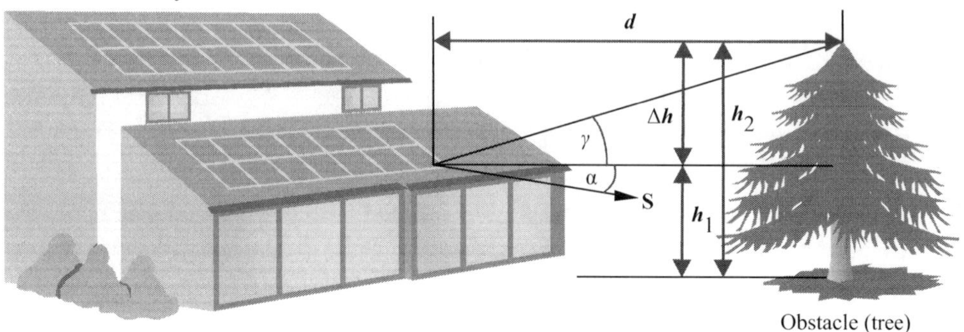

Figure 2.21 Determining the height and azimuth angle of an obstacle relative to a reference point

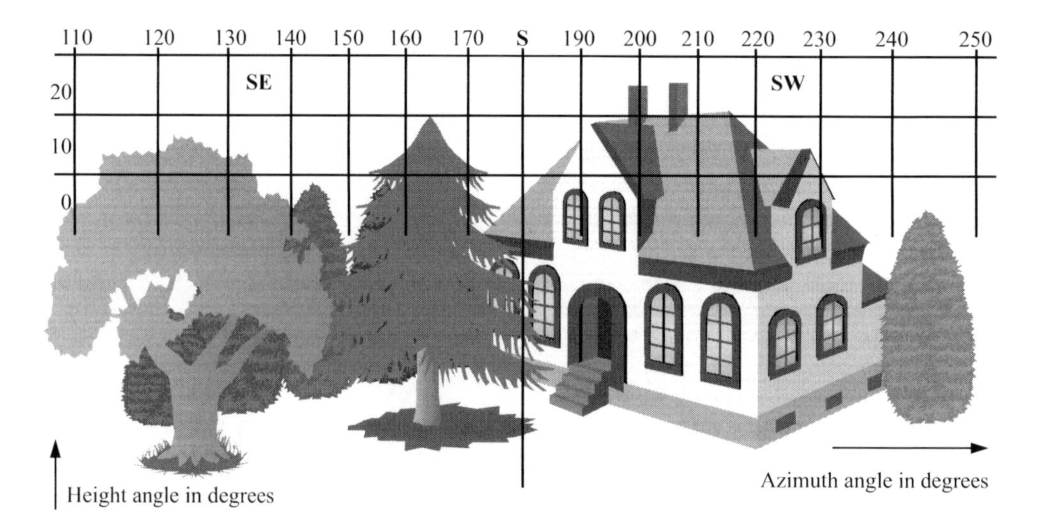

Figure 2.22 Surroundings within an angular grid

The height angle γ is calculated thus:

$$\gamma = \arctan\left(\frac{h_2 - h_1}{d}\right) = \arctan\left(\frac{\Delta h}{d}\right) \tag{2.44}$$

These calculations must be made for all obstacles close to the solar energy system, so each object's height and distance from the reference point must be known. Optical tools – which can be bought in stores or made at home – can be used to determine the height and azimuth angles of objects.

Figure 2.22 shows an example of surroundings in an angular grid with spherical coordinates. Figure 2.23 shows a silhouette of the surroundings as a polygon chain transferred

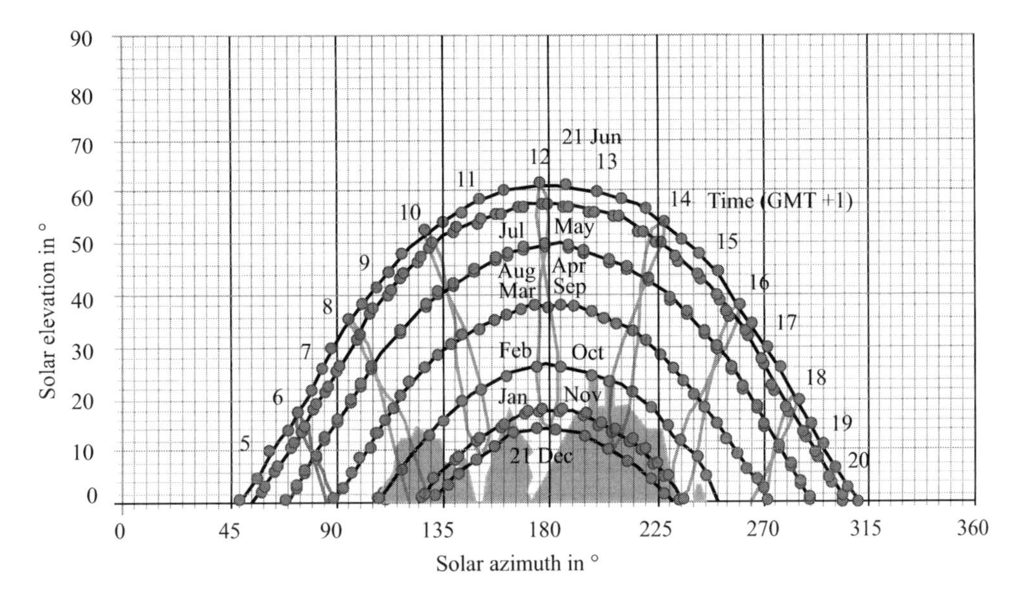

Figure 2.23 Ecliptic diagram for Berlin with silhouette of surroundings

to an ecliptic diagram. We can now see when shading will occur at this location. In the example above, the system is nearly completely shaded on December 21, with sunlight only reaching the system for around an hour at the reference point in the morning and at midday. On February 21, no shading is expected, however, after 9 a.m. From March to September, no shading at all can be seen.

Determining the degree of direct shading

In the additional calculations below, it is assumed that a description of the surroundings is available as a polygon chain, as shown in Figure 2.23. To see whether the sun's current position is within or outside of the object's polygon in the silhouette, a line is drawn from the sun's position horizontally towards the left (Figure 2.24).

We then count how many times this line overlaps the edges of the polygon. All edges of the polygon with at least one point to the left of the sun's position must be investigated to see if there is any overlapping with the line. If the number of points of intersection is even, the test point (the sun's position) is outside of the object's polygon. If the number of points is odd, the test point is within the polygon chain, so direct sunlight is shaded.

Not all objects in the environment completely block sunlight. For instance, trees let part of the irradiance through, and the amount is calculated using the **degree of transmissivity** τ. [Sat87] provides the following degrees of transmissivity for deciduous trees in various seasons:

- Without foliage (winter) $\tau = 0.64$
- With foliage (summer) $\tau = 0.23$

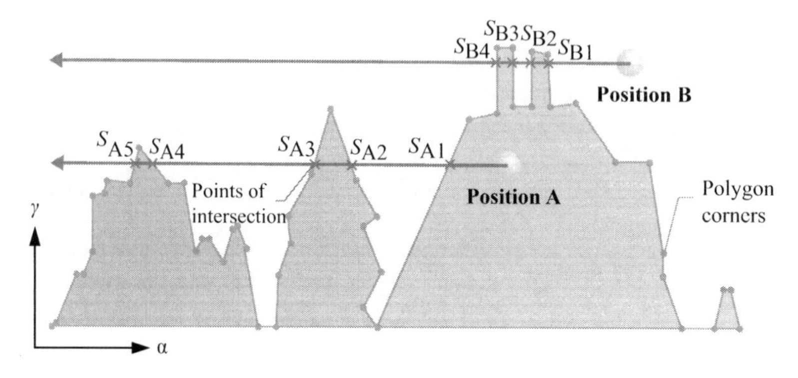

Figure 2.24 Shading test for two different positions of the sun A and B (Position A has 5 points of intersection; Position B, 4)

The direct degree of shading S_{dir} is then used to describe the share of direct irradiance blocked by shading:

$$S_{dir} = \begin{cases} 0 & \text{, if } (\gamma_S, \alpha_S) \text{ outside the object polygon} \\ 1 & \text{, if } (\gamma_S, \alpha_S) \text{ within the non-transparent object polygon} \\ 1-\tau & \text{, if } (\gamma_S, \alpha_S) \text{ within the transparent object polygon} \end{cases} \quad (2.45)$$

Determining the diffuse degree of shading

To determine the diffuse degree of shading, the object's polygon is projected onto a hemisphere. The objects represented in the object's polygon now block the part of diffuse irradiance incident on the hemisphere through the object's polygon. To take a simple example, a polygon with four corners has two on the horizontal plane. See Figure 2.25.

The line connecting the two points $\mathbf{p}_1$ and $\mathbf{p}_2$ under the assumption that $\alpha_1 \neq \alpha_2$ and $\gamma_1 > 0$ and $\gamma_2 > 0$ is described by the straight line $\gamma(\alpha) = m \cdot \alpha + n$

with $m = \dfrac{\gamma_2 - \gamma_1}{\alpha_2 - \alpha_1}$ and $n = \dfrac{\gamma_1 \cdot \alpha_2 - \gamma_2 \cdot \alpha_1}{\alpha_2 - \alpha_1}$

In an isotropic distribution of radiation, the diffuse share of sunlight $E_{diff,hor,Ai}$ through area A_i on a horizontal surface is:

$$E_{diff,hor,A_i}(p_1, p_2) = L_{e,iso} \int_{\alpha_1}^{\alpha_2} \int_{0}^{(m\cdot\alpha+n)} \sin\gamma \cos\gamma \, d\gamma \, d\alpha = \tfrac{1}{2} \cdot L_{e,iso} \int_{\alpha_1}^{\alpha_2} \sin^2(m\cdot\alpha+n) \, d\alpha$$

$$= \begin{cases} \tfrac{1}{2} \cdot L_{e,iso} \cdot (\alpha_2 - \alpha_1) \cdot \sin^2 \gamma_1 & \text{for } m = 0 \\ \tfrac{1}{2} \cdot L_{e,iso} \cdot (\alpha_2 - \alpha_1) \cdot (\tfrac{1}{2} + \tfrac{1}{4} \cdot \dfrac{\sin 2\gamma_1 - \sin 2\gamma_2}{\gamma_2 - \gamma_1}) & \text{for } m \neq 0 \end{cases} \quad (2.46)$$

Here, $L_{e,iso}$ is the isotropic radiance, which can be eliminated in the calculation of the diffuse degree of shading by using the least common denominator. In an anisotropic distribution of radiation, the integral also spreads across the radiance function L_e, which

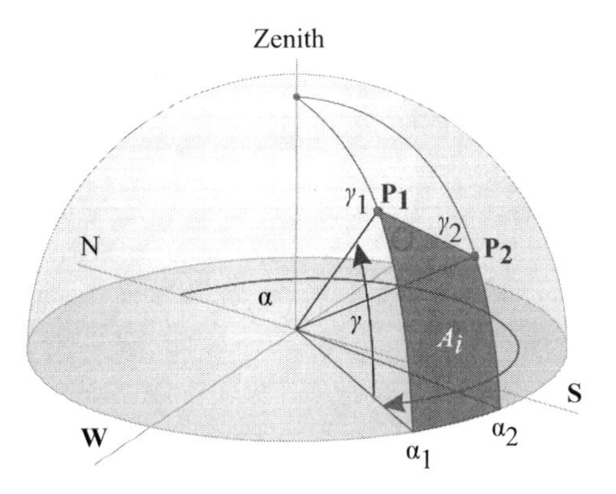

Figure 2.25 An area demarcated by the connection of two points and the horizon

is irrespective of direction, and can generally only be resolved with numeric methods. The incident angle θ must be taken into account for a slanted plane in accordance with formula 2.32. For the isotropic case, there is a similar analytical solution for the integral [Qua96].

For an object's polygon chain with n polygon chain points $\mathbf{p}_1 = (\alpha_1, \gamma_1)$ to $\mathbf{p}_n = (\alpha_n, \gamma_n)$, the diffuse share of irradiance $E_{\text{diff,P}}$ can be determined from the polygon chain's area $E_{\text{diff,A}i}$ using the sum of the amount of irradiance $E_{\text{diff,A}i}$ through surfaces A_i:

$$E_{\text{diff,P}} = \left| \sum_{i=1}^{n-1} E_{\text{diff,A}_i} (p_i, p_{i+1}) \right| \tag{2.47}$$

The **diffuse degree of shading** $S_{\text{diff,hor}}$ for a horizontal surface is now derived from the ratio of reduced diffuse irradiance and total diffuse irradiance. For an isotropic radiation distribution, the following applies:

$$S_{\text{diff,hor}} = \frac{E_{\text{diff,hor,P}}}{E_{\text{diff,hor}}} = \frac{E_{\text{diff,hor,P}}}{\pi \cdot L_{\text{e,iso}}} \tag{2.48}$$

Likewise, the following applies for the diffuse degree of shading of a slanted surface:

$$S_{\text{diff,gen}} = \frac{E_{\text{diff,gen,P}}}{E_{\text{diff,gen}}} = \frac{E_{\text{diff,gen,P}}}{L_{\text{e,iso}} \cdot \frac{\pi}{2} \cdot \left(1 + \cos \gamma_{\text{E}}\right)} \tag{2.49}$$

For transparent shading, the degree of shading S must be weighted using the degree of transmissivity τ.

Overall determination of shading

The total horizontal irradiance $E_{\text{G,hor}}$ including shading can now be calculated from the direct irradiance $E_{\text{dir,hor}}$ and diffuse irradiance $E_{\text{diff,hor}}$ on the horizontal plane along with the degrees of shading S_{dir} and $S_{\text{diff,hor}}$:

$$E_{\text{G,hor}} = E_{\text{dir,hor}} \cdot (1 - S_{\text{dir}}) + E_{\text{diff,hor}} \cdot (1 - S_{\text{diff,hor}}) \tag{2.50}$$

The overall irradiance $E_{G,gen}$ on a slanted plane is calculated in similar fashion. Change the albedo A to take account of possible detrimental effects of reflection from the ground when determining reflected irradiance $E_{refl,gen}$:

$$E_{G,gen} = E_{dir,gen} \cdot (1 - S_{dir}) + E_{diff,gen} \cdot (1 - S_{diff,gen}) + E_{refl,gen} \quad (2.51)$$

[Qua96] provides a more detailed description of how to calculate shading losses.

Optimal distance for solar modules on stands

Often, solar thermal and photovoltaics systems are installed on the ground or flat roofs to collect solar energy. The systems are then not horizontal, but installed at an angle on **stands**. Installation at an angle improves energy yield considerably; **optimal orientation** can increase energy yield by 12.5 % in Berlin at an angle of 30° relative to the horizontal plane, as shown in Table 2.13.

In addition, horizontal systems have greater losses from dirt. Airborne particles, bird droppings, and other types of dirt collected on the surface. When the modules are installed at an angle, rain and snow wash the dirt off, keeping the degree of soiling limited. The more acute the tilt angle, the higher the cleaning effect of rain and snow. At an angle of around 30°, losses from dirt range from 2 % to 10 % under German conditions depending on the location. As the tilt angle decreases, the degree of soiling rises considerably. Under extreme climatic conditions, the losses from dirt can be much greater, such as during long dry spells with a lot of dust. In such cases, it can make sense to clean the modules regularly.

One drawback of stands is that one row of modules might shade another. As the discussion below shows, an optimal distance between the rows can minimize the losses from shading.

The distance of each row d along with the length l is used to calculate the **degree of area usage** as shown in Figure 2.26:

$$f = \frac{l}{d} \quad (2.52)$$

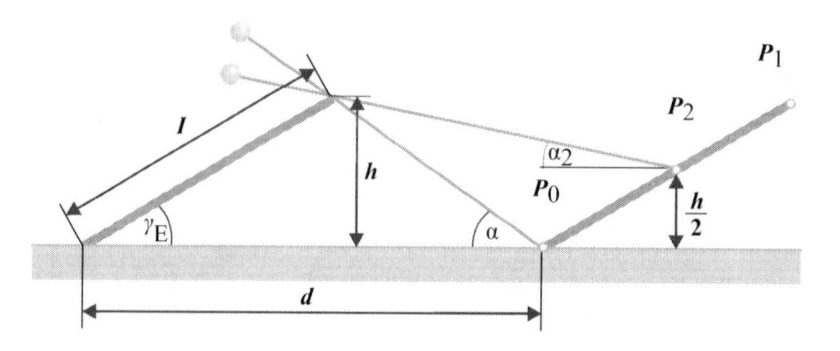

Figure 2.26 The geometry of solar modules on stands

Various points on the slanted plane of the system are shaded to varying degrees. Point P_0 is the most detrimentally affected. The greater the degree of surface usage *f*, the greater the shading, as represented by the **shading angle**:

$$\alpha = \arctan\left(\frac{f \cdot \sin \gamma_E}{1 - f \cdot \cos \gamma_E}\right) \tag{2.53}$$

In addition to the degree of surface utilization *f*, the tilt angle γ_E also affects the shading angle (Figure 2.27).

As the shading angle increases, so do the irradiation losses. Figure 2.28 shows the relative shading losses *s* at point P_0 relative to shading angle α and tilt angle γ_E. **Relative shading losses** *s* are calculated from the irradiation $H_{G,gen}$ on slanted, unshaded modules and the irradiation $H_{G,gen,red}$ when the modules shade each other:

$$s = 1 - \frac{H_{G,gen,red}}{H_{G,gen}} \tag{2.54}$$

As the tilt angle increases, the system becomes increasingly susceptible to shading.

Because photovoltaic modules are especially sensitive to shading, irradiation at point P_0 can be used as a rough reference value for the entire system. Table 2.14 shows the shading angle α and the resulting shading losses *s* for various degrees of area utilization at tilt angles of 10° and 30°. The angle results in relative irradiation gains, which can be taken into account via a factor *g*.

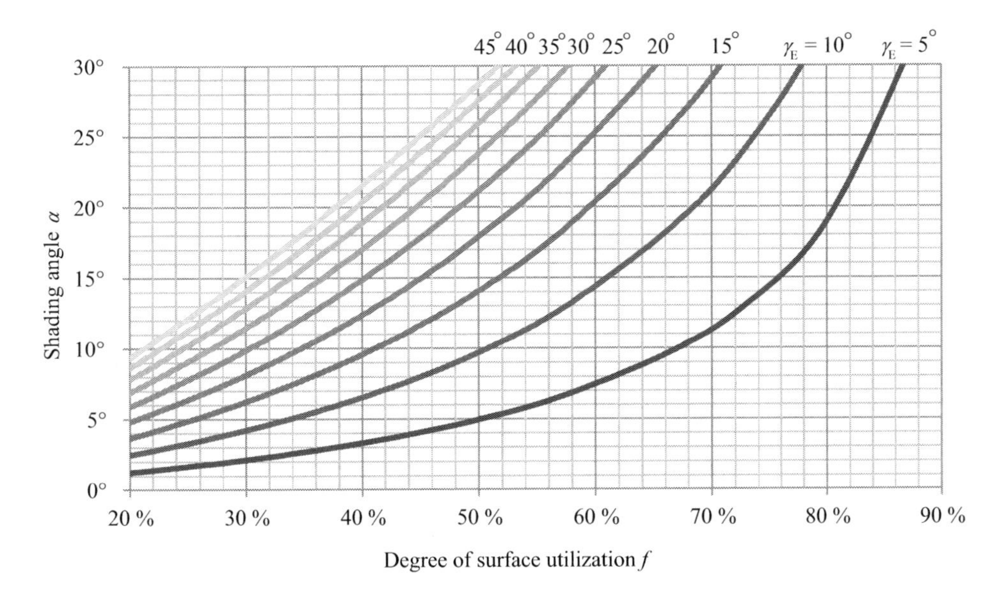

Figure 2.27 The shading angle α relative to the degree of surface utilization *f* and tilt angle γ_E

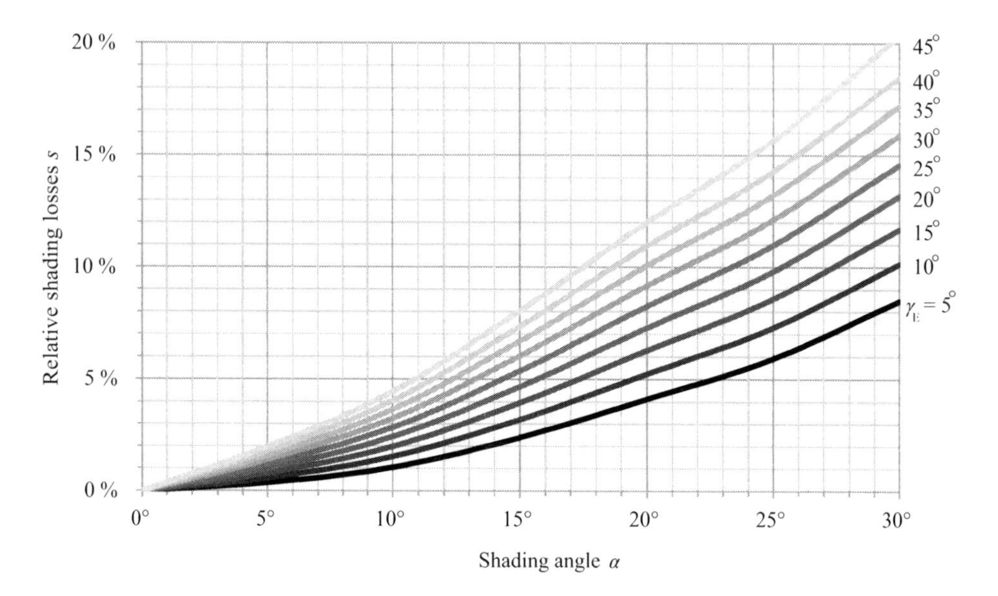

Figure 2.28 Relative shading losses s relative to shading angle α and tilt angle γ_E in Berlin

Table 2.14 Shading losses s and overall correction factor k for point P_0 under different degrees of surface utilization and tilt angles in Berlin

	$\gamma_E = 30°$				$\gamma_E = 10°$			
F	α	s	g	k	α	s	g	k
1:1.5	38.8°	0.246	1.125	0.848	18.6°	0.048	1.064	1.013
1:2.0	23.8°	0.116	1.125	0.995	9.7°	0.015	1.064	1.048
1:2.5	17.0°	0.074	1.125	1.042	6.5°	0.009	1.064	1.054
1:3.0	13.2°	0.048	1.125	1.071	4.9°	0.006	1.064	1.058
1:3.5	10.7°	0.035	1.125	1.086	3.9°	0.004	1.064	1.060
1:4.0	9.1°	0.029	1.125	1.092	3.3°	0.004	1.064	1.060

The factor g is calculated from the ratio of annual irradiation $H_{G,gen}$ on the slanted plane to the horizontal irradiation $H_{G,hor}$:

$$g = \frac{H_{G,gen}}{H_{G,hor}} \tag{2.55}$$

The gains from the tilt angle and the losses from shading add up to produce correction factor k, which indicates the ratio of irradiation at point P_0 to the horizontal plane:

$$k = (1 - s) \cdot g \tag{2.56}$$

Table 2.15 Shading losses and overall correction factor on average at points P_0, P_1, and P_2 with different degrees of surface utilization and tilt angles in Berlin

f	$\gamma_E = 30°$			$\gamma_E = 10°$		
	s_m	g	k	s_m	g	k
1:1.5	0.098	1.125	1.015	0.018	1.064	1.045
1:2.0	0.048	1.125	1.071	0.006	1.064	1.058
1:2.5	0.032	1.125	1.089	0.004	1.064	1.060
1:3.0	0.021	1.125	1.101	0.003	1.064	1.061
1:3.5	0.016	1.125	1.107	0.002	1.064	1.062
1:4.0	0.013	1.125	1.110	0.002	1.064	1.062

By reducing the degree of surface utilization, values around $f < 0.33$ are possible but only represent an insignificant improvement of the ratio of row length l to the distance between rows d of 1:3. If the degree of surface utilization is greater than 0.4 (1:2.5), a lower tilt angle increases energy yield. When we consider that a tilt angle of 10° produces a 5 % loss relative to 30° due to greater dirt accumulation on the modules, then a smaller tilt angle is only recommended if the surface utilization exceeds 0.5 (1:2).

The shading losses are not as great with solar thermal collectors as with photovoltaic modules. For the former, the average level of irradiation can be used when calculating yield. For further calculations, the average shading losses were calculated at points P_0, P_1 and P_2, as shown in Figure 2.26. The losses at point P_0 were previously calculated. No shading occurs at point P_1. The shading angle at point P_2

$$\alpha_2 = \arctan\left(\frac{f \cdot \sin \gamma_E}{2 - f \cdot \cos \gamma_E}\right) \tag{2.57}$$

in the middle of the slanted surface is relevant. Table 2.15 shows the average relative shading losses s_m along with the overall correction factor k for the three points P_0, P_1 and P_2. These losses are much lower than at P_0 alone. Further improvements are hardly possible here at surface utilization degrees $f < 0.5$ (1:2). A smaller tilt angle is generally no longer recommended.

Irradiance measurement devices and solar simulators

Measuring global irradiance

Two basic approaches are commonly pursued for the measurement of global irradiance: semiconductor sensors and thermal sensors (Figure 2.29). Devices that use one of these sensor types are called **pyranometers**.

A **semiconductor sensor**, which usually consists of silicon as with photovoltaics solar cells, generates short-circuit current that increases proportional to irradiance. Relatively small measurement resistance allows the sensor to run the close to the short circuit, and the current is converted into a voltage signal. Semiconductor sensors are sensitive to temperature changes, however, so that the sensor's temperature also needs to be recorded for

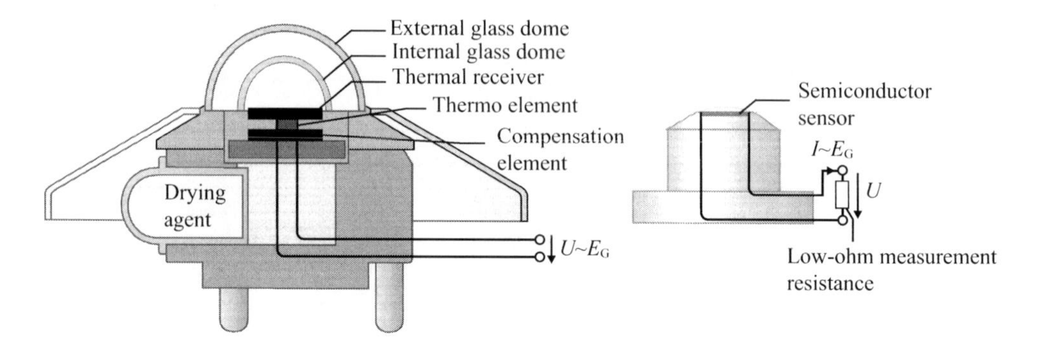

Figure 2.29 Pyranometer for the measurement of global irradiance. Left: pyranometer with a thermal sensor; right: pyranometer with a silicon semiconductor

corrections to be made to the measurement values. The measurement current generally increases by 0.15 %/°C. Another drawback of semiconductor sensors is spectral sensitivity (cf. Chapter 5, Photovoltaics), meaning that not all wavelengths of sunlight are equally converted into a power signal. If, for instance, the spectral composition of sunlight changes when the sun is low in the sky, the sensor may provide a slightly different reading.

Thermal sensors have a black receiver that heats up in the sun. Two glass domes thermally isolate it. The difference between the receiver's surface temperature and the ambient temperature is proportional to the irradiance. A thermal element on the back of the receiver converts the temperature difference into a voltage of signal. The benefit of this sensor type is constant spectral sensitivity across a wide range of wavelengths. The sensor is therefore more exact under constant radiance conditions. The drawback is that the sensor has a bit of inertia. Depending on the specific design, it can take up to a minute for the correct final values to be reached. This type of sensor is therefore not suited for quality assurance in the manufacture of solar cells, where measurements are taken with a flash of light.

The requirements for thermal sensors are defined in ISO 9060. There, a distinction is made between secondary standard, first-class, and second-class sensors. Secondary standard sensors are the most accurate. But even then, there is a tolerance of ±1 to 3 % in determining daily irradiation. Pyranometers are generally very sensitive to dirt. They therefore have to be cleaned regularly. If they are not, measurement errors can quickly reach 10 % and higher.

Measuring direct and diffuse irradiance

Global irradiance consists of two parts: direct and diffuse irradiance. To measure each, the share of radiance from the other must be eliminated from the measurement.

A sphere is permanently positioned between the sensor and the sun to cast a shadow on the sensor's surface and absorb the direct part of the sunlight so that diffuse irradiance can be measured.

Pyrheliometers are used to measure direct-normal irradiance. The sensor, usually thermal, is attached to the end of a cylinder. Direct sunlight can only enter the beginning

Figure 2.30 Left: A station that measures global, diffuse, and direct irradiance. Top right: pyranometer with sphere casting shadow. Bottom right: pyrheliometer

of this cylinder through a small opening. The cylinder is constantly focused on the sun, so the sensor measures direct irradiance. Diffuse irradiance enters at an angle and is absorbed by the cylinder.

Artificial suns

In practice, it is not possible to conduct all tests and measurements under real ambient conditions. In photovoltaics in particular, solar cells and modules must be measured for classification during production. To this end, constant irradiance conditions not possible under fluctuating real-world conditions are needed. Artificial light sources are therefore used that give off light that is as much like the sun as possible.

Usually, xenon flashes and arc lamps are used. Various filters can reduce individual parts of the spectrum so that the light produced more closely mimics the spectrum of natural sunlight. **Flashers** are frequently used in photovoltaics to simulate the sun.

Three factors determine the quality of solar simulators. First, there must not be any considerable spatial fluctuations across the surface exposed to the light. Proper optics can ensure evenness. Second, the solar simulator must be able to produce constant conditions over long time frames. Finally, the main criterion is the spectrum of light close to the solar spectrum *AM1.5*g (cf. Figure 2.4.). To this end, even good solar simulators still deviate considerably. There are quality requirements for the categorization of solar simulators in classes A, B, and C according to these three criteria. Table 2.16 provides an overview of the criteria in IEC 904–9 [IEC95].

Table 2.16 Allowable tolerances and specifications for solar simulators in compliance with IEC 904–9

Specification	*Class A*	*Class B*	*Class C*
Spectral overlapping (fraction of the ideal percentage)	0.75–1.25	0.6–1.4	0.4–2.0
Spatial irregularity	±2 %	±5 %	±10 %
Temporal instability	±2 %	±5 %	±10 %

Table 2.17 Ideal distribution of total irradiance across wavelengths in IEC 904–9

Wavelength range in nm	*400–500*	*500–600*	*600–700*	*700–800*	*800–900*	*900–1,100*
Share	18.5 %	20.1 %	18.3 %	14.8 %	12.2 %	16.1 %

For a review of spectral overlapping, the spectrum of light was subdivided into six spectral regions. Table 2.17 shows the shares that each region should make up in total irradiance. Deviations from the defined shares must not exceed the values indicated in Table 2.16 for inclusion in that class.

3 Non-concentrated solar thermal

Fundamentals

Solar thermal is especially important for the use of solar energy. The history of solar thermal is very long. For instance, in 214 BC Archimedes used a solar mirror to boil water.

Essentially, **solar thermal** means that solar energy is used to create heat. A very wide range of technologies can be used for this purpose. In addition to using solar heat for space heating, hot water supply, and process heat in industry, solar thermal can also be used for cooling and to generate electricity in solar thermal power plants. A distinction can roughly be made between the following applications:

- solar pool heating
- solar hot water supply
- solar low-temperature heat for space heating
- solar cooling
- solar process heat
- solar thermal power generation.

Because these applications cover a lot of area, only the fundamentals can be covered here. The focus is on the most common systems currently in use – closed collector systems for hot water supply and open collector systems (absorbers) for swimming pool heating.

The following sections include a number of calculations from thermodynamics. The parameters used are introduced and explained in this section. Table 3.1 contains the most important parameters along with symbols and units. Energy in the form of **heat** Q is related to **heat flow** $\dot{Q}$ as follows:

$$Q = \int \dot{Q} \, \mathrm{d}t \tag{3.1}$$

If the temperature changes $\Delta\vartheta$, the **heat also changes** ΔQ, as can be calculated via the **heat capacity** c and mass m of the material in question:

$$\Delta Q = c \cdot m \cdot \Delta\vartheta \tag{3.2}$$

In descriptions of thermodynamic events, the use of temperatures ϑ in degrees Celsius or absolute temperatures T in Kelvin often leads to confusion. When absolute values are used, the following applies:

$$T = \vartheta \cdot \tfrac{\mathrm{K}}{^\circ\mathrm{C}} + 273.15\ \mathrm{K} \tag{3.3}$$

Table 3.1 For heat calculations, symbols and units are used

Name	*Symbol*	*Unit*
Heat, energy	Q	Ws = J
Heat flow	$\dot{Q}$	W
Temperature	ϑ	°C
Absolute temperature	T	K (Kelvin, 0 K = −273.15 °C)
Heat capacity	C	J/(kg K)
Heat conductivity	λ	W/(m K)
Heat transmission ratio	U'	W/(m K)
Thermal transmittance	U (formerly k)	W/(m K)
Heat transfer coefficient	α	W/(m K)

If a temperature difference $\Delta\vartheta$ is indicated in degrees Celsius (°C), it corresponds to the number for the temperature difference ΔT in Kelvin (K). To ensure that the calculation is correct, the temperature in the equation above for a calculation of heat change should be indicated in Kelvin. The same applies for numerous formulae used below. But because the Celsius scale is far more common than the Kelvin scale, most of the temperature differences and formulae that involve them are based on Celsius values below.

The heat flow $\dot{Q}$ that occurs during a heat change is as follows if heat capacity c is constant:

$$\dot{Q} = \frac{dQ}{dt} = c \cdot \frac{dm}{dt} \cdot \Delta\vartheta + c \cdot m \cdot \frac{d\Delta\vartheta}{dt} = c \cdot \dot{m} \cdot \Delta\vartheta + c \cdot m \cdot \Delta\dot{\vartheta} \tag{3.4}$$

Table 3.2 shows the heat capacities of various materials.

Table 3.2 Heat capacity c of various substances ϑ = 0–100 °C

Name	*c in Wh/(kg K)*	*c in kJ/(kg K)*	*Name*	*c in Wh/(kg K)*	*c in kJ/(kg K)*
Aluminium	0.244	0.879	Air (dry, 20 °C)	0.280	1.007
Concrete	0.244	0.88	Brass	0.107	0.385
Ice (−20 °C to 0°C)	0.58	2.09	Paraffin	0.582	2.094
Iron	0.128	0.456	Molten salt ($NaNO_3$/KNO_3)	0.43	1.55
Ethanol (20 °C)	0.665	2.395	Sand, dry	0.22	0.8
Plaster	0.31	1.1	Silicon	0.206	0.741
Glass, glass wool	0.233	0.840	Oil (as heat carrier) VP1 (200 °C)	0.569	2.048
Wood (spruce)	0.58	2.1	Tyfocor (antifreeze) 55 % (50 °C)	0.96	3.45
Copper	0.109	0.394	Water	1.163	4.187

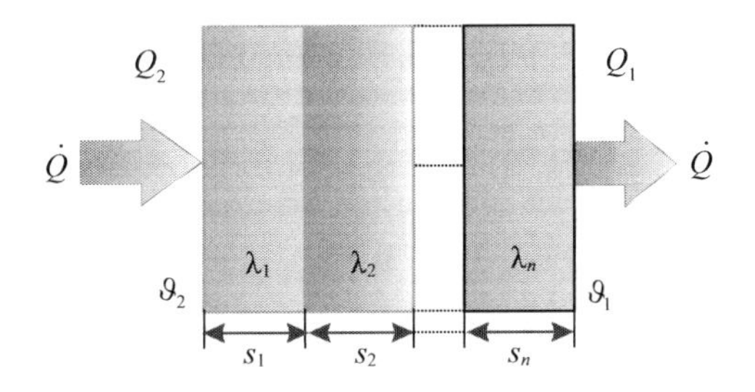

Figure 3.1 Heat transmissivity of a barrier consisting of n layers with the same cross-section A

If these materials are arranged in layers with a cross-section area A consisting of n layers as shown in Figure 3.1 with the temperature of ϑ_1 on one side and of ϑ_2 on the other, the formula for **heat transmissivity** is:

$$\dot{Q} = U \cdot A \cdot (\vartheta_2 - \vartheta_1) \tag{3.5}$$

The heat flow $\dot{Q}$ increases heat on the one side and decreases it on the other until the two sides have the same temperature level. Because the heat content is often much greater on one side than the other, the temperature change on the side with the much greater heat content is usually negligible. But the side with the greater heat content is not necessarily the one with the higher temperature, as the example of an exterior building wall shows. The formula for the **heat transmissivity coefficient**

$$U = \left(\frac{1}{\alpha_1} + \frac{1}{\alpha_2} + \sum_{i=1}^{n} \frac{s_i}{\lambda_i} \right)^{-1} \tag{3.6}$$

is the product of the heat transfer coefficients α_1 and α_2 on the two sides, heat transmissivity $\lambda_{1,}$ and layer thickness s_i of the n layers. Heat conductivity λ is shown in Table 3.3 for various substances.

Table 3.3 The heat conductivity of various substances

Material	*Heat conductivity λ in W/(m K)*	*Material*	*Heat conductivity λ in W/(m K)*
Ice (0 °C)	2.23	Wood (spruce)	0.14
Water (20 °C)	0.60	Clay	0.5–0.9
Air (dry, 20 °C)	0.026	Glass	0.76
Hydrogen	0.186	Plaster	0.45
Copper	380	Straw bale	0.038–0.067
Iron	81	Isofloc (paper fibres)	0.045
Steel	52–58	EPS insulation	0.03–0.05
Granite	2.8	PUR insulation	0.024–0.035
Concrete	2.1	Glass wool	0.032–0.05
Masonry	0.3–1.4	Mineral wool	0.035–0.045
Autoclaved aerated concrete	0.08–0.25	Evacuated insulation panel	0.008

Solar thermal systems

Solar pool heating

Just because we talk about solar pool heating first does not mean that heated pools are an especially good idea ecologically. Swimming pools, after all, always consume a lot of water and energy. Germany has around 8,000 public swimming pools (indoor and outdoor). Another approximately 500,000 private swimming pools collectively have a water surface larger than all of the public pools combined. These figures show that private pools are not just a luxury, but also a waste of energy and water for a privileged few. Nonetheless, all pools are especially suited for the use of solar thermal energy. Back in 1988, an estimated €400 million in energy costs accrued at all of the public pools in West Germany alone. Because energy prices have generally risen since then, the costs are probably much higher today. The use of solar energy is a way of saving both energy and costs because this technology is so straightforward.

Outdoor pools in Germany generally follow the supply of solar energy. In winter, when there is little sun, outdoor pools are usually not open. In contrast, during the warm season it is easy to use solar energy to heat up the water. Figure 3.2 shows how solar swimming pool heating works.

In and around Germany, the temperature of open bodies of water is generally around 20 °C during the outdoor swimming season. It therefore generally suffices to raise the temperature just a few degrees. For such low-temperature applications, very simple absorbers can be used to convert solar energy into useful heat. A pump system pumps the pool water through the absorber. The water heats up inside the absorber before being fed back into

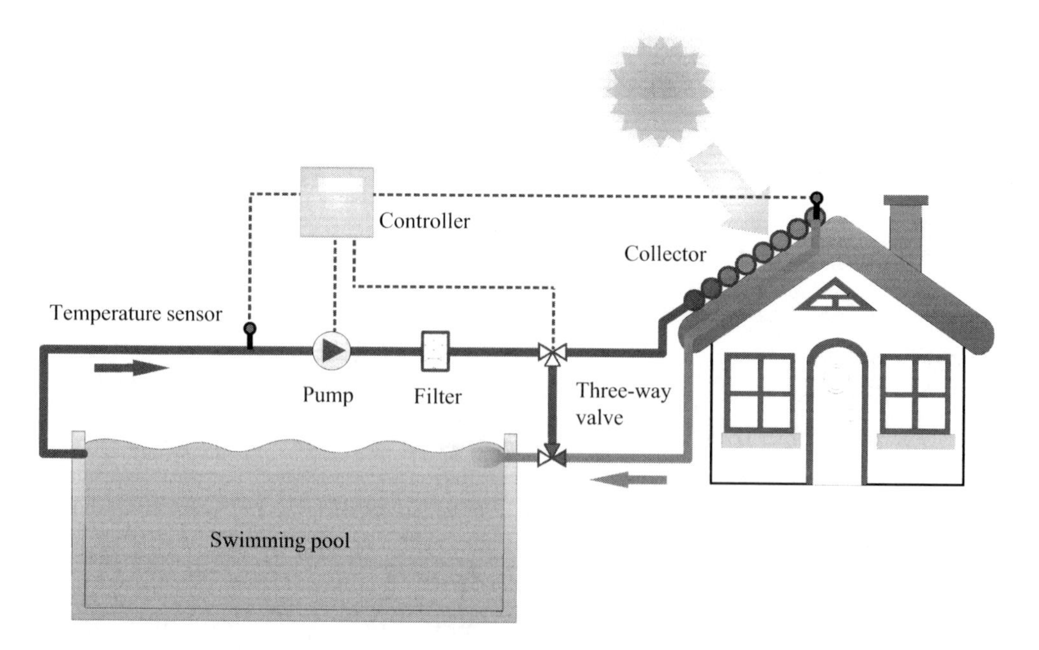

Figure 3.2 How solar swimming pool heating works

the pool. No hot water storage tank – like the ones used in homes – is needed. The pool itself serves as the storage tank.

For the solar circuit, the **pump** has to be set so that it only runs when the absorber can be expected to increase the water's temperature. If the pump sent pool water through the absorber under a cloudy sky, the water might give off heat to the ambient air, thereby cooling down the pool. To prevent this from happening, a two-point control system with hysteresis used. Here, one temperature sensor measures the pool's temperature; another, the absorber's. The pump switches on if the differential temperature exceeds a set threshold. A conventional auxiliary heater can also be integrated in the system if need be.

If the system is based solely on solar heat, the temperature of the pool will fluctuate depending on the weather. If the weather is bad, the pool's temperature drops a bit. Generally, however, pools are not used as much on such days, so lower temperatures are usually considered acceptable.

The **heat demand of outdoor pools** generally ranges from 150 kWh to 450 kWh per m^2 of surface area. When solar is used to heat up the water to 23 °C, no additional heat from fossil fuels is needed at all. Over a surface area of 2,000 m^2, 75 000 l of heating oil equivalent to 150 000 kg CO_2 (if the boiler is used 80 % of the time) can then be saved in a season [Fac93]. Covering the pool at night to minimize heat losses also saves energy.

The **absorber's area** should be equal to around 50–80 % of the pool's surface. The absorber costs around €100 per square metre. Over the year, the cost of solar heating is generally lower than the cost of fossil heating. The cost of a heating system based exclusively on solar is only greater if the pool is to have a higher temperature or if the pool is outdoors and used all year round; then, the absorber size needed could exceed the cost of a fossil heating system.

The electricity to power the pump used for the pool's heating system and for other equipment can also come from photovoltaics. In such cases, the temperature control system may not even be needed because the PV system will only provide power when the sun is shining. Chapter 5 deals with photovoltaic systems in more detail.

Solar hot water

For hot water used in kitchens and showers, higher temperatures are needed than for swimming pools. The simple absorbers used at swimming pools are therefore no good for hot water supply. At high temperatures, the losses from convection – the influence of wind, rain, and snow – would be unacceptably great, as would the heat losses from thermal radiation. Simple absorber systems would therefore only be able to cover a small part of hot water supply. Collectors are therefore used to produce much higher temperatures and reduce losses. Here, a distinction is made between flat-plate collectors, evacuated collectors, evacuated-tube collectors, and collectors with integrated tanks. For space heating, flat-plate collectors sometimes also use air, not just water, as the heat carrier. We will come back to solar collectors in more detail later on.

A complete system for hot water supply not only includes a closed collector, in which an absorber heats up the water, but also other components ranging from storage tanks to pumps and complex control systems to ensure that users feel that their supply of solar hot water is just as convenient as conventional hot water supply.

Figure 3.3 Solar thermal systems for hot water supply. Left: gravity system; right: systems with pumps (photos: left: Cornelia Quaschning, right: Viessmann Werke)

A simple solar hot water supply system can be a primitive absorber consisting of a black container filled with water and facing the sun at midday. If the container is higher than the tap, hot water will flow out without further ado. One example is a solar shower used for camping. Such devices are basically just a black bag filled with water and hung from a branch. If the bag has been in the sun long enough, you can take a hot shower with the water.

Naturally, such a system does not provide the convenience people are used to in everyday life; after all, once the bag is empty, you have no more hot water and have to refill it manually. That's not what most people think of as convenience. But if the bag and the tap have an airtight connection, a water hose can be connected to immediately refill the bag. And instead of the bag, you can use a collector. An off-the-shelf collector probably does not contain enough water for a hot shower, however, and you might also burn yourself because the water gets so hot. The solar thermal systems used in homes are therefore a bit more sophisticated, as described below.

Gravity and thermosiphon systems

The kind of system shown in Figure 3.4 exploits the greater specific density water has when it is cold. Cold water is heavier than warm and therefore sinks. The collector is then always lower than the heat storage tank. Cold water from the storage tank then sinks down into the collector through a pipe. When the water warms up in the collector, it rises again, eventually passing through a pipe at the top of the collector before heading on to the storage tank. In this way, a circuit is created between the collector and the storage tank. The hot water is always available from the storage tank, where it is taken from the top. Any hot water consumed is replaced immediately by fresh, but cold water from the

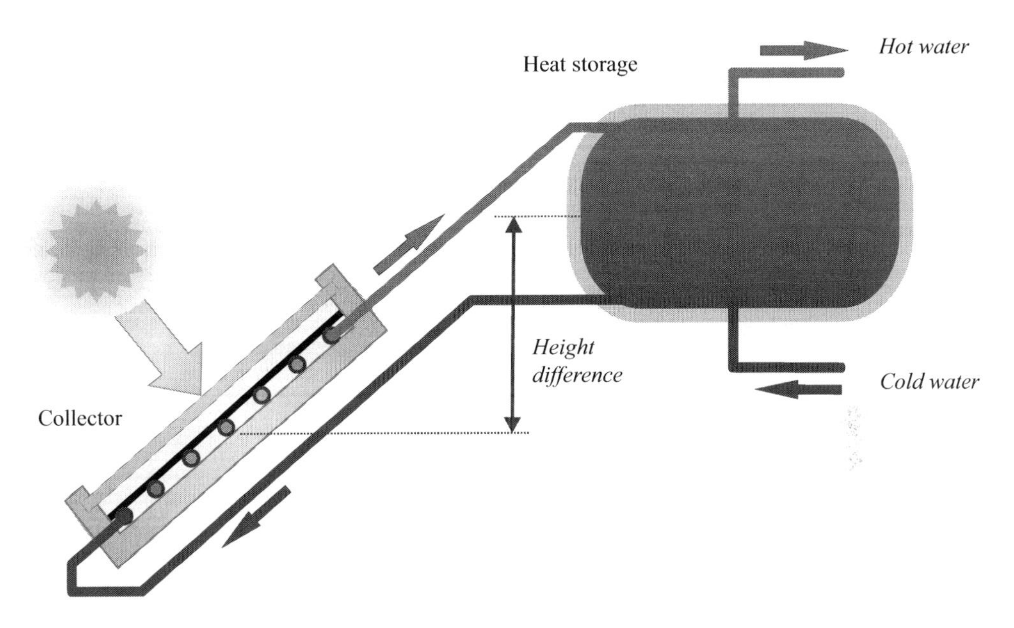

Figure 3.4 How a solar thermal system with gravity feed works (thermosiphon)

plumbing system. This fresh cold water then starts heating up in the collector. When more sunlight is available, hot water rises up faster because of the greater temperature differences. In this way, the flow rate within the circuit adapts to the amount of sunlight available.

In such a gravitational system, it is crucial that the storage tank be at least 0.6 m to one metre above the collector. Otherwise, the circuit might run backwards, for instance at night, cooling off the hot water from the tank via the collector in the process. Likewise, if the height difference is too small, water does not properly flow through the circuit. In Mediterranean countries and similar locations with a lot of irradiation, storage tanks are therefore usually installed on flat roofs just beside the collector.

On pitched roofs, the storage has to be as far above the roof as possible if the collector is also to be installed on the roof. The storage tank's great mass may, however, exceed the building's load-bearing capacity. Furthermore, it can be hard to connect the system to a conventional heating system located, generally, in the basement.

If drinking water directly enters the collector in a single-circuit system, the system might freeze in the winter and be damaged. In areas with frost, **dual-circuit systems** are therefore used. Potable water flows directly into storage in such systems and is taken from there. In contrast, the water that flows through the collector has antifreeze added to it; this section is called the solar circuit. Heat is taken from the storage tank via the heat exchanger; the solar circuit and the drinking water circuit are separate. Glycols are often used as antifreeze in the solar circuit. Whatever antifreeze is used, it should be non-toxic in case the content of the solar circuit leaks into the drinking water. Therefore, ethylene glycol – used in automobiles – is not permitted in drinking water systems. To prevent corrosion, the antifreeze has to be tailored to the materials used in the system.

Gravity solar thermal systems have a number of drawbacks, however. First, such systems have a lot of inertia; they cannot properly react to quick changes in irradiance. Gravity systems are therefore less suited for collector surfaces larger than 10 m^2. Second, the storage tank always has to be higher than the collector in such systems, which is not always easy to implement. Finally, high temperatures in the storage tank and collector circuit can reduce the collector's efficiency. In return, gravity solar thermal units have simple structures. Unlike the systems described in the section below, no control system or pump is needed to circulate water. The system therefore cannot fail when one of these elements is defective. Finally, electricity is needed to power a pump, so gravity systems save this energy in addition.

Solar thermal with forced circulation

Unlike solar thermal systems that use gravity to circulate water, those with forced circulation have an electric pump. Here, the collector and the storage tank can be installed anywhere, though the connecting lines should be kept as short as possible because pipes always entail losses. The storage tank does not, however, need to be located above the collector. Figure 3.5 shows a solar thermal system with forced circulation.

One **temperature sensor** is located in the storage tank; another, in the collector. If the temperature in the collector is higher than in the storage tank by a specified amount, the control system switches on a pump to circulate the heat carrier medium within the

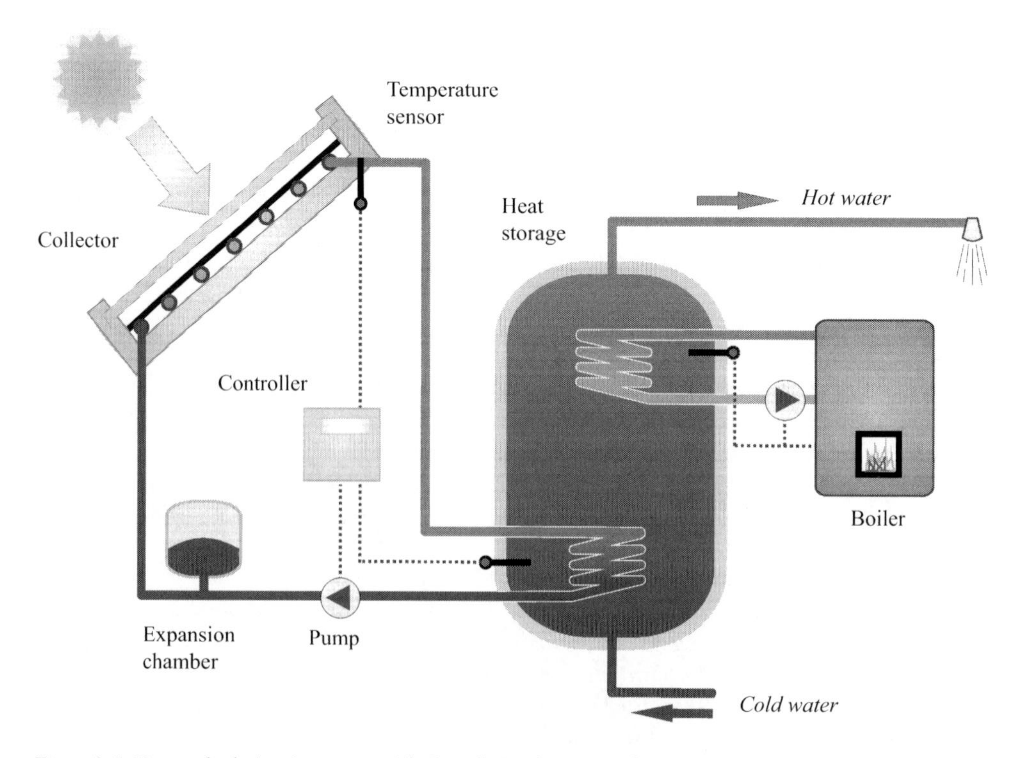

Figure 3.5 How a dual-circuit system with forced circulation works

collector circuit. Usually, the temperature difference is set between 5 °C and 10 °C. If the differential temperature falls below a minimum value, the pump switches off again. These two threshold values should be selected to prevent the pump from repeatedly switching on and off again when there is little irradiance.

The **pumps** used in the solar circuit can be the same as those used to circulate water in heating systems. Such pumps are generally technically advanced – and hence both reliable and relatively inexpensive. Often, variable-speed pumps are used. They allow the flow rate to be adapted to the amount of irradiance. The pumps are then set to ensure that the heat carrier medium has a **throughput** of 30 l/h to 50 l/h per square metre of collector surface. The flow rate is faster for swimming pool absorbers. If it is too low, the temperature in the collector rises, thereby reducing efficiency. If the flow rate is too great, the pump consumes too much energy.

In Germany, such pumps usually run on 230 V alternating current from the grid. It is also possible, however, to power the pump with direct current, which can come from a small photovoltaic module. In this case, the energy needed to power the pump also comes from the sun.

Large systems that cover much higher consumption levels than in single-family homes use **dual storage systems**. Here, one storage tank is behind the other. The first storage tank is heated up from the solar collector, while the second storage tank is connected to a conventional auxiliary heating system. Here, solar covers a relatively small share of heat supply, but the system is affordable. To increase the coverage from solar, a circulation line can be installed from the second tank into the first one. When there is enough irradiation,

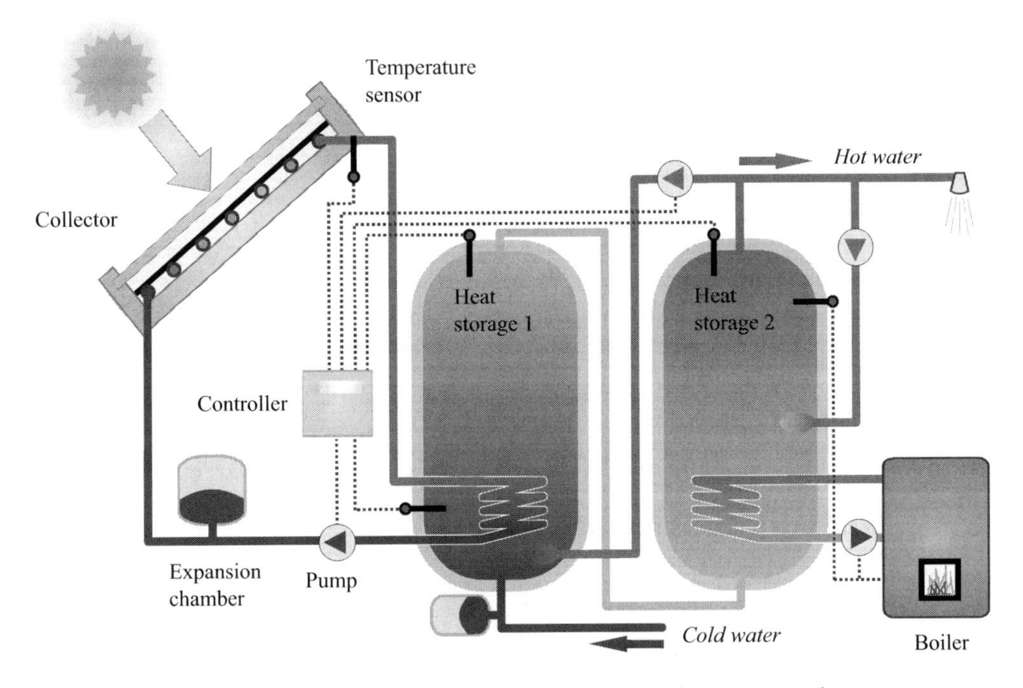

Figure 3.6 How a solar supply of hot drinking water works with two heat storage tanks

solar energy can heat up both of the tanks then. Figure 3.6 shows such a system with two storage tanks for a solar supply of hot drinking water.

Solar support for space heating

Solar thermal cannot only easily be used to supply hot water, but also for space heating. In principle, the collector and the storage tank merely need to be larger – and connected to the heater circuit. Drinking water is generally not used in the heater circuit, however, so separate storage tanks are needed for space heating and hot water supply. A combined storage tank is an elegant option that also reduces heat losses (Figure 3.7).

In Germany, these systems are usually designed to cover only part of space heating. Solar energy suffices in particular to cover most of this demand in spring and autumn. In contrast, the collectors generally do not provide enough energy to cover all of the demand in winter.

Purely solar heating

In principle, it is also possible to meet the demand for heating energy entirely from the sun. In such cases, **seasonal storage** is needed to provide heat from the summer during the winter. This approach increases the overall cost of solar heat. For economic reasons, smaller storage tanks sufficing for a few days are generally used instead. In climates like Germany's, solar energy can then meet only 20 to around 70 % of heat demand, depending on the building's insulation standard.

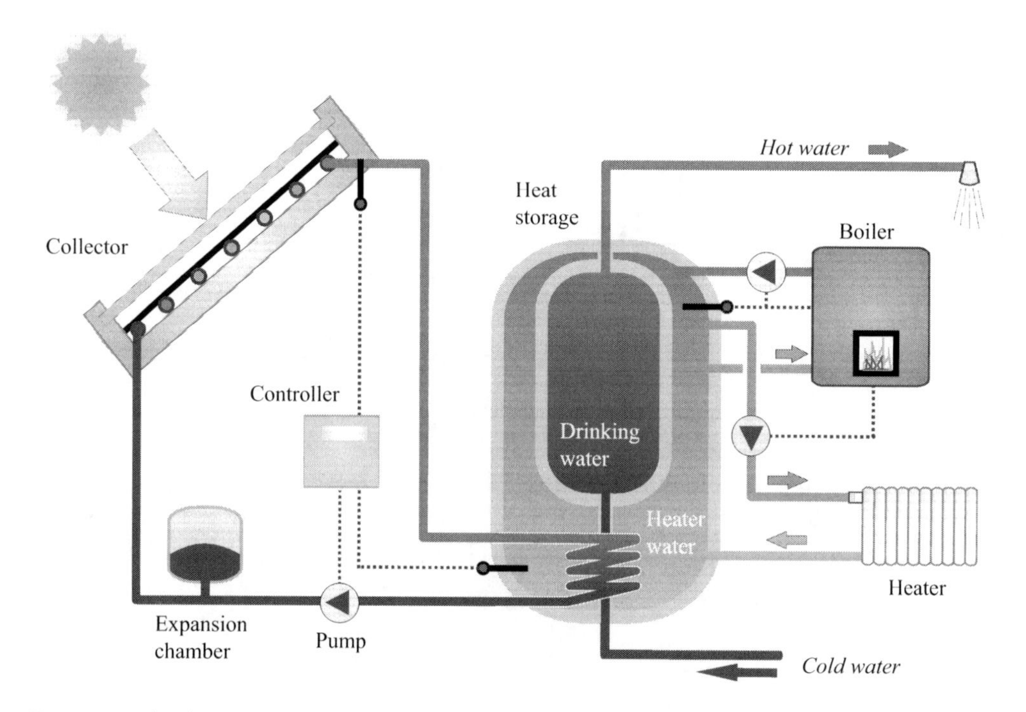

Figure 3.7 Solar thermal system for hot water supply and space heating support

Optimal building insulation is essential if solar is to meet all of the demand for heat. Here, the minimum insulation standard would be a Passive House, one that requires zero heating energy. The storage tank would then generally need to be 30 to 50 m³ to provide enough heat even through the winter for a single-family home. The storage tank then generally has **layers**. Hot water is not as dense as cold water. If the hot water is above the cold and the water is prevented from circulating, the layers remain stable. More hot water can then be taken from the top of the storage tank at the desired temperature than would be the case if the water were mixed within the tank. To keep the layers from mixing, water can be added to and taken from the tank at different levels, such as when heat from the collectors is added or taken out. Figure 3.8 shows a system with a three-way valve leading to different connections used relative to storage temperature.

Numerous homes have been built in Germany and Switzerland with purely solar space heating. One example is the residential complex in Oberburg, Switzerland, which was completed in 2007 (Figure 3.9). Here, a 205 m³ combined storage tank is integrated in the building. The 276 m² part of the roof facing south is covered with flat-plate collectors. They heat up the storage tank to around 90 °C in the summer. In the winter, the top end of the storage tank does not fall below 60 °C, which suffices for space heating and hot water. Excess heat can even be exported to neighbouring buildings through connecting lines.

Solar district heat

If a neighbourhood has a lot of buildings in it with solar collectors, they can be connected to create a district heat network (Figure 3.10). In addition, a large central collector system

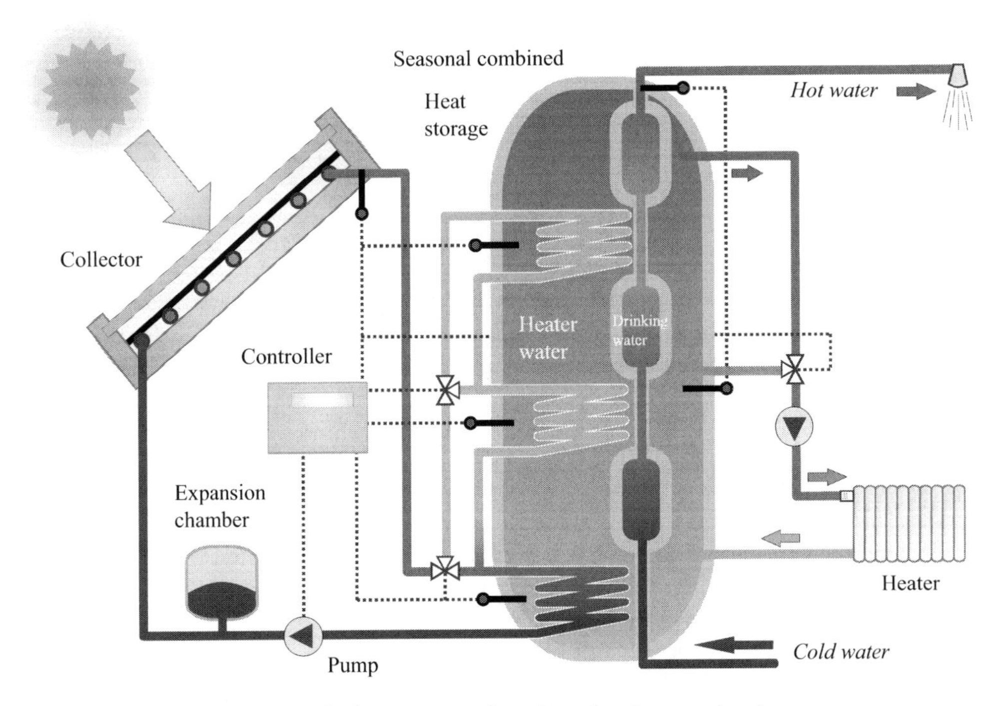

Figure 3.8 Solar thermal system for hot water supply and purely solar space heating

Figure 3.9 A residential complex with solely solar space heating in Oberburg, Switzerland. Left: installation of a 205 m³ combined storage tank; right: view of the 276 m² of collector on the south-facing roof.

Photos: Jenni Energietechnik AG, www.jenni.ch

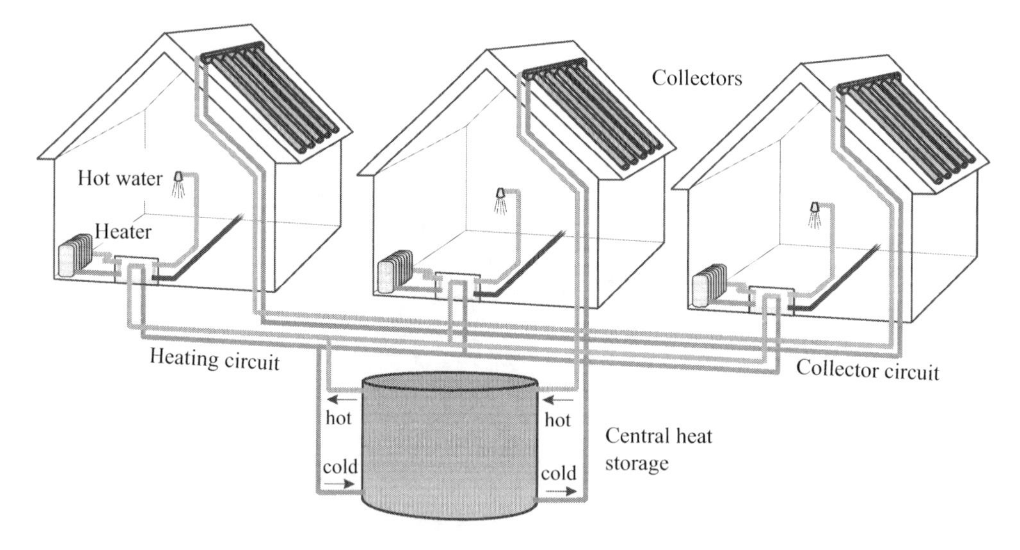

Figure 3.10 How a solar district heat network works

can be built. The centrepiece of such a district heat network is central heat storage. The larger the storage facility, the lower the heat losses can be, thereby allowing more heat to be stored over a longer time frame. But because connections in district heat networks are longer, costs are higher, as are line losses.

A number of solar district heat neighbourhoods have already been successfully built, though few of them are truly large. Table 3.4 shows data from a select group of such

Table 3.4 Data from existing solar district heat projects

	Marstal (Denmark)	*Brædstrup (Denmark)*	*Kungälv (Sweden)*	*Neckarsulm (Germany)*
Completed in	2003	2007	2001	1997/2001
Collector surface in 1,000 m^2	19	8	10	2.7/6.5
Collector output in MW	12.85	5.6	7	1.89/4.55
Storage type	Tank/gravel/pond	Water tank	Water tank	Geothermal
Storage size in 1,000 m^3	2/3.5/10	2	1	20/63.3
Solar heat yield in MWh/a	8,824	4,000	3,500	832/2,018
Auxiliary heater	Fuel oil	Natural gas	Biomass/oil	Natural gas
Auxiliary heat output in MW	18	29.8	25	n/a
Share of solar	32 %	9 %	4 %	50 %
Cost of solar facility in million €	7.333	1.57	2.218	1.483, n/a

projects. The largest solar district heat project in Europe is currently located in Marstal, Denmark. There, around 19,000 m² of collector surface area has been installed over numerous project stages. The district heat network serves more than 1,400 customers.

Solar cooling

Though it may sound like a contradiction, solar heat is also a great way to cool buildings. In sunny, hot parts of the world, a large number of energy-hungry air-conditioners provide pleasant indoor temperatures. The sunnier and hotter it gets, the greater the need for cooling. But as irradiance increases, so does the output of a thermal collector. While demand for heat is disproportionate to irradiance, demand for cooling is nearly exactly in line with it.

Along with a large, powerful collector, the heart of a solar cooling unit is generally an absorption chiller. But here, absorption is unrelated to the solar absorber. Rather, the absorption in this chiller is related to the chemical process of **sorption**. Chemists speak of sorption or absorption when a fluid takes up another gas or fluid. One common example is carbon dioxide in carbonated water.

Absorption chillers use an absorbable coolant with a low boiling point – such as ammonia – that is later dissolved in water. In addition to ammonia, water itself can be used as the coolant if it is under negative pressure. Then lithium bromide is a good solvent.

In the evaporator, the coolant boils at low temperatures. In the process, it takes heat out of the cooling system. The coolant then needs to be turned back into a fluid so it can either operate once again in order to draw off more heat. If we include a few more tricks and implement sorption, liquefaction is also possible from solar heat.

First, the chiller's absorber mixes the evaporated coolant with the solvent. During the absorption process, heat is released. This heat is either used to heat up water, etc., or given off as waste heat. A pump then transports the solvent mixture to the regenerator, which heats up the coolant and the solvent and separates them thanks to their different boiling

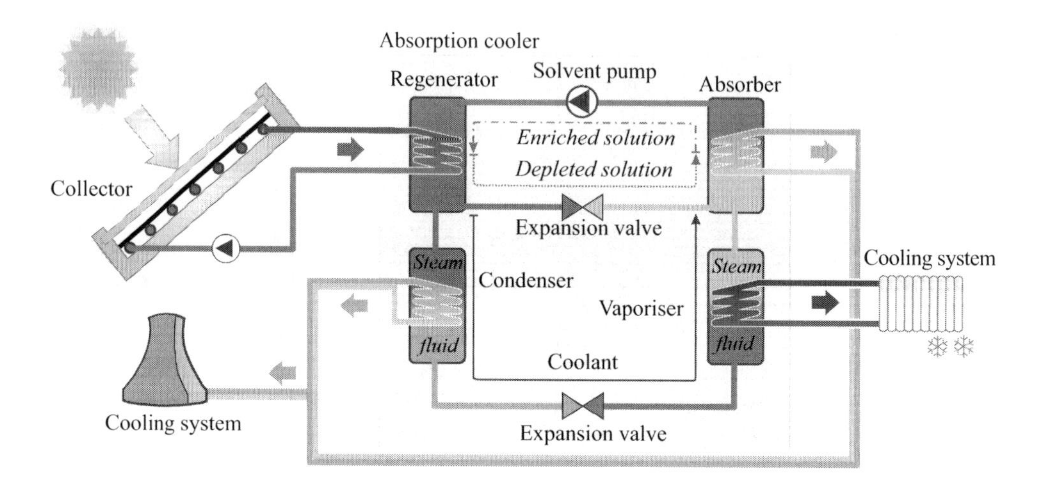

Figure 3.11 How a solar cooling system works

points. The heat used for this purpose can come from powerful solar collectors. Temperatures from 100 to 150 °C are optimal. The coolant that has now been separated is still a vapour; it is passed on to a heat exchanger (condenser), where it gives off its heat and liquefies again. The condensation heat can also be used as useful heat or given off as waste heat. The fluid coolant passes through a thermal expansion valve to return back to the evaporator, while the solvent goes back to the chiller's absorber. When it expands in the thermal expansion valve, the coolant cools down considerably; it can then once again pass on cold to the cooling system via the evaporator.

Solar collectors

To produce higher temperatures, special collectors are needed with more sophisticated technology than is used to heat pools. A distinction is made between:

- collectors with integrated tanks
- flat-plate collectors
- evacuated-tube collectors.

The flat-plate and evacuated-tube collectors described below require a storage tank because the use of the collector alone is not fruitful. The collectors themselves have very little space for the heat medium. Under great irradiance, the content of the collector would quickly exceed 100 °C if this content were not constantly refreshed. Such high temperatures would increasingly lead to heat losses – not an efficient way of exploiting incident sunlight. Indeed, all drinking water systems have storage tanks to collect large amounts of the heat absorbed by the collector for later hot water supply when the sun is not shining (as much). Again, with swimming pool heating systems, the pool itself serves as the storage tank.

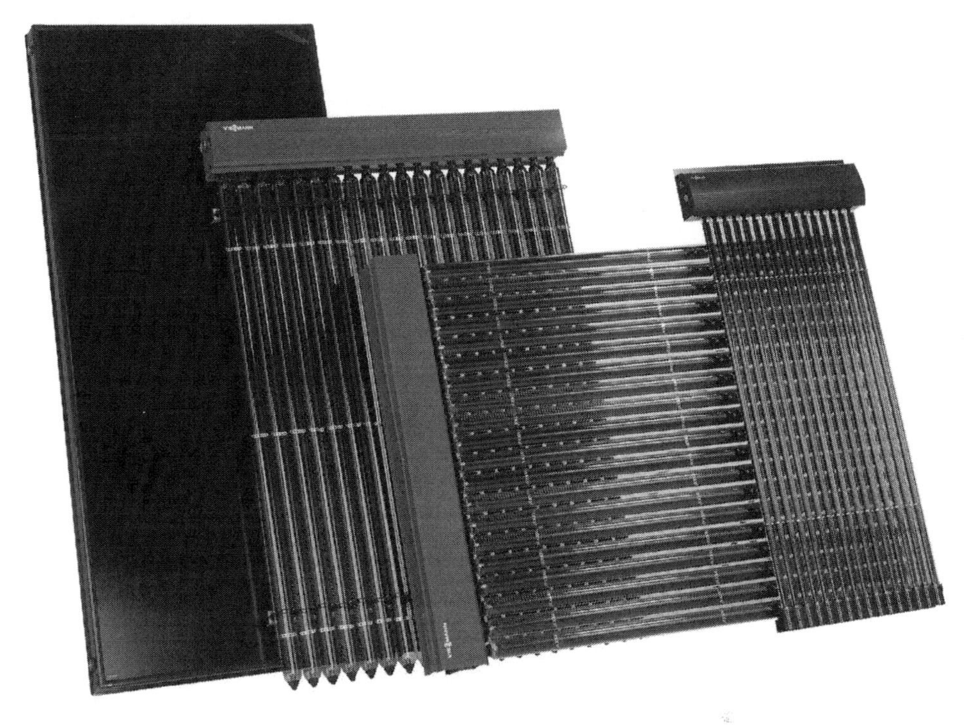

Figure 3.12 Flat-plate and evacuated-tube collectors

(Image: Viessmann Werke)

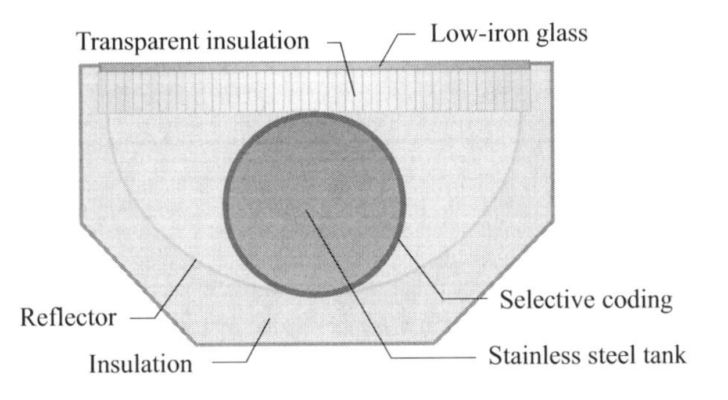

Figure 3.13 Cross-section of a collector with integrated storage

Collectors with integrated tanks

A special way of providing hot water from solar heat is to have the storage tank integrated in the collector. Here, the system runs the risk of freezing over in the winter.

Transparent insulation is one way of solving the problem. Here, the materials used have a transmissivity not quite as good as low-iron solar glass, but a much better thermal transmittance, meaning that heat losses are much lower. Various conventional cladding materials and transparent insulation products are compared in Table 3.5. The use of transparent insulation can reduce heat losses from collectors with integrated storage tanks to acceptable levels. Figure 3.13 shows a collector with an integrated stainless steel tank.

Table 3.5 Thermal transmittance U and total energy transmissivity (g-value) of various conventional and transparent insulation materials

Conventional glazing			*Transparent insulation between low-iron glass with air gap*		
Material	*U-value in W/(m² K)*	*g-value*	*Material*	*U-value in W/(m² K)*	*g-value*
Single glazing (4 mm)	5.9	0.86	Aerogel granular (20 mm)	0.85	0.4
Double glazing (4 mm, 12 mm air, 4 mm)	3.0	0.77	Polycarbonate honeycomb (100 mm)	0.7	0.66
Triple glazing with IR coating (36 mm)	0.5 . . . 1.2	0.53 . . . 0.62	Polycarbonate capillary (100 mm)	0.7	0.64

(Data: [Fac90, Hum91, Bun92])

The collector itself is well insulated on the rear. Reflectors on the inside wall send incident light back on to the storage tube. A layer of transparent insulation is on the top underneath the glass cover.

Such a storage collector, of course, does not need heat storage outside of the collector, as other collectors do. As a result, the entire system has a simple design and is quite inexpensive because so few components are needed. If the temperature in the stored water is too low, a downstream instantaneous gas heater can be set to bring the water up to the desired temperature via a thermostat.

Collectors with integrated storage have a much greater mass than other collectors, however, which can be a drawback. This collector is therefore not usable on all roof constructions. In addition, the heat losses are greater than with an optimally insulated external storage tank during long periods of bad weather. Collectors with integrated storage have therefore not yet managed to conquer a large share of the market.

Flat-plate collectors

Flat-plate collectors are currently the most common type for a supply of hot water in Europe. Such units mainly consist of three components:

- a transparent cover
- a collector case
- an absorber.

An absorber inside the flat-plate collector converts sunlight into heat, which is passed on to the water flowing through the absorber in tubes. The absorber is in a housing with the best possible insulation on the back of the collector to reduce heat losses as much as possible on the back.

Nonetheless, the collector has losses that depend on the temperature difference between the absorber and its environment. Here, a distinction is made between losses from convection and radiation. Moving air causes convection losses.

Figure 3.14 Cross-section of a flat-plate collector
(Source: © Bosch Thermotechnik GmbH)

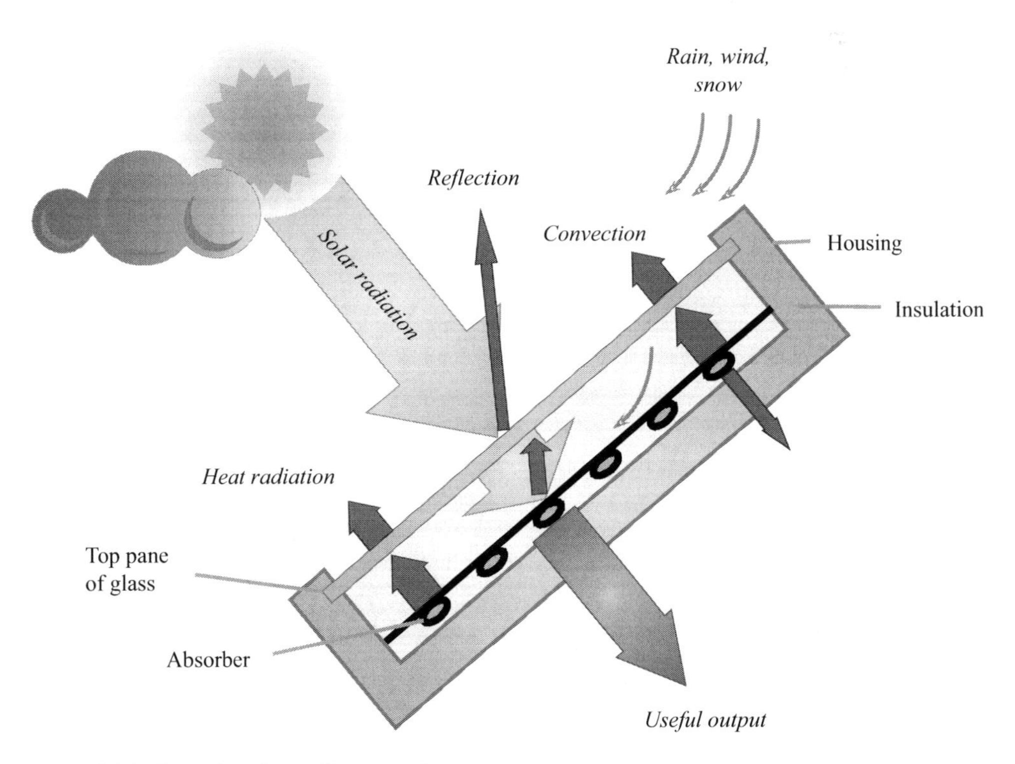

Figure 3.15 How a flat-plate collector works

A pane of glass covers the collector facing the sun. This glass pane prevents most of the convection losses. The pane also prevents the absorber from radiating heat into the collector's ambient atmosphere as in a greenhouse. But the pane of glass has a drawback: reflection losses when part of the sunlight is reflected rather than reaching the absorber. Figures 3.15 and 3.16 show how a flat-plate collector works.

As Figure 3.17 shows, the **front pane** reflects part of the incident radiant flux Φ_e, absorbs part of it, and lets most of it through. The degree of reflection ρ, the degree of absorption α and the degree of transmissivity τ express these events. The sum of these three degrees must add up to 1:

$$\rho + \alpha + \tau = 1 \tag{3.7}$$

For the various types of radiant flux, the following applies:

$$\Phi_e = \Phi_\rho + \Phi_\alpha + \Phi_\tau = \rho\,\Phi_e + \alpha\,\Phi_e + \tau\,\Phi_e \tag{3.8}$$

When solar radiation is absorbed, the pane heats up. In line with the thermal balance, the absorbed share then has to be emitted as radiation Φ_ε because the pane would otherwise be heated up endlessly. The degree of emissivity ε then corresponds to the degree of absorption α :

$$\alpha = \varepsilon \tag{3.9}$$

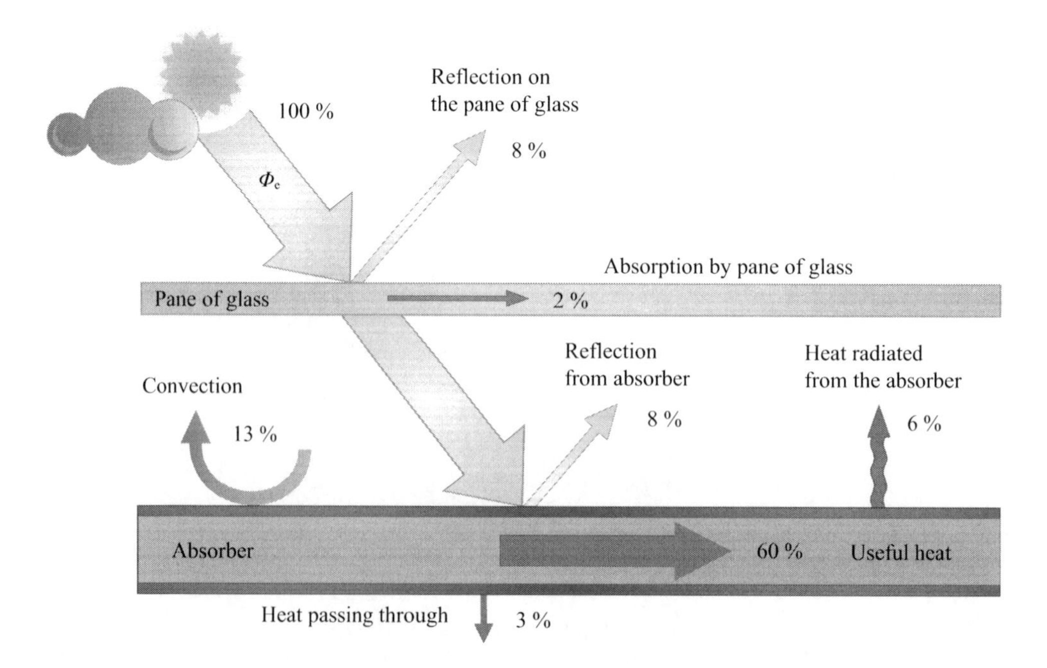

Figure 3.16 Energy conversion in a solar collector and the various types of losses

(Based on [Wag95])

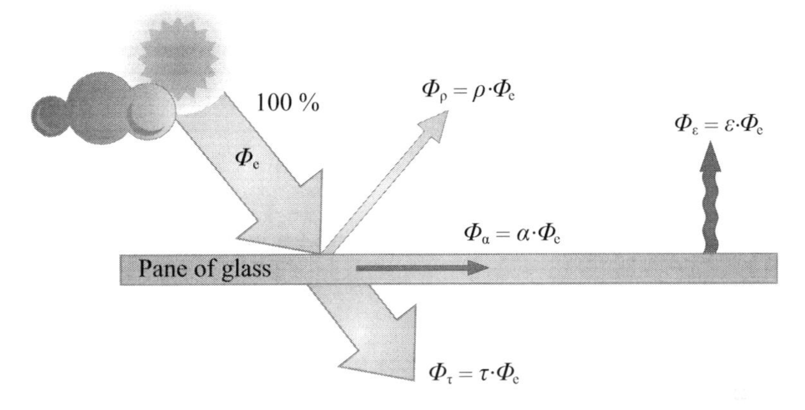

Figure 3.17 Sunlight incident on the front pane

While the front pane is intended to let sunlight through as much as possible, it is also designed to keep back the absorber's heat radiation as in a greenhouse – and reduce convection losses to the environment. A simple cover suffices to serve this purpose in many collectors. Usually, low-iron solar glass that has been thermally treated is used as the front pane; it lets through a lot of light and weathers the elements well. Modern solar glass has a transmissivity degree of around 0.92, a level that special antireflection glass can increase up to 0.96. Plastics have not proven to be a worthwhile option because they are not sufficiently resistant to ultraviolet light and do not weather the elements nearly as well; after some time, they 'go blind'.

Double glazing reduces the heat losses considerably, but at the expense of the amount of light let through.

The **collector casing** can be made of plastic, metal, and even wood. The pane of glass should be fitted into the casing so as to be as airtight as possible; that way, less heat escapes, and dirt, bugs, and moisture cannot so easily enter the collector. A lot of collectors have controlled ventilation so that any condensation that builds up within the collector or enters it can be blown out, instead of building up on the glass pane.

The material used for **rear ventilation** has to insulate well and withstand temperature fluctuations. Polyurethane hard foam boards are suitable, as are mineral fiber boards. Any adhesives or other materials used along with the insulation, etc., must not give off vapors when heated up lest they also produce condensation that reduces the amount of sunlight entering the collector.

In most cases, solar collectors heat up water. For space heating, however, you eventually need to heat up air, not water. In conventional heating systems, radiators and pipes installed in the floor pass the heat from water on to indoor air. Air can be used directly in solar collectors instead of water. Because air does not convey heat as well as water does, absorbers then have to have a much larger cross-section. Otherwise, **air collectors** do not differ much from flat-plate collectors that use water as the heat carrier. Figure 3.18 shows an air collector with a ribbed absorber. An integrated solar module can power the ventilator. Air collectors are an interesting alternative to support space heating in particular. In contrast, heat storage is more complicated in combination with air collectors.

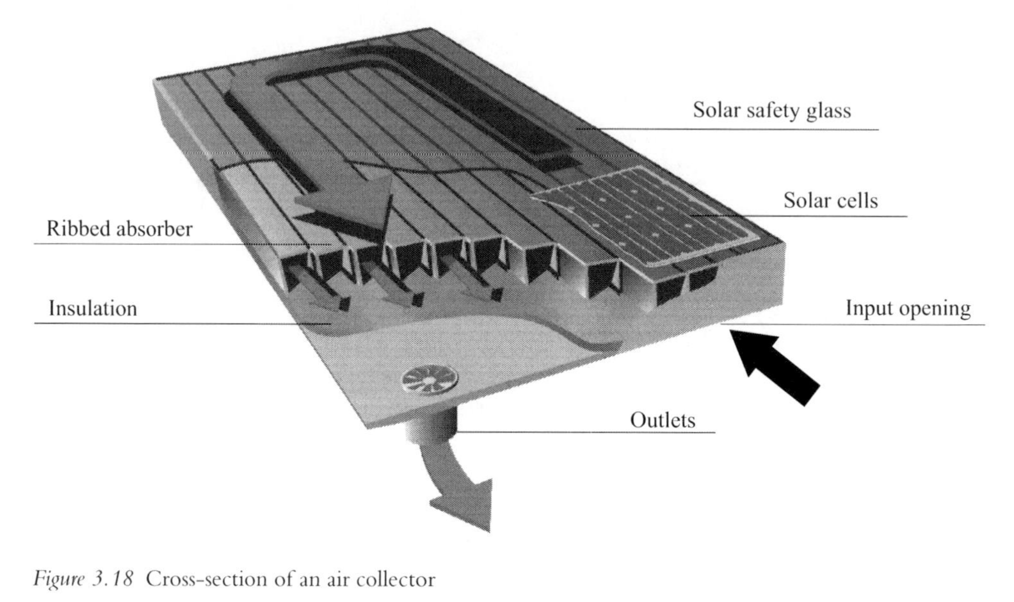

Figure 3.18 Cross-section of an air collector

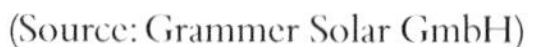
(Source: Grammer Solar GmbH)

If there is a vacuum between the absorber and the front pane, heat losses from **convection** – air moving within the collector – can be reduced considerably. This principle is applied in evacuated collectors. Support struts are needed to keep the back of the collector from touching the glass cover in such units because external air pressure would otherwise press them together. The vacuum cannot be maintained over long periods because air cannot be completely prevented from entering the gaps between the glass and the collector casing. Therefore, flat evacuated collectors need to be newly evacuated at regular intervals. The collector has a special valve to connect with a vacuum pump for this purpose. This regular maintenance is not necessary for a different kind of collector: the evacuated tube collector.

Evacuated tube collectors

Evacuated tube collectors exploit the principle that a high vacuum is much easier to maintain in a completely closed **glass tube** than in a flat collector over a long time frame. Because of their shape, glass tubes withstand external air pressure better so that no support struts are required. In evacuated tube collectors, a distinction is made between:

- heat pipes
- end-to-end heat medium tubes.

Collectors that use a heat pipe have a closed glass tube with a diameter of several centimetres containing a flat metal absorber sheet with a heat pipe in the middle. The heat pipe itself contains the heat carrier, which can be methanol. When the sun heats up this medium, it evaporates. The gas rises into the condenser at the top of the heat pipe – sticking out of the side of the glass tube. Here, the heat medium re-condenses, giving off the energy to another heat carrier flowing by through a **heat exchanger**. Once it has condensed again, the fluid medium sinks back down into the heat pipe. Such tubes have to be installed at a slight angle in order to work. Figure 3.19 shows a cross-section of how such a collector works.

Evacuated tube collectors with an **end-to-end heat pipe** have the heat carrier running directly through the collector. It is therefore not necessary to have a heat exchanger within the collector, and the collector does not necessarily have to be at a slight angle.

It is not possible to completely prevent atmospheric hydrogen from entering the vacuum because hydrogen molecules are extremely small. Over time, the penetration of hydrogen destroys the vacuum. Hydrogen 'getters' are therefore installed in the collector to absorb hydrogen over the collector's service life.

Evacuated tube collectors produce much more energy, especially in the cold season. Evacuated tube collectors take up much less space than normal flat-plate collectors to provide the same energy. In return, evacuated tube collectors are much more expensive. If it is not possible to heat up the water by using a conventional heating system, evacuated tube collectors are the best choice in areas with little irradiation. For this reason, the technology has conquered a very large part of the market in China.

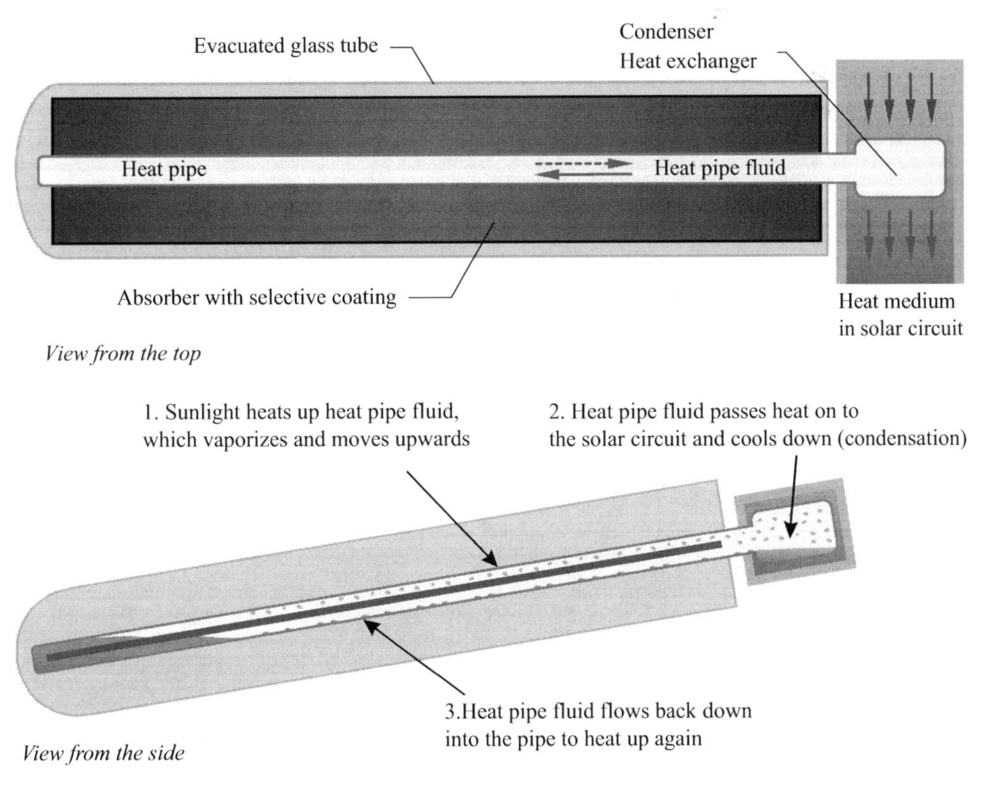

Figure 3.19 How an evacuated tube collector works

Collector absorbers

The centrepiece of any collector is the solar absorber. Different absorber types are used depending on the specific collector and purpose. If a dark hose is spread out on a lawn in the summer, the water within the hose will heat up quickly. Swimming pool absorbers are hardly more complicated. They can consist of black tubes spread permanently across a large surface, such as a roof.

The tubes used are made of plastics that are UV-resistant and can withstand the chlorinated swimming pool water. The **materials** used include polyethylene (PE), polypropylene (PP), and ethylene propylene diene monomer (EPDM), with the latter having a longer service life, but also being more expensive. For ecological reasons, PVC should not be used because hydrochloric acids and dioxins can be emitted when the material is burned in waste facilities. In addition, toxic heavy metals and plasticizers are added during the processing of PVC.

In flat-plate and evacuated tube collectors, temperatures can reach 200°C. All of the materials used have to withstand these temperatures. Therefore, copper, steel, and aluminium are used almost exclusively for these **absorbers**. Figure 3.20 shows some of the types that are used. Today, absorbers with copper tubes that are pressed in, soldered on, welded on, or glued in are used.

It is well-known that black materials absorb sunlight very well and heat up in the process. Metallic materials are not naturally black and therefore have to be coated. Black paint is an obvious option. While temperature-resistant black paint does the job, there are far better materials for **absorber coatings**. When a black surface heats up, it gives off some of that heat as thermal radiation. The effect can be observed with an electric stove. When the stove is on, you can feel the heat on your skin. The same effect occurs when an absorber is painted black. Only part of the heat is passed on to the water flowing inside the collector, with the other part emitted to the surroundings as undesired thermal radiation.

Selective coatings absorb sunlight roughly as well as a black surface does, but they give off far less thermal radiation. Cermet, Tinox, and electroplated nickel can be used as selective coatings. Cermet is a composite material consisting of ceramics in a metallic matrix, whereas Tinox is short for titanium nitride oxide. Selective coatings cannot be painted or sprayed on; rather, more complex processes are required, such as galvanic separation,

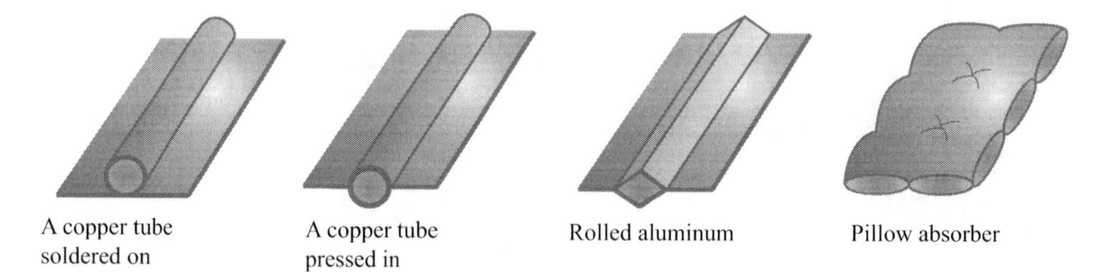

Figure 3.20 Designs of solar absorbers (left: flag absorbers, right: pillow absorber)

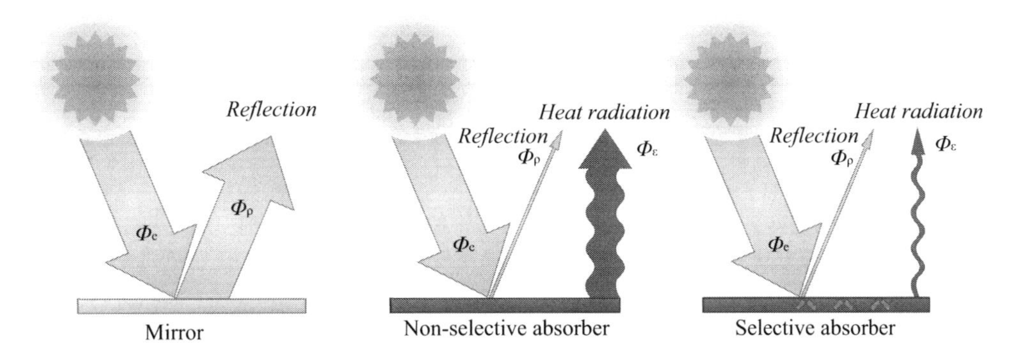

Figure 3.21 Losses from various absorber coatings

electron beam of physical vapour deposition, and sputtering in a vacuum. Figure 3.21 shows the properties of various absorbers.

A selective absorber's properties can be better described with an analysis of the thermal radiation spectrum. As explained in Chapter 2, Planck's Law can be used to calculate spectral radiance relative to an emitter's temperature.

The temperature of the sun is 5,777 K. Most of the solar spectrum is wavelengths shorter than 2 μm. Here, the absorber should have a high absorption coefficient. Sunlight heats up the absorber to around 350 K, the maximum for the spectrum being far above wavelengths smaller than 2 μm. Kirchhoff's law of thermal radiation states that absorption coefficient α is identical with emission coefficient ε, which should be as low as possible above 2 μm to ensure that a hot absorber does not give off much energy to its surroundings as thermal radiation.

Figure 3.22 shows the relative radiance for the maximum of the spectrums at 350 K and 5,777 K along with the degree of absorption of a non-selective absorber and a selective absorber. Table 3.6 shows the data for the degree of absorption of non-selective and several selective absorber materials. Most collectors sold today have selective absorber coatings.

Table 3.6 Absorption degree α, transmissivity degree τ, and reflection degree ρ for various absorber materials

	Visible			*Infrared*		
	$\alpha = \varepsilon$	τ	ρ	$\alpha = \varepsilon$	τ	ρ
Non-selective absorber	0.97	0	0.03	0.97	0	0.03
Black chrome	0.95	0	0.05	0.09	0	0.91
Cermet	0.95	0	0.05	0.05	0	0.95
Tinox	0.95	0	0.05	0.05	0	0.95

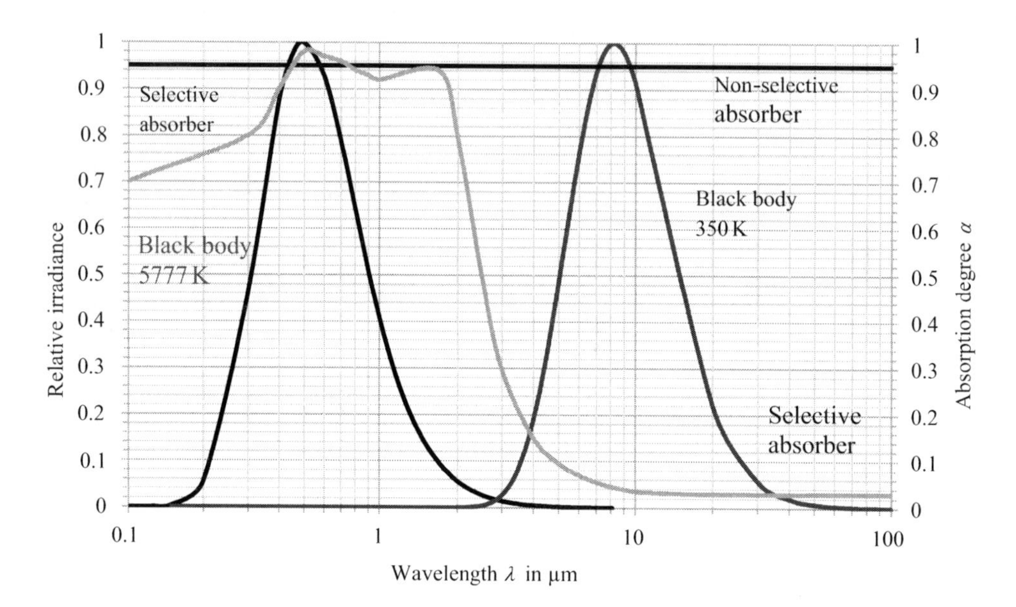

Figure 3.22 Spectrums of black bodies at 5,777 K and 350 K along with the absorption degrees of selective and non-selective absorbers

Collector output and efficiency

Above, we discussed collector design and the materials required; below, we will focus on calculating collector output and efficiency, both of which are crucial towards determining a system's yield.

The collector converts into usable heat the irradiance E that passes through the pane of glass and reaches collector surface A_K weighted with the glass pane's transmissivity degree τ. In order to calculate the collector's useful output $\dot{Q}_{KN}$, losses from reflection $\dot{Q}_R$, convection $\dot{Q}_K$ and thermal radiation $\dot{Q}_S$ must be subtracted from the incident irradiance. The formula for the **collector's useful output** is:

$$\dot{Q}_{KN} = \tau \cdot E \cdot A_K - \dot{Q}_K - \dot{Q}_S - \dot{Q}_R \tag{3.10}$$

The losses from convection $\dot{Q}_K$ and thermal radiation $\dot{Q}_S$ are combined in $\dot{Q}_V$. As described above, the losses from thermal radiation $\dot{Q}_S$ are far lower with selective absorbers than with non-selective ones. The vacuum between the absorber and the pane of glass reduces convective heat losses $\dot{Q}_K$ in evacuated and evacuated tube collectors. Reflection losses $\dot{Q}_R$ can be calculated from the amount of insolation that passes through the front pane using the reflection degree ρ.

$\dot{Q}_V = \dot{Q}_K + \dot{Q}_S$ and $\dot{Q}_R = \tau \cdot \rho \cdot E \cdot A_K$ produces

$$\dot{Q}_{KN} = \tau \cdot E \cdot A_K \cdot (1 - \rho) - \dot{Q}_V \tag{3.11}$$

The absorber's absorption degree $\alpha = 1 - \rho$ is used to produce

$$\dot{Q}_{KN} = \tau \cdot \alpha \cdot E \cdot A_K - \dot{Q}_V = \eta_0 \cdot E \cdot A_K - \dot{Q}_V \quad (3.12)$$

with

$$\eta_{0i} = \alpha \cdot \tau \quad (3.13)$$

η_{0i} is the ideal optical efficiency here. In practice, this value is not reached because the suboptimal heat transfer from the absorber to the heat carrier medium within the absorber causes the absorber material to heat up. The temperature differences lead to additional losses that can be taken into account via the **collector's efficiency factor** F':

$$\eta_0 = F' \cdot \alpha \cdot \tau = F' \cdot \eta_{0i} \quad (3.14)$$

The corrected value is called the **conversion factor**. It reflects the collector's **optical efficiency**, meaning its efficiency when there are no thermal losses from convection or thermal radiation. The collector efficiency factor is relative to the collector's design and the flow rate of the heat carrier but generally ranges between 0.8 and 0.97.

Thermal losses $\dot{Q}_V$ are relative to the average collector temperature ϑ_K, the ambient temperature ϑ_U, and the coefficients a, a_1, and a_2:

$$\dot{Q}_V = a_1 \cdot A_K \cdot (\vartheta_K - \vartheta_U) + a_2 \cdot A_K \cdot (\vartheta_K - \vartheta_U)^2 \approx a \cdot A_K \cdot (\vartheta_K - \vartheta_U) \quad (3.15)$$

The collector's efficiency η_K can be calculated from the collector's feasible useful output $\dot{Q}_{KN}$ and the irradiance E incident on the collector's surface A_K:

$$\eta_K = \frac{\dot{Q}_{KN}}{E \cdot A_K} = \eta_0 - \frac{\dot{Q}_V}{E \cdot A_K}$$ describes the **collector's efficiency**:

$$\eta_K = \eta_0 - \frac{a_1 \cdot (\vartheta_K - \vartheta_U) + a_2 \cdot (\vartheta_K - \vartheta_U)^2}{E} \approx \eta_0 - \frac{a \cdot (\vartheta_K - \vartheta_U)}{E} \quad (3.16)$$

Figure 3.23 shows a calculation of collector efficiency for a flat-plate collector with the parameters $\eta_0 = 0.83$, $a_1 = 3.41$ W/(m² K), and $a_2 = 0.014$ W/(m² K²). Thermal losses increase considerably along with the differential temperature between the collector and its surroundings. At low irradiance, efficiency drops more quickly, so that this collector cannot provide any energy at an irradiance of $E = 200$ W/m² if the differential temperature is above 40 °C.

The diagram also shows how to calculate the **collector's standstill temperature**. During standstill, no heat is taken from the collector, so its efficiency is therefore $\eta_K = 0$. At an irradiance of 400 W/m², the collector's standstill temperature is around 75 °C above the ambient temperature. At an irradiance of 1,000 W/m², the standstill temperature can exceed 200 °C depending on the type of collector. The materials the collectors are made of must be designed to withstand such temperatures, lest the collector be destroyed by overheating.

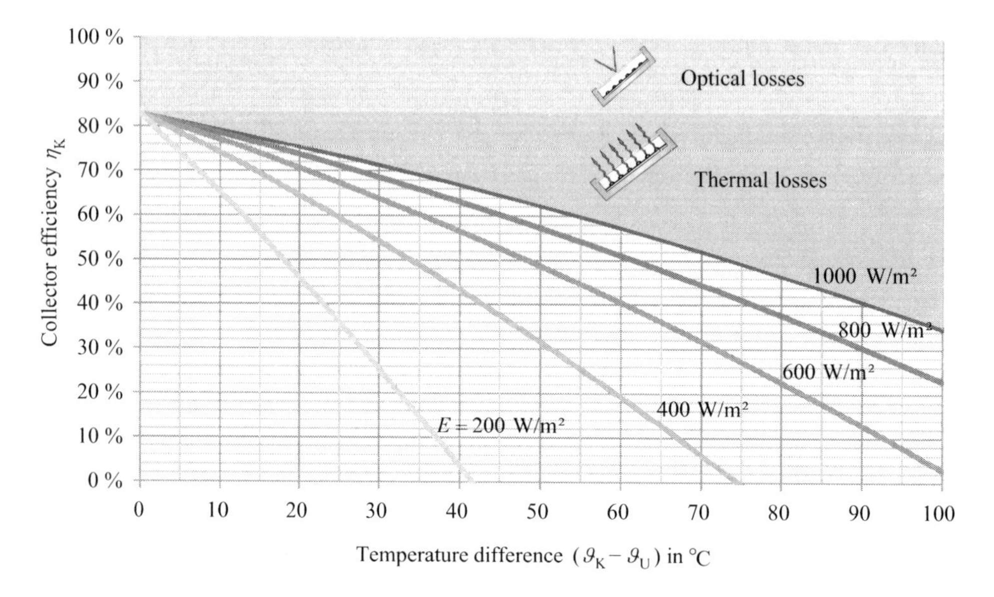

Figure 3.23 Collector efficiencies η_K at various levels of irradiance E and temperature differences ($\vartheta_K - \vartheta_U$)

The calculation of efficiency here assumes there is no wind. As the wind increases, so do the thermal losses. They can be accounted for by means of modified loss factors.

Various parameters – η_0, a_1, and a_2 – are expected for different types of collector designs. For instance, a flat-plate collector with double glazing has lower heat losses than one with single glazing (loss coefficients a_1 and a_2). In return, the double glazing lets through less sunlight, which reduces optical efficiency.

Table 3.7 shows typical conversion factors η_0 along with loss coefficients a_1 and a_2 for various collector types. Figure 3.24 shows collector efficiencies. When the temperature differences are very slight – which is the case in solar swimming pool systems, for example – inexpensive absorbers without glazing have good efficiency ratings. Evacuated tube collectors have far lower thermal losses than flat-plate collectors do, making them especially efficient when temperature differences are great – in other words, when outdoor temperatures are low but water temperatures are high.

Table 3.7 Example conversion factors and loss coefficients

Collector type	η_0	a_1 *in W/(m² K)*	a_2 *in W/(m² K²)*
Unglazed absorber	0.91	12.0	0
Flat-plate collector, single glazing, non-selective absorber	0.86	6.1	0.025
Flat-plate collector, single glazing, selective absorber	0.81	3.8	0.009
Flat-plate collector, double glazing, selective absorber	0.73	1.7	0.016
Evacuated tube collector	0.80	1.1	0.008

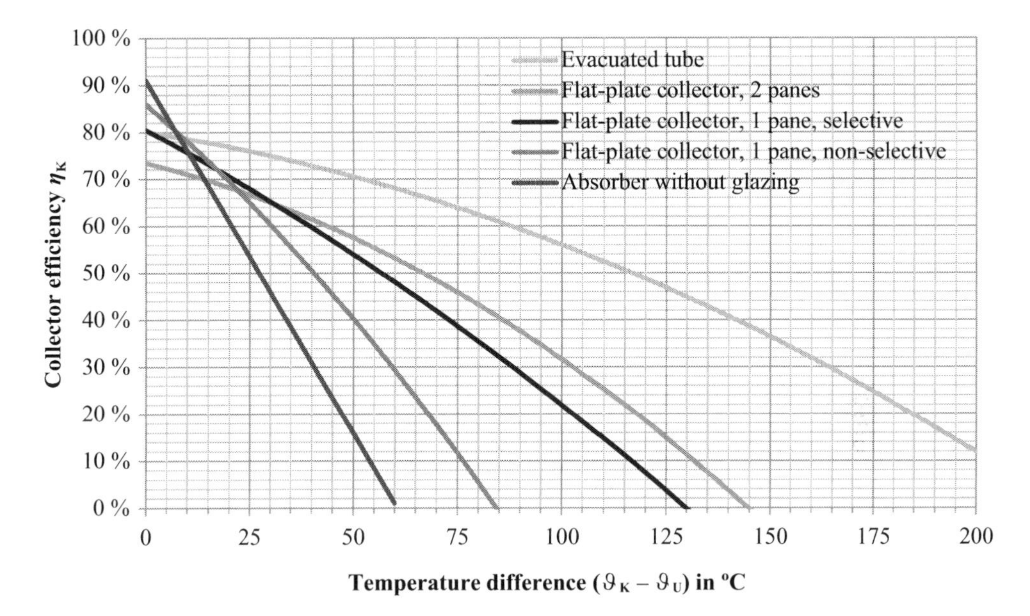

Figure 3.24 Collector efficiencies η_K relative to the differential temperature (ϑ_K–ϑ_U) for various collector types at an irradiance E of 800 W/m²

Collector parameters are generally determined based on measurements. The collector's useful output can be calculated from a measurement of the mass flow $\dot{m}$ through the collector under stationary conditions, the collector's input temperature ϑ_{KE}, its output temperature ϑ_{KA}, and irradiance E

$$\dot{Q}_{KN} = \dot{m} \cdot c \cdot \left(\vartheta_{KA} - \vartheta_{KE}\right) \tag{3.17}$$

along with collector efficiency.

$$\eta_K = \frac{\dot{Q}_{KN}}{E \cdot A_K} \tag{3.18}$$

The heat medium is then heated up in stages for the measurement to produce various average collector temperatures.

$$\vartheta_K = \frac{\vartheta_{KA} + \vartheta_{KE}}{2} \tag{3.19}$$

If the ambient temperature ϑ_U is also measured, efficiencies can be calculated relative to the differential temperature $\vartheta_K - \vartheta_U$. The parameters η_0, a_1, and a_2 are then entered in a quadratic equation for a parameter estimation

$$\eta_K = \eta_0 - \frac{a_1}{E} \cdot \left(\vartheta_K - \vartheta_U\right) - \frac{a_2}{E} \cdot \left(\vartheta_K - \vartheta_U\right)^2 = f_0 - f_1 \cdot x - f_2 \cdot x^2 \tag{3.20}$$

such as the MS-EXCEL trendline function.

Often, the function of the collector's efficiency is indicated by the individual loss factor a, not with the two loss factors a_1 und a_2, to simplify the process. In this case, however, a should be determined by means of a linear function in a parameter estimation. If the factor a_2 is left out, significant inaccuracies result. DIN EN 12975–2 [DIN06] includes comprehensive tips on inspecting solar collectors.

The angle of incidence of direct sunlight θ (see Chapter 2) also influences conversion factor η_0. The conversion factor drops as the angle of incidence increases because reflections off the pane of glass increase. For more exact calculations, the conversion factor is therefore modified by using the **incidence angle modifier** K (IAM), which includes angle of incidence θ and the collector's efficiency factor F', which is irrespective of the angle:

$$\eta_0(\theta) = \eta_{0i} \cdot F' \cdot K(\theta) \tag{3.21}$$

Usually, a distinction is made here between IAM K_{dir} for direct irradiance $E_{Kol,dir}$ and correction factor K_{diff} for diffuse irradiance $E_{Kol,diff}$. For the conversion factor, we then have the following formula:

$$\eta_0(\theta) = \eta_{0i} \cdot \frac{F' \cdot K_{dir}(\theta) \cdot E_{Kol,dir} + F' \cdot K_{diff} \cdot E_{Kol,diff}}{E_{Kol,dir} + E_{Kol,diff}} \tag{3.22}$$

For a simple estimation, the IAM $K_{dir}(50°)$ can be used for flat-plate collectors. For an incident angle of 0°, $K_{dir}(0°) = 1$. For an angle range for θ of 0 to around 70°, the function

$$K_{dir}(\theta) = 1 - b_0 \cdot \left(\frac{1}{\cos\theta} - 1 \right) \tag{3.23}$$

with

$$b_0 = \frac{1 - K_{dir}(50°)}{\cos^{-1} 50° - 1} = \frac{1 - K_{dir}(50°)}{0{,}5557} \tag{3.24}$$

describes the curve. The IAM K_{diff} for diffuse irradiance is only used in general, not relative to incidence angle θ.

For evacuated tube collectors, a distinction has to be made between incident angles parallel to or crossing the collector's axis. Here, the incident angle θ is divided into a longitudinal share θ_l and a transversal share θ_t. [The85] states that

$$\theta_l = \left| \gamma_E + \arctan\left(\tan(90° - \gamma_S) \cdot \cos(\alpha_S - \alpha_E)\right) \right| \text{ and} \tag{3.25}$$

$$\theta_t = \left| \frac{\arctan\left(\cos\gamma_S \cdot \sin(\alpha_S - \alpha_E)\right)}{\cos\theta} \right| \tag{3.26}$$

The angles α_S and γ_S describe the position of the sun; the angles α_E and γ_E, the orientation of the collector (see Chapter 2).

The incident angle modifier K_{dir} for the evacuated tube collector also consists of a longitudinal share $K_{dir,l}$ and a transversal share $K_{dir,t}$:

$$K_{dir}(\theta) = K_{dir,l}(\theta_l) \cdot K_{dir,t}(\theta_t) \tag{3.27}$$

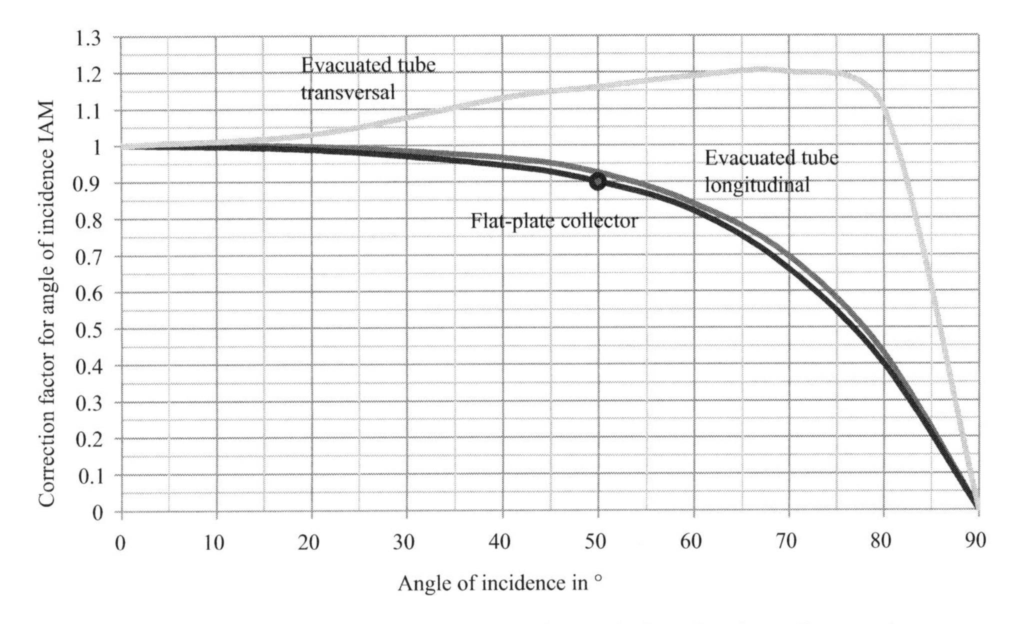

Figure 3.25 Incident angle modifier relative to the incident angle for a flat-plate collector and an evacuated tube collector

The longitudinal incident angle modifier $K_{dir,l}$ along the length of the tubes has a curve similar to a flat-plate collector. The transversal incident angle modifier $K_{dir,t}$ is very different and generally even increases over a large angle area. Figure 3.25 shows the typical curve of incident angle modifiers for a flat-plate and evacuated tube collector.

Pipes and tubes

A solar hot water system always needs pipes and tubes. Lines are needed to connect the collector to the storage tank and the storage tank to the taps.

In terms of energy, such lines are merely undesired sources of losses; in the best case, they make a minor contribution to space heating in the winter. If these lines are not (properly) insulated, line losses can be very great, and a majority of the heat can even be lost before it reaches its intended destination.

The collector's flow rate – the amount of the heat medium that flows through the collector each hour – is closely linked to line length and diameter. The collector's flow rate should be set to prevent the temperature from rising too much within the collector, which would reduce efficiency; in return, the flow rate should not be too great, lest too much energy be used to drive the pump.

The **collector's flow rate** can be calculated from the collector's useful output $\dot{Q}_{KN}$, the heat carrier's heat capacity c, and the desire differential temperature $\Delta\vartheta$ of the heat medium between the collector's input and output:

$$\dot{m} = \frac{\dot{Q}_{KN}}{c \cdot \Delta\vartheta} = \frac{\dot{Q}_{KN}}{c \cdot (\vartheta_{KA} - \vartheta_{KE})} \quad (3.28)$$

If the collector's previously calculated useful output is added, the collector's flow rate is

$$\dot{m} = \frac{\eta_0 \cdot E \cdot A_K - a_0 \cdot A_K \cdot (\vartheta_K - \vartheta_U) - a_2 \cdot A_K \cdot (\vartheta_K - \vartheta_U)^2}{c \cdot \Delta\vartheta} \tag{3.29}$$

which can also be expressed as follows in terms of the collector's surface area:

$$\dot{m}' = \frac{\dot{m}}{A_K} = \frac{\eta_0 \cdot E - a_0 \cdot (\vartheta_K - \vartheta_U) - a_2 \cdot (\vartheta_K - \vartheta_U)^2}{c \cdot \Delta\vartheta} \approx \frac{\eta_0 \cdot E - a \cdot (\vartheta_K - \vartheta_U)}{c \cdot \Delta\vartheta} \tag{3.30}$$

For instance, if a flat-plate collector with $\eta_0 = 0.8$ and $a = 4$ W/(m² K) is to heat up a medium with a heat capacity of $c = 0.96$ Wh/(kg K) by $\Delta\vartheta = 10$ K from $\vartheta_{KE} = 35$ °C to $\vartheta_{KA} = 45$ °C, the required collector flow rate at an ambient temperature of $\vartheta_U = 20$ °C, an average collector temperature of $\vartheta_K = 40$ °C and an irradiance of $E = 800$ W/m² a $\dot{m}' = 58{,}3$ kg/(m²h).

If the collector's input temperature ϑ_{KE} is known, we can use

$$\Delta\vartheta = \vartheta_{KA} - \vartheta_{KE} \quad (3.31) \quad \text{and} \quad \vartheta_K = \frac{\vartheta_{KA} + \vartheta_{KE}}{2} \tag{3.32}$$

in the equation for the collector's flow rate $\dot{m}'$ to calculate the **collector's output temperature**

$$\vartheta_{KA} = \frac{\eta_0 \cdot E + \dot{m}' \cdot c \cdot \vartheta_{KE} + a \cdot (\vartheta_U - \frac{1}{2}\vartheta_{KE})}{\dot{m}' \cdot c + \frac{1}{2}a} \tag{3.33}$$

If the collector's flow rate is increased in the example above to 18 kg /(m² h), the collector's output temperature ϑ_{KA} increases to 65 °C at the same collector input temperature. The average collector temperature ϑ_K then increases to 50 °C, thereby reducing the collector's efficiency from 70 % to 65 %. Such low flow rates – and even lower ones – are common in gravity solar thermal systems, also called low-flow systems.

Often, the collector's flow rate is given in l/h or l/(m² h). This **volumetric flow rate** $\dot{V}$ is a product of the density ρ of the **mass flow rate** $\dot{m}$:

$$\dot{V} = \frac{1}{\rho} \cdot \dot{m} \tag{3.34}$$

Water has a density of slightly less than 1 kg/l. If the heat carrier medium including antifreeze is 1.06 kg/l, the figure for volumetric flow rate is roughly equivalent to the mass flow rate.

From the cross-section area A_R of the lines in the collector circuit and the flow rate v_R of the heat medium, the necessary **line diameter** d_R can be determined via $\dot{V} = A_R \cdot v_R = \pi \cdot \frac{1}{4} \cdot d_R^2 \cdot v_R$.

$$d_R = \sqrt{\frac{4 \cdot \dot{m}}{\rho \cdot v_R \cdot \pi}} = \sqrt{\frac{4 \cdot \dot{m}' \cdot A_K}{\rho \cdot v_R \cdot \pi}} \tag{3.35}$$

With the collector surface of $A_K = 5$ m², a flow rate of $v_R = 1$ m/s, a mass flow rate relative to the surface area of $\dot{m}' = 50$ kg/(m²h) and a density of $\rho = 1060$ kg/m³, the line diameter d_R needs to be at least slightly less than 10 mm. Table 3.8 shows the properties of off-the-shelf copper lines. A 12 x 1 copper pipe with an interior diameter of 10 mm thus suffices.

Table 3.8 Properties of standard copper pipes

Designation	*Exterior Ø in mm*	*Interior Ø in mm*	*Specific mass in kg/m*	*Content in l/m*	*Flow rate in l/h (v=1 m/s)*
12 x 1 (DN10)	12	10	0.31	0.079	280
15 x 1 (DN12)	15	13	0.39	0.133	480
18 x 1 (DN15)	18	16	0.51	0.201	720
22 x 1 (DN20)	22	20	0.59	0.314	1,130
28 x 1.5 (DN25)	28	25	1.12	0.491	1,770
35 x 1.5 (DN32)	35	32	1.41	0.804	2,900
42 x 1.5 (DN40)	42	39	1.71	1.195	4,300

In practice, slightly larger diameters are used. Friction within the lines slows down the heat medium, leading to pressure losses, which are not as severe if the pipe diameters are greater. Tables 3.9 and 3.10 show the **recommended pipe diameters** for copper pipes used in systems with pumps and gravity systems.

The location of the collector and storage tank generally determines line lengths. Once we have selected a line diameter and insulated the pipe, we can calculate the line's thermal losses. Here, a distinction is made between the losses during the heat-up phase and during circulation.

Losses during the heat-up phase

If the collector circuit is not used, such as overnight, the lines cool off and eventually reach the outdoor temperature. When the collector circuit begins to flow again, the lines need to be heated up first before the heat can be passed on to the storage tank via the heat exchanger. Energy is needed not only to heat up the heat medium, but also the lines themselves.

If lines with the mass m_L, heat capacity c_L, and the mass's heat carrier m_W with a capacity of c_W are heated up from temperature ϑ_1 to temperature ϑ_2, the amount of heat required Q_{LA} for a number n_A of heating events is calculated as follows:

$$Q_{LA} = n_A \cdot (m_L \cdot c_L + m_W \cdot c_W) \cdot (\vartheta_2 - \vartheta_1) = n_A \cdot (m \cdot c)_{eff} \cdot (\vartheta_2 - \vartheta_1) \tag{3.36}$$

Table 3.9 Recommended diameters for copper lines in pumped facilities filled with a mixture of water and antifreeze [Wag95]

	Line length				
Collector surface	*10 m*	*20 m*	*30 m*	*40 m*	*50 m*
Up to 5 m²	15 x 1	15 x 1	15 x 1	15 x 1	15 x 1
6 to 12 m²	18 x 1	18 x 1	18 x 1	18 x 1	22 x 1
13 to 16 m²	18 x 1	22 x 1	22 x 1	22 x 1	22 x 1
17 to 20 m²	22 x 1	22 x 1	22 x 1	22 x 1	22 x 1
21 to 25 m²	22 x 1	22 x 1	22 x 1	22 x 1	28 x 1.5
26 to 30 m²	22 x 1	22 x 1	28 x 1.5	28 x 1.5	28 x 1.5

Table 3.10 Recommended copper line diameters for gravity systems with a mixture of water and antifreeze [Lad95]

	Height difference between collector and storage				
Collector surface	*0.5 m*	*1 m*	*2 m*	*4 m*	*6 m*
Up to 4 m²	22 x 1	22 x 1	18 x 1	18 x 1	15 x 1
Up to 10 m²	28 x 1.5	28 x 1.5	22 x 1	22 x 1	18 x 1
Up to 20 m²	42 x 2	35 x 1.5	28 x 1.5	28 x 1.5	22 x 1

For example, the heat-up losses of a 20 m long copper line with a cross-section of 15 x 1 mm filled with a heat carrier containing antifreeze are to be calculated for a temperature increase of $\vartheta_1 = 20$ °C to $\vartheta_2 = 50$ °C. This line's mass with a heat capacity of $c_L = 0.109$ Wh/(kg K) is $m_L = 7.8$ kg, while the heat carrier's mass has a heat capacity of $c_W = 0.96$ Wh/(kg K) and a density of $\rho_W = 1.06$ kg/l at $m_W = 2.82$ kg. During one heat-up phase ($n_A = 1$) with $(m\ c)_{eff} = 3.6$ Wh/K, the line has losses during the heat-up phase of $Q_{LA} = 108$ Wh.

In addition, there are heat-up losses for stop valves, temperature sensors, the pump, and other components integrated in the line circuit. If the mass and heat capacity of these components are known, the losses can also be calculated as above.

Circulation losses

Once the line has heated up, there are constantly losses as heat escapes outwards. For line length l and heat transmission ratio U', the circulation losses from heat transmission during circulation time t_Z at an ambient temperature of ϑ_A and a heat carrier temperature of ϑ_W are:

$$Q_Z = U' \cdot l \cdot t_Z \cdot (\vartheta_W - \vartheta_A) \qquad (3.37)$$

The **heat transmission ratio**

$$U' = \frac{\pi}{\frac{1}{2\lambda} \ln \frac{d_{LA}}{d_{RA}} + \frac{1}{\alpha \cdot d_{LA}}} \qquad (3.38)$$

can be calculated from the insulation layer's heat conductivity λ for a line with an external diameter d_{RA} and an insulated line's heat conductivity with external diameter d_{LA}. The heat conductivity λ of various materials is shown in Table 3.3. For the **heat transmission ratio** α between the insulation layer and the air, values between $\alpha = 10$ W/(m² K) at $U' = 0.2$ W/(m K) and $\alpha = 15.5$ W/(m²K) at $U' = 0,5$ W/(m K) are interpolated linearly.

To take the example above, the circulation losses of a 20 m long 15 x 1 mm ($d_{RA} = 15$ mm) copper line are to be calculated. The heat transmission ratio is $\alpha = 10$ W/(m² K). If the insulation thickness is 30 mm ($d_{LA} = 0.075$ m) and the insulation's heat conductivity $\lambda = 0.040$ W/(m K), the heat transmission ratio is $U' = 0.1465$ W/(m K). At an ambient temperature of $\vartheta_A = 20$ °C and a heat carrier temperature of $\vartheta_W = 50$ °C, the circulation losses for circulation time $t_Z = 8$ h are

$$Q_Z = 703 \text{ Wh}$$

The circulation losses for a 20 m long line with a diameter of 22 mm and an insulation thickness of only 10 mm are calculated at a differential temperature between the heat carrier and ambient air of 40 °C and a circulation time of 10 h at $Q_Z = 2500$ Wh. Proper insulation is crucial to prevent heat losses on the path to the storage tank. Lines should therefore be kept as short as possible, and clamps are to be attached without creating unavoidable thermal bridges. When laying the pipes, keep in mind that they expand and contract due to great temperature fluctuations. Use pipe elbows to account for these expansions and contractions.

If the controller stops circulation, the line and the heat carrier within cool down. The heat within the line

$$Q(t_1) = (c \cdot m)_{\text{eff}} \cdot (\vartheta_W(t_1) - \vartheta_A) \tag{3.39}$$

at time t_1 is a factor of the ambient temperature ϑ_A and the heat carrier's temperature $\vartheta_W(t_1)$. This heat is reduced by the heat flow

$$\dot{Q} = U' \cdot l \cdot (\vartheta_W(t) - \vartheta_A) \tag{3.40}$$

At time t_2, the **stored heat** is calculated with

$$Q(t_2) = Q(t_1) - U' \cdot l \cdot (\vartheta_W(t_1) - \vartheta_A) \cdot (t_2 - t_1) \tag{3.41}$$

from which the heat carrier's temperature is calculated:

$$\vartheta_W(t_2) = \frac{Q(t_2)}{(c \cdot m)_{\text{eff}}} + \vartheta_A = \left(1 - \frac{U' \cdot l \cdot (t_2 - t_1)}{(c \cdot m)_{\text{eff}}}\right) \cdot (\vartheta_W(t_1) - \vartheta_A) + \vartheta_A \tag{3.42}$$

This formula only applies, however, for short periods of time because the temperature of the heat carrier changes, requiring the calculation of heat flow to be redone as it changes. For larger intervals of time, the heat carrier's temperature $\vartheta_W(t_2)$ can be calculated at a time t_2 by dividing the large time interval $t_2 - t_1$ into n smaller time intervals Δt. With

$$\Delta t = \frac{t_2 - t_1}{n} \tag{3.43}$$

we get

$$\vartheta_W(t_2) = \left(1 - \frac{U' \cdot l \cdot \Delta t}{(c \cdot m)_{\text{eff}}}\right)^n \cdot (\vartheta_W(t_1) - \vartheta_A) + \vartheta_A \tag{3.44}$$

$\Delta t \to 0$, $t_1 = 0$ and $t_2 = t$ then produce

$$\vartheta_W(t) = \exp\left(-\frac{U' \cdot l}{(c \cdot m)_{\text{eff}}} \cdot t\right) \cdot (\vartheta_{W0} - \vartheta_A) + \vartheta_A \tag{3.45}$$

A 20 m long 15 x 1 copper line with 30 mm of insulation (U' = 0.15 W/(m K)) and $(c\ m)_{\text{eff}}$ = 3.6 Wh/K cools down at an ambient temperature of ϑ_A = 20 °C within one hour from $\vartheta_W(t_1)$ = 50 °C to $\vartheta_W(t_2)$ = 33 °C. If the collector then restarts circulation, the losses during the heat-up phase described above begin again.

Storage

For solar thermal systems, there are various types of storage depending on the application. The purpose of heat storage is to ensure the desired level of heat across a fluctuating supply of solar energy. A distinction is made between

- short-term storage
- long-term storage.

Short-term storage is designed to bridge just a few hours or days, whereas long-term storage is specially designed to compensate for seasonal heat differences. The latter therefore has a much greater volume. The following options are available for large storage facilities:

- artificial storage basins
- caverns in rock
- aquifers (storage and groundwater)
- storage underground and in rock.

Storage can also be further divided into temperature ranges:

- low-temperature storage below 100 °C
- medium-temperature storage from 100 °C and 500 °C
- high-temperature storage exceeding 500 °C.

Finally, there are various types of heat storage, such as:

- sensible heat storage
- latent heat storage (storage through phase changes)
- thermochemical/sorption storage.

This book only takes a closer look at **low-temperature storage for the storage of sensible heat**. Table 3.11 shows the properties of various low-temperature storage media.

Table 3.11 Properties of various low-temperature storage media [Gie89, Kha95]

	Density ρ in kg/m³	*Heat conductivity λ in W/(m K)*	*Heat capacity c in kJ/(kg K)*
Water (0 °C)	999.8	0.5620	4.217
Water (20 °C)	998.3	0.5996	4.182
Water (50 °C)	988.1	0.6405	4.181
Water (100 °C)	958.1	0.6803	4.215
Granite	2,750	2.9	0.89
Soil (coarse gravel)	2,040	0.59	1.84
Clay	1,450	1.28	0.88
Storage concrete	2,370	1.18	0.98

Drinking water storage

Drinking water storage facilities can be precisely dimensioned to suit the desired degree of solar fraction and hot water demand based solely on simulations. To start with, the required **storage volume** is 1.5 to 2 times greater than daily consumption. In addition to the storage volume for daily demand, there is a 50 % additional standby volume and 20 l/m² of collector surface as the pre-heating volume. Pressurized hot water storage starts at systems smaller than 100 l and ends exceeding 1,000 l. For a single-family home with 4 to 6 persons, the recommended storage size is 300 l to 500 l.

Generally, heat storage has two heat exchangers (see Figure 3.5 on page 84). The **heat exchanger** for the solar circuit is in the lower half of the storage tank, while the heat exchanger for the conventional backup heat system is at the top. There is an opening in the middle of each heat exchanger for the installation of the control system's temperature sensor. The cold water connection is at the bottom; the hot water outlet, at the top – to ensure that water of different temperatures is properly layered.

Figure 3.26 shows a horizontal cylindrical heat storage tank with a rounded end, which we will now investigate for further calculations. Within the storage tank, **heat losses** occur as heat passes through the tank's insulation. To ensure that the tank is properly insulated, at least 100 mm of insolation should be applied with a heat conductivity of $\lambda = 0.04$ W/(m K). The latest super-insolation products – such as evacuated fibre optics – have a residual pressure smaller than 10^{-3} mbar to achieve heat conductivities of $\lambda = 0.005$ W/(m K) [Fac95].

The **amount of heat that can be stored** in the hot water tank

$$Q = c \cdot m \cdot (\vartheta_{SP} - \vartheta_A) \tag{3.46}$$

is the product of the differential temperature between the average storage temperature ϑ_{SP}, the ambient temperature ϑ_A, heat capacity c and the storage content's mass m. At a water temperature of 50 °C, the heat capacity $c_{H2O} = 4.181$ kJ/(kg K) = 1.161 Wh/(kg K) and the density $\rho_{H2O} = 0.9881$ kg/l. With 300 l of storage volume, the differential temperature is 50 °C, so 17.2 kWh of heat can be stored.

The **storage losses** $\dot{Q}_{SP}$ of a cylindrical, round tank are composed of the losses $\dot{Q}_{SP,Z}$ in the cylindrical area and the losses $\dot{Q}_{SP,K}$ in the two spherical domes:

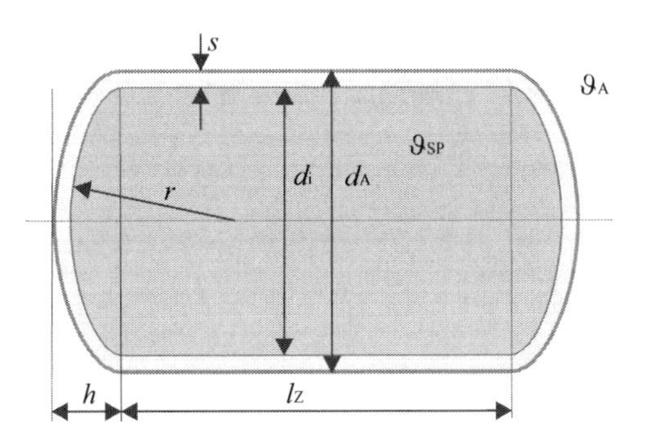

Figure 3.26 A cylindrical, rounded heat storage tank

$$\dot{Q}_{SP} = \dot{Q}_{SP,Z} + 2 \cdot \dot{Q}_{SP,K} \tag{3.47}$$

The losses

$$\dot{Q}_{SP,Z} = U' \cdot l_Z \cdot (\vartheta_{SP} - \vartheta_A) \tag{3.48}$$

within the cylindrical area are calculated similarly to the process for the line losses above from the heat transmission ratio U', length l_Z and the difference between the average storage temperature ϑ_{SP} and the ambient temperature ϑ_A.

The **heat transmission ratio** U' also depends upon the insulation's heat conductivity λ, the heat transmission ratio α between the insulation and the air, and the external diameter d_A and internal diameter d_i of the cylindrical part of the insulation:

$$U' = \frac{\pi}{\frac{1}{2 \cdot \lambda} \cdot \ln \frac{d_A}{d_i} + \frac{1}{\alpha \cdot d_A}} \tag{3.49}$$

For the heat transmission ratio α, values are used in linear interpolation between 10 W/(m² K) for U' = 0.2 W/(m K) and 15.5 W/(m² K) for U' = 0.5 W/(m K).

The heat losses

$$\dot{Q}_{SP,K} = U \cdot A_K \cdot (\vartheta_{SP} - \vartheta_A) \tag{3.50}$$

of the round part are also calculated based on the differential temperature between storage and the ambient temperature, the heat transmissivity coefficient U and the spherical dome's surface A_K.

Assuming that the tank wall's temperature is the same as storage temperature ϑ_{SP}, the heat transmission ratio α_1 from the tank wall to the insulation layer, the heat transmission ratio α_2 of the insulation layer to the air, insulation thickness s, and the insolation's heat conductivity λ can be used to calculate **thermal transmittance**:

$$U = \frac{1}{\frac{1}{\alpha_1} + \frac{1}{\alpha_2} + \frac{s}{\lambda}} \tag{3.51}$$

The heat transmission ratio from the tank wall to the insulation layer can be estimated as α_1 = 300 W/(m² K). The **heat transmission ratio** α_2 from the insulation layer to air depends on the wall's orientation [VDI2067]:

horizontal wall with heat rising:

$$\alpha_2 = 2{,}3 \frac{\mathrm{W}}{\mathrm{m}^2\mathrm{K}} \cdot \sqrt[4]{(\vartheta_{SP} - \vartheta_A) / {}^\circ\mathrm{C}}$$

horizontal wall with heat falling:

$$\alpha_2 = 1{,}7 \frac{\mathrm{W}}{\mathrm{m}^2\mathrm{K}} \cdot \sqrt[4]{(\vartheta_{SP} - \vartheta_A) / {}^\circ\mathrm{C}}$$

vertical wall (spherical) with heat given off to the side:

$$\alpha_2 = 2{,}2 \frac{\mathrm{W}}{\mathrm{m}^2\,\mathrm{K}} \cdot \sqrt[4]{(\vartheta_{\mathrm{SP}} - \vartheta_{\mathrm{A}})/{}^\circ\mathrm{C}}$$

The **surface**

$$A_{\mathrm{K}} = 2\pi \cdot r \cdot h \tag{3.52}$$

of the spherical dome can be calculated with r and h, as shown in Figure 3.26.

In the following **example**, the heat losses of a 300 l storage tank are calculated in accordance with DIN 4803. The ambient temperature is $\vartheta_{\mathrm{A}} = 20$ °C; the storage temperature, $\vartheta_{\mathrm{SP}} = 60$ °C. The dimensions are $l_{\mathrm{Z}} = 1.825$ m, $d_{\mathrm{A}} = 0.7$ m, $d_{\mathrm{R}} = d_{\mathrm{i}} = 0{,}5$ m, $r = 0.45$ m, $h = 0.11$ m, and $s = 0.1$ m. The insulation's heat conductivity $\lambda = 0.035$ W/(m K) and the heat transmission ratio $\alpha = 15.5$ W/(m² K) in the cylindrical section produce the heat transmission ratio $U' = 0.64$ W/(m K). The spherical dome surface $A_{\mathrm{K}} = 0.311$ m² and the spherical dome's thermal transmissivity $U = 0.33$ W/(m² K) are used to calculate heat storage losses as follows:

$$\dot{Q}_{\mathrm{SP}} = (U' \cdot l_{\mathrm{Z}} + 2 \cdot U \cdot A_{\mathrm{K}}) \cdot (\vartheta_{\mathrm{SP}} - \vartheta_{\mathrm{A}}) = 55\ \mathrm{W}$$

As time t passes, the storage temperature ϑ_{SP} drops when storage is at rest. The losses are also reduced in the process. If heat is added to or taken from storage, the **storage temperature** is calculated as follows:

$$\vartheta_{\mathrm{SP}}(t) = \exp\left(-\frac{U' \cdot l_{\mathrm{Z}} + 2 \cdot U \cdot A_{\mathrm{K}}}{c \cdot m} \cdot t\right) \cdot (\vartheta_{\mathrm{SP0}} - \vartheta_{\mathrm{A}}) + \vartheta_{\mathrm{A}} \tag{3.53}$$

similar to the formula for temperature lines. The value

$$\tau = \frac{c \cdot m}{U' \cdot l_{\mathrm{Z}} + 2 \cdot U \cdot A_{\mathrm{K}}} \tag{3.54}$$

represents **storage's time constant**. It indicates how long it takes for the differential temperature to drop to 1/e = 36.8 % of the initial value. In this example, the time constant $\tau = 250$ h = 10.4 days.

Figure 3.27 shows the curve of storage temperature in storage at rest. Here, it is clear that the storage tank gives off a considerable amount of heat to its surroundings. In only a bit more than a week, the storage temperature has been cut in half. Such a storage facility is thus only good if the heat needs to be stored just for a few days. In larger storage facilities, the ratio of the volume to the surface decreases, thereby considerably reducing relative heat losses. When the volume reaches 1,000 m³, the time constant can almost stretch out to six months. Such facilities are suitable for seasonal storage, specifically from the summer into the winter. Proper insulation is especially important in keeping heat losses to a minimum for seasonal heat storage.

A storage facility connected to a solar energy system will, however, rarely be at rest; rather, the solar collector will constantly feed it with heat, and consumers will take heat from it. The storage temperature curve becomes a bit harder to calculate then. Various simulation programs are available for this purpose.

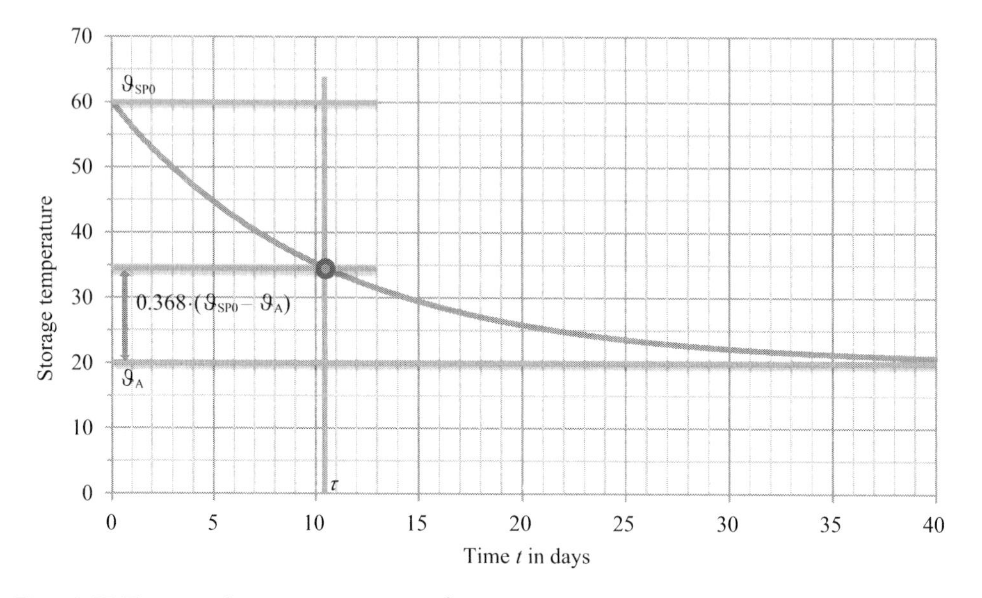

Figure 3.27 The curve for storage temperature ϑ_{SP} across time with a 300 l tank at rest

All of these storage temperatures calculated above are averages. In most storage systems, however, a desired **temperature layering** will occur. At the top of the storage tank where hot water is taken out, the temperature is higher than at the bottom, where cold water enters. The storage tank therefore needs to be divided into multiple layers, with the heat flow being calculated between the individual layers.

Swimming pools

The swimming pool itself stores heat when connected to a solar collector. Here, heat is consumed within the storage facility itself, not drained off to serve another application. Convection losses $\dot{Q}_K$, radiation losses $\dot{Q}_S$, evaporation losses $\dot{Q}_{Vd}$, and transmission losses $\dot{Q}_{Tr}$ to the ground can be used to calculate the pool's losses $\dot{Q}_{SB}$:

$$\dot{Q}_{SB} = \dot{Q}_K + \dot{Q}_S + \dot{Q}_{Vd} + \dot{Q}_{Tr} \tag{3.55}$$

In a heated outdoor swimming pool used seasonally, **transmission losses** $\dot{Q}_{Tr}$ to the ground are negligible.

Convection losses

$$\dot{Q}_K = \alpha_K \cdot A_W \cdot (\vartheta_W - \vartheta_L) \tag{3.56}$$

can be calculated from air temperature ϑ_L, water temperature ϑ_W, the water surface A_W, and thermal transmissivity:

$$\alpha_K = \left(3.1 + 4.1 \cdot v_W \cdot \frac{\text{s}}{\text{m}}\right) \cdot \frac{\text{W}}{\text{m}^2\text{K}} \tag{3.57}$$

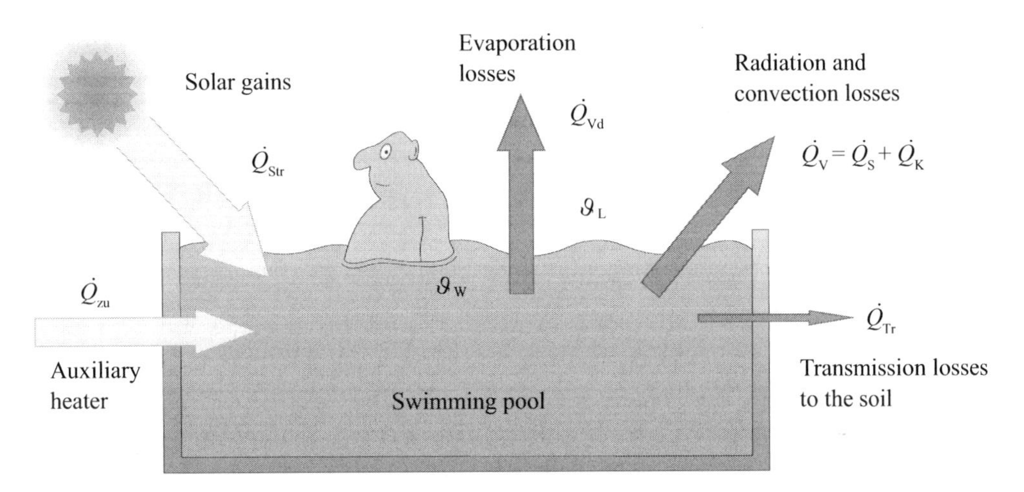

Figure 3.28 A swimming pool's energy balance

Here, v_W is wind velocity 0.3 m above the surface of the water. Values at other heights can be converted to this height using the formulae in Chapter 6.

Radiation losses from thermal radiation are the result of radiation exchanged between the surface of the pool and the sky. The Stefan–Boltzmann constant $\sigma = 5.67051 \cdot 10^{-8}$ W/(m^2 K^4), the degree of emissivity ε_W of the water ($\varepsilon_W \approx 0.9$), the water's surface A_W, and the absolute temperature of the water T_W and the sky T_H can be used to calculate **radiation losses**:

$$\dot{Q}_S = \sigma \cdot \varepsilon_W \cdot A_W \cdot \left(T_W^4 - T_H^4\right) \tag{3.58}$$

The **temperature of the sky** can be calculated as follows:

$$T_H = T_L \cdot \left(0.8 + \frac{\vartheta_{tau}(\vartheta_L)}{250°C}\right)^{0.25} \tag{3.59}$$

with T_L in K representing the absolute air temperature and ϑ_{tau} representing dew point [Smi94].

Dew point

$$\vartheta_{tau} = 234.175\ °C \cdot \frac{\ln\left(\frac{\phi \cdot p}{0.61078\ \text{kPa}}\right)}{17.08085 - \ln\left(\frac{\phi \cdot p}{0.61078\ \text{kPa}}\right)} \tag{3.60}$$

can be calculated from relative humidity φ and **vapour pressure:**

$$p = 0.61078\ \text{kPa} \cdot \exp\left(\frac{17.08085 \cdot \vartheta_L}{234.175\ °C + \vartheta_L}\right) \tag{3.61}$$

Table 3.12 Vapour pressure p of water and dew point ϑ_{tau} at 70 % relative humidity relative to air temperature ϑ_l

ϑ_l in °C	10	12	14	16	18	20	22	24	26	28	30
p in kPa	1.23	1.40	1.60	1.82	2.07	2.34	2.65	2.99	3.37	3.79	4.25
ϑ_{tau} in °C	4.8	6.7	8.6	10.5	12.5	14.4	16.3	18.2	20.1	22.0	23.9

Relative to air temperature ϑ_L, vapour pressure p is measured in pascals (1 Pa = 1 N/m² = 0.01 mbar). In Germany, the average relative humidity φ during the swimming season is around 70 %.

Table 3.12 shows vapour pressure and dew point for a number of temperatures for the formulae above.

Evaporation losses

$$\dot{Q}_{Vd} = h_V \cdot \dot{m}_V \tag{3.62}$$

can generally be calculated from the evaporating volumetric flow rate $\dot{m}_V$ and the evaporation heat h_V = 2257 kJ/kg of water. Usually, however, empirical formula are used. At wind velocity v_W 0.3 m above the water's surface, vapour pressure p at water temperature ϑ_W and air temperature ϑ_L, at relative humidity φ for swimming pool surface A_W, evaporation losses can be calculated as follows [Hah94]:

$$\dot{Q}_{Vd} = A_W \cdot (0{,}085\frac{\text{m}}{\text{s}} + 0{,}0508 \cdot v_W) \cdot \left(p(\vartheta_W) - \varphi \cdot p(\vartheta_L)\right) \tag{3.63}$$

In a swimming pool with a surface of A_W = 20 m², wind velocity v_W = 1 m/s, ambient temperature ϑ_L = 20 °C, water temperature ϑ_W = 24 °C and relative humidity φ = 0.7 the pool's losses are as follows, assuming transmission losses to the ground are negligible:

$$\dot{Q}_{SB} = \dot{Q}_K + \dot{Q}_S + \dot{Q}_{Vd} = 576\text{ W} + 1{,}493\text{ W} + 3{,}672\text{ W} = 5{,}741\text{ W}$$

It would therefore require a lot of energy to heat up the pool if there were no additional gains from irradiation.

The **solar radiation gains** $\dot{Q}_{Str}$ can be calculated from irradiance E, the water surface A_W and the degree of absorption α_{abs}:

$$\dot{Q}_{Str} = \alpha_{abs} \cdot E \cdot A_W \tag{3.64}$$

The degree of absorption α_{abs} is around 0.8 for white tiles, 0.9 for light blue tiles, and more than 0.95 for dark blue tiles; it increases with the water depth. Hence, the losses of a 20 m² pool in the example above with an absorption degree of 0.9 are compensated for at an irradiance of only E = 319 W/m².

If the **pool is covered overnight**, heat losses can be reduced by around 40 to 50 %.

The heating energy demand needed to keep a swimming pool at 23 °C without an overnight cover is around 300 kWh/m² for the entire season, though that figure can vary greatly from one location to another. A solar energy system can cover all of this demand for heating energy. This **area of the solar absorber** should be equivalent to 50 to 80 % of the pool's surface.

System design

Demand for useful energy

A hot water supply system's demand for useful energy Q_N can be calculated from the amount of energy consumed. The heat capacity of water (c_{H2O} = 1.163 Wh/(kg K)), the amount of water consumed m, the cold water temperature ϑ_{KW}, and the hot water temperature ϑ_{WW} produce:

$$Q_N = c \cdot m \cdot (\vartheta_{WW} - \vartheta_{KW}) \tag{3.65}$$

Table 3.13 shows rough values for **domestic hot water demand**. If no specific data are available, the cold water temperature can be assumed to be ϑ_{KW} = 10 °C.

Table 3.14 shows estimates for hot water demand in buildings with relatively **high occupancy**, such as hotels and homes for senior citizens. Demand for useful energy in restaurants can be estimated at 230 to 460 Wh/per meal and at 2,500 to 5,000 Wh/user in saunas.

If washing machines have direct hot water connections, demand for hot water will be accordingly higher.

Because the tables only allow consumption to be roughly estimated, a **closer analysis** is recommended when a system is planned. Table 3.15 provides an overview of values that can serve as starting points.

In addition to demand for useful heat for the year as a whole, demand varies over time. If there are great differences between days and between the summer and the winter, the hot water supply system will have to be dimensioned to account for these fluctuations. For the precise dimensioning and forecast of output, simulation software is recommended.

Table 3.13 Hot water demand, domestic [VDI82]

	Hot water demand in litres per day and per person		*Specific useful heat in Wh/(per day per person)*
	ϑ_{WW}= 60 °C	*ϑ_{WW}= 45 °C*	
Low demand	10–20	15–30	600–1200
Medium demand	20–40	30–60	1,200–2,400
High demand	40–80	60–120	2,400–4,800

Table 3.14 Hot water demand in hotels, etc. [VDI82]

	Hot water demand in litres per day and per person		*Specific useful heat in Wh/(per day per person)*
	ϑ_{WW}= 60 °C	*ϑ_{WW}= 45 °C*	
Room with bath	95–138	135–196	5,500–8,000
Room with shower	50–95	74–135	3,000–5,500
Other rooms	25–35	37–49	1,500–2,000
Retirement homes, B&Bs, etc.	25–50	37–74	1,500–3,000

Table 3.15 Hot water consumption for various applications

	Consumption	*Temperature*	*Useful heat*
Dishwashing per person	12–15 l/day	50 °C	550–700 Wh/day
Handwashing	3–5 l	37 °C	95–160 Wh
Bathtub	150 l	40 °C	5,200 Wh
Shower	30–45 l	37 °C	940–1,400 Wh
Hair wash	10–15 l	37 °C	310–470 Wh

Degree of solar cover and degree of usage

The share of energy required Q_{zu} for conventional backup heat depends on the amount of useful heat Q_N, energy losses (line heat-up losses Q_{LA}, circulation losses Q_Z in the collector circuit, and storage losses Q_{SP}) and the collector's useful energy Q_{KN}:

$$Q_{zu} = Q_N + Q_{LA} + Q_Z + Q_{SP} - Q_{KN} \tag{3.66}$$

In addition to the line losses in the collector's circuit, there are also line losses between the storage tank and the taps, which can be calculated in the same way. Simulation programs are generally calculate and add up all of these factors over a year. The result is annual cumulative amounts for each share of energy.

One important parameter for solar energy systems is the **degree of solar fraction** (SD). It indicates how much of total energy demand comes from the solar energy system alone. Specifically, it is defined as the share of solar energy from the storage tank in total energy demand, consisting of useful energy Q_N and storage losses Q_{SP}:

$$SD = \frac{Q_N + Q_{SP} - Q_{zu}}{Q_N + Q_{SP}} = \frac{Q_{KN} - Q_{LA} - Q_Z}{Q_N + Q_{SP}} \tag{3.67}$$

For Middle-European climate regions, systems for solar hot water supply are generally dimensioned so that solar meets roughly 50 to 60 % of demand. This level represents a compromise between the goal of high energy output and affordability.

Figure 3.29 shows the solar fraction relative to collector area for various storage sizes as calculated in a simulation program. In all simulations, consumption is assumed to be constant. When the collector area is very small, the solar fraction increases quickly as the collector's size grows. In contrast, the collector surface area has to be roughly doubled to increase the solar fraction from 60 to 75 %. Larger storage volumes are not worthwhile if the collector area is small because the storage losses would be too great. Large storage only makes sense if the collector area is also large.

Another important parameter in describing a solar thermal system is the **collector cycle efficiency** η_{KN}. It represents the overall efficiency of the solar thermal system and is the product of the ratio of energy from the solar energy system to the energy incident on the collector. The annual amount of irradiation H_{Solar} within the collector and the collector surface A_K produce:

$$\eta_{KN} = \frac{SD \cdot (Q_N + Q_{SP})}{H_{solar} \cdot A_K} = \frac{Q_{KN} - Q_{LA} - Q_Z}{H_{solar} \cdot A_K} \tag{3.68}$$

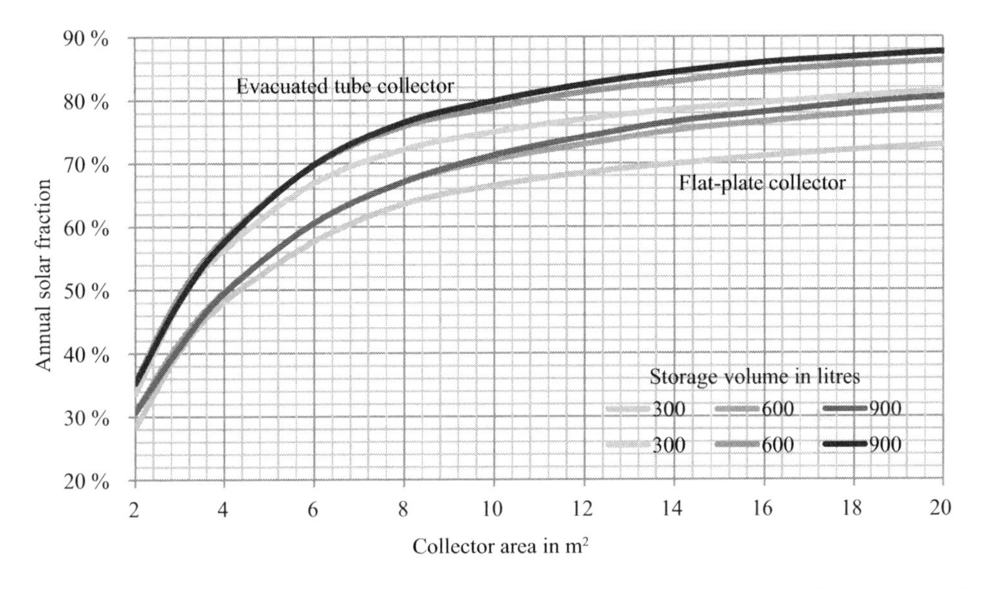

Figure 3.29 Solar fraction for drinking water relative to collector area for various collector types and storage sizes. Location: Berlin, collector angle: 30° south, heat demand: 7.4 kWh/day

In practice, the collector's cycle efficiency generally ranges from 20 to 50 %. In addition to the system's quality, the solar fraction influences cycle efficiency. The higher the solar fraction, the lower cycle efficiency is.

In **calculating the amount of primary energy saved** E_{PE}, the electricity E_{pump} for the collector's circulation pump, the primary energy efficiency η_{zu} of the conventional backup heater, and the efficiency η_{E} of primary energy consumption for electricity from the grid have to be taken into account. The amount of primary energy saved is thus

$$E_{\mathrm{PE}} = \frac{Q_{\mathrm{KN}} - Q_{\mathrm{LA}} - Q_{\mathrm{Z}}}{\eta_{\mathrm{zu}}} - \frac{E_{\mathrm{pump}}}{\eta_{\mathrm{E}}} \tag{3.69}$$

if the energy used to manufacture the solar energy system is not taken into account. The primary energy consumed by the collector's circulation pump is between 2 and 15 % of the collector's yield. Pipe diameters and the pump's output must be optimally coordinated in order to keep the pump's energy consumption to a minimum.

Solar drinking water supply

As explained in the previous section, solar thermal systems for drinking water supply are generally designed in climates like Germany's so that solar meets 50 to 60 % of hot water demand over the year. Because the amount of sunlight fluctuates greatly in Germany, the solar energy system nearly meets all of the demand for hot water in the summer months. In the winter, however, the share of solar can fall below 10 % (Figure 3.30). A conventional heating system then has to provide the rest.

A solar thermal system for hot water supply can be designed relative to the number of people in the household according to a simple rule of thumb:

- Collector size: 1–1.5 m² (flat-plate) collector surface per person
- Storage size: 80–100 l per person

When evacuated tube collectors are used, the collector can be roughly 30 % smaller. The minimum, however, is 3 to 4 m² because losses in the lines would otherwise be too great.

For a detailed design, hot water demand first has to be determined. The storage volume V_{Sp} should be twice the total demand of the *P* people with a daily demand per person of V_{person}:

$$V_{Sp} = 2 \cdot P \cdot V_{person} \tag{3.70}$$

The hot water demand Q_{Person} per person per day can be used to calculate the annual demand for useful heat Q_N for hot water supply:

$$Q_N = 365 \cdot P \cdot Q_{person} \tag{3.71}$$

A family of four with an average consumption of 45 l or 1.8 kWh of hot water per day thus requires:

$$V_{Sp} = 2 \cdot 4 \cdot 45\,\text{l} = 360\,\text{l}$$

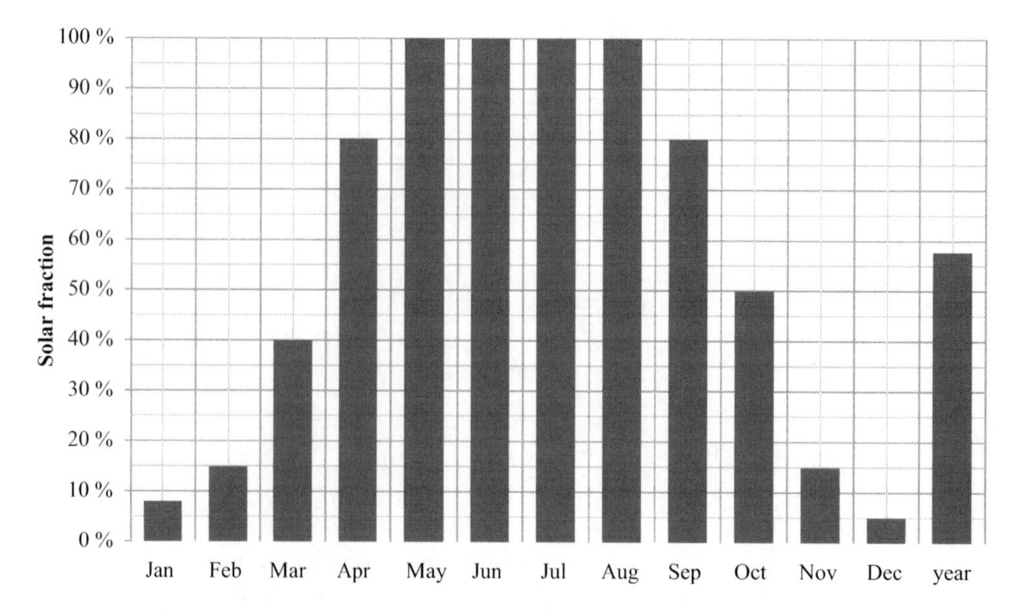

Figure 3.30 Characteristic curve of the solar fraction for solar thermal hot water systems in Germany

Typical storage tanks are around 300 to 400 l. A 400-litre tank would be a bit too large. A tank with a volume of 300 litres is a bit smaller than the calculated demand but should suffice. The annual demand for heat energy is:

$$Q_N = 365 \cdot 4 \cdot 1.8 \text{ kWh} = 2{,}628 \text{ kWh}$$

Once the storage has been dimensioned, we can determine the size of the collector area A_K. To do so, we need to know the annual irradiation H_{solar} incident on the collector (cf. Chapter 2). If we leave out storage losses for the moment, the collector's optimal size can be calculated as follows:

$$A_K \approx \frac{SD}{\eta_{KN}} \cdot \frac{Q_N}{H_{solar}} \tag{3.72}$$

At an annual solar fraction of $SD = 60$ % and an average collector cycle efficiency of $\eta_{KN} = 30$ %, the required collector area for flat-plate collectors to meet the demand for useful energy above at an irradiation of 1,100 kWh/m² is:

$$A_K \approx \frac{60\ \%}{30\ \%} \cdot \frac{2{,}628 \text{ kWh}}{1{,}100 \frac{\text{kWh}}{\text{m}^2}} = 4.8 \text{ m}^2$$

Here, it is crucial to remember that the solar fraction and the collector cycle efficiency are related to each other and cannot vary independently. It is only possible to produce more detailed plans with the use of complex computer programs. In addition to the size of the collectors and storage, the design of other components needs to be taken into account, such as pumps, control systems, and pipes.

Solar thermal for space heating support

A larger collector area is needed if the solar thermal system is to not only provide hot water, but also support space heating. Here, optimal insulation of the building is the first step towards ensuring that the sun can cover as much of the demand for space heat as possible. While demand for hot water is generally constant across the year, demand for space heat is concentrated in the winter months. In the winter, however, solar collectors produce the least amount of energy. When solar thermal systems are to support space heating, they are therefore generally designed to serve this function only in the transitional months of the spring and fall in addition to providing hot water. In the deep winter, a conventional heating system then covers most of the heat demand (Figure 3.31).

As the collector area and storage tank increase in size, so does the solar fraction. In return, the share of demand for heat that the conventional heating system has to meet is reduced. If the conventional system is fired with oil or gas, a larger solar energy system reduces carbon emissions. A very large system, however, produces more excess energy that cannot be used. Larger systems are therefore generally less affordable than small ones. When designing the system, you therefore have to decide whether the priority is on the largest

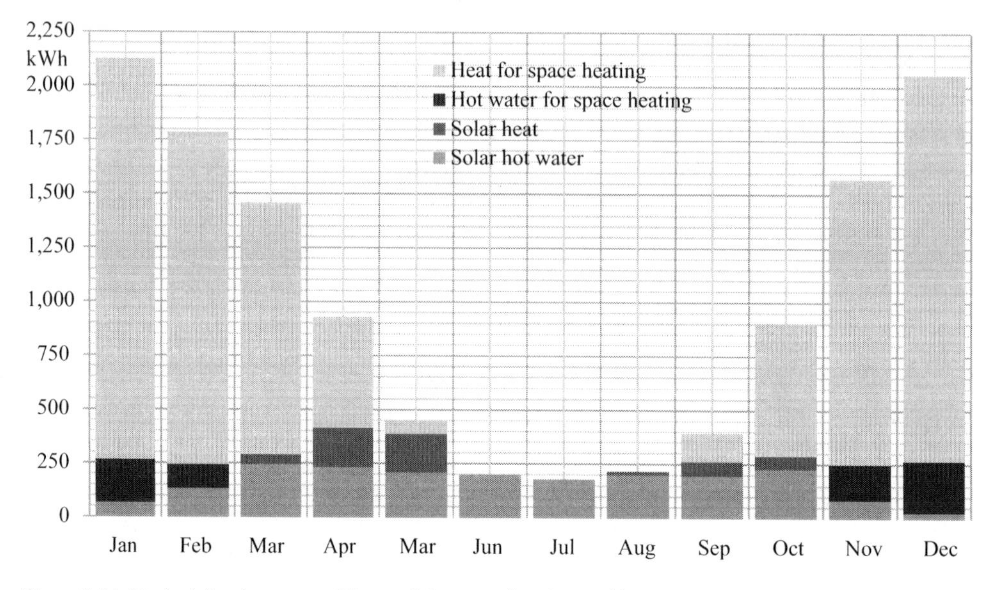

Figure 3.31 Typical development of demand for space heating and hot water in Germany and the share of solar energy and conventional heat in coverage in an old building with 20 % solar coverage

possible share of solar coverage or affordability. The following two design versions are good starting points:

Variant 1: Small system with focus on affordability

- Collector area for flat-plate collectors: 0.8 m² per 10 m² of floor space
- Collector area for evacuated tube collectors: 0.6 m² per 10 m² of floor space
- Storage size: at least 50 litres per m² of collector area

Variant 2: Medium-size array for higher degree of solar coverage

- Collector area for flat-plate collectors: 1.6 m² per 10 m² of floor space
- Collector area for evacuated tube collectors: 1.2 m² per 10 m² of floor space
- Storage size: 100 litres per m² of collector area

Optimal design naturally also takes into account the actual demand for heating energy. And of course, that demand is very different in an unrenovated building from in a modern nearly zero-energy home. Table 3.16 shows the simulation results for optimal systems dimensioned according to the two variants above.

Even though the midsize system has a collector twice as big and storage four times bigger, the degree of solar coverage does not necessarily double. The building's insulation standard plays a much greater role in the solar coverage rate than the size of the solar thermal array does. Those looking to get as much of their heat as possible from solar to protect the climate should therefore look into optimal insulation. And once again, detailed planning is only possible with sophisticated computer programs.

Table 3.16 Solar fraction

	Old building	*Standard new building*	*Advanced new building*	*Passive House/nearly zero-energy building*
Heat demand for hot water in kWh	2,700	2,700	2,700	2,700
Demand for space heat in kWh	25,000	11,500	3,900	1,950
Solar fraction variant 1 (small system)	13 %	22 %	40 %	51 %
Solar fraction variant 2 (midsize system)	22 %	36 %	57 %	68 %

Assumptions: Berlin, orientation 30° south, unshaded, 130 m² floor area, optimal flat-plate collector with combined storage

Table 3.17 Data from completed residential buildings in Germany and Switzerland with 100 % solar heat supply

	Oberburg (Switzerland)	*Niederwinkling*	*Kappelrodeck*	*Regensburg*	*Oberburg (Switzerland)*
Building type	*Single-family home*	*Single-family home*	*Single-family home*	*Single-family home*	*Complex (8 units)*
Completed in	1989	1998	2006	2006	2007
Treated floor area in m²	130	170	147	186	851
Heat output in kW	2.8	2.7	3.5	5	12
Collector surface in m²	84	75	112	83	276
Storage volume in m³	118	27	42.8	38	205

Purely solar heating

Likewise, computer programs are generally needed to design systems with a 100 % solar heat supply. It should be kept in mind, however, that a lot of standard simulation programs for solar thermal systems are not fully able to map large systems with seasonal storage.

Optimal building insulation with a low building heat load is a requirement for purely solar heat supply. The goal is then to optimally coordinate the size of the storage tank and collector. Experience from previous projects is a good starting point (Table 3.17).

Solar updraft tower

Solar updraft towers are fundamentally different from the solar thermal systems discussed above. In these towers, a large collector converts solar heat into electricity. But unlike in concentrating solar power plants discussed in the next chapter, sunlight is not concentrated here.

Solar updraft towers heat up air. The collector field is a large area covered by a glass or plastic. In the middle of this covered area is a tall chimney. The cover over the collector

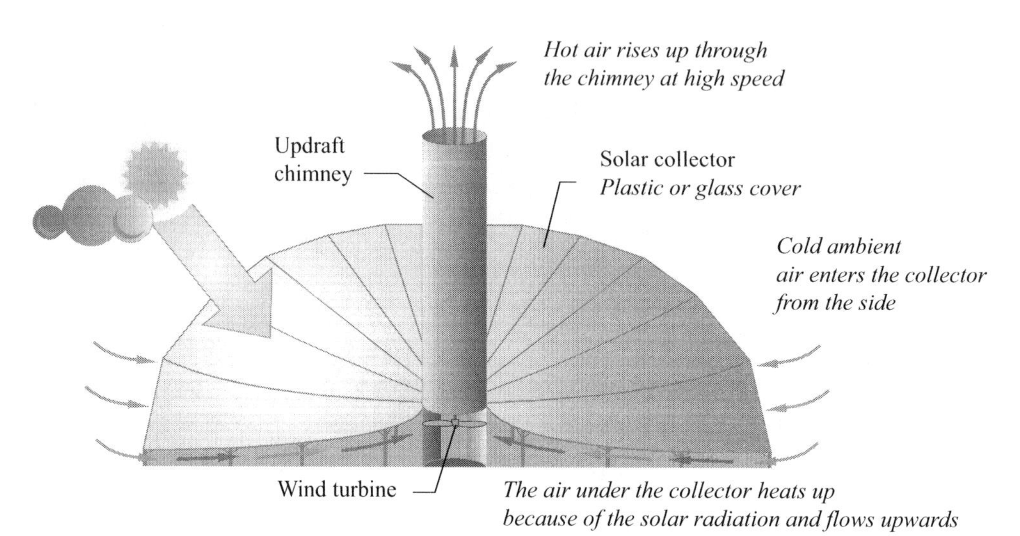

Figure 3.32 How a solar updraft tower works

area rises slightly towards the chimney in the middle. Around the edges of the covered area, air can stream inwards. The sun heats up the air under the cover. The air rises along the slightly pitched cover towards the chimney, where it shoots up quickly. The rising air inside the chimney is used to drive wind turbines that drive a generator to produce electricity. The heat under the covered area suffices to produce electricity for a while after the sun has gone down. Figure 3.32 shows how a solar updraft tower works.

The overall efficiency of a solar updraft tower

$$\eta_{\text{total}} = \eta_{\text{tower}} \cdot \eta_{\text{colletor}} \cdot \eta_{\text{turbine}} \cdot \eta_{\text{generator}} \tag{3.73}$$

results from the tower's efficiency η_{tower}, the collector's efficiency $\eta_{\text{collector}}$, the wind turbine's efficiency η_{turbine}, and the generator's efficiency $\eta_{\text{generator}}$. The tower converts the air's thermal energy from the collector into kinetic energy. Without consideration of losses, the maximum efficiency is:

$$\eta_{\text{tower,max}} = \frac{P_{\text{kin,air}}}{\dot{Q}_{\text{therm,air}}} = \frac{\dot{m} \cdot g \cdot h \cdot \dfrac{\Delta\rho}{\rho_0}}{\dot{m} \cdot c_{\text{p}} \cdot \Delta T} = \frac{g \cdot h \cdot \dfrac{\Delta\rho}{\rho_0}}{c_{\text{p}} \cdot \Delta T} \tag{3.74}$$

The kinetic energy is directly related to the height of the tower h and the differential pressure $\Delta\rho$ caused by the upwind within the tower [Ung91]. The thermal energy is the product of the air's heat capacity c_{p} and temperature increase ΔT. At a constant volume, the following applies for ideal gases

$$\frac{\Delta\rho}{\rho_0} = \frac{\Delta T}{T_0} \tag{3.75}$$

If the two previous formulae are combined, we have

$$\eta_{\text{tower,max}} = \frac{g \cdot h}{c_p \cdot T_0} \tag{3.76}$$

The tower's efficiency thus increases linearly along with its height h. A tower with a height of $h = 1000$ m at a reference temperature of $T_0 = 300$ K and a heat capacity of air at $c_p = 1$ kJ/(kg K) would have a maximum efficiency of 3.3 %. If the collector's efficiency is 60 % and the turbine's 80 %, maximum efficiency is further reduced to around 1.5 %.

At the beginning of the 1980s, a small demonstration plant with a nominal capacity of 50 kW was built near the southern Spanish town of Manzanares. The collector had an average diameter of 122 m and an average height of 1.85 m. The chimney was 195 m tall with a diameter of 5 m. The system was dismantled, however, in 1988 when the chimney toppled over during a storm; all of the planned tests had, however, already taken place during the system's projected service life.

New power updraft towers were recently discussed in Australia, for instance. One project could have a generator capacity of 200 MW; the tower would be 1,000 m tall, with a diameter of 180 m – and the collector would have a diameter of 6,000 m. The project failed to get financing, however. But in desert regions, solar updraft towers have the potential to be competitive with conventional generators.

4 Concentrated solar power

Introduction

The operating temperatures of non-concentrating collectors are limited. As shown in Chapter 3, good evacuated tubes allow temperatures exceeding 200 °C to be reached easily. However, higher temperatures are hardly possible, and efficiency decreases quickly at temperatures above 100 °C because the heat losses increase so much. Much higher temperatures are needed for process heat and power generation from thermal circulation processes. The only way to reach such temperatures and keep efficiency high is to concentrate the irradiance.

Although there are a wide range of options for concentrated solar thermal systems, only a few had been tried by the 1990s. And while non-concentrating solar thermal systems and photovoltaics grew tremendously in the 1990s, concentrated solar power remained largely an issue for researchers in that decade. In the 1970s and 1980s, ample research funding was made available in the wake of the oil crises, but those sources began to dry up in the 1990s.

The renaissance for commercial systems did not begin until 2006 with the planning and construction of concentrated solar power plants primarily in Spain and the US. Yet, the benefits of concentrated solar power systems are tremendous for a renewable energy supply, so the importance of the technology will continue to increase rapidly.

Concentrating irradiance

To concentrate sunlight, a device – the concentrator – with an aperture area of A_K generally focuses light on a receiver with an area of A_R (Figure 4.1).

For the solar radiation to be concentrated greatly, the light has to be incident at one angle. Specifically, diffuse irradiance (see Chapter 2) cannot be used in concentrated solar plants. Only direct sunlight can be concentrated, so this technology is especially interesting for sunny regions of the world.

The **concentration factor** C can be generally described by using the formula:

$$C = \frac{A_K}{A_R} \tag{4.1}$$

Because sunlight does not reach the earth absolutely in parallel, there is a limit to how much it can be concentrated. Because the sun is so much larger than the earth, the former does not appear as an infinitely small dot, but rather as a sphere under the half cone angle:

$$\varphi_{S/2} = \arcsin\left(\frac{r_S}{r_{SE}}\right) = \arcsin\left(\frac{6,963 \cdot 10^8\,\mathrm{m}}{1.5 \cdot 10^{11}\,\mathrm{m}}\right) = 0.27° = 16' \tag{4.2}$$

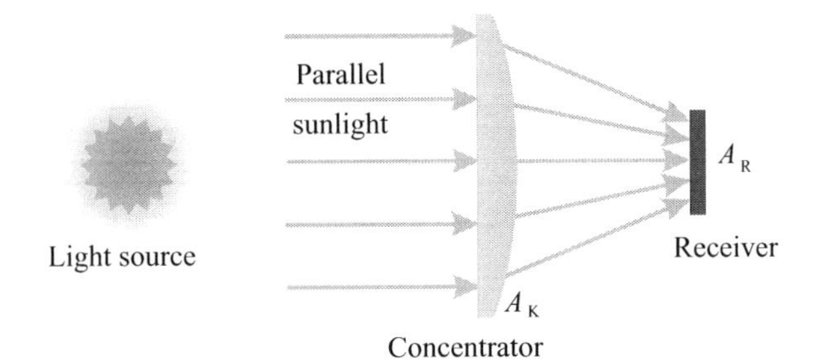

Figure 4.1 How light is concentrated

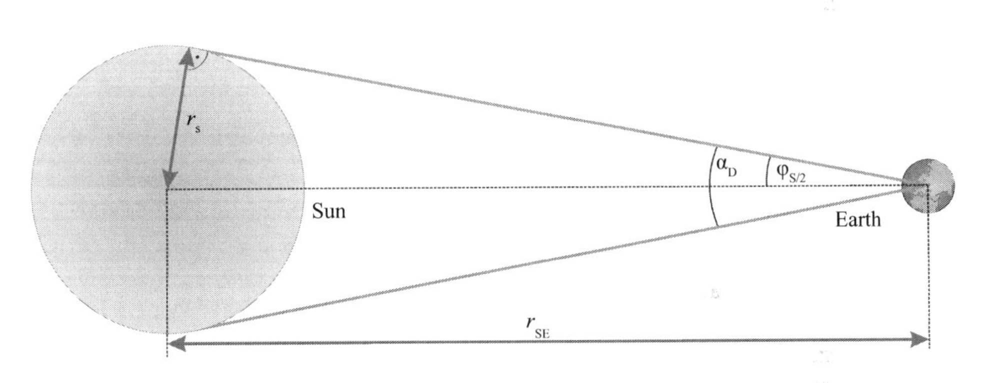

Figure 4.2 Half cone angle for the sun

It is a factor of the sun's radius r_S and the sun's distance r_{SE} to the earth; it varies slightly over the course of the year as this distance varies. Figure 4.2 shows the geometry for a determination of the solar half cone angle.

The shadow cast by a high-rise shows that sunlight is not absolutely parallel. At a distance, it no longer has clear boundaries, but blends into a semi-shaded area.

The entire half cone angle is also called divergence α_D. The diversions of sunlight $\alpha_D = 0.0093$ rad can be used to calculate the theoretical **maximum concentration** of a dual-axis tracking point concentrator:

$$C_{max} = \frac{4}{\alpha_D^2} = \frac{1}{\varphi_{S/2}^2} = 46,211 \tag{4.3}$$

The maximum concentration of single-axis linear concentrators is reduced to:

$$C_{max,linear} = \sqrt{C_{max}} \approx 215 \tag{4.4}$$

These calculations only apply for the half cone area of sunlight on earth. The maximum concentration is greater on planets further from the sun, which does not mean, however, that higher temperatures are possible.

Because the sun's energy only reaches the earth as heat radiation, the laws for a black body's radiation from Chapters 2 and 3 are used. If no heat is taken from an ideal absorber with $\varepsilon = 1$, it has to radiate the energy it absorbs, aside from the effects of convection. Without consideration of the atmosphere's influence, the Stefan–Boltzmann constant $\sigma = 5.67051 \cdot 10^{-8}$ W m^{-2} K^{-4} and the solar constant E_0 are used as follows:

$$C \cdot E_0 = \sigma \cdot T_A^4 \tag{4.5}$$

If we then include the **absorber's temperature** T_A, we get:

$$T_A = \sqrt[4]{\frac{C \cdot E_0}{\sigma}} \tag{4.6}$$

The maximum concentration C_{max} thus produces the maximum absorber temperature:

$$T_{A,max} = \sqrt[4]{\frac{C_{max} \cdot E_0}{\sigma}} = \sqrt[4]{\frac{46211 \cdot 1367 \ \text{W m}^{-2}}{5{,}67 \cdot 10^{-8} \ \text{W m}^{-2} \ \text{K}^{-4}}} = 5777 \ \text{K} = T_{Sun} \tag{4.7}$$

It corresponds exactly to the sun's surface temperature T_{sun}, which we know from Chapter 2. If we then add

$$\frac{E_0}{\sigma} = \frac{T_{Sun}^4}{C_{max}} \tag{4.8}$$

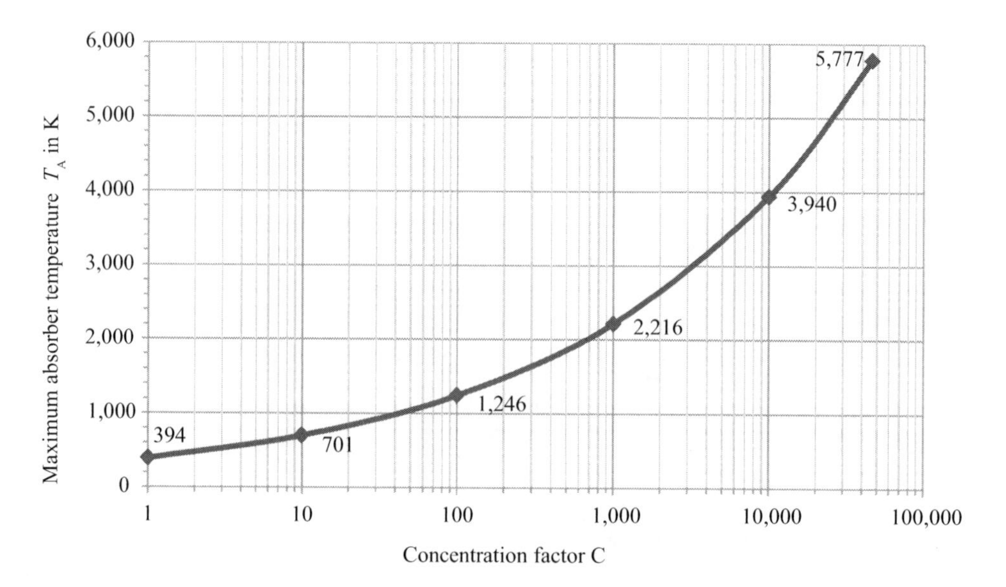

Figure 4.3 Maximum absorber temperature relative to concentration factor

in (4.6) the theoretical maximum absorber temperature T_A can be determined for other concentration factors C:

$$T_A = T_{sun} \sqrt[4]{\frac{C}{C_{max}}} \tag{4.9}$$

Figure 4.3 shows the theoretical absorber temperature for various concentration factors. The standstill temperature of good non-concentrating collectors ($C = 1$) exceeds 470 K, far above the theoretical maximum of 394 K found here (see Chapter 3). The glass cover and selective absorber coatings reduces longwave radiation from non-concentrating collectors. The figures for real-world collectors therefore do not conflict with the theoretical considerations in this chapter.

In general, greater concentrations are required to reach higher temperatures. In practice, for instance, the solar concentrator at Odeillo (France) has reached around 4,000 K by concentrating more than 10,000 times.

Concentrating collectors

Lenses are currently only used to a great extent with concentrating photovoltaics. They are generally too expensive to be used with concentrated solar thermal systems, in which mirrors are generally used.

A reflector that concentrates sunlight along a line or on a point generally has a parabolic shape. If the reflector is circular, sunlight is not concentrated on a single point. Glass mirrors have proven to be useful reflectors because of their long service lives.

The reflector has to track the sun so that sunlight is always at a 90° angle to the aperture. A distinction is made between single-axis and dual-axis trackers. Single-axis trackers concentrate sunlight on an absorber tube, while dual-axis trackers generally have a central absorber close to the focal point. The tracker uses either a sensor to determine the optimal orientation to the sun or a computer to calculate the sun's position.

Linear collectors

Collector types and geometry

Linear collectors are generally parabolic troughs that focus sunlight on an absorber tube. The concentrator can be either a closed unit or distributed units. If the concentrator consists of distributed mirrors, each of which has to be optimally positioned for the absorber tube, it is called a **Fresnel collector** (Figure 4.4).

Up to now, most commercial systems have been parabolic trough collectors, but Fresnel collectors are also in serial production. A large number of parabolic trough collectors have been made especially for concentrated solar power plants. Figure 4.6 shows parabolic trough collectors in Spain, while Table 4.1 provides technical data for various commercial parabolic trough collectors.

The shape of a parabolic trough collector with focal length f and collector length along the x and y-axis can be described with

$$y = \frac{x^2}{4f} \tag{4.10}$$

as shown in Figure 4.7.

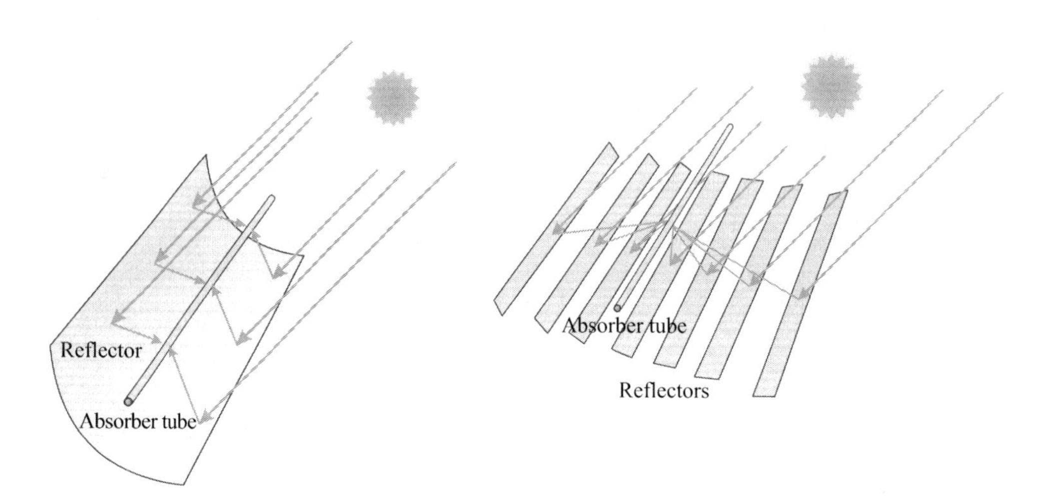

Figure 4.4 Concentrating sunlight with linear collectors (left: parabolic trough, right: Fresnel collector)

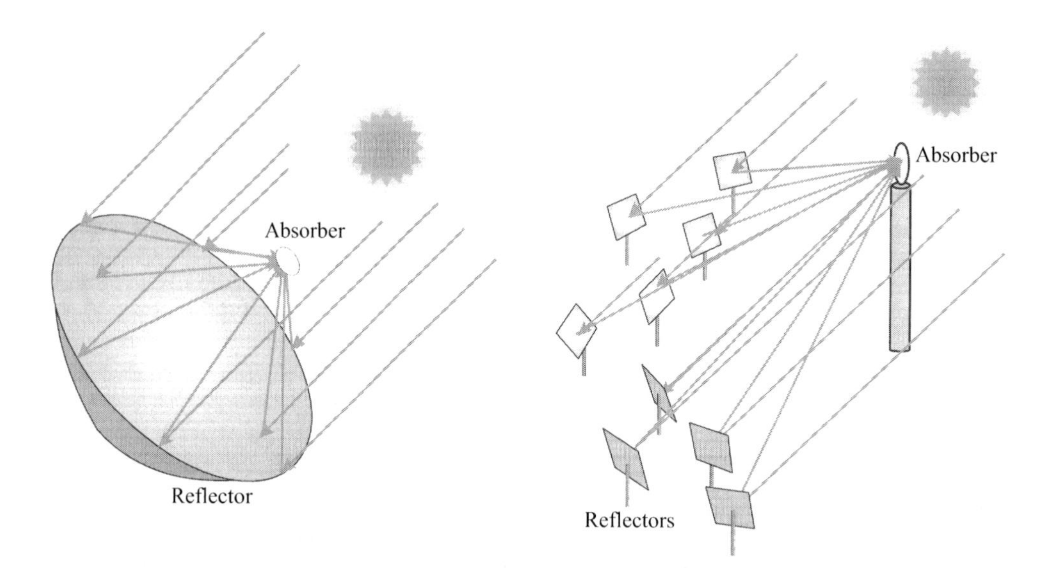

Figure 4.5 Concentrating sunlight with point concentrators (left: parabolic dish, right: array of heliostats)

Table 4.1 Technical data for parabolic trough collectors

Collector type	*LS-1*	*LS-2*	*LS-3*	*Euro Trough*	*SGX-2*	*Astro*
First installation (year)	1984	1986	1988	2001	2006	2007
Concentration factor C	61	71	82	82	82	82
Aperture width d_K in m	2.5	5.0	5.76	5.76	5.76	5.76
Collector length l_K in m	50	48	99	150	100	150
Aperture area A_K in m²	128	235	545	825	470	833
Absorber tube diameter in mm	42.4	70	70	70	70	70

Figure 4.6 Parabolic trough collectors at the PSA research facility in Almería, Spain

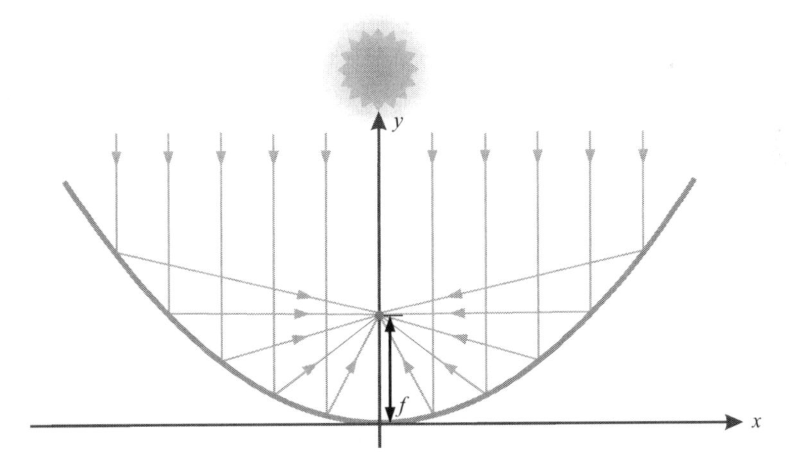

Figure 4.7 The geometry of a parabolic trough collector

Collector output and efficiency

The direct irradiance $E_{dir,K}$ on the collector, the collector's aperture area A_K, optical efficiency η_{opt}, and thermal collector losses $\dot{Q}_V$ are used to calculate the collector's output:

$$\dot{Q}_{KN} = E_{dir,K} \cdot A_K \cdot \eta_{opt} - \dot{Q}_V \tag{4.11}$$

Because parabolic trough collectors generally only track the sun along one axis, direct sunlight generally is not incident on the collector at a 90° angle. Cosine losses reduce the direct-normal irradiance $E_{\text{dir,s}}$ incident on a surface perpendicular to the sun. The losses are determined through multiplication with the incident angle's cosine θ. Because the absorber tube is outside of the reflector at the focal point, part of the solar radiation at the end of the collector does not reach the absorber tube. These additional final collector losses are taken into account with $f_{\text{end losses}}$. The direct irradiance on the collector is then:

$$E_{\text{dir,K}} = E_{\text{dir,s}} \cdot \cos\theta \cdot (1 - f_{\text{end losses}}) \tag{4.12}$$

The solar azimuth α_S, solar elevation γ_S (see Chapter 2), the deviation α of the tracking axis from North/South, and the angle β of the tracking axis can be used to calculate the **incident angle** θ for all geometries and times [Sti85]:

$$\theta = \arccos\left(\sqrt{1 - \left(\cos\left(\gamma_S - \beta\right) - \cos\beta \cdot \cos\gamma_S \cdot \left(1 - \cos\left(\alpha_S - \alpha\right)\right)\right)^2}\right) \tag{4.13}$$

For collectors with horizontal axis ($\beta = 0$), the formula to calculate the angle of incidence is even simpler:

$$\theta = \arccos\left(\sqrt{1 - \cos^2\gamma_S \cdot \cos^2\left(\alpha_S - \alpha\right)}\right) \tag{4.14}$$

For collectors with horizontal axis facing North/South ($\alpha = 0, \beta = 0$), we then have:

$$\theta = \arccos\left(\sqrt{1 - \cos^2\gamma_S \cdot \cos^2\alpha_S}\right) \tag{4.15}$$

The **collector's end losses** $f_{\text{end losses}}$ – the part of the solar radiation that does not reach the absorber tube at the end of the collector because of its angle – can be calculated with focal length f and collector length l_K:

$$f_{\text{end losses}} = \frac{f \cdot \tan\theta}{l_K} - f_{\text{end gains}} \tag{4.16}$$

If n collectors are then set up in a row, part of the irradiance incident at a slant from one collector can hit the absorber tube of a neighboring collector. The share of the final energy gains

$$f_{\text{end gains}} = \frac{(n-1)}{n} \cdot \frac{\max\left(0;\, f \cdot \tan\theta - d_D\right)}{l_K} \tag{4.17}$$

increases as the collector distance d_D decreases. Figure 4.8 shows the end losses and gains along with the underlying geometries.

With $n = 4$ EuroTrough collectors (see Table 4.1) with a length l_K of 150 m, a focal length of $f = 1.71$ m and a collector distance of $d_D = 0.5$ m, the end gains $f_{\text{end gains}} = 0.24$ % while the end losses $f_{\text{end losses}} = 0.41$ % at an incident angle of $\theta = 30°$. The example shows that end losses are more or less negligible if collectors are sufficiently long.

Evacuated tubes are generally used as absorbers and parabolic trough collectors; they are described in Chapter 3 in the context of non-concentrating collectors. These collector

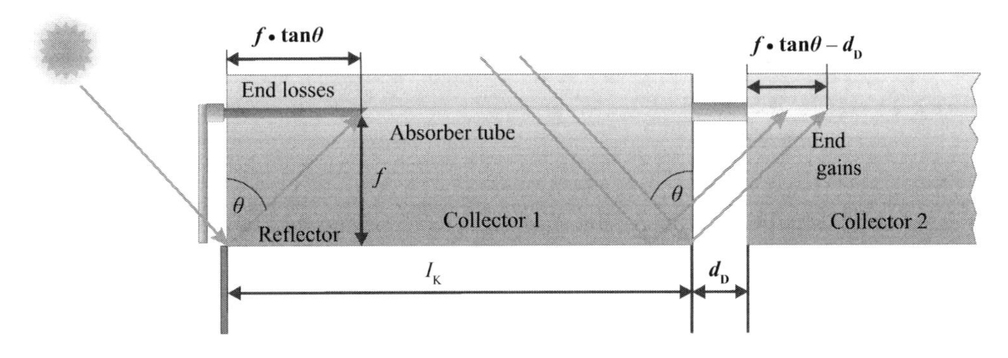

Figure 4.8 End losses and gains with parabolic trough collectors arranged in a row

tubes are largely identical except that they have to be able to withstand higher temperatures in concentrating collectors. The **optical efficiency**

$$\eta_{opt} = (\rho \cdot \gamma \cdot \tau \cdot \alpha) \cdot K \cdot \eta_{cleanliness} = \eta_0 \cdot K \cdot \eta_{cleanliness} \tag{4.18}$$

of parabolic trough collectors is the product of the mirror's degree of reflection ρ, the glass encasement tube's degree of transmissivity τ , and the absorber tube's degree of absorption α (Figure 4.9). Not all of the reflected light reaches the absorber tube because the mirror reflects the light in different directions. The intercept factor γ takes account of this reduction. The optical nominal efficiency η_0, which includes the aforementioned parameters, is generally measured.

If the light is not incident at a right angle, the optical losses increase. The **incidence angle modifier** K (IAM) takes account of the rising optical losses relative to angle of incidence $\boldsymbol{\theta}$. With

$$K = \max\left(1 - c_1 \cdot \frac{\theta}{\cos\theta} - c_2 \cdot \frac{\theta^2}{\cos\theta}; 0\right) \tag{4.19}$$

we have an empirical formula to describe the curve. The parameters c_1 and c_2 are determined by taking measurements at different incident angles.

In practice, the optical efficiency η_{opt} is below the level under ideal conditions because of dirt on the equipment. The parameter $\eta_{cleanliness}$ takes account of this reduction. Dirt on the reflector refracts light so that a lower share reaches the absorber. Concentrating systems are therefore much more sensitive to dirt than non-concentrating ones. If the system is regularly cleaned every few days, cleanliness can be maintained at 90 to 95 %. If the intervals between cleaning are longer, the losses increase significantly.

As with non-concentrating collectors, heat losses $\dot{Q}_V$ can be divided into convection losses $\dot{Q}_K$ and radiation losses $\dot{Q}_S$ (Figure 4.9):

$$\dot{Q}_V = \dot{Q}_K + \dot{Q}_S = \alpha_K \cdot A_{abs} \cdot (T_{glass} - T_U) + A_{abs} \cdot \varepsilon \cdot \sigma \cdot (T_A^4 - T_U^4) \tag{4.20}$$

With the thermal transmissivity coefficient α_K and absorber tube surface A_{abs}, we can roughly calculate the convection losses from the temperature difference between the glass

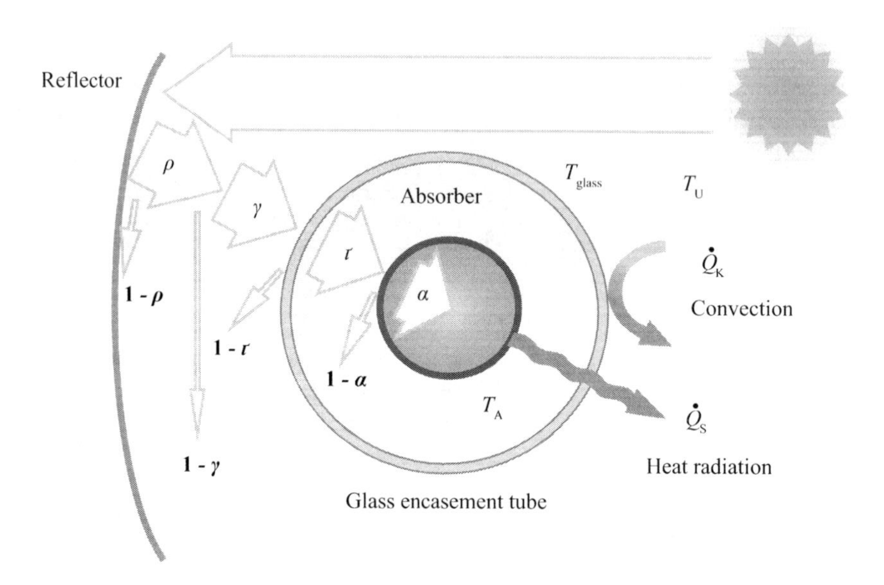

Figure 4.9 Optical and thermal events in the absorber tube

encasement tube and the ambient temperature. In contrast, the absorber tube's surface temperature T_A and the ambient temperature T_U are decisive for radiation losses. To reduce the emission coefficient ε, selective surface coatings are generally applied, as described in Chapter 3.

The collector length l_K, the absorber tube's diameter d_A, the collector's aperture area A_K, and the concentration factor C are used to calculate the absorber tube's surface:

$$A_{abs} = l_K \cdot d_A \cdot \pi = \frac{A_K}{C} \cdot \pi \tag{4.21}$$

Heat losses

$$\dot{Q}_V = A_K \cdot \frac{\pi}{C} \cdot \left(\alpha_K \cdot \left(T_{glass} - T_U\right) + \varepsilon \cdot \sigma \cdot (T_A^4 - T_U^4)\right) \tag{4.22}$$

decrease as the concentration factor increases.

The **collector's efficiency**

$$\eta_K = \frac{\dot{Q}_K}{E_{dir,s} \cdot A_K} = \eta_{opt} - \frac{\dot{Q}_V}{E_{dir,s} \cdot A_K} \tag{4.23}$$

usually refers to the direct-normal irradiance $E_{dir,s}$ incident on collector area A_K. As concentration factor C increases, so does the collector's efficiency along with heat losses. Figure 4.10 shows calculations for $\eta_{opt} = 0.75$, $\varepsilon = 0.1$, $\alpha_K = 2$ W/(m² K), $E_{dir,S} = 1000$ W/m² under the assumption that the glass encasement tube's differential temperature relative to the ambient temperature is 30 % of the difference between the absorber's temperature and the ambient temperature.

Here, it is clear that high concentration factors are required to reach high temperatures; in general, concentration factors exceeding 100 require dual-axis point concentrators.

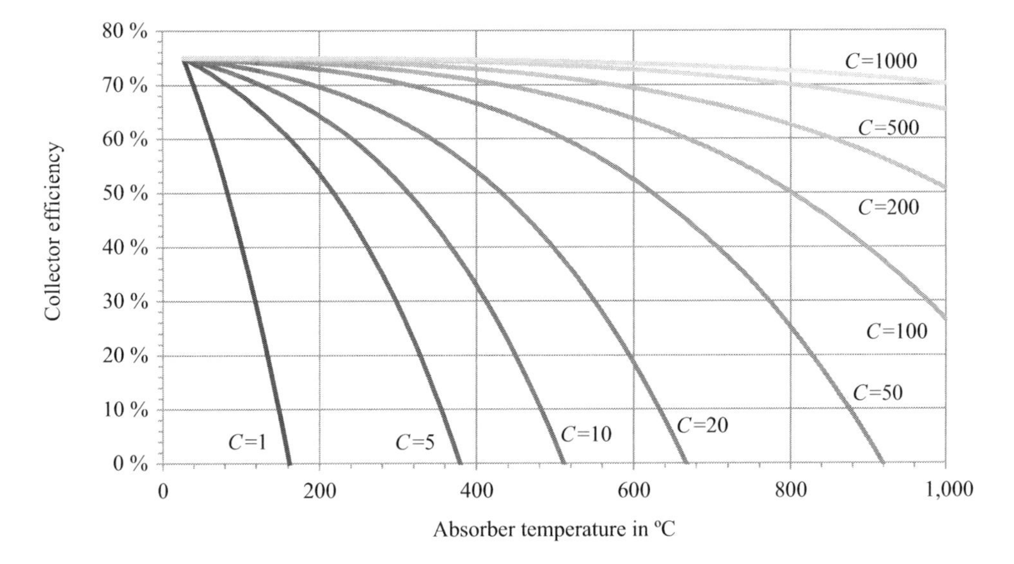

Figure 4.10 The collector's efficiency relative to the absorber's temperature at various concentrations

Table 4.2 Typical parameters to determine the efficiency of parabolic trough collectors

η_0	$c_1\ (1)^{-1}$	c_2 in $(1)^{-2}$	b_0 in K^{-1}	b_1 in $K^{-1} W m^{-2}$	b_2 in $K^{-2} W m^{-2}$
0.75	–0.000884	0.00005369	$7.276 \cdot 10^{-5}$	$4.96 \cdot 10^{-3}$	$6.91 \cdot 10^{-4}$

In practice, heat losses not accounted for – such as those caused by the absorber's fixtures – generally lead to higher heat losses than calculated above. Therefore, collectors are usually measured, and heat losses determined empirically by means of the factors b_0, b_1, and b_2 with $\Delta T = (T_A - T_U)$ [Dud94]:

$$\dot{Q}_V = \left(b_0 \cdot K \cdot E_{\mathrm{dir,K}} + b_1 + b_2 \cdot \Delta T\right) \cdot A_K \cdot \Delta T \tag{4.24}$$

Table 4.2 shows typical parameters for the calculation of efficiency.

Expansion under heat

The great temperature differences between standstill and operating temperatures mean that concentrating systems have to be able to withstand considerable expansion. A body's thermal expansion when it heats up to temperature ϑ_2 with an expansion coefficient α and length l_1 at temperature ϑ_1 is:

$$\Delta l = l_2 - l_1 = l_1 \cdot \alpha \cdot (\vartheta_2 - \vartheta_1) = l_1 \cdot \alpha \cdot \Delta T \tag{4.25}$$

A 100 m long steel pipe then expands at a differential temperature of $\Delta T = 350$ K by around 0.5 m.

Table 4.3 Linear expansion coefficients for various bodies at a temperature range of 0 to 500 °C [Her12]

Material	*α in 10^{-6} K^{-1}*	*Material*	*α in 10^{-6} K^{-1}*
Aluminium	27.4	Stainless steel	18.2
Copper	17.9	Fused quartz	0.61
Steel C60	13.9	Common glass	10.2

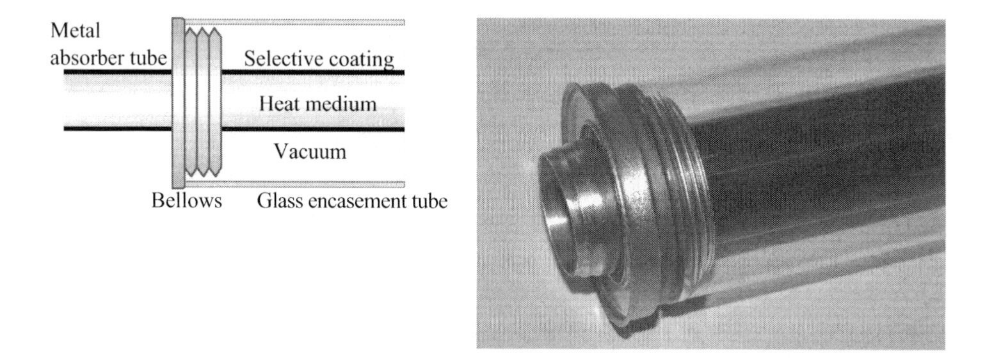

Figure 4.11 Left: bellows to absorb the different expansion rates of glass and metal. Right: high-temperature absorber tube made by Schott for parabolic trough power plants

Flexible elements used to attach and connect the absorber tube must be able to absorb these expansions. The absorber tube's steel pipe and glass encasement expand at different rates, causing special problems. Special folds are therefore used to compensate for these differences in length so that the glass does not break during expansion.

Parabolic trough collector fields

In concentrating solar power plants, three or four parabolic trough collectors are generally looked up in rows. A number of these rows are then lined up parallel to each other to create a field of collectors. When the sun is low in the sky, one row of collectors can shade another.

Relative to solar azimuth α_S, solar elevation γ_S, and the angle β and deviation α of the tracking axis from North/South, the tracking angle and orientation of the collectors can be determined:

$$\rho = \arctan\left(\frac{\cos\gamma_S \cdot \sin(\alpha_S - \alpha)}{\sin(\gamma_S - \beta) + \sin\beta \cdot \cos\gamma_S \cdot (1 - \cos(\alpha_S - \alpha))}\right) \tag{4.26}$$

The distance between the rows of collectors d_R and the collector's aperture width d_K (see Figure 4.12) are used to calculate the marginal angle

$$\rho_{\text{threshold}} = \arccos\left(\frac{d_K}{d_R}\right) \tag{4.27}$$

at which shading is expected.

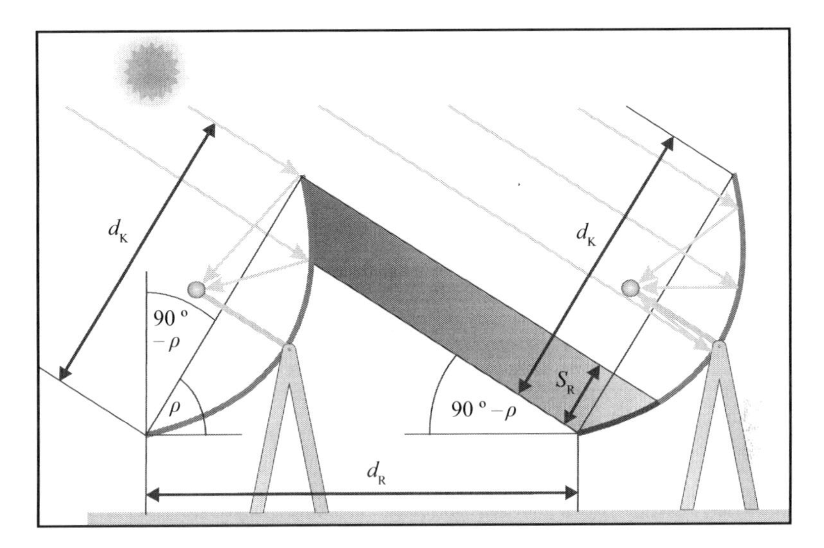

Figure 4.12 The shading of one row of parabolic trough collectors by another

In this case, direct irradiance $E_{dir,K}$ incident on the collector with m parallel collector rows is reduced relative to the ratio of the width of the shading S_R to the aperture width d_K:

$$f_{\text{row to row shading}} = \max\left(0;\quad \frac{m-1}{m} \cdot \frac{S_R}{d_K}\right) = \max\left(0;\quad \frac{m-1}{m} \cdot \frac{d_K - d_R \cdot \cos\rho}{d_K}\right) \tag{4.28}$$

At a typical collector aperture width of $d_K = 5.76$ m and a row distance of $d_R = 17.3$ m, the boundary angle is $\rho_{\text{threshold}} = 70.6°$. If the tracking angles are greater, the rows can be expected to shade each other. At $m = 50$ parallel collector rows and a tracking angle of $\rho = 80°$, the row losses are, for instance, $f_{\text{row to row shading}} = 47$ %.

However, when the sun is low in the sky and tracking angles are large, shading between rows can be tolerated generally because there is generally also not enough irradiance to produce a lot of energy. On the other hand, if there is not enough space between the rows, shading losses will occur even at higher levels of irradiance. Likewise, if the rows are too far apart, the heat and pump losses increase in the lines connecting the rows. A row distance of d_R that is around three times the aperture width d_K is generally a good compromise with nearly maximum energy output.

The circulation heat losses Q_Z through the pipes in the collector field can be calculated as above with the pipe losses for non-concentrating collector systems (see the section on 'Circulation losses' in chapter 3). In practice, there are other losses at valves, faucets, and a poorly insulated sections of piping in addition to thermal transmissivity losses. In a collector field with $m \cdot n$ collectors, an average field temperature of T_{field}, and an ambient temperature of T_U, the field losses

$$\dot{Q}_{V,\text{field}} = f \cdot m \cdot n \cdot A_K \cdot (T_{\text{field}} - T_U) = f \cdot A_{\text{field}} \cdot (T_{\text{field}} - T_U) \tag{4.29}$$

can be calculated nearly empirically. In existing systems, a thermal field loss factor of $f = 0.0583$ W/(m² K) has been calculated [Lip95].

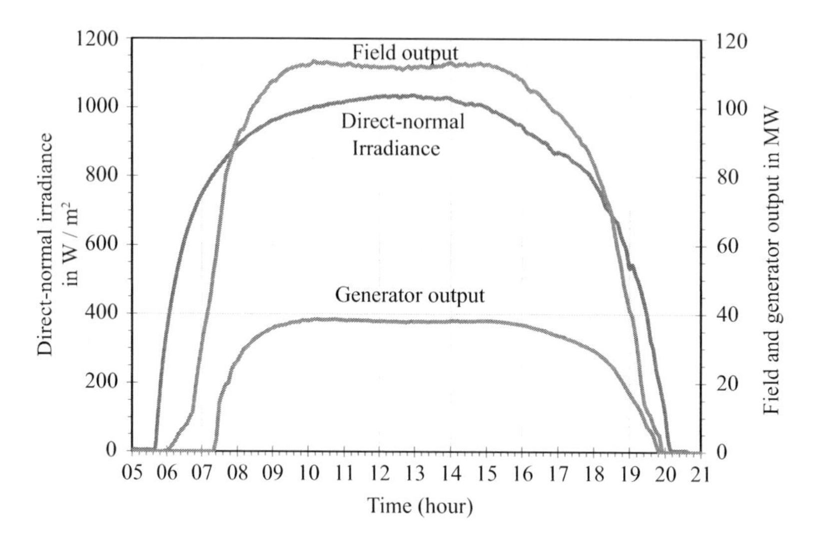

Figure 4.13 Typical time curve for direct-normal irradiance and the output of the field and generator over the course of the day for a parabolic trough power plant

The line losses during the heat-up phase Q_{LA} can also be calculated as we did above with non-concentrating systems. Because of the great mass of the piping and carrier, large amounts of energy are needed during the heat-up phase. If the entire collector field's aperture area is $A_{field} = 250{,}000\ m^2$, some 43 km of absorber tubes are needed with a total metal pipe mass of 164 t. In addition, there are 5 km of connection lines in the field with a mass of 325 t along with around 600 t of heat carrier oil. Heating up the collector field by around 200 °C results in heat losses during the heat-up phase of $Q_{LA} = 85$ MWh. Even at high levels of irradiance, the collectors would need far more than half an hour to produce this energy. The field therefore takes some time each morning to warm up (Figure 4.13).

Point concentrators

To reach higher operating temperatures, greater concentrations than those provided by parabolic trough collectors are needed. Point concentrators are required; they focus sunlight along two axes. With symmetrical dishes, sunlight can be focused on a single focal point (Figure 4.5). However, the size of parabolic dishes is limited in terms of cost. Most systems have a diameter of less than 10 m.

Heliostats fields are used for bigger sizes. Here, a number of mirrors with dual-axis trackers (the heliostats) are arranged in a circle or semicircle around the tower. They focus sunlight on a receiver at the top of a tower.

Depending on the sun's position, the heliostats can also shade each other. Special computer programs are therefore needed to determine the optimal distance between the mirrors and shading losses relative to time. The further the heliostats are from the tower, the greater the distance between the heliostats needs to be. Roughly, the distance between the mirrors

$$\Delta x \approx x \cdot \frac{z_S}{z_T} \tag{4.30}$$

Figure 4.14 Heliostats at the PSA research facility in Almería, Spain

is relative to the mirror's distance x from the tower, the height of the mirror z_S, and the height of the tower z_T [Kle93].

In addition to shading losses, inaccurate tracking, reflection losses, wind loads, cosine losses, and technical failures reduce the field's efficiency. A mirror field's overall efficiency generally ranges from 55 to 80 %.

Heat engines

In addition to providing high-temperature process heat, concentrating systems are mainly used to generate power. Concentrating photovoltaic cells and heat engines can be used to convert concentrated sunlight into electricity. The first large commercial concentrating PV systems were built in 2008, but heat engines have been used since the 1980s in commercial concentrating solar power plants.

Carnot process

Carnot proposed an ideal circular process in which heat and work are periodically input and output. The useful mechanical work that is output

$$|W| = Q_{in} - |Q_{out}| \tag{4.31}$$

is a product of the difference between the heat input Q_{in} and the heat output Q_{out}. The thermal efficiency of the Carnot process

$$\eta_{th,C} = \frac{|W|}{Q_{in}} = \frac{T_{in} - T_{out}}{T_{in}} = 1 - \frac{T_{out}}{T_{in}} \tag{4.32}$$

is defined by the temperatures of the heat source T_{in} and the heat sink T_{out}. To achieve high efficiency, heat has to be input at a high temperature level and output at a low temperature level. An efficiency of 100 % is theoretically only possible at a heat sink temperature of 0 K. The efficiencies of existing circular processes are always below the Carnot efficiency.

Clausius–Rankine process

The **Clausius–Rankine process** is a thermodynamic circular process (Figure 4.15). It is used in **thermal power plants** and is therefore the most important process for power generation in such plants. The working medium in this process is water. Under pressure, it is preheated (a–b) and then turned into steam (b–c). The temperature at which it turns into steam depends on the pressure. At a pressure of 100 bar (equivalent to 10,000 kPa), for instance, the boiling point is 311 °C or 584.1 K. In the next stage of the process, the water vapour is then further superheated (c–d). In the steam turbine, the water vapour produces mechanical energy (d–e). The mechanical energy then drives a generator that produces electricity. Generators are described in greater detail in Chapter 6 (wind power).

In this closed circuit, the steam pressure can be reduced below the ambient pressure, thereby increasing efficiency. Typical condensation pressures are below 100 mbar or 10 kPa. At 100 mbar, the condensation temperature of water pressure is 46 °C or 319 K.

The condenser takes off low-temperature heat Q_{out}, thereby allowing the water vapour to condense back into fluid water (e–f). Heat can dissipate either through fresh water cooling (fresh or seawater) or by means of dry or wet cooling towers. The lower the cooling temperature, the lower condensation pressures need to be, thereby increasing efficiency. The feed water pump brings the fluid water back up to the boiler's pressure (f–a), thereby closing

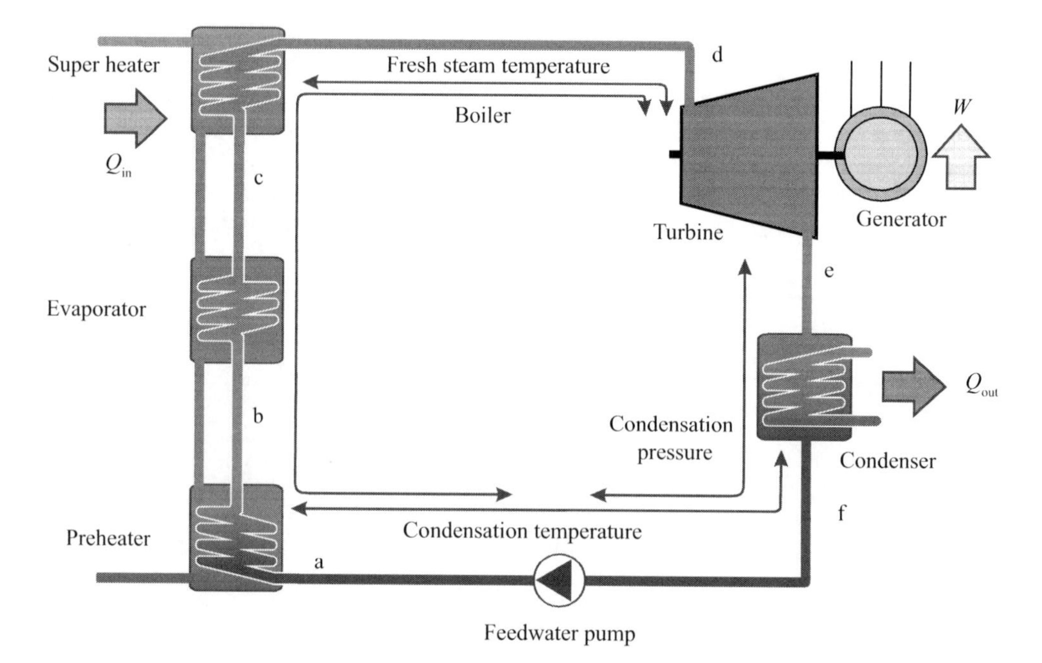

Figure 4.15 Diagram of the Clausius–Rankine process (steam turbine process)

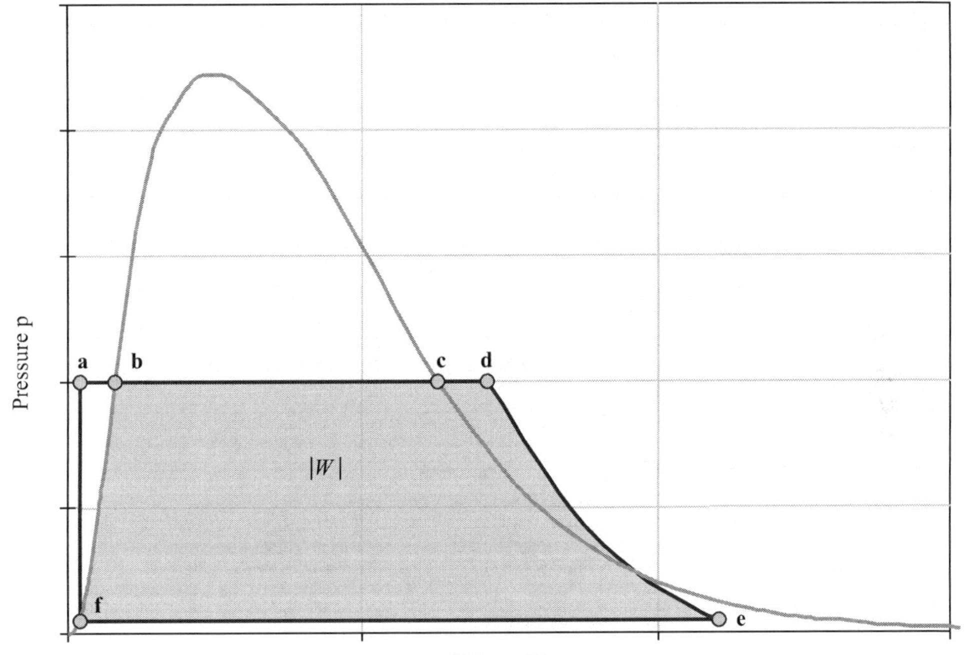

Figure 4.16 Pressure–volume diagram for the Clausius–Rankine process

the circuit. To produce this pressure, the pump has to consume a considerable amount of the electricity generated in the process.

Figure 4.16 shows a pressure-volume diagram of the Clausius–Rankine process. The boiler feed water pump produces the pressure (f–a). At the same pressure p, the water is preheated with additional heat in an isobaric process Q_{in} (a–b) so that it vaporizes (b–c) and super heats (c–d), thereby increasing the volume V considerably. The area where the fluid becomes gas – called the 'wet steam area' – is indicated by the curve in the diagram. When the water vapour expands again in the turbine, the pressure drops as the volume increases further (d–e). In the condenser, the vapour re-condenses at the same pressure to become a fluid again (e–f), thereby reducing the volume considerably.

In addition to such known physical properties as pressure, volume, and temperature, another parameter is used in thermodynamics: **entropy** S. Entropy cannot be directly measured. It describes the extent to which a process is reversible. If energy or heat has to be added to a system, the system's entropy increases. In contrast, when heat leaves a system, entropy drops. A change in entropy dS is therefore always related to a change in heat dQ at a certain temperature T:

$$dS = \frac{dQ}{T} \tag{4.33}$$

Entropy is measured as J/K; the reference point can be selected freely. In thermodynamics, one often also speaks of the specific entropy relative to mass s. Figure 4.17 shows a T-s

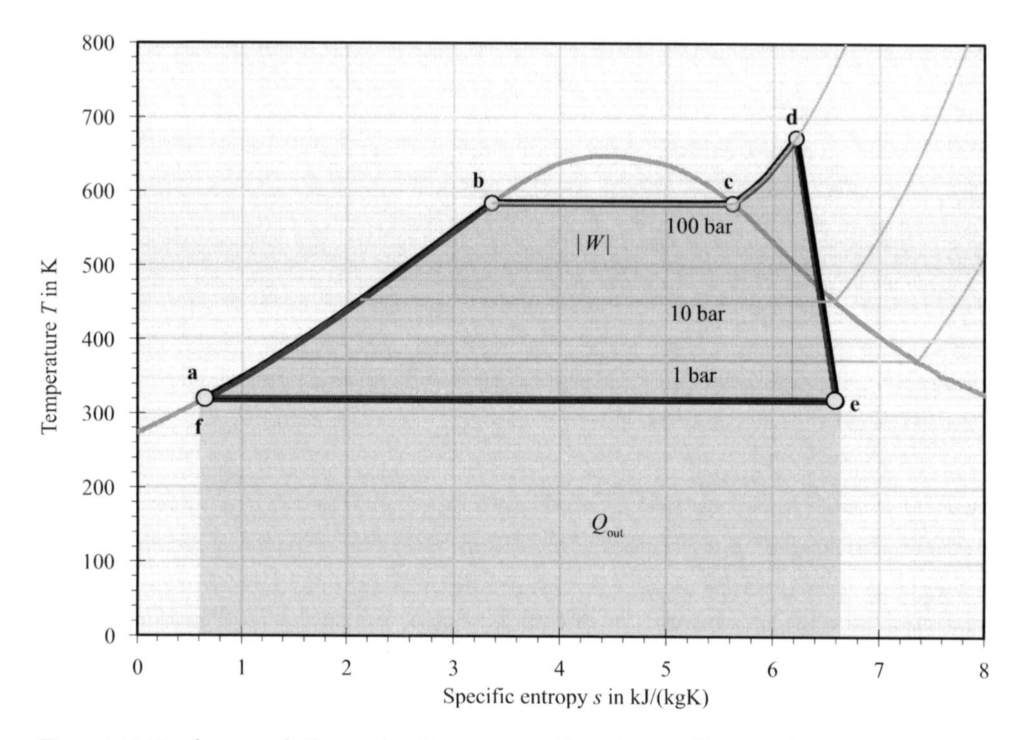

Figure 4.17 T–s diagram of Clausius–Rankine process without intermediate superheating

diagram for the Clausius–Rankine process. The wet vapour area (where the fluid becomes gas) is also represented here by the curve.

Preheating (a–b) follows the limit curve. Here, heat is added, and entropy S increases. Vaporization (b–c) occurs at constant temperature. In the subsequent superheating phase (c–d), the temperature increases further. Here, heat is also input, increasing entropy in the process. In the ideal case, the steam within the turbine undergoes an isotropic process, meaning that entropy remains the same. If this were the case here, the diagram above would have a vertical line, but entropy also slightly increases under real-world conditions due to non-reversible expansion (d–e). In the subsequent condensation phase, Q_{out} is given off, so that entropy reduces again (e–f). The feedwater pump produces pressure in an isentropic process in which neither the temperature nor the entropy change, so that a and f overlap in the diagram.

The enclosed area at the top of the T-s diagram represents the mechanical energy provided $|W|$; the area at the bottom, the heat given off Q_{out}. The energy input into the process Q_{in} is the sum of both of these areas, so the efficiency of the process can be derived by dividing the enclosed area at the top from the total area. To increase efficiency, the enclosed area has to become bigger.

The cooling process limits the lower temperature level, which cannot be changed much, so efficiency improvements will mainly come from higher temperatures and pressures. At higher pressures, the boiling point (b–c) occurs at higher temperatures. At 200 bar, for instance, the temperature reaches 365.7 °C or 639 K. Above 220 bar, vaporization takes place under supercritical conditions. In other words, the water vapour turns directly into steam without the water boiling. Superheating to higher temperatures also increases efficiency. However, extremely high pressures and temperatures require very expensive types of steel and alloys. Improvements in efficiency thus are not cheap. At a pressure of 167 bar and a temperature of 538 °C, the net efficiency can be around 42 %; at 300 bar and 720 °C, it has only increased to 47.5 %. In conventional power plants, increases in efficiency in order to reduce carbon emissions can only have a limited impact.

Concentrating solar parabolic trough power plants currently run at far lower temperatures. At a pressure of 100 bar and a temperature of 371 °C, turbine efficiencies of 38 % can be reached. A number of additional technical measures have been taken to improve efficiency further. For instance, waste heat from the turbine can be used to preheat the feed water in multiple stages. Furthermore, the turbine can have a high-pressure and a low-pressure stage. Between the two stages, an intermediate superheating phase occurs to further increase the enclosed area in the T-s diagram. Greater process temperatures from direct vaporization in the trough collectors and the use of solar towers are expected to produce further improvements. Nonetheless, the heat process is the weakest link in a concentrating solar power plant's efficiency chain despite all of the technological advances.

Brayton cycle

In **open gas turbines**, the Brayton cycle (also known as the Joule cycle) is used (Figure 4.18). Here, a compressor draws in ambient air, compresses it in an isentropic process, and feeds it into a combustion chamber. In conventional systems, fuels (such as kerosene and natural gas) heat up the air in an isobar process. The heat required can also come from solar thermal processes, however, as explained below in relation to solar power towers. The hot air then expands in the gas turbine, which then drives the compressor and a generator for power production. A significant part of the turbine's work is required to drive the compressor.

To reach higher efficiencies and the conversion of thermal energy into electricity, gas turbines can be combined with steam turbines. Here, a gas turbine process first occurs at high temperatures. The waste heat from the gas turbine vaporizes and superheats water in a downstream boiler. The steam produced in the process then drives a steam turbine. Such 'combined-cycle gas turbines' (**CCGTs**) have overall efficiencies of up to 60 %. This principle can be used in solar power towers with high-pressure receivers (see the section 'Pressurized receiver' later in this chapter).

Stirling process

Another type of heat engine used in concentrated solar power plants is the Stirling engine, also known as an external combustion engine. It uses air, helium, or hydrogen as the working medium. The gas is compressed in an isothermal process without increasing the temperature, then heated up at the same volume in an isochoric process before it expands in an isothermal process and then cools down again in an isochoric process. In practice, one or two pistons are used in such systems. They move the working gas between a hot and a cold area. Concentrated irradiance can be used as a heat source for the hot section. A regenerator separates the two sections. The expansion and compression of the working gas moves the pistons. Stirling engines have not yet been commercial successes to a major extent. Up to now, there have only been small batch productions, though they have proven quite successful in practice.

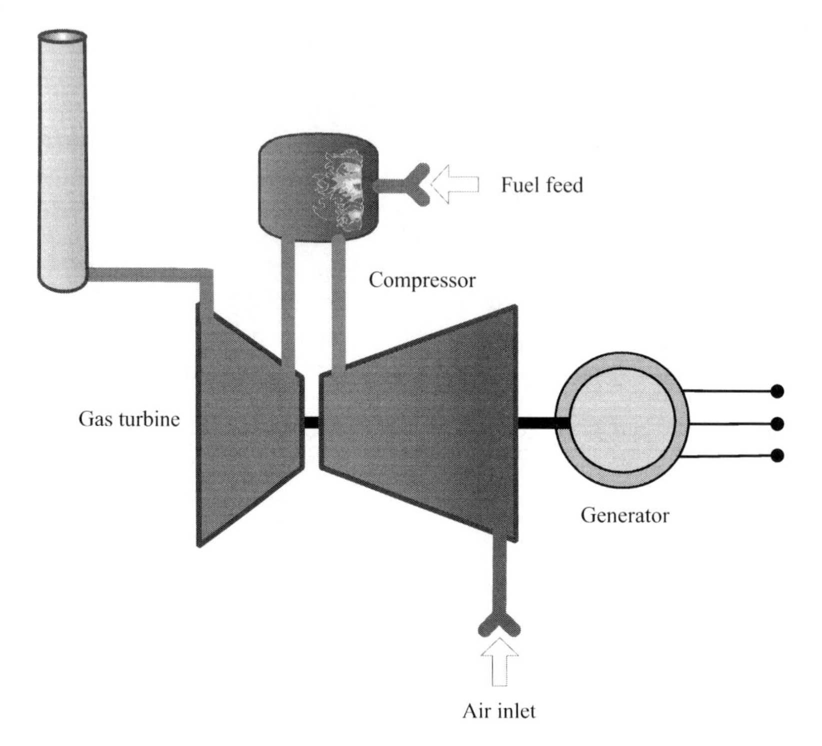

Figure 4.18 How an open gas turbine works (Brayton cycle)

Concentrating solar thermal systems

Parabolic trough power plants

In 1906, the development of solar thermal power plants began in the United States. In the US and in Cairo, Egypt, (a British colony at the time), demonstration plants were built and tested successfully. The systems visually resemble the ones we have today to a surprising extent. But in 1914, just before World War I broke out, material problems and other technical difficulties put an end to the first attempts at large-scale solar power production.

In 1978, however, the technology was reborn in the United States. The Public Utilities Regulatory Policy Act (PURPA) required public utilities to purchase power from independent producers at clearly defined costs. The oil crises more than doubled power costs, so Southern California Edison (SCE) offered long-term Standard Offer Contracts for power from renewable energy. In combination with tax incentives, these SOCs made building these plants financially attractive. In 1984, the first parabolic trough power plant was built in the Mojave Desert. Other power plants were also completed in intervals of less than a year, each one larger and more technically advanced. But in the mid-1980s, energy prices plummeted. The tax incentives expired at the end of 1990, putting an end to construction of parabolic trough power plants in California.

Up to 2007, California's **SEGS parabolic trough power plants** (Solar Electric Generation Systems) were the only power plants in the world of this kind. Here, a million mirrors with a total aperture area of 2.3 million m^2 concentrate sunlight. These power plants have an electric output of 354 MW and produce 800 GWh each year, enough to cover

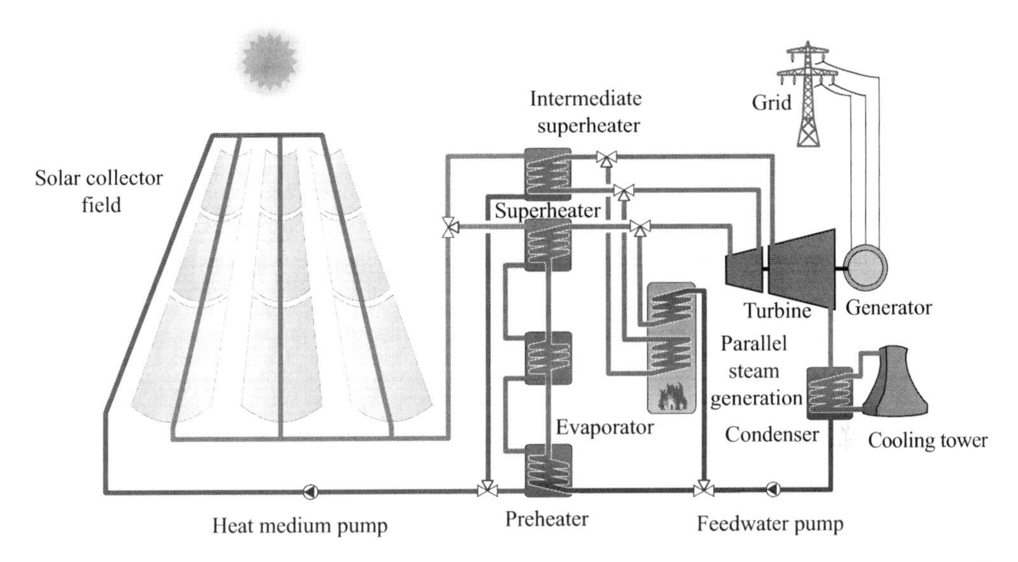

Figure 4.19 A parabolic trough power plant with parallel steam generation and a separate collector and steam turbine circuit

the total needs of a full 60,000 Americans. Eight of the plants can also run on fossil fuels to provide electricity at night or during bad weather. The share of natural gas consumed annually is, however, limited to 25 % of the thermal energy input.

The SEGS systems built in California in the 1980s and 1990s have a thermal oil running through the absorber tube within the trough collectors; the sun heats up the oil to nearly 400 °C. Heat exchangers transfer the heat to a steam turbine process with intermediate superheating. In addition to the solar collector field, a parallel fossil-fired steam generator is used to produce heat in the steam turbine circuit (Figure 4.19). Table 4.4 summarizes the technical data for these American parabolic trough power plants.

Greater political support led to a revival of concentrated solar power technology. The second time around, new power plant concepts were investigated.

In numerous power plants, a small amount of backup power from natural gas is allowed to prevent the system from freezing. Too much natural gas consumption would, however, increase carbon emissions. In this case, thermal storage should be integrated (Figure 4.20).

Here, the solar field is a bit bigger, and it fills up a storage tank during the day. In the evenings and at night, heat is taken from storage to power the turbine. If the storage tank is large enough, the solar power plant can provide electricity around the clock (Figure 4.21).

The storage tank must be designed to withstand high temperatures. Water is not a good option as a storage medium because it has a large volume when it becomes a gas. In addition to soil, concrete, and stone, molten salt is an alternative. In practice, such systems have two tanks. When heat is being built up, molten salt is taken from the cold tank to the hot tank; when heat is consumed, the process is reversed. The hot tank has a temperature of around 380 °C. The cold tank's temperature must be above the molten salt's freezing point. Temperatures for the cold tank generally are around 280 °C.

At present, the thermal oil's thermal properties limit the temperatures for parabolic trough power plants. Tests in Spain showed that water can also be used directly as a heat

Table 4.4 Technical data for the SEGS parabolic trough power plants in California [Pil96]

System	*I*	*II*	*III*	*IV*	*V*	*VI*	*VII*	*VIII*	*IX*
Year of completion	1984	1985	1986	1986	1987	1988	1988	1989	1990
Capacity in M, W	13.8	30	30	30	30	30	30	80	80
Land area in 1,000 m^2	290	670	800	800	870	660	680	1,620	1,690
Aperture area in 1,000 m^2	83	165	233	233	251	188	194	464	484
Collector type	LS1–2	LS1–2	LS-2	LS-2	LS-2	LS-2	LS2–3	LS-3	LS-3
Oil input temp. in °C	241	248	248	248	248	293	293	293	293
Oil output temp. in °C	307	321	349	349	349	391	391	391	391
Steam (solar)									
– Pressure in bar	38	27	43.4	43.4	43.4	100	100	100	100
– Temperature in °C	247	300	327	327	327	371	371	371	371
Efficiency in %									
– Steam turbine (solar)	31.5	29.4	30.6	30.6	30.6	37.6	37.6	37.6	37.6
– Steam turbine (gas)	–	37.3	37.3	37.3	37.3	39.5	39.5	37.6	37.6
– Solar field (thermal)	35	43	43	43	43	43	43	53	50
– Solar-electric (net)	9.3	10.7	10.2	10.2	10.2	12.4	12.3	14.0	13.6
Sales price in US$/kW	4,490	3,200	3,600	3,730	4,130	3,870	3,870	2,890	3,440

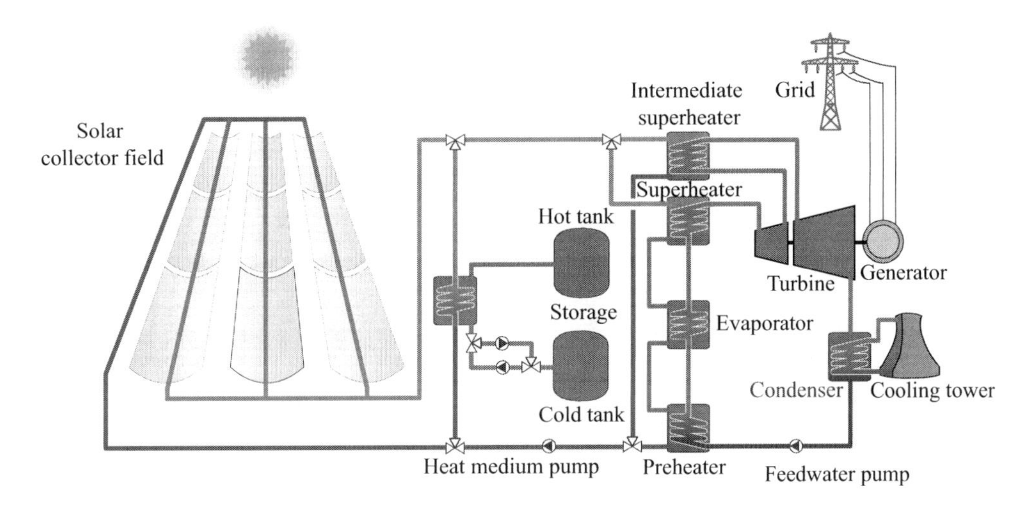

Figure 4.20 A parabolic trough power plant with thermal storage

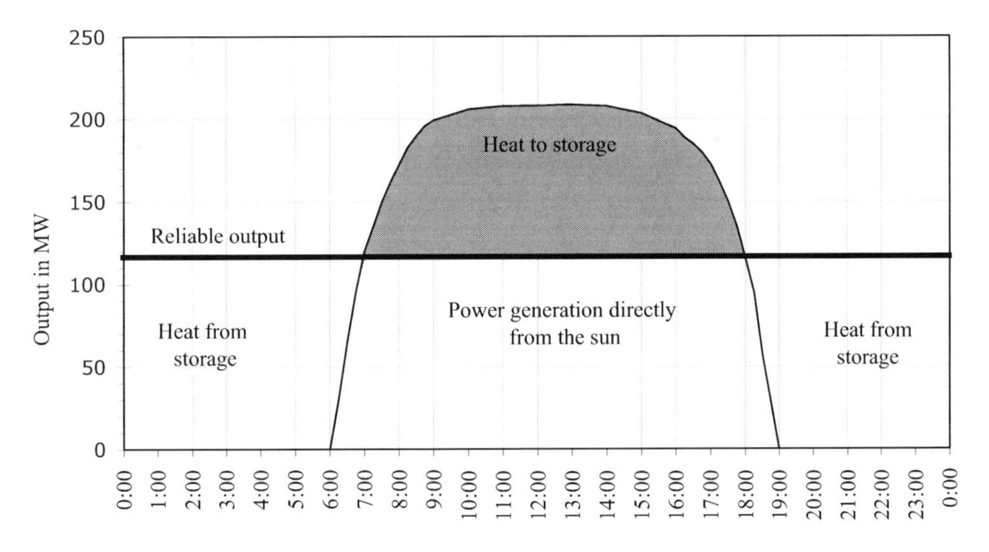

Figure 4.21 Guaranteed output of a concentrated solar power plant with thermal storage

medium. It can vaporize in the absorber tube at a pressure of 100 bar and superheats to temperatures of 500 °C. The result is greater efficiency for the circuit. At the same time, costs are lower when the thermal oil and heat exchanger between the oil and the water vapour circuit are not needed. However, for direct solar vaporization the integration of thermal storage has not been solved satisfactorily.

The World Bank has provided funding for the market launch of concentrating solar power in developing countries. In these projects, a combined-cycle gas turbine is to be combined with a solar field (Figure 4.22).

In an **ISCCS plant** (Integrated Solar Combined Cycle Power Station), a solar field is combined with a traditional combined-cycle gas turbine. When there is not enough

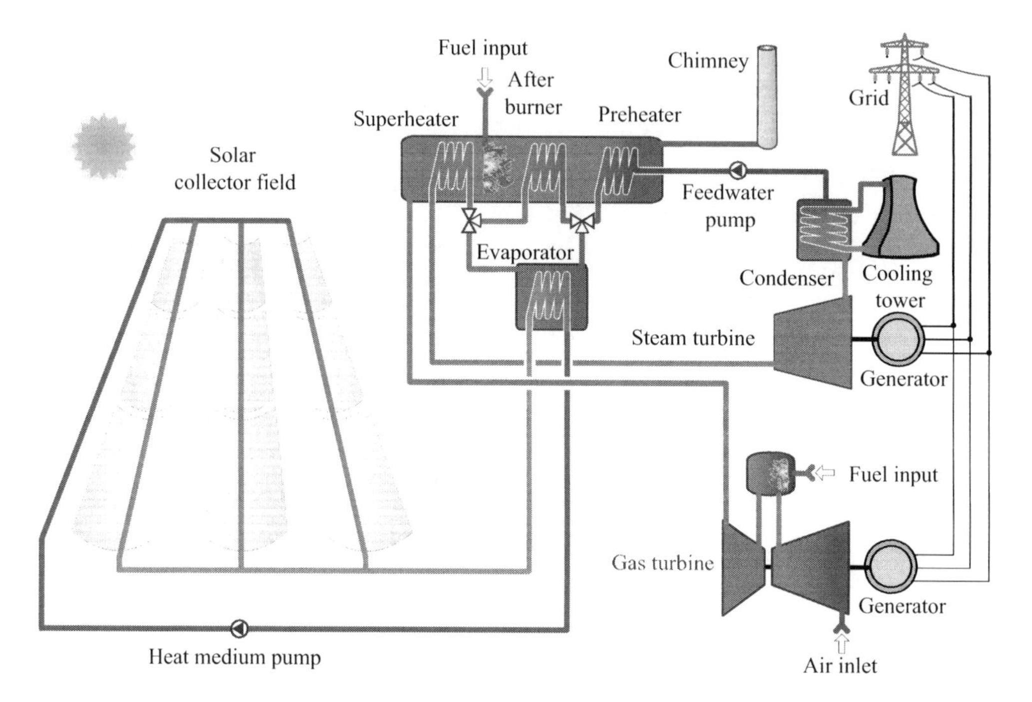

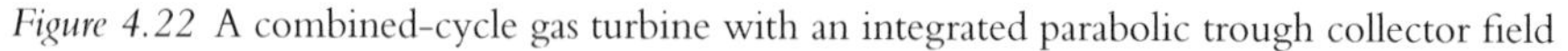
Figure 4.22 A combined-cycle gas turbine with an integrated parabolic trough collector field

solar heat to drive the gas turbine process, waste heat is taken from the heat recovery steam generator. In this process, however, solar only makes up a small share of the energy input at far below 10 %. For long-term climate protection, this power plant concept is thus only an option if biogas or green hydrogen is used instead of natural gas. No such plans exist yet.

Solar thermal systems are also possible at a smaller scale. At low temperatures, power can be generated in the ORC process, which is explained in the section on geothermal power generation.

The most interesting markets for concentrating solar power are currently the US and Spain. But other countries are also getting on board. Table 4.5 shows the technical data for some new parabolic trough power plants that have gone into operation since 2007.

In practice, parabolic trough power plants always have an axis facing North/South so that as much energy as possible can be generated. When the axis is East/West, energy yield drops by around 20 %. On the other hand, the average monthly output from systems facing East/West is much smoother across the year. In very sunny areas, solar thermal power plants can therefore run for over 7,000 full-load hours per year (a year has roughly 8,760 hours) in solar-only generators, thereby completely offsetting conventional power plants. If the feasible cost-reduction potential is tapped, parabolic trough power plants will be a realistic alternative to conventional plants.

Solar power towers

Solar power towers are another type of concentrating solar power plant. Here, hundreds or even thousands of mirrors are arranged around a tower. The mirrors (also called heliostats)

Table 4.5 Technical data for a group of new parabolic trough power plants

System	*Nevada Solar Onexe*	*Andasol 1*	*Ain Beni Mathar*	*Shams 1*
Country	USA	Spain	Morocco	UAE
Year of completion	2007	2008	2010	2012
Capacity in MW	64	50	472	100
Power plant type	SEGS	SEGS	ISCCS	SGES
Share of solar	100 %	88 %	< 5 %	100 %
Land area in 1,000 m²	1,400	2,000	800	2,500
Aperture area in 1,000 m²	357	510	183	628
Collector type	SGX-2	EuroTrough	ASTRO	ASTRO
Electricity generation in GWh/a	129	180	3,538	210
Full-load hours in h/a	2,015	3,600	7,500	2,100
Storage duration in full-load hours (h)	0.5	7.5	0	0
Investment costs in million €	230	300	420	470

track the sun and focus its light on top of the tower. Each individual mirror has to be focused exactly within a fraction of a degree so that the reflected sunlight truly reaches the focal point.

An absorber at the focal point heats up the highly concentrated sunlight to temperatures exceeding 1,000 °C. Water vapour, air, or molten salt can be used as the heat medium. A gas or steam turbine drives a generator that converts the heat into electricity.

There is less commercial experience with solar power towers than with parabolic trough power plants. In Almería (Spain), Barstow (USA) and Rehovot (Israel), there have been test systems for some time now where components are optimized and new equipment tested (Figure 4.23). In 2006, the first commercial solar power tower (the PS10) with a capacity of 11 MW went into operation near Seville, Spain. Table 4.6 shows the technical data of recent solar power towers. The first commercial plants (the PS10 and PS20) used conventional receiver technologies to produce saturated steam, but later receiver concepts promise significantly higher efficiencies.

Open volumetric receiver

In the tower concept with open volumetric receivers (Figure 4.24), a fan draws in ambient air through the receiver, which is also the focal point of the heliostats. A wire mesh, ceramic foam, and metallic/ceramic honeycomb structures can be used as receiver materials. The receiver heats up from the solar energy and gives off the heat to the ambient air passing through it. The air cools down the front side, so very high temperatures are only reached inside the receiver. The result is lower radiation losses. At temperatures ranging from 650 °C to 850 °C, the air enters a heat recovery steam generator, which vaporizes and super heats water to drive a steam turbine circuit. If need be, this type of power plant can receive backup heat from other fuels via a duct burner. In 2008, a demonstration power plant for this technology went into operation in Jülich, Germany.

Figure 4.23 Solar power tower at the PSA research facility in Almería, Spain

Table 4.6 Technical data for a select group of solar power towers

System	*PS10*	*PS20*	*Jülich*	*Gemasolar*
Country	Spain	Spain	Germany	Spain
Year of completion	2006	2008	2008	2011
Capacity in MW	11	20	1,5	19,9
Receiver type	Saturated steam	Saturated steam	Air, open	Salt
Share of solar	>85 %	>85 %	100 %	>85 %
Land area in 1,000 m²	600	900	160	1950
Heliostat area in 1,000 m²	75	151	18	305
Number of heliostats	624	1255	2150	2650
Electricity generation in GWh/a	24.3	48.6	1	110
Full-load hours in h/a	2,200	2,430	670	5,500
Tower height in m	100	160	60	140
Storage duration in full-load hours	0.5	1	1	15
Investment costs in million €	48.5	95.4	21.7	230

Pressurized receiver

A further development of the tower concept has a pressurized receiver; in the midterm, it looks promising (Figure 4.25).

Here, the air is heated up in the volumetric pressurized receiver to around 15 bar at temperatures up to 1,100 °C. A transparent fused quartz dome separates the absorber from the surrounding area. The hot air produced then drives a gas turbine. The turbine's waste heat is then used to drive a downstream steam turbine process.

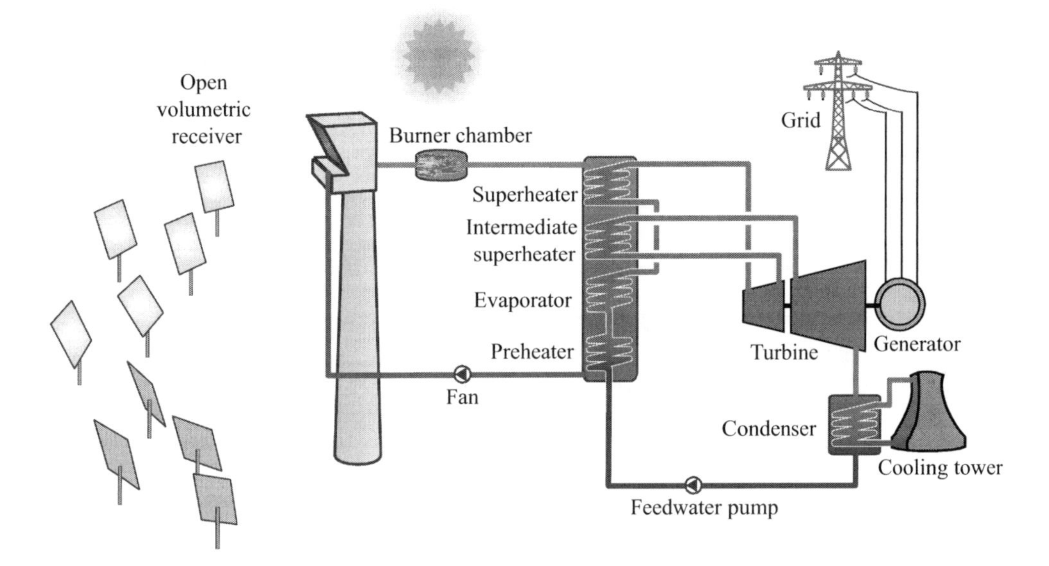

Figure 4.24 Solar power tower with open volumetric receiver

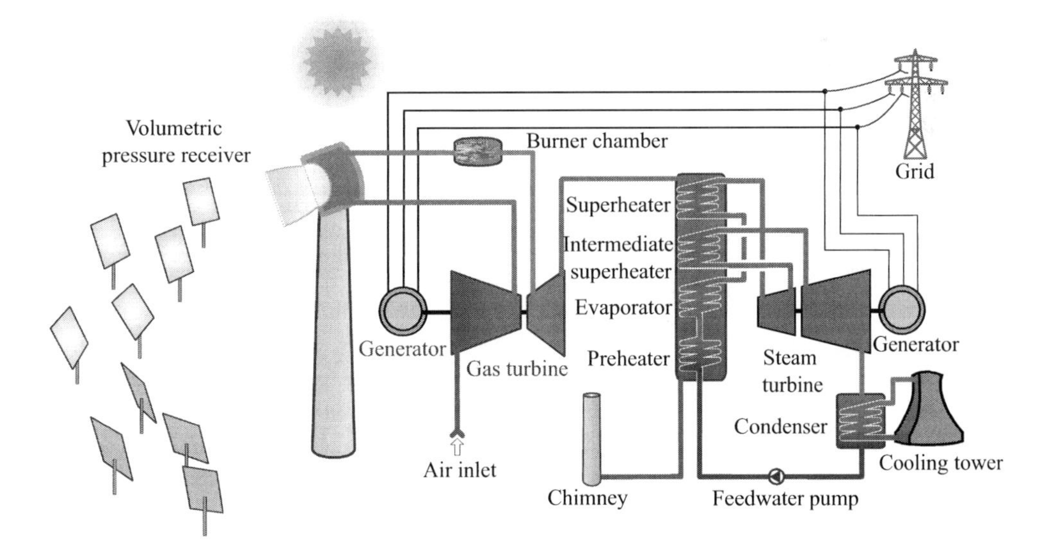

Figure 4.25 Solar power tower with volumetric pressurized receiver to drive gas and steam turbines with solar energy

The combined gas and steam turbine process increases the efficiency of the conversion of heat into electricity from around 35 % in the steam turbine process to more than 50 %. Overall, the conversion of solar energy into electricity can then have an efficiency exceeding 20 %. These prospects justify the more complicated, more expensive receiver technology.

Dish Stirling systems

While parabolic trough and power tower plants only make sense economically at outputs of at least a few megawatts, dish Stirling engines can be smaller and serve off-grid locations, for instance. In these systems, a mirror dish concentrates light on a focal point. To ensure that the dish concentrates light as much as possible, the dish tracks the sun very accurately along two axes.

A receiver is located at the focal point. It passes heat on to the centrepiece of the system: the Stirling engine. The engine turns the heat into kinetic energy to drive a generator that produces electricity.

Stirling engines can not only run on solar energy, but also from combustion heat. In combination with a biogas burner, these systems can therefore also run at night or during periods of bad weather. When biogas is used, the system is also carbon-neutral.

A number of prototypes of purely solar units have been built in Saudi Arabia, the US, and Spain (see Table 4.7 and Figure 4.26). The price of power from dish Stirling systems per kilowatt-hour is, however, relatively high compared with power towers and parabolic troughs. However, serial production could reduce the cost of such systems significantly.

Solar furnaces and solar chemistry

In addition to providing process heat and generating electricity, concentrated solar energy can also be used for material tests and in solar chemistry.

A large **solar furnace** is located in Odeillo, France. There, a large number of small mirrors have been installed on a slope to reflect sunlight onto a concave mirror with a diameter of 54 m; a research centre is at the focal point. Temperatures of 4,000 °C power are reached – great enough for experiments and industrial processes. Other solar furnaces have been set up in Almería, Spain (Figure 4.27) and in Cologne, Germany.

In addition to producing chemicals at high temperatures, solar thermal can also be used to produce hydrogen. In such processes, electricity is not used for electrolysis. Rather, hydrogen can be produced in chemical processes at high temperatures provided by solar energy. For example, the chemical process can take place within a solar tower's receiver. Hydrogen is considered an important energy carrier, especially in transport and fuel cells (see Chapter 10). If the vision of a hydrogen economy is to become reality someday, concentrating solar chemical systems could play a major role in the production of climate-friendly hydrogen.

Power imports

The description of solar thermal power plants above mentions that they are rarely economically useful in areas with relatively low annual irradiation, such as Germany.

Table 4.7 Technical data for the EuroDish Stirling unit

Concentrator diameter	8.5 m	Reflection degree	94 %
Aperture area	56.7 m²	Working gas	Helium
Focal length	4.5 m	Gas pressure	20–150 bar
Average concentration factor	2,500	Receiver's gas temperature	650 °C
Gross electric output	9 kW	Max. wind velocity for operation	65 km/h
Net electric output	8.4 kW	Maximum wind velocity for survival	160 km/h

(Data: [Sch02])

Figure 4.26 EuroDish prototypes at the PSA research facility in Almería, Spain

Figure 4.27 Solar furnaces at the PSA research facility in Almería, Spain. Top right: secondary concentrator within a building

Table 4.8 Key data for overhead power lines for HVAC und HVDC [Hos88]

Type of transmission *Nominal voltage*	*HVAC* *380 kV*	*HVAC* *750 kV*	*HVAC* *1,150 kV*	*HVDC* *±600 kV*
Image of pylon				
Line cross-section Al/St in mm^2	805/102	805/102	805/102	805/102
Number of strands	4	4	6	4
Number of lines	2 × 3	2 × 3	2 × 3	2 × 2
Specific resistance in Ω/km	0.009	0.009	0.006	0.009
Thermal maximum in MW	2 ×3,812	2 ×7,015	2 ×16,120	2 × 6,500
Transmission capacity in MW[1])	2 ×2,121	2 × 4,187	2 × 9,630	2 × 3,860
Losses in kW/km[1])	2 × 280	2 × 280	2 × 421	2 × 187
Relative losses per 1,000 km[1])	13.2 %	6.7 %	4.4 %	4.8 %

1) At 1 A/mm^2

The annual amount of direct-normal irradiance in Germany is less than half the level in the sunniest parts of the world; in addition, the efficiency of concentrating solar power plants drops considerably when they run at partial capacity. Power generation costs are therefore expected to be around three times higher than at the best locations. From an economic perspective, it thus makes more sense to build concentrating solar power plants in, say, northern Africa and transport the electricity to Germany than to build them in Germany.

There is also enough space for such systems in northern Africa. Even if we set aside unsuitable areas – such as sand dunes, nature conservation zones, farmland, and mountains – only 1 % of the remaining area in northern Africa would suffice to theoretically meet all of civilization's entire power demand with concentrated solar power plants. In addition to conventional high-voltage alternating current lines (HVAC), high-voltage direct-current lines (HVDC) are also an option. Table 4.8 covers the key data for both types of transmission at various voltage levels.

In addition to transmission losses, there are also losses from the conversion of alternating current into direct current and back for HVDC lines. The technology is already available, and it is competitive. A 5,000 km 600 kV HVDC line would have transmission losses of around 18 %, a figure that would drop below 14 % at 800 kV.

At a possible power cost from the site of a utility-scale concentrated solar power plant of 3 to 4 cents/kWh, the losses would add on a cost of around 0.5 cents/kWh. The line itself would add on around 0.5 to 1 cents/kWh. Overall, renewable power would then be available at 4 to 5 cents/kWh from Africa to Germany to ensure a reliable power supply from a combination of photovoltaics, wind power, and concentrated solar power [Qua05]. Such

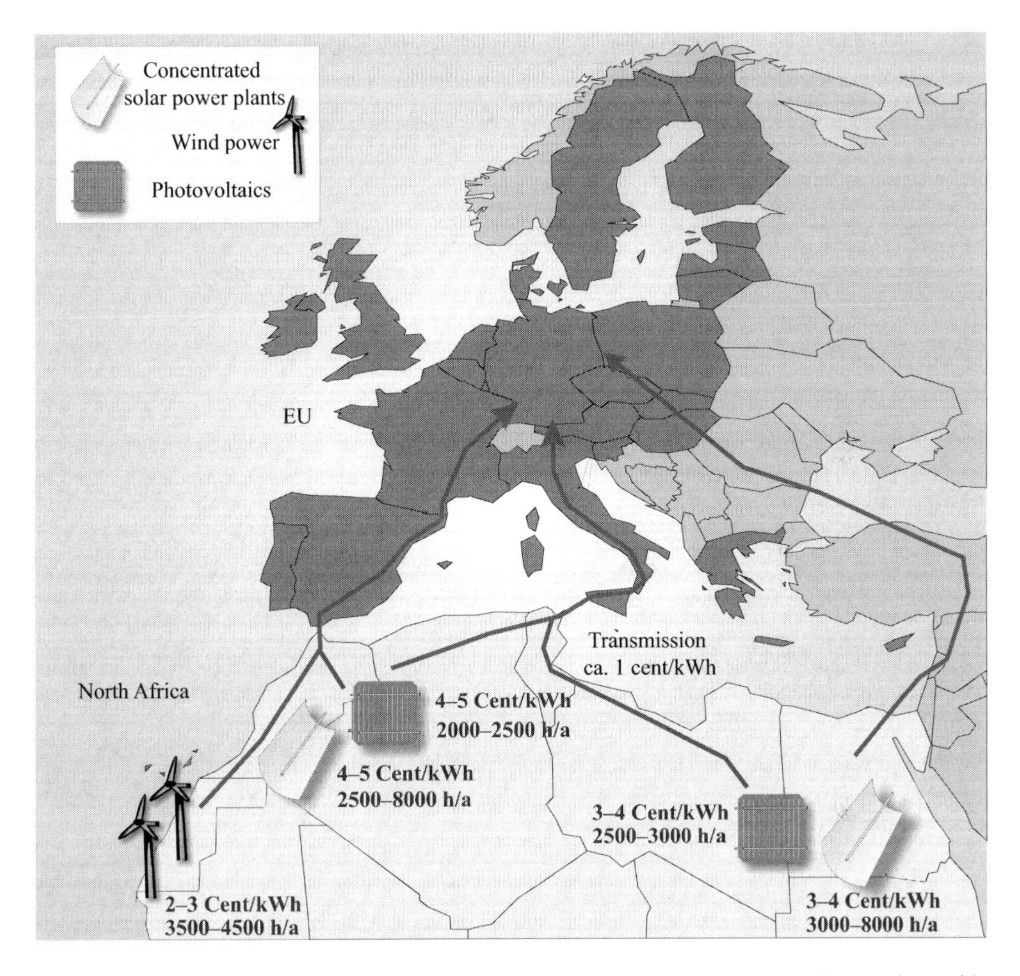

Figure 4.28 Options for renewable power imports from northern Africa to the EU along with possible power costs and capacity factors for the midterm

long power lines, however, will take quite some time to build because local residents often oppose proposed routes. Critics of this import strategy therefore believe small, distributed solutions are better because they can be built much more quickly and therefore start helping protect the climate faster.

These low costs will probably be possible when 30 GW of concentrated solar power plants and 200 GW of photovoltaics have been built, which is possible in less than 10 years. Germany would play an especially important role in this scenario in terms of the development and export of solar technology, thus making this solution very attractive for import and export countries.

In contrast, the use of hydrogen to transport electricity is only interesting in exceptional cases. When hydrogen is made by means of electrolysis, the losses exceed 50 % when we include transport and the reconversion of hydrogen into electricity. Production of hydrogen therefore only makes sense when it can be used directly for transport or to store electricity seasonally.

5 Photovoltaics

Introduction

Photovoltaics consists of the two words *photo* and *volta*. Photo (Greek phõs, photós) means 'light', while volta (Count Volta, 1745-1827, Italian physicist) is a unit for electrical voltage. In other words, photovoltaics takes sunlight and converts it directly into electricity.

The **history of photovoltaics** started in 1839, when Becquerel discovered the photoelectric effect. But it took another hundred years for the age of semiconductors to start. In 1949, Shockley developed a model for p-n transitions, and in 1954 the first silicon solar cell literally saw the light of day at Bell Labs in the US. At the time, the first cell had an efficiency of around 5 %. Costs were not an issue; after all, the cell was to provide electricity to vessels and satellites in space.

Since then, the efficiency of solar cells has continued to rise. In the lab, silicon cells easily have efficiencies above 20 % these days. New materials have also been tested, and some of them are now used. The cost of solar cells has been brought down considerably. In off-grid applications, photovoltaics is already competitive. But even on the grid, the cost of power from photovoltaics is already equal to or below power from the grid (grid parity) in a number of countries – and counting. As more and more PV is installed, costs will come down even further. In a few years, photovoltaics will be competitive with power from the grid worldwide.

Of all sources of renewable electricity, photovoltaics has the widest range of applications. It is especially **modular**. In other words, the power generator can come in almost any size – from a few milliwatts for pocket calculators and watches to multi-megawatts for utility-scale power plants.

For a long time, photovoltaics was mainly used in consumer applications, toys, and off-grid applications, such as telecommunications facilities. Germany's **1,000 Roofs Program** launched at the beginning of the 1990s led to the construction of 2,250 PV systems connected to the grid, demonstrating for the first time that photovoltaics was a good source of power on a significant scale. In central Europe, photovoltaics poses tremendous potential. It can be installed on roofs and façades without taking up any space. Building-integrated PV has the potential to cover a third of power demand in Germany in the midterm.

Figure 4.1 shows a roof-integrated photovoltaic system built in the United States back in the 1980s. At the time, capacities exceeding 100 kW_p broke records, but today the largest PV power plants have more than 100 megawatts. Thanks to the priority that renewable power has on the grid and to feed-in tariffs specified in the German **Renewable Energy Act**, Germany became a leader in photovoltaics at the end of the 1990s. Until then, Japan had dominated the market.

Figure 5.1 Roof-integrated photovoltaic system (Georgetown University in Washington, DC, built in 1984, 4,464 solar modules, 337 kW_p total capacity)

Table 5.1 Overview of key electrotechnical parameters

Name	*Symbol*	*Unit*	*Name*	*Symbol*	*Unit*
Electrical energy	W, E	W s, J	Specific resistance	ρ	Ω m
Electrical output	$P = U \cdot I$	W	Electrical field strength	$E = -dU/ds$	V / m
Electrical voltage	U	V	Inductance (coil)	L	H = V s/ A
Electrical current	I	A	Capacity	C	F = A s/ V
Electrical resistance	$R = U / I$		Electrical charge	$Q = \int I dt$	C = A s
Electric conductance	$G = 1 / R$	S	Force in electrical field	$F = E \cdot Q$	N

In Germany, installation figures have skyrocketed in recent years. Since 2011, Germany has had more installed PV capacity than nuclear capacity. Photovoltaics now plays a crucial role in German power supply.

This chapter presents the basics, such as how photovoltaics works, how to calculate a photovoltaic system, and what its applications are. Table 5.1 shows an overview of the key electrotechnical parameters used in such descriptions.

How solar cells work

Bohr's atomic model

Starting with the Bohr model of individual atoms, the energy band model for solids will be discussed to describe how semiconducting solar cells work. In Bohr's atomic model, electron mass orbits the nucleus

$$m_e = 9{,}1093897 \cdot 10^{-31}\ \mathrm{kg} \tag{5.1}$$

with a radius of r_n at orbital frequency ω_n. The result is the **centrifugal force**:

$$F_Z = m_e \cdot r_n \cdot \omega_n^2 \tag{5.2}$$

Between the nucleus, which consists of Z positively charged protons and additional neutrons without a charge, and the electron with the **elementary charge**

$$e = 1{,}60217733 \cdot 10^{-19} \text{ A s} \tag{5.3}$$

the **coulomb charge** holds things together:

$$F_C = \frac{1}{4\neq \cdot \varepsilon_0} \cdot \frac{Z \cdot e^2}{r_n^2}, \text{ with} \tag{5.4}$$

$$\varepsilon_0 = 8{,}85418781762 \cdot 10^{-12} \frac{\text{A s}}{\text{V m}} \tag{5.5}$$

being the **electric force constant** or the dielectric constant. The coulomb force and the centrifugal force must balance each other out for the electron to stay within its orbital. Here, orbits must have an angular momentum that is an integer multiple of

$$\hbar = \frac{h}{2 \cdot \neq} \tag{5.6}$$

with $\hbar$ calculated from **Planck's constant**

$$h = 6{,}6260755 \cdot 10^{-34} \text{ Js} \tag{5.7}$$

Quantization of the angular momentum produces:

$$n \cdot \hbar = m_e \cdot r_n^2 \cdot \omega_n \tag{5.8}$$

With that and the equilibrium $F_Z = F_C$, we get the following equation for the orbital's radius

$$r_n = \frac{n^2 \cdot \hbar^2 \cdot 4 \cdot \neq \cdot \varepsilon_0}{Z \cdot e^2 \cdot m_e} \tag{5.9}$$

and for the **orbital frequency**:

$$\omega_n = \frac{1}{(4 \cdot \neq \cdot \varepsilon_0)^2} \cdot \frac{Z^2 \cdot e^4 \cdot m_e}{\hbar^3 \cdot n^3} \tag{5.10}$$

The integral index n indicates the number of each orbital. The energy that an electron carries along with it on its orbital is:

$$E_n = \tfrac{1}{2} \cdot m_e \cdot v_e^2 = \tfrac{1}{2} \cdot m_e \cdot r_n^2 \cdot \omega_n^2 = \frac{1}{n^2} \cdot \frac{Z^2 \cdot e^4 \cdot m_e}{32 \cdot \neq^2 \cdot \varepsilon_0^2 \cdot \hbar^2} \tag{5.11}$$

The energy of a hydrogen atom with atomic number $Z = 1$ on the first orbital ($n = 1$) is $E_{1,Z=1} = 13.59$ eV.

If an electron is to be lifted to the next orbital up, the amount of external energy required is $\Delta E = E_n - E_{n+1}$. Keep in mind here that the number of electrons that can be within a single orbital is limited. The first orbital ($n = 1$) can have two electrons; the second, 8, then 18, 32, 50, etc. As n increases, the orbital's energy decreases, eventually reaching $E_\infty = 0$ at $n = \infty$. In addition to the Bohr model, there are other models describing the atom not discussed here (see [Her12]).

Photoelectric effect

The energy of photons in the light provides the energy to move an electron into a higher orbital. The **photon's energy** at wavelengths λ is calculated as

$$E = \frac{h \cdot c}{\lambda} \tag{5.12}$$

with the speed of light $c = 2.99792458 \cdot 10^8$ m/s. If a photon with 13.59 eV of energy impacts an electron in the first orbital of a hydrogen atom, the energy suffices to lift the electron on to orbital E_∞, thereby completely separating it from the nucleus. This energy is also called ionization energy, while the process of photons completely removing electrons from the nucleus is called the external **photoelectric effect**. In the example of hydrogen, the photon has to have a wavelength λ smaller than 90 nm, which only x-rays have.

In photovoltaics, visible, ultraviolet, and infrared radiation is used, and the protons here have far less energy; therefore, the external photoelectric effect does not apply as it requires radiation with high energy. Instead, the internal photoelectric effect applies, as explained below.

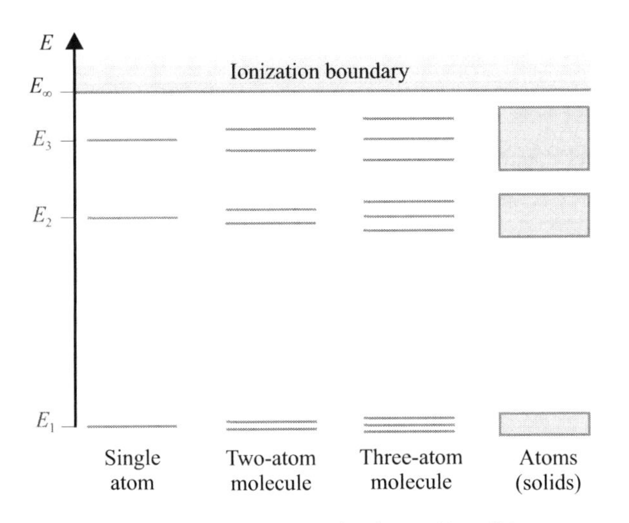

Figure 5.2 Energy states of electrons in an atom, in molecules, and in solids

In individual atoms, electrons have precisely defined energy states, but in the molecules consisting of multiple atoms electrons interact and break up identical energy levels into closely related neighbouring levels. In a solid with k atoms, the individual levels are so close that they cannot be separated. Energy bands are created in the process for the individual energy levels in the electron orbitals (Figure 5.2). As explained above, only a limited number of electrons can inhabit these energy bands.

In the band model, the bands are filled up with electrons from the first one to the last in order. The first completely filled band is called the **valence band** VB. The next band up is the **conduction band** CB, which can be either partly filled or completely empty. The space between the valence band and the conduction band cannot contain energy states; this area is called the band gap E_g.

Depending on how the bands are arranged and how many electrons they contained, solids are divided into the categories of electric conductors, semiconductors, and non-conducting isolators (Figure 5.3). Conductors either have a conduction band only partly filled with electrons or overlapping conduction and valence bands. If the conduction band partly contains electrons, the electrons can move within the solid, therefore allowing electrons to be conducted. The specific electrical resistance of conductors is very low at $\rho < 10^{-5}\ \Omega$ m. Conductors are generally metallic materials.

Isolators have a large specific electrical resistance at $\rho > 10^{7}\ \Omega$ m. Their conduction band has no electrons, and the great band gap ($E_g \geq 5$ eV) means that electrons have a hard time moving from the valence band into the conduction band.

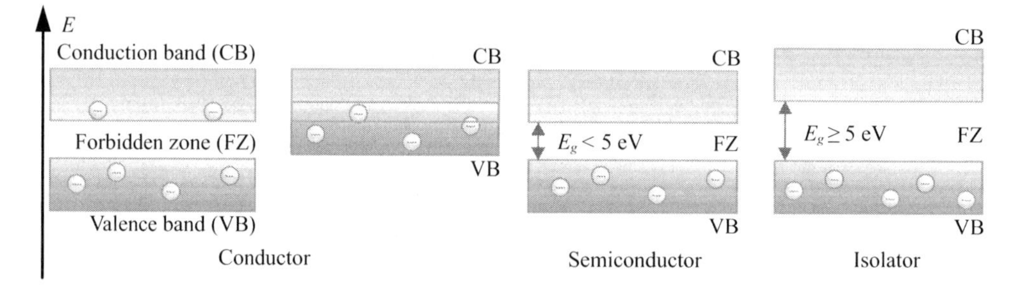

Figure 5.3 Energy bands for conductors, semiconductors, and isolators

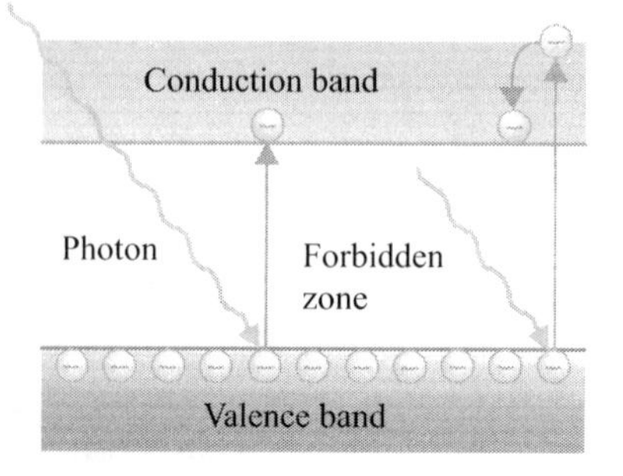

Figure 5.4 Photons moving electrons from the valence band into the conduction band (internal photo-electric effect)

For photovoltaics, **semiconductors** are crucial. Their conductivity ranges from $10^{-5}\,\Omega$ m to $10^{7}\,\Omega$ m. Like isolators, their conductance band has no electrons, but the relatively small band gap ($E_g < 5$ eV) means that solar energy can move electrons into the conduction band (Figure 5.4). When photons move electrons into the conduction band, we speak of the **internal photoelectric effect**.

If the photon's energy is smaller than the band gap, the electron cannot get through to the conduction band. If the photon's energy is too great, the electron does enter the conduction band, but part of the energy is lost because the electron falls back to the edge of the conduction band.

Technically, the internal photoelectric effect can be used for photoelectric resistance, which changes relative to irradiance, or to generate electricity – as in photovoltaics.

How a solar cell works

Photovoltaics also uses semiconductors. They generally have four electrons – called valence electrons – in the outermost electron shell. Elements from Group IV of the periodic table are the main semiconductors; examples include silicon (Si), germanium (Ge), and tin (Sn). Compounds of two elements from Group III and V (called III-V compounds) also have four valence electrons, as do II-VI compounds and combinations of various elements.

Table 5.2 Band gaps of various semiconductors at 300 K

IV semiconductors		*III-V semiconductors*		*II-VI semiconductors*	
Material	E_g	*Material*	E_g	*Material*	E_g
Si	1.107 eV	GaAs	1.35 eV	CdTe	1.44 eV
Ge	0.67 eV	InSb	0.165 eV	ZnSe	2.58 eV
Sn	0.08 eV	InP	1.27 eV	ZnTe	2.26 eV
		GaP	2.24 eV	HgSe	0.30 eV

(Data: [Lec92])

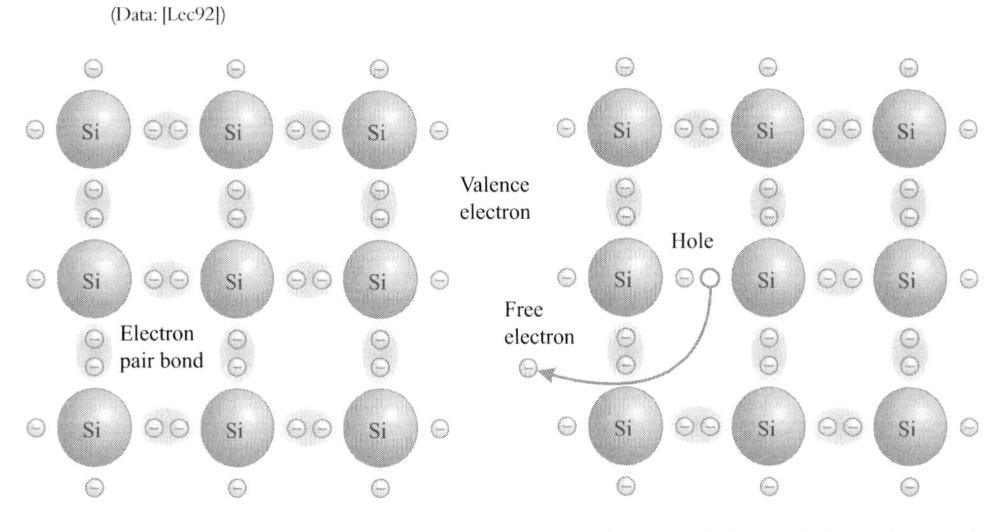

Figure 5.5 The crystal structure of silicon (left); intrinsic conductivity of electron holes in the crystal lattice (right)

An example of a III–V semiconductor is gallium arsenide (GaAs); a II–VI semiconductor is cadmium telluride (CdTe). Table 5.2 shows the various band gaps of a few semiconductors.

Silicon is commonly used in photovoltaics. It is the second most common element in the earth's crust after oxygen, but is usually only found in chemical bonds.

Silicon (Si) is a semiconductor from Group IV of the periodic table of elements; it has four valence electrons in the outermost shell. Two electrons from neighbouring atoms form a covalent bond in the silicon crystal lattice to form a stable electron configuration. In a way, the two atoms now share the electrons. The covalent bond with four neighbours gives silicon the stable electron configuration of the noble gas argon (Ar). In the band model, the valence band is full and the conduction band empty. Light and heat can lift an electron from the valence band into the conduction band. The electron is then free to move around in the crystal lattice. An electron hole is left behind in the valence band, as shown in Figure 5.5. These electron holes produce the semiconductor's **intrinsic conductivity**.

For each electron, there is an electron hole. In other words, there are always the same number of electrons as electron holes. We then speak of **electron density** n and **hole density** p:

$$n = p \tag{5.13}$$

The product of electron density and electron hole density is called the **intrinsic carrier density** n_i, and it is relative to absolute temperature T and band gap E_g:

$$n \cdot p = n_i^2 = n_{i0}^2 \cdot T^3 \cdot \exp(-\frac{E_g}{k \cdot T}) \tag{5.14}$$

The **Boltzmann constant** applies:

$$k = 1.380658 \cdot 10^{-23}\ \mathrm{J/K} \tag{5.15}$$

For silicon, $n_{i0} = 4.62 \cdot 10^{15}\ \mathrm{cm^{-3}\ K^{-3/2}}$. At absolute zero ($T = 0$ K), there are therefore no free electrons or holes, but their number increases quickly as temperatures rise.

If electric voltage is connected to the silicon lattice, negatively charged electrons flow towards the anode. Bound electrons next to a hole can switch places and enter the hole, which then moves in the opposite direction – when voltage is connected, it moves towards the cathode. The **mobility** μ_n and μ_p of the electrons and holes in the semiconductor also depends on the temperature. For silicon, μ_n and μ_p can be calculated with $\mu_{0n} = 1350\ \mathrm{cm^2/(V\ s)}$ and $\mu_{0p} = 480\ \mathrm{cm^2/(V\ s)}$ at $T_0 = 300$ K via:

$$\mu_n = \mu_{0n} \cdot \left(\frac{T}{T_0}\right)^{-3/2} \ (5.16) \text{ and } \mu_p = \mu_{0p} \cdot \left(\frac{T}{T_0}\right)^{-3/2} \tag{5.17}$$

The **electric conductivity**

$$\kappa = \frac{1}{\rho} = e \cdot (n \cdot \mu_n + p \cdot \mu_p) = e \cdot n_i \cdot (\mu_n + \mu_p) \tag{5.18}$$

of the semiconductor is the sum of the electron and hole current. At low temperatures, conductivity drops considerably. This effect is used to make temperature sensors for low temperatures.

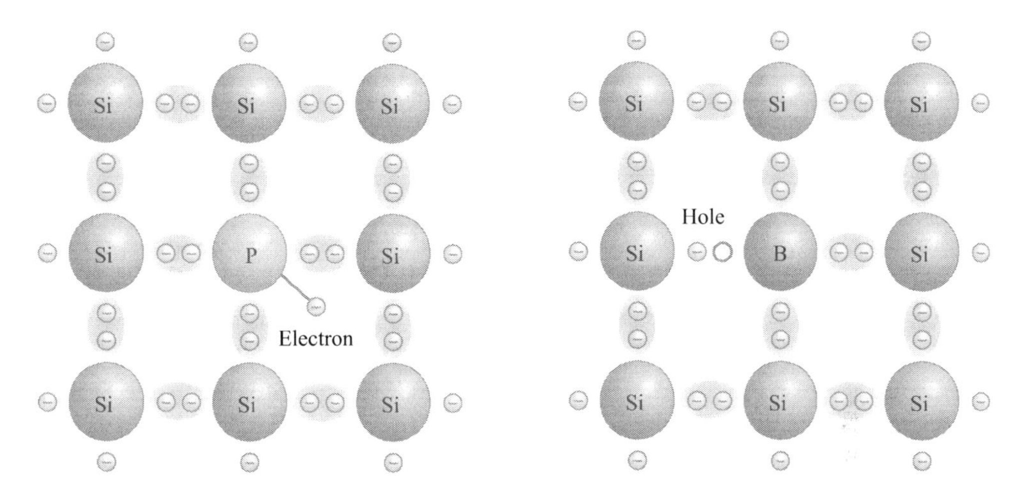

Figure 5.6 n- and p-doped silicon

Light also affects electrical conductivity. It is used to make light-sensitive photoelectric resistors. However, external electric voltage has to be applied in the process. Intrinsic conductivity cannot be used to generate electrical current. Here, another effect is used: **doping**.

Atoms from Group V, such as phosphorus (P) and antimony (Sb) have five valence electrons, unlike silicon. If these atoms are put into the silicon lattice, the fifth electron cannot bind with a neighbouring electron. Because that electron is only loosely bound, it takes very little energy to separate it from the atom compared with a properly bound electron. It is then available as a free electron. The addition of atoms from Group V is called n-doping. The foreign atoms are called **donors**.

In **n-doping, the density of free electrons is calculated**

$$n = \sqrt{\frac{n_D \cdot N_L}{2}} \cdot \exp\left(-\frac{E_D}{2 \cdot k \cdot T}\right) \tag{5.19}$$

from the density n_D of donor atoms, the effective density state N_L in the conduction band, and the donors' ionization energy E_D which is the energy needed to separate the electrons. In silicon at a temperature of $T = 300$ K, $N_L = 3.22 \cdot 10^{19}$ cm^{-3}, and with phosphorus atoms as donors $E_D = 0.044$ eV.

With n-doping, there are far more free electrons than electron holes, so the electrons are called the majority carriers. Electric conductivity is mainly based on the transport of electrons, and the semiconductor is n-conductive.

If atoms from Group III are used instead of from Group V – such as boron (B) and aluminium (Al) – with three valence electrons, a valence electron is then missing in the silicon lattice, so there is an extra electron hole. It then takes little energy E_A relative to an n-type conductor to separate the whole so it can 'move freely'. Electric conductivity is mainly based on the transport of positive charge carriers – the electron holes. By means of p-doping, the semiconductor becomes p-conductive. The foreign atoms are called **acceptors**.

The acceptor density n_A, the effective density state N_V in the valence band, and the ionization energy E_A ($N_V = 1.83 \cdot 10^{19}$ cm^{-3} for silicon at $T = 300$ K and $E_A = 0{,}045$ eV for boron) produces the following **density of extra holes** for a **p-type semiconductor**:

$$p = \sqrt{\frac{n_A \cdot N_V}{2}} \cdot \exp\left(-\frac{E_A}{2 \cdot k \cdot T}\right) \qquad (5.20)$$

If a p-type and an n-type semiconductor come into contact, a p–n **junction** is created. As explained above, the n-type semiconductor has excess free electrons; the p-type semiconductor, excess free electron holes. The electrons then move from the n-type semiconductor to the p-type semiconductor, whereas the holes move in the opposite direction (Figure 5.7).

An area with few free charge carriers results at the transitional zone. Where electrons have moved into the p-zone, they leave behind positively ionized doping atoms. The depletion region then has a positive charge. Where holes have diffused into the n-zone, they leave behind negatively iodized doping atoms, resulting in a negatively charged depletion region.

An electric field is thus created between the n-side and the p-side, and it opposes the motion of the charge carriers so that diffusion does not continue infinitely. The result is a **differential potential**.

$$U_d = \frac{k \cdot T}{e} \cdot \ln \frac{n_A \cdot n_D}{n_i^2} \qquad (5.21)$$

Because of the charge neutrality, the widths d_n and d_p of the depletion regions in each semiconductor zone are:

$$d_n \cdot n_D = d_p \cdot n_A \qquad (5.22)$$

The following applies for the **total width of the depletion zone**:

$$d = d_n + d_p = \sqrt{\frac{2 \cdot \varepsilon_r \cdot \varepsilon_0 \cdot U_d}{e} \cdot \frac{n_A + n_D}{n_A \cdot n_D}} \qquad (5.23)$$

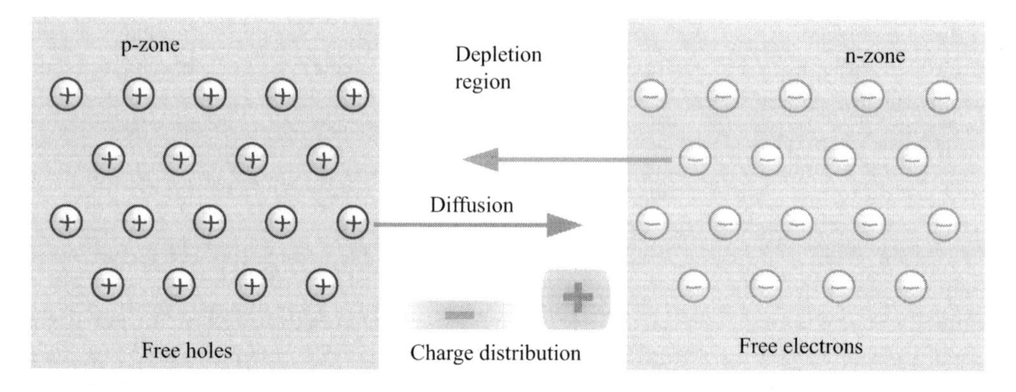

Figure 5.7 Depletion region at the p–n junction caused by the diffusion of electrons and electron holes

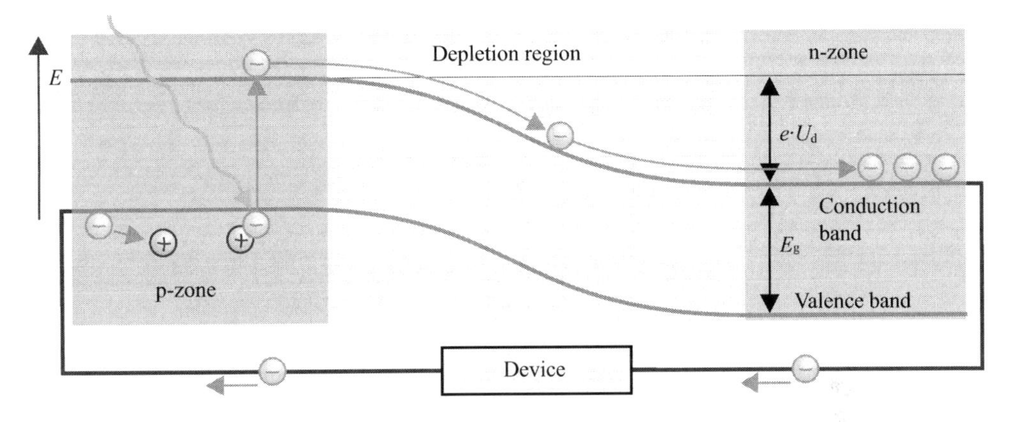

Figure 5.8 How a solar cell works in the energy band model

For silicon with a hole concentration of $n_D = 2{\cdot}10^{16}$ cm^{-3} and $n_A = 1{\cdot}10^{16}$ cm^{-3} at a temperature of $T = 300$ K, the differential potential is $U_d = 0.73$ V. At $\varepsilon_r = 11.8$, $d_n = 0.13$ µm and $d_p = 0.25$ µm.

If photons then move electrons from the valence band into the conduction band within the depletion zone, thereby separating them from the atom, they then move through the electric field into the n-side, and the holes created in the process move into the p-side. In the energy band model, the process can be illustrated by bending the bands into the depletion zone. The current is closed if connected to a device that consumes electricity (Figure 5.8).

As explained above, a solar cell only uses part of the energy from the photons. If the energy from the photons is less than the band gap, their energy will not suffice to move an electron from the valence band into the conduction band. This is the case at wavelengths greater than:

$$\lambda_{\mathrm{max}} = \frac{h \cdot c}{E_g} = \frac{1{,}24\ \mu\mathrm{m\ eV}}{E_g} \tag{5.24}$$

But at wavelengths near the band gap, not all of the incident irradiance is converted into electric energy. Part of the solar radiation is reflected, while another part passes through the semiconductor unused. In addition, electrons can be recombined with electron holes within the semiconductor, meaning that they fall back into the valence band before they reach the point of power consumption. Figure 5.9 shows these events.

Likewise, only part of the energy from radiation at short wavelengths with high energy is used because the energy that exceeds the band gap is lost when electrons fall back into the edge of the conduction band. The amount of energy that can be used therefore greatly depends upon the wavelength of incident light. The part that can be used is expressed as the external quantum efficiency $\eta_{\mathrm{ext}}(\lambda)$ relative to the wavelength. This efficiency indicates how many of the incident photons within a given wavelength actually produce current.

As explained above, solar cells are only able to optically absorb part of the photons. Because of reflection and transmission, some of the photons remain unused. The degree of absorption $\alpha(\lambda)$, reflection $\rho(\lambda)$, and transmission $\tau(\lambda)$ can be used along with the external quantum efficiency η_{ext} to calculate the internal quantum efficiency.

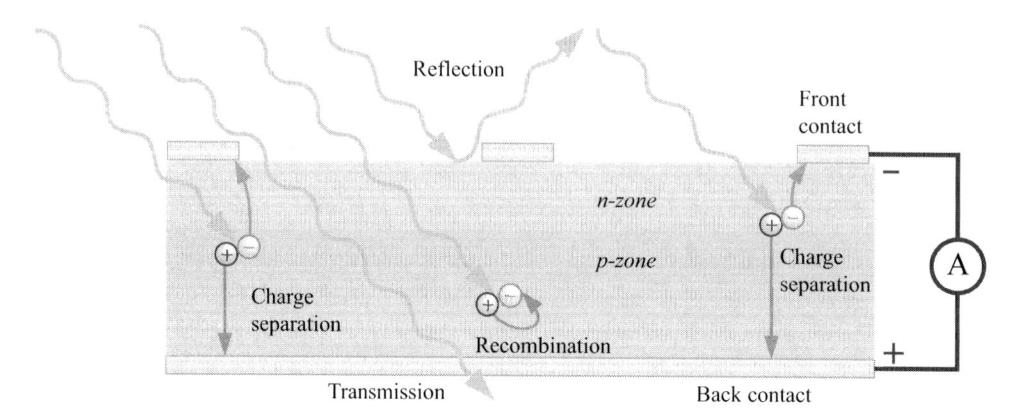

Figure 5.9 Events within a solar cell exposed to sunlight

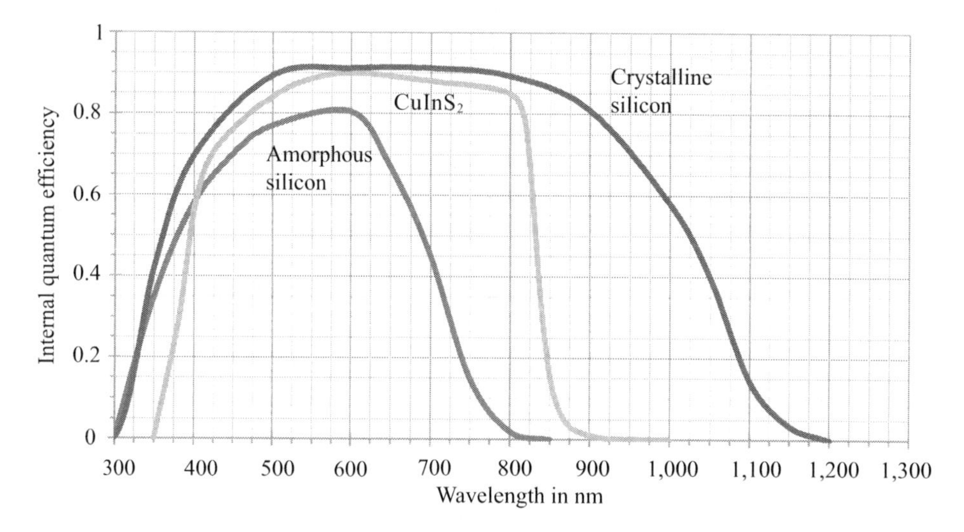

Figure 5.10 Typical curve for the internal quantum efficiency of various solar cell types

$$\eta_{\text{int}}(\lambda) = \frac{\eta_{\text{ext}}(\lambda)}{\alpha(\lambda)} = \frac{\eta_{\text{ext}}(\lambda)}{1-\rho(\lambda)-\tau(\lambda)} \tag{5.25}$$

Figure 5.10 shows the typical **internal quantum efficiency** for various solar cells. Depending on the solar cell's specific design and quality, its characteristics may differ within the same type. The higher and wider the curve of quantum efficiency is, the greater the solar cell's efficiency.

Spectral sensitivity is another major parameter:

$$S(\lambda) = \frac{e \cdot \lambda}{h \cdot c} \cdot \eta_{\text{ext}}(\lambda) = \frac{\lambda}{1{,}24\ \mu\text{m}} \cdot \frac{\text{A}}{\text{W}} \cdot \eta_{\text{ext}}(\lambda) \tag{5.26}$$

It is calculated from external quantum efficiency η_{ext} and wavelength λ. Figure 5.11 shows spectral sensitivity S relative to wavelength λ. It is relatively easy to measure spectral

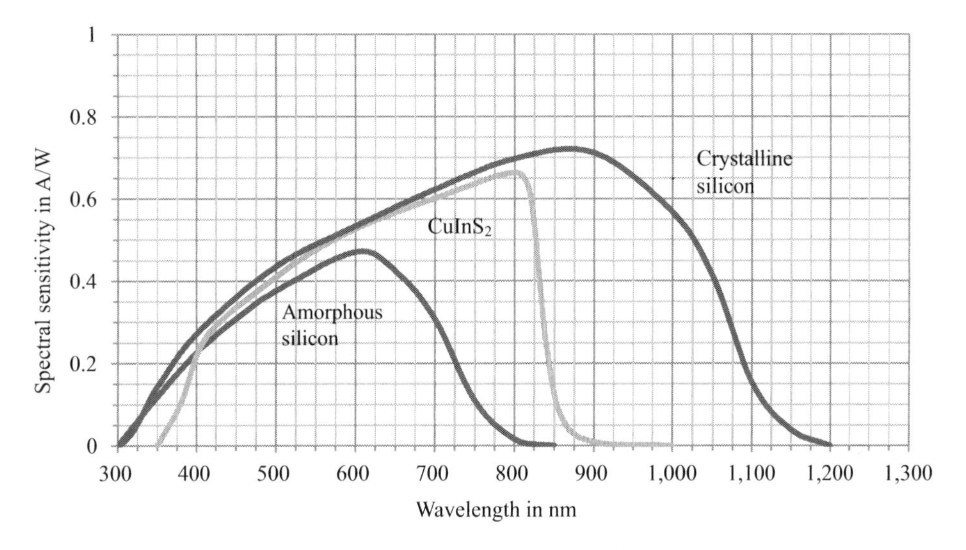

Figure 5.11 Typical curve of a solar cell's spectral sensitivity for various solar cell types

sensitivity. A solar cell with area A is exposed to a known monochromatic irradiance $E(\lambda)$, and the short circuit current $I_K(\lambda)$ is measured:

$$S(\lambda) = \frac{I_K(\lambda)}{A \cdot E(\lambda)} \tag{5.27}$$

Often, relative spectral sensitivity

$$S_{rel}(\lambda) = \frac{S(\lambda)}{S_{max}} \tag{5.28}$$

is indicated at a value between 0 and 1. It is based on spectral sensitivity $S(\lambda)$ and maximum spectral sensitivity S_{max}.

If both ends of the solar cell are short-circuited, the short circuit current or, approximately, the **photoelectric current** I_{Ph} starts flowing. This current can be calculated based on the solar cell's area A, spectral sensitivity S, and the spectrum of sunlight (such as the AM1.5g spectrum mentioned in Chapter 2):

$$I_{Ph} = \int S(\lambda) \cdot E(\lambda) \cdot A \cdot d\lambda \tag{5.29}$$

The share of irradiance E in incident irradiance E_0, absorbed by the semiconductor depends on semiconductor's thickness d and the material's **absorption coefficient** α:

$$E = E_0 \cdot (1 - \exp(-\alpha \cdot d)) \tag{5.30}$$

A distinction is made between direct and indirect semiconductors. The absorption coefficient of indirect semiconductors, such as silicon, is much lower. In indirect semiconductors, the photons also need to transmit crystal momentum to the electron in addition to the energy required. The electrons thus have a harder time crossing the band gap in indirect semiconductors.

For the direct semiconductor GaAs, the absorption coefficient α(GaAs) ≈ 630 mm^{-1}, but the value drops for silicon to α(Si) ≈ 7.2 mm^{-1} for radiation at a wavelength of around 1 μm. If both semiconductors are to absorb the same share of sunlight, the silicon semiconductor has to be 87.5 times thicker than the GaAs. For more exact calculations, the absorption coefficient has to be calculated relative to wavelengths. For crystalline silicon solar cells to have high absorption, the crystal has to be at least 200 μm thick. Light trapping – when the light reflects inside the material – extends the path of light, so the material can be thinner.

There are also other effects in photovoltaics and other cell technologies, such as MIS cells, not discussed here; see the bibliography ([Wag07, Goe05, Lew01]).

Manufacturing solar cells and solar modules

Solar cells made of crystalline silicon

Various semiconductor materials can be used to make solar cells. The most common one is silicon. Below, we therefore only discuss how to make solar cells from silicon.

Silicon is the second most common element on earth after oxygen; it occurs, for instance, in silicon dioxide (SiO_2), commonly known as sand. To get silicon out of silicon dioxide, temperatures of around 1800 °C are needed for the following reduction process:

$$SiO_2 + 2C \xrightarrow{1800\,^{\circ}C} Si + 2CO \tag{5.31}$$

The product is called **metallurgical silicon** (MG-Si), which is still has a lot of impurities at a purity of 98 %. In an aluminothermic reduction, silicon can also be produced:

$$3SiO_2 + 4Al \xrightarrow{1100\,^{\circ}C \ldots 1200\,^{\circ}C} 2Al_2O_3 + 3Si \tag{5.32}$$

But this process also produces silicon with great impurities. To make semiconductors for the computer industry, electronic-grade silicon (EG-Si) with impurities of only 10^{-10} % is required. The requirements are not quite so high for solar-grade silicon (SOG-Si), though purification processes are also needed.

In the Siemens process, hydrogen chloride (HCl) is added to silicon. In an exothermic reaction, trichlorosilane ($SiHCl_3$) and hydrogen (H_2) are created:

$$Si + 3HCl \longrightarrow SiHCl_3 + H_2 \tag{5.33}$$

Triclorosilane is fluid at 30 °C, and distillation systems remove impurities from it. The silicon is then recovered by means of **chemical vapour deposition** (CVD). At temperatures exceeding 1,000 °C, silicon is deposited on a thin silicon ingot inside the reactor when the trichlorosilane is put into the reactor along with highly pure hydrogen:

$$4SiHCl_3 + 2H_2 \xrightarrow{1200\,^{\circ}C} 3Si + SiCl_4 + 8HCl \tag{5.34}$$

The result is highly pure silicon ingots up to 30 cm thick and 2 m long. These ingots can then be used to make **polycrystalline solar cells**. A lot of manufacturers melt down the silicon once again and pour it into square blocks. The crystals of polycrystalline silicon are irregularly arranged. These crystals meet at 'grain boundaries', where losses occur within the solar cell.

One recent poll appointed research has been on directly cleaning metallurgical silicon during manufacture without the relatively complex silane process. Called 'upgraded metallurgical silicon' (UMG-Si), this product does not have the purity of electronics-grade silicon from the silane process. It is therefore also called 'dirty silicon'. Nonetheless, UMG silicon can be used to make solar cells with quite good efficiencies.

Table 5.3 provides an overview of the abbreviations already used and a few others to come so you can keep an overview.

To increase the efficiency of solar cells, **monocrystalline material** can be made from polycrystalline silicon. Two methods are used: seed crystal is pulled out of molten silicon in the Czochralski process; the other process is called zone melting. The seed crystal is dipped into molten polycrystalline silicon to produce the desired monocrystalline precursor material. The grain boundaries disappear, thereby reducing losses within the cell.

The pure polycrystalline and monocrystalline rods and blocks are called 'ingots'. The crystalline silicon ingots are then cut, usually by wire saws, into wafers some 150 μm to 250 μm thick. Some 30 to 50 % of the material is lost as waste in the process. Further technical developments have managed to reduce the thickness of the wafers over the past few years, thereby reducing the amount of material needed – and lowering costs. The silicon waste produced during the sawing of wafers combines with grains of silicon carbide and oil or glycol (which are used in the sawing process) to produce **slurry**. Until recently, the recovery of pure silicon from this slurry was prohibitively expensive. Now, recycling processes are affordable.

The alternative to a wire saw is zone melting, which used to be used commercially but is now rare. In the **string ribbon process** (SR process) silicon is first melted. In this process, two thin ribbons are pulled through the molten silicon. The molten silicon forms a thin film between the two ribbons similar to bubbles of soap and hardens.

The **EFG process** (edge-defined film-fed growth) also starts with molten silicon. But here, an octagonal mold with a thin gap inside it is dipped into the molten silicon. Capillary forces pushed the silicon into the gap. A foil is lowered onto this silicon, which then forms

Table 5.3 Overview of common abbreviations related to silicon

Abbreviation	*Full name*	*Note*
Si	Silicon	General term
a-Si, α-Si	Amorphous silicon	Used in thin-film photovoltaics
Cz-Si	Czochralski silicon	Mono-Si from the Czochralski process
EG-Si	Electronic-grade silicon	For computer processors > 99.999999999 %
FZ-Si	Float-zone silicon	Mono-Si
EFG-Si	Edge-defined film-fedgrowth Si	Octagonal silicon
MG-Si	Metallurgical-grade silicon	Purity > 98 %
Mono-Si	Monocrystalline silicon	Generally black in appearance
Poly-Si	Polycrystalline silicon	The most common type on the market
μc-Si	Microcrystalline silicon	Niche product
PVG-Si	Photovoltaic-grade silicon	Essentially the same as solar silicon
SOG-Si	Solar-grade silicon	Solar silicon, purity > 99.99 %
SR-Si	String ribbon silicon	Created by growth along a ribbon
UMG-Si	Upgraded MG-Si	Purity > 99.99 %

Figure 5.12 Polycrystalline silicon for solar cells. Left: raw silicon; middle: silicon ingots; right: silicon wafers (photos: PV Crystalox Solar plc.)

a fluid film on the foil. The foil is lifted, and the fluid silicon solidifies on its underside. The result is a long octagon that can be separated at the edges to make silicon wafers.

Whether the sawing or the pulling process is used, the result is always silicon **wafers**. In various stages, the wafers then have to be cleaned, doped, and have contacts added to them. Any damage caused during the sawing process is first removed with hydrofluoric acid. Otherwise, the wafers would break too easily in further processing.

Atoms of phosphorus or boron are added to dope the silicon for the p–n transition. Here, the vapour deposition process is used. At temperatures between 800 °C and 1200 °C, gaseous dopants are combined with a carrier gas, such as nitrogen (N_2) or oxygen (O_2). This gas then flows over the wafers, and the dopant atoms are deposited on the silicon semiconductor relative to the gas mixture, the temperature, and the flow rate. The doped semiconductor's surface is then cleaned by means of etching.

A screen printing process then adds contacts to the front and back of the cell. Metal and alloys based on aluminium or silver are used as material for the contacts. On the back, the contact usually covers the entire area, while the front has thin contact fingers so that as little of the solar cell as possible is shaded.

Finally, an **antireflection coating** is added, giving the solar cell its characteristic blue colour. Metallic silicon reflects light well, but the coaching reduces the reflection of light dramatically. Generally, a roughly 70 nm thick layer of silicon nitride (Si_3N_4) is used. Silicon dioxide (SiO_2) and titanium dioxide (TiO_2) can also be used as antireflection layers. It is now also possible to make antireflection coatings in other colours than blue so that photovoltaic modules can be better integrated aesthetically in buildings. Figure 5.13 shows a cross-section of a crystalline solar cell.

There are various ways to increase a solar cell's efficiency further. For instance, the solar cell's surface can have microscopic pyramids added to it in order to increase the surface so that more sunlight can be absorbed. The front contacts can also be buried. Lasers were used to give BP's Saturn cell its **laser grooved buried contacts**, **LGBC**). The lasers cut small grooves into the front of the cell, into which the front contacts are chemically deposited.

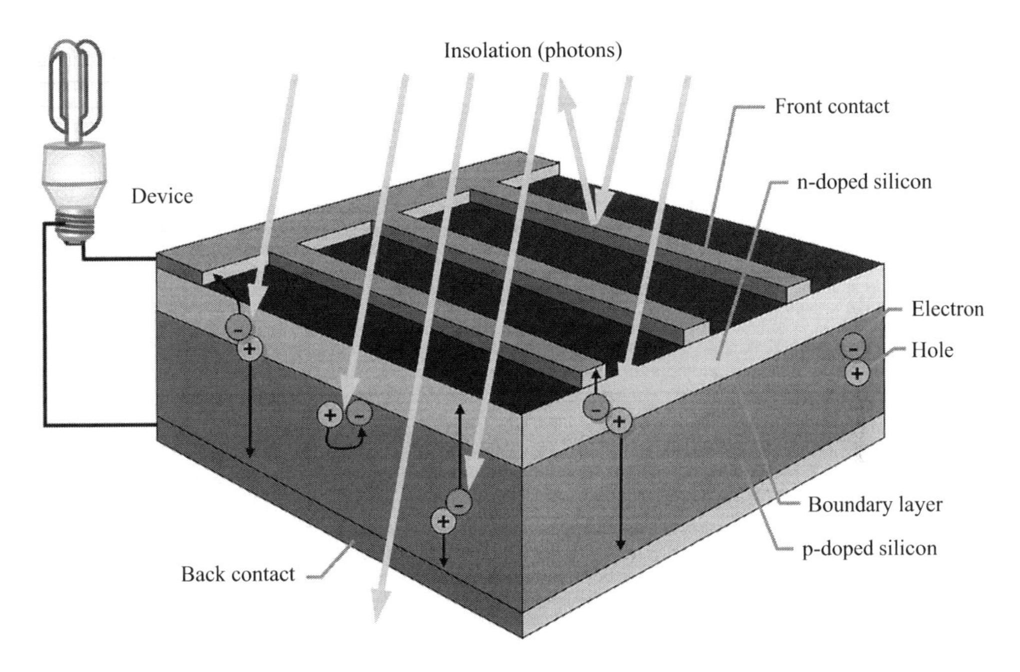

Figure 5.13 The structure of a crystalline solar cell

These 'buried' contacts are only an eighth as large, but deeper. As a result, reflection losses from the front contacts are reduced.

Reflections are reduced even further if the front contacts are moved to the back. **Back-contact solar cells** have both contacts on the back. They are electrically isolated from each other.

If monocrystalline solar cell material is combined with amorphous silicon, the results can also be very high efficiency. The **HIT cell** (HIT: heterojunction with intrinsic thin-layer) is made by Sanyo.

All of these efforts increase the solar cell's efficiency. In serial production, efficiencies can exceed 20 %. But the extra effort also makes manufacturing more complicated.

Solar modules made of crystalline cells

Without protection from the elements, individual solar cells would quickly be destroyed, so they are combined in **modules**. A crystalline solar cell typically has an edge length of 10 cm to 20 cm. Solar modules used in battery systems generally contain 32 to 40 cells. There are, however, other module sizes with far fewer or far more cells for grid-connected systems.

As with solar thermal flat-plate collectors, low-iron panes of glass are used as covers. The back generally also has either a pane of glass or plastic foil (usually Tedlar). The solar cells are embedded in plastic between the cover and the rear foil or rear glass pane. Generally, EVA (ethylene vinyl acetate) hardened at temperatures up to 150 °C at low pressure for 10 to 15 minutes is used. The process is called **lamination**. To prevent the glass from breaking and facilitate installation, the modules are then put into a frame. Connections are usually made with sockets and plugs. Bypass diodes are also installed to provide protection from

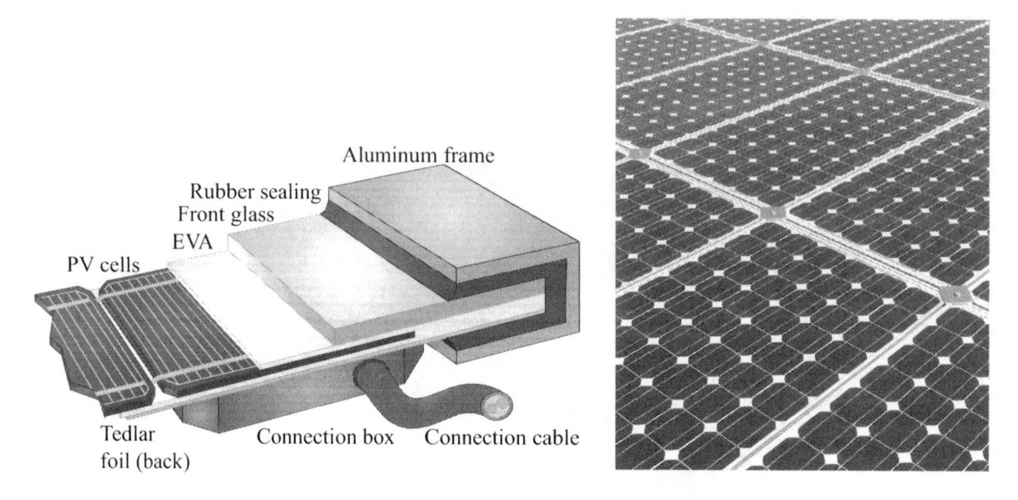

Figure 5.14 Left: the basic structure of a solar module; right: installed solar modules

dangerous operating conditions. The section 'Technical data for solar modules' later in this chapter provides the technical data for a few solar modules.

Solar cells made of amorphous silicon

In addition to cells made of crystalline silicon, thin-film cells are also now common. Such cells can either be made of amorphous silicon or other materials, such as cadmium telluride (CdTe) or copper-indium-diselenide (CIS). Thin-film cells have the benefit of making do with very little material, and they are expected to be very affordable. For these reasons, thin-films cells are held to have great development potential. The efficiency of thin-films cells is, however, currently far below that of crystalline silicon cells. In other words, you need more modules taking up more space and entailing greater insulation costs for the same output.

Solar cells made of amorphous silicon have a carrier substrate, which is generally glass. A transparent conducting oxide (**TCO**), such as tin oxide, is sprayed onto the glass. A laser then cuts this surface into strips for the later application of front contacts. Silicon and the dopant are then applied to the substrate by means of vapour deposition at high temperatures. First, a 10 nm p-layer is applied, followed by 10 nm of buffer layer. Then comes an intrinsic layer of 500 nm amorphous silicon, followed by a 20 nm n-layer. The rear contacts made of aluminium powder are then screen-printed on, and the cells are embedded in a polymer layer for protection. Figure 5.15 shows the basic structure of an amorphous silicon cell.

When the silicon is applied by means of vapour deposition, its crystalline structure is lost, so the atoms are amorphous. Amorphous silicon solar cells can be roughly 100 times thinner than crystalline cells because the former has a much higher absorption coefficient than the latter. The band gap of amorphous silicon is a bit greater at 1.7 eV. A lot less material is needed, and manufacturing can be streamlined. In the lab, efficiencies of up to 12 % have been reached. Nonetheless, solar cells made of amorphous silicon are mainly used for small applications, such as pocket calculators and watches, because their efficiency is far lower than that of crystalline cells at 6–7 %, compared with 20 %, respectively. In addition, amorphous solar cells undergo a degradation process that reduces efficiency by 1–2 % in the first few months of use before it finally stabilizes at a lower level.

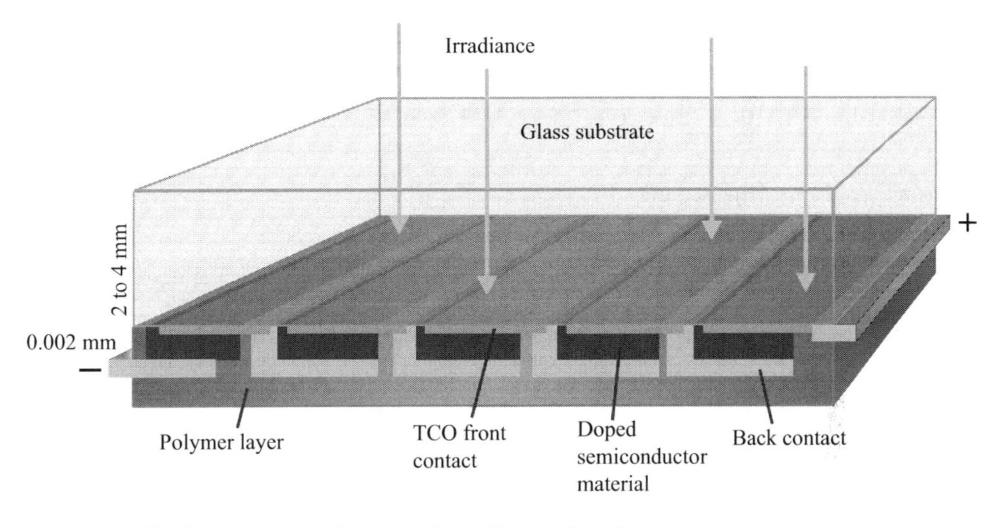

Figure 5.15 The basic structure of an amorphous silicon solar cell

Solar cells made of other materials

Increasingly, other thin-film materials are used instead of amorphous silicon. A very wide range of materials and technologies are also being tested. The following materials are now used in serial production:

- cadmium telluride (CdTe)
- cadmium sulphide (CdS)
- copper-indium-sulphide (CIS)
- copper-indium-diselenide ($CuInSe_2$ or CIS)
- copper-indium-gallium-diselenide/sulphide (CIGS).

The manufacturing process is similar to the one for amorphous silicon solar cells. While global silicon reserves are practically inexhaustible, other materials, such as tellurium, are only available to a limited extent. CdTe and CIS can reach higher efficiencies than amorphous silicon can. Nonetheless, the efficiency of thin-film solar cells is noticeably lower than that of crystalline solar cells. Because more module surface is therefore needed to produce the same amount of power, installation costs are therefore higher. On the other hand, less material is required and production processes can be streamlined, so thin-film cells are less expensive than crystalline cells to manufacture. It is too soon to tell which option will prevail.

Thin-film cells can also be made of polycrystalline silicon. **Microcrystalline cells** consist of extremely small polycrystalline areas on a substrate. Microcrystalline and amorphous cell material is combined to produce micromorph solar cells.

Other promising technological developments include solar cells containing pigment based on titanium dioxide (TiO_2) and organic solar cells. Both of these technologies are, however, still in the research stage. The main progress needs to be made in efficiency and the long-term stability of the cells.

In space, other materials, such as gallium arsenide (GaAs), are used. GaAs is especially resistant to cosmic radiation and has high efficiencies exceeding 20 %. It is also too expensive to use on earth. In contrast, more expensive materials can be used in concentrating photovoltaics because these cells are so small. Tandem and triple cells combine two or three materials, respectively, of different band gaps. The uppermost layer of cell material absorbs shortwave light. Long waves pass through this layer and are absorbed by those underneath. In this way, the spectrum of light is better exploited, and efficiency levels can exceed 40 %.

Module tests and quality assurance

Under real-world conditions, solar modules are designed to run for at least 20 to 30 years. Various test methods have been developed to check whether a serially manufactured solar module is likely to run properly for that long. IEC 61215 [DIN05b] and IEC 61646 [DIN08] are two standards that specify comprehensive test procedures for crystalline and thin-film solar modules. First, the manufacturer sends eight samples to a certified test lab, which conducts the following tests:

- visual inspection and output measurement
- wet insulation resistance
- long-term test under outdoor conditions (60 kWh/m²)
- long-term hotspot test with cell shading
- temperature test of bypass diodes
- UV pretreatment test (15 kWh/m²)
- damp humidity test (10 cycles from –40 °C to +85 °C at 85 % rel. humidity)
- thermal cycling (50 and 200 cycles from –40 °C to +85 °C)
- damp heat test (1000 h at 85 °C and 85 % rel. humidity)
- hail impact (25 mm balls of ice shot at a velocity of 23 m/s)
- static load test and resistance test of connections.

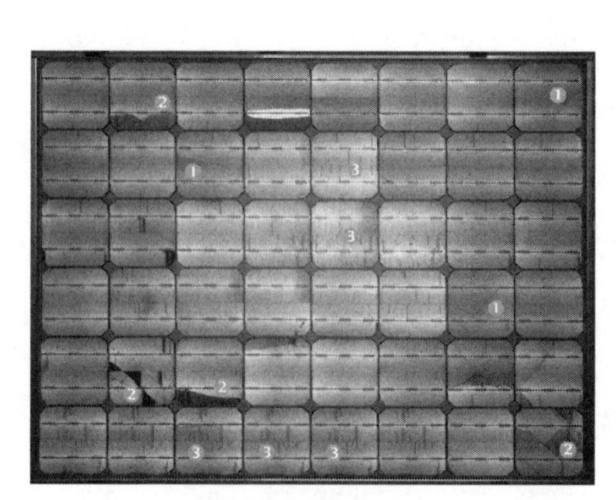

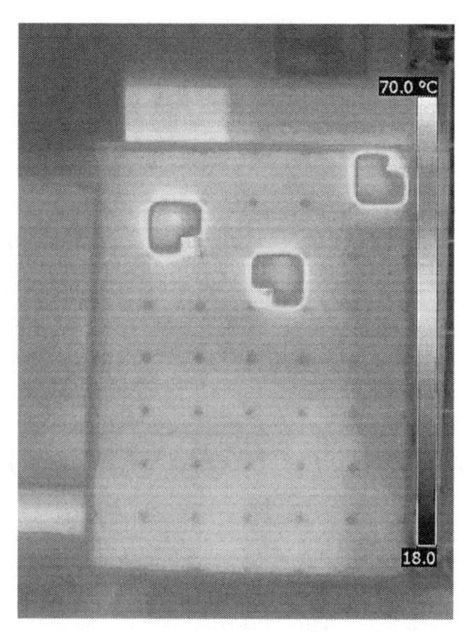

Figure 5.16 Left: an image of electroluminescence reveals damaged cells (1: micro-fissures, 2: broken cells, 3: contact degradation/flawed screen printing). Right: thermography of a module with three partly shaded solar cells.

Photos: Oliver Suchaneck, HTW Berlin

If all of the tests are passed, the modules are considered reliable for the long term. Keep in mind, however, that this standard merely concerns type certification. Quality can easily fluctuate during serial production. Subsequent sample tests therefore make sense. If the modules are to be built in special locations, such as close to the sea or on farms, additional tests, such as salt mist tests and tests of ammonia resistance, can help detect whether problems might occur in advance.

Other test methods provide additional information about production quality and possible flaws. A **lamination test** can reveal whether the lamination process went optimally or whether the EVA foil can be expected to age prematurely. **Electroluminescence (EL)** can reveal cell flaws within the module. Here, the module is connected to an external power source in the dark. The solar cells then emit infrared radiation, which can be recorded with a special camera. The images show great differences in brightness that cannot be seen with the naked eye, indicating differences in cell production, poor contacts, and micro-fissures. **Thermography** is also useful. Here, an infrared camera visualizes temperature differences while a PV system is running. Shaded cells and individual cells with low outputs heat up, thereby reducing power production; in the worst case, the entire system fails. Flawed contacts and improperly connected modules can also be detected in this way.

Electrical description of solar cells

Simple equivalent circuit

A solar cell has the same physical structure as a **diode**. It consists of an n-doped and a p-doped semiconductor with a depletion region, meaning that a solar cell not exposed to sunlight acts quite like a diode and can be described as one in simplified fashion.

The saturation current in reverse bias I_S and diode factor m are used in the following formula for cell current I relative to cell voltage (here $U = U_D$):

$$I = -I_D = -I_S \cdot \left(\exp\left(\frac{U_D}{m \cdot U_T}\right) - 1\right) \quad (5.35)$$

The temperature factor U_T at a temperature of 25 °C is $U_T = 25.7$ mV. The saturation current I_S is roughly 10^{-10} A. The diode factor m is 1 in an ideal diode. If the diode factor is chosen between 1 and 5, we can better describe a solar cell.

When a solar cell is exposed to sunlight, a power source can be connected in parallel to the diode in the simplified equivalent circuit. The power source produces photocurrent I_{Ph}, which is relative to irradiance E via the coefficient c_0:

$$I_{Ph} = c_0 \cdot E \quad (5.36)$$

Kirchhoff's first law gives us the equation for the solar cell's **current–voltage characteristic** in the simplified equivalent circuit, as shown in Figure 5.17. Figure 5.18 shows the curve at different levels of irradiance.

$$I = I_{Ph} - I_D = I_{Ph} - I_S \cdot \left(\exp\left(\frac{U}{m \cdot U_T}\right) - 1\right) \quad (5.37)$$

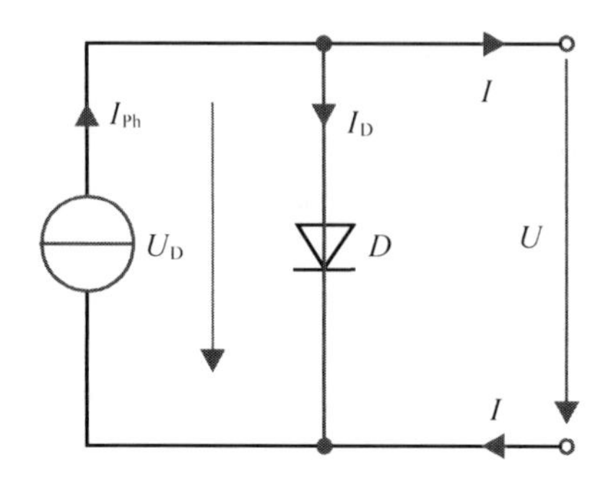

Figure 5.17 A solar cell's simplified equivalent circuit

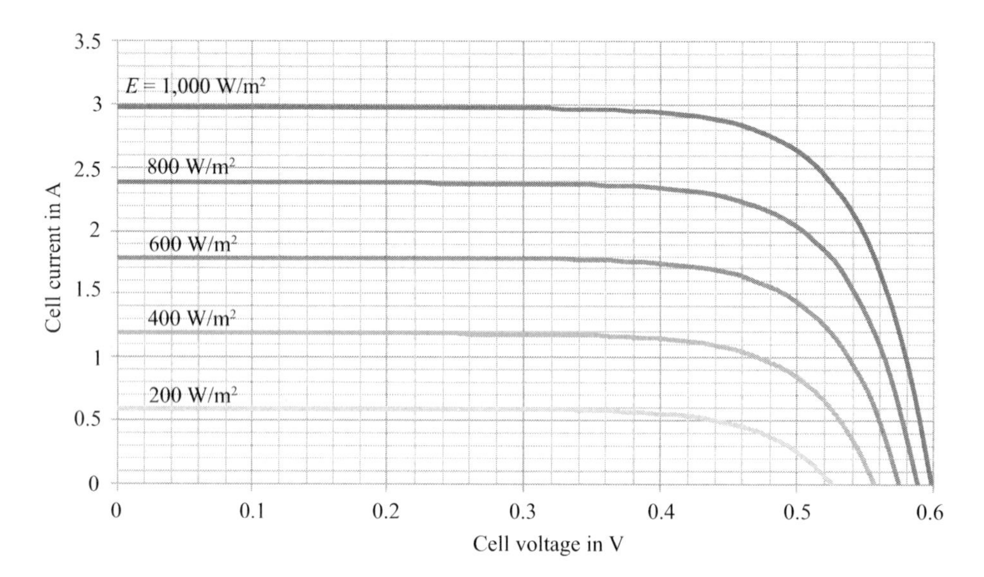

Figure 5.18 Influence of irradiance E on a solar cell's current-voltage characteristic

Expanded equivalent circuit (single-diode model)

For most applications, the simple equivalent circuit suffices to describe solar cells. The current-voltage characteristic only differs from actual values by a few percent. For a more exact description, however, the simple equivalent circuit has to become more sophisticated. In a real solar cell, voltage drops on the path that charge carriers take from the semiconductor to the external contacts. Series resistance R_S causes this reduction in voltage. Leak current along the edge of the solar cell can also be observed and described as parallel resistance R_P. Figure 5.19 shows both types of resistance in an equivalent circuit.

In real cells, series resistance R_S comes in at a few milliohms, while parallel resistance R_P is generally much greater than 10 Ω. Figures 5.20 and 5.21 show the influence of these two types of resistance on the current-voltage characteristic, also called the I–V curve.

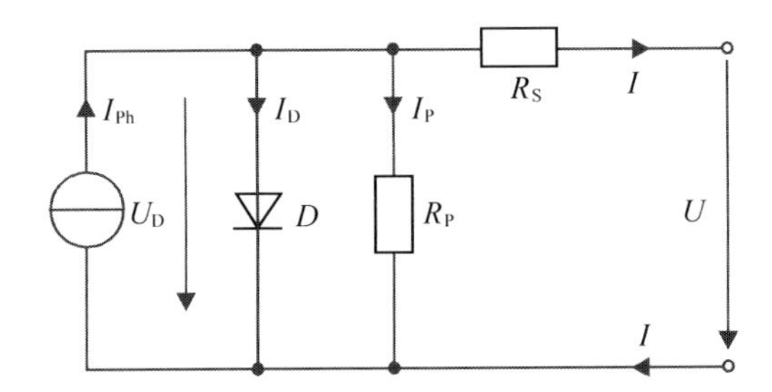

Figure 5.19 Expanded equivalent circuit for a solar cell (single-diode model)

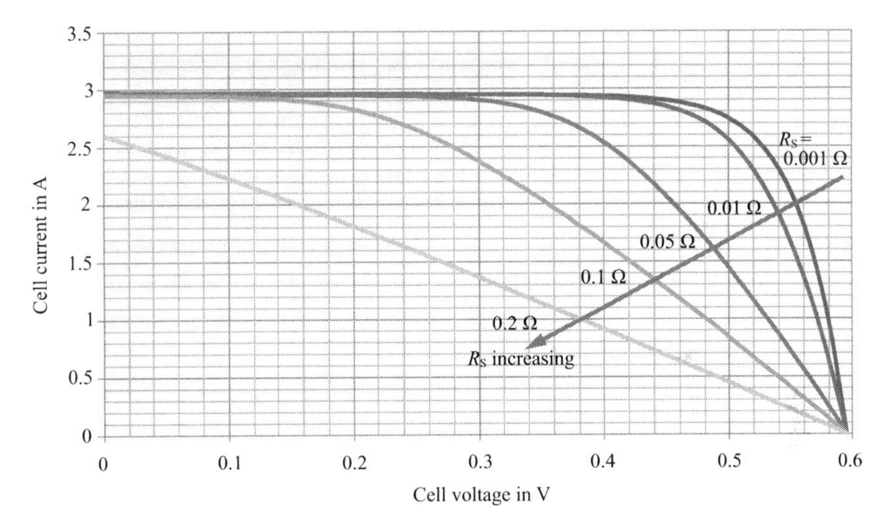

Figure 5.20 Influence of series resistance R_S on a solar cell's current–voltage characteristic

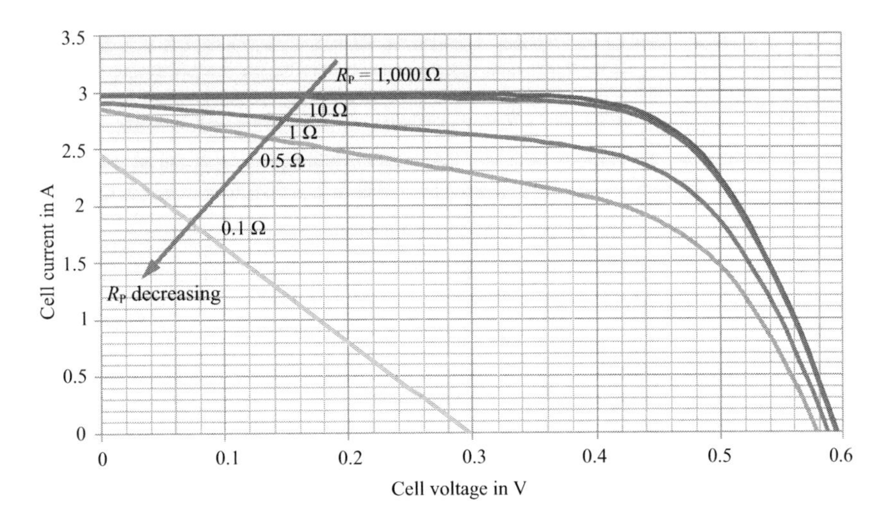

Figure 5.21 Influence of parallel resistance R_p on a solar cell's current–voltage characteristic

Kirchhoff's node law $0 = I_{Ph} - I_D - I_P - I$ is used with the formula $I_P = \frac{U_D}{R_P} = \frac{U + I \cdot R_S}{R_P}$ for the I–V curve of the solar cell's expanded equivalent circuit:

$$0 = I_{Ph} - I_S \cdot \left(\exp\left(\frac{U + I \cdot R_S}{m \cdot U_T} \right) - 1 \right) - \frac{U + I \cdot R_S}{R_P} - I \tag{5.38}$$

However, this implicit formula cannot be broken down into I and U as easily as the formula for the simplified equivalent circuit could (5.37). Rather, we need numeric methods.

One of the most common options is the Newton Method to find the root of a function. A solar cell's I–V curve can be expressed in a closed formula:

$$f(U,I) = 0. \tag{5.39}$$

The current I of a given voltage U_V and the voltage U of a given current I_V are to be determined. The solution is the root of the function $f(U,I)$. To reach this solution, the following formulae are calculated in an iterative process:

$$U_{i+1} = U_i - \frac{f(U_i, I_V)}{\frac{df(U_i, I_V)}{dU}} \quad \text{and} \quad I_{i+1} = I_i - \frac{f(U_V, I_i)}{\frac{df(U_V, I_i)}{dI}} \tag{5.40}$$

Starting from U_0 and I_0, a solution for the implicit formula for a given current I_V or voltage U_V can be found by repeating the iteration until the difference between the iterated steps falls below a previously selected value ε. The iteration is discontinued when: $|U_i - U_{i-1}| < \varepsilon$ or $|I_i - I_{i-1}| < \varepsilon$.

The Newton Method quickly produces convergence depending on the starting values U_0 and I_0. For diode breakdowns, pre-iteration with another method can be useful.

The iteration to determine a solar cell's current I met a given voltage U_V is as follows in accordance with 5.38:

$$I_{i+1} = I_i - \frac{I_{Ph} - I_S \cdot \left(\exp\left(\frac{U_V + I_i \cdot R_S}{m \cdot U_T} \right) - 1 \right) - \frac{U_V + I_i \cdot R_S}{R_P} - I_i}{-\frac{I_S \cdot R_S}{m \cdot U_T} \cdot \exp\left(\frac{U_V + I_i \cdot R_S}{m \cdot U_T} \right) - \frac{R_S}{R_P} - 1} \tag{5.41}$$

Two-diode model

The two-diode model is an even better way of describing a solar cell (Figure 5.22). Here, a second diode is parallel to the first. Each has a different saturation current and diode factor. For crystalline silicon solar cells, this model provides a nearly optimal description. It is, however, only of limited use for thin-film solar cells. The deviations are especially great under partial loads.

The implicit formula for the two-diode model is:

$$0 = I_{Ph} - I_{S1} \cdot \left(\exp\left(\frac{U + I \cdot R_S}{m_1 \cdot U_T} \right) - 1 \right) - I_{S2} \cdot \left(\exp\left(\frac{U + I \cdot R_S}{m_2 \cdot U_T} \right) - 1 \right) - \frac{U + I \cdot R_S}{R_P} - I \tag{5.42}$$

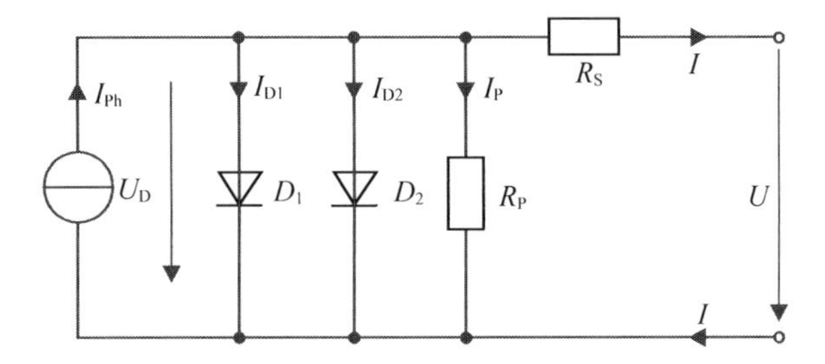

Figure 5.22 Two-diode model of a solar cell

Table 5.4 Parameters for various PV modules for the two-diode model [PRE94]

Parameter	c_0	I_{S1}	I_{S2}	m_1	m_2	R_S	R_P
Unit	m²/V	nA	μA	–	–	mΩ	Ω
AEG PQ 40/50	$2.92 \cdot 10^{-3}$	1.082	12.24	1	2	13.66	34.9
Siemens M50	$3.11 \cdot 10^{-3}$	0.878	12.71	1	2	13.81	13.0
Kyocera LA441J59	$3.09 \cdot 10^{-3}$	1.913	8.25	1	2	12.94	94.1

The first diode is generally an ideal diode ($m_1 = 1$), while the diode factor of the second diode is $m_2 = 2$. This calculation also has to be performed in a numeric method. Table 5.4 shows the parameters that can be used to simulate some PV modules well.

Two-diode model with expansion term

The solar cell's equivalent circuit has to be expanded further so we can describe the curve in the negative breakdown area, where there are great negative voltages. An expansion term, shown as an additional current source in Figure 5.23 $I(U_D)$, describes the diode breakdown in the negative voltage area. This source of current generates current relative to diode voltage U_D, thus allowing us to describe the solar cell's electrical behaviour when the voltage is considerably negative.

Breakdown voltage U_{Br}, the landslide breakdown exponent n and adjustment value for conductance b are used in the following formula for the I–V curve:

$$0 = I_{Ph} - I_{S1}\left(\exp\left(\frac{U + I \cdot R_S}{m_1 \cdot U_T}\right) - 1\right) - I_{S2}\left(\exp\left(\frac{U + I \cdot R_S}{m_2 \cdot U_T}\right) - 1\right) - \frac{U + I \cdot R_S}{R_P} - I \quad (5.43)$$

$$\underbrace{- b \cdot (U + I \cdot R_S) \cdot \left(1 - \frac{U + I \cdot R_S}{U_{Br}}\right)^{-n}}_{\text{Exparsion term}}$$

Figure 5.24 shows a polycrystalline solar cell's I–V curve. The parameters $I_{S1} = 3 \cdot 10^{-10}$ A, $m_1 = 1$, $I_{S2} = 6 \cdot 10^{-6}$ A, $m_2 = 2$, $R_S = 0.13\ \Omega$, $R_P = 30\ \Omega$, $U_{Br} = -18$ v, $b = 2.33$ mS and $n = 1.9$ are used to determine the curve.

The great series resistance comes about in this case when the resistance of the connection lines is taken into account. The positive voltage and current range represents the cell's

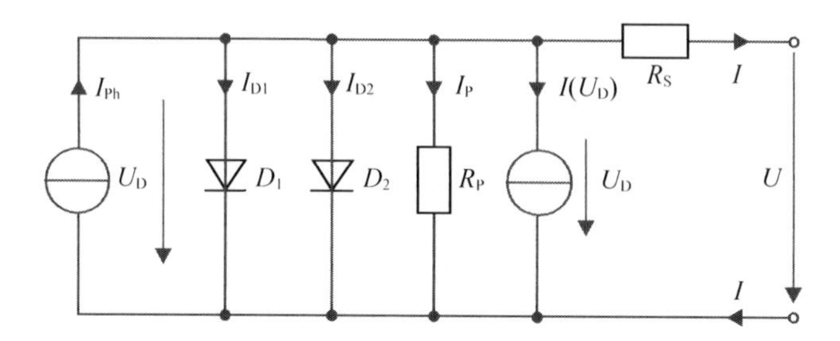

Figure 5.23 A two-diode equivalent circuit with a second source of current for a breakdown under negative voltage

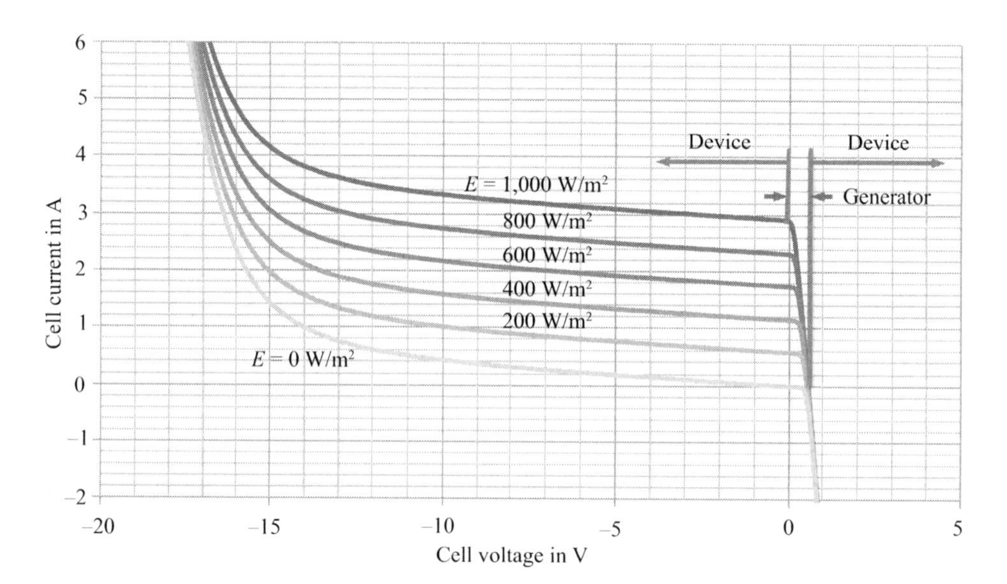

Figure 5.24 A polycrystalline cell's I–V curve over the entire voltage range

generator range. If the voltage or current is negative, the cell consumes power. An external voltage source or other solar cells must provide the electrical output for this to happen.

Other electrical cell parameters

Up to now, we have only spoken about the voltage and current of solar cells. Here, we discuss other electrical cell parameters. Table 5.5 provides an overview of the main solar cell parameters.

If the cell is short-circuited, the terminal voltage at the solar cell is zero. The short circuit current I_{sc}, which is roughly equivalent to the photocurrent I_{Ph}, then flows. Because the photocurrent is proportional to irradiance E, this also applies for **short circuit current**:

$$I_{SC} \approx I_{Ph} = c_0 \cdot E$$

Table 5.5 Electrical solar cell parameters

Designation	*Symbol*	*Unit*	*Note*
Open-circuit voltage	U_{SC}	V	$U_{OC} \sim \ln E$
Short-circuit voltage	I_{SC}	A	$I_{OC} \approx I_{Ph} \sim E$
MPP voltage	U_{MPP}	V	$U_{MPP} < U_{OC}$
MPP current	I_{MPP}	A	$I_{MPP} < I_{SC}$
MPP output	P_{MPP}	W or W_p	$P_{MPP} = U_{MPP}\ I_{MPP}$
Fill factor	FF		$FF = P_{MPP}\ /\ (U_{OC}\ I_{SC}) < 1$
Efficiency	η	%	$\eta = P_{MPP}\ /\ (E\ A)$

If the cell has an open circuit with current I at zero, the cell has **open-circuit voltage** U_{OC}. In the I–V curve of the simplified equivalent circuit, U_{OC} can be calculated with $I = 0$:

$$U_{OC} = m \cdot U_T \cdot \ln\left(\frac{I_{SC}}{I_S} + 1\right) \quad (5.44)$$

Because short-circuit current I_{SC} has a linear relation to irradiance E, the relation of open-circuit voltage is as follows:

$$U_{OC} \sim \ln(E) \quad (5.45)$$

A solar cell's maximum output can be determined at a specific voltage. Figure 5.25 shows the output-voltage characteristic along with the current-voltage characteristic. The output curve clearly has a maximum point. One therefore speaks of a **maximum power point (MPP)**.

The voltage U_{MPP} at the *MPP* is less than the open-circuit voltage $U_{OC,}$ and the current I_{MPP} is less than the short-circuit voltage I_{SC}. The current and voltage both depend upon irradiance and temperature, as with short-circuit current and open-circuit voltage. The output P_{MPP} at the *MPP* is calculated as follows:

$$P_{MPP} = U_{MPP} \cdot I_{MPP} < U_{OC} \cdot I_{SC} \quad (5.46)$$

The share of current is more important in the interdependence on irradiance, so the MPP output roughly increases proportional to irradiance E.

The MPP output of solar cells and modules is usually determined under **standard test conditions (STC)** (E = 1000 W/m², ϑ = 25 °C, spectral AM1.5g) to facilitate comparisons. Because solar modules almost always have a lower output under natural conditions, W_p (**watts-peak**) is often the designation used for the finding from the measurement.

The parameters of solar modules are therefore also determined under **Normal Operating Test Conditions (NOTCs)**. Here, measurements are taken at a **Normal Operating Cell Temperature (NOCT)** at an irradiance of 800 W/m², an ambient temperature of 20 °C, spectrum AM1.5g, and a wind velocity of 1 m/s. Generally, the NOCT is around 45 °C.

The fill factor *FF* is another parameter:

$$FF = \frac{P_{MPP}}{U_{OC} \cdot I_{SC}} = \frac{U_{MPP} \cdot I_{MPP}}{U_{OC} \cdot I_{SC}} \quad (5.47)$$

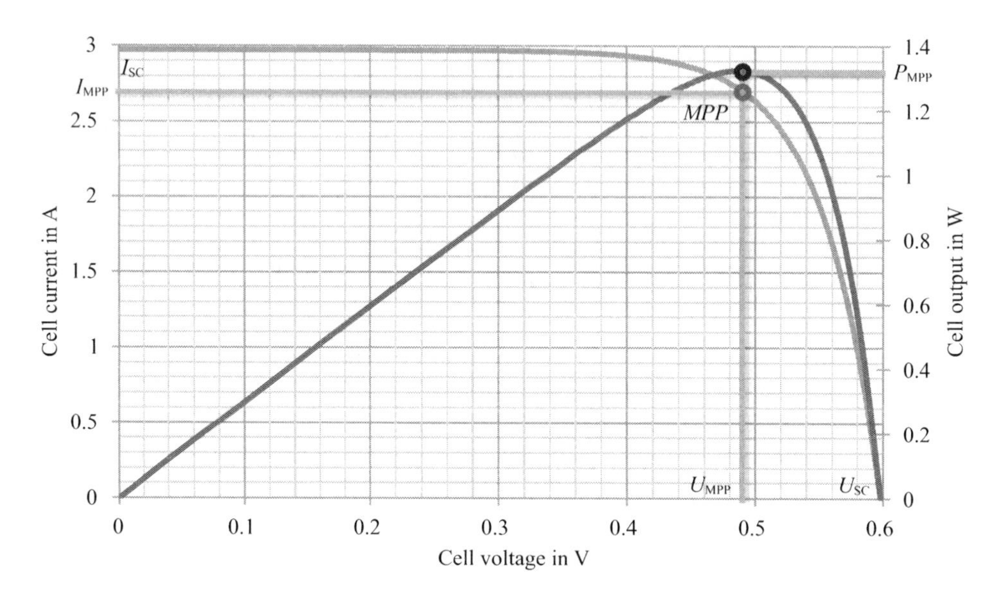

Figure 5.25 A solar cell's I–V and P–U curves with MPP

Table 5.6 Maximum efficiencies and fill factors for various cell technologies

Cell/module type	η_{max} *cell, lab*	η_{max} *cell, series*	η_{max} *module, series*	*FF module, series*
Mono-Si	25.0 %	22.9 %	20.4 %	0.80
Poly-Si	20.4 %	17.8 %	16.0 %	0.78
SR-Si	17.8 %	15.6 %	14.1 %	0.74
EFG-Si	18.2 %	14.4 %	12.8 %	0.72
HIT	23.0 %	21.6 %	19.0 %	0.78
μc-Si / a-Si	11.9 %	10.4 %	10.1 %	0.68
a-Si	12.5 %	7.6 %	7.4 %	0.67
CdS / CdTe	17.3 %	12.8 %	11.8 %	0.74
CIS / CIGS	20.3 %	15.1 %	14.5 %	0.73
Concentrator	43.5 %	40.0 %	29.5 %	0.85
GaAs	26.4 %	— [1]	— [1]	— [1]
Dye-sensitized	11.8 %	— [2]	— [2]	— [2]
Organic	10.6 %	— [2]	— [2]	— [2]

Note: As of 07/2012 [1] Only special applications, such as outer space [2] Not yet in serial production

It serves as a quality criterion for solar cells by describing how close the solar cell's I–V curve is to the rectangle of U_{OC} and I_{SC}. This value is always less than one.

A solar cell's **efficiency** η is the product of the MPP output P_{MPP}, irradiance E, and the solar cell's area A:

$$\eta = \frac{P_{MPP}}{E \cdot A} = \frac{FF \cdot U_{OC} \cdot I_{SC}}{E \cdot A} \tag{5.48}$$

Usually, efficiency is indicated under standard test conditions. Table 5.6 shows the maximum efficiencies and typical fill factors for various types of cells and modules. The module's efficiency is always less than the cell's because the module's frame and the space between the cells do not have any output.

Temperature factor

The formulae used to describe solar cell models always assume a constant temperature of 25 °C. But as described above, a solar cell's curve changes relative to temperature. Now, we discussed how the formulae used for this calculation can be changed to take account of different temperatures.

First, the temperature voltage is no longer a fixed value, but has to be calculated for each specific temperature.

The **temperature voltage** is calculated from the absolute temperature in Kelvin ($T = \vartheta$ K/°C + 273.15 K), the Boltzmann constant $k = 1.380658 \cdot 10^{-23}$ J/K, and the elementary charge $e = 1.60217733 \cdot 10^{-19}$ As:

$$U_T = \frac{k \cdot T}{e} \tag{5.49}$$

The temperature factor of **saturation currents** I_{S1} and I_{S2} can be calculated from the coefficients c_{S1} and c_{S2} and the band gap E_g from Table 5.2 [Wol77]:

$$I_{S1} = c_{S1} \cdot T^3 \cdot \exp\left(-\frac{E_g}{k \cdot T}\right) \tag{5.50}$$

$$I_{S2} = c_{S2} \cdot T^{5/2} \cdot \exp\left(-\frac{E_g}{2 \cdot k \cdot T}\right) \tag{5.51}$$

Open-circuit voltage drops as saturation currents increase along with the temperature. To simplify further calculations, we will leave out consideration of how series resistance R_S, parallel resistance R_P, and diode factors change according to temperatures.

In formulae (5.50) and (5.51), the role of band gaps relative to temperature is left out, but it is crucial in determining **photocurrent** I_{Ph}. The band gap decreases as temperatures increase, so photons need less energy to move electrons into the valence band, thereby increasing photocurrent. The temperature dependence of photocurrent is calculated as follows using the coefficients c_1 and c_2:

$$I_{Ph}(T) = (c_1 + c_2 \cdot T) \cdot E \tag{5.52}$$

Table 5.7 Parameters for the temperature factors of various PV modules [PRE94]

Parameter *unit*	c_{S1} A/K^3	c_{S2} $A\ K^{-5/2}$	c_1 m^2/V	c_2 $m^2/(V\ K)$
AEG PQ 40/50	210.4	$18.1 \cdot 10^{-3}$	$2.24 \cdot 10^{-3}$	$2.286 \cdot 10^{-6}$
Siemens M50	170.8	$18.8 \cdot 10^{-3}$	$3.06 \cdot 10^{-3}$	$0.179 \cdot 10^{-6}$
Kyocera LA441J59	371.9	$12.2 \cdot 10^{-3}$	$2.51 \cdot 10^{-3}$	$1.932 \cdot 10^{-6}$

Table 5.7 shows the parameters used to calculate the temperature factors of various solar modules.

A number of calculations are simplified by assuming that the current, voltage, and output of solar cells and modules change in line with voltage. We then only need three temperature coefficients for the description.

As explained above, short-circuit current increases with temperatures. Normally, short-circuit current I_{SC} is indicated under standard test conditions (ϑ = 25 °C). The temperature coefficient α_{ISC} of short-circuit current is used to calculate short-circuit current at another temperature:

$$I_{SC}(\vartheta_2) = I_{SC}(\vartheta_1) \cdot (1 + \alpha_{ISC} \cdot (\vartheta_2 - \vartheta_1)) \tag{5.53}$$

The temperature coefficient of MPP current generally only slightly deviates from that of short-circuit current. If the coefficient is given as a relative parameter, the same coefficient can be used for short-circuit current and MPP current.

The temperature coefficient α_{UOC} of open-circuit voltage is calculated in the same way as short-circuit current, but in the negative. As the temperature increases, open-circuit voltage drops faster than short-circuit current increases.

Because the temperature coefficient of voltage is greater than that of current, there is also a negative temperature coefficient α_{PMPP} for MPP output. For silicon solar cells, it is generally around -4 %/°C = $-4 \cdot 10^{-3}$/°C. In other words, when the temperature increases by 25 °C, output decreases by 10 %.

Figure 5.26 shows I–V curves relative to a changing temperature ϑ. Clearly, open-circuit voltage drops considerably as the temperature increases. In contrast, short-circuit current only increases slightly. As a result, MPP output is reduced as temperatures rise, as described above.

The temperature response of different solar cell materials varies. Above all, amorphous silicon, micromorph silicon, and cadmium-telluride have relatively constant power generation

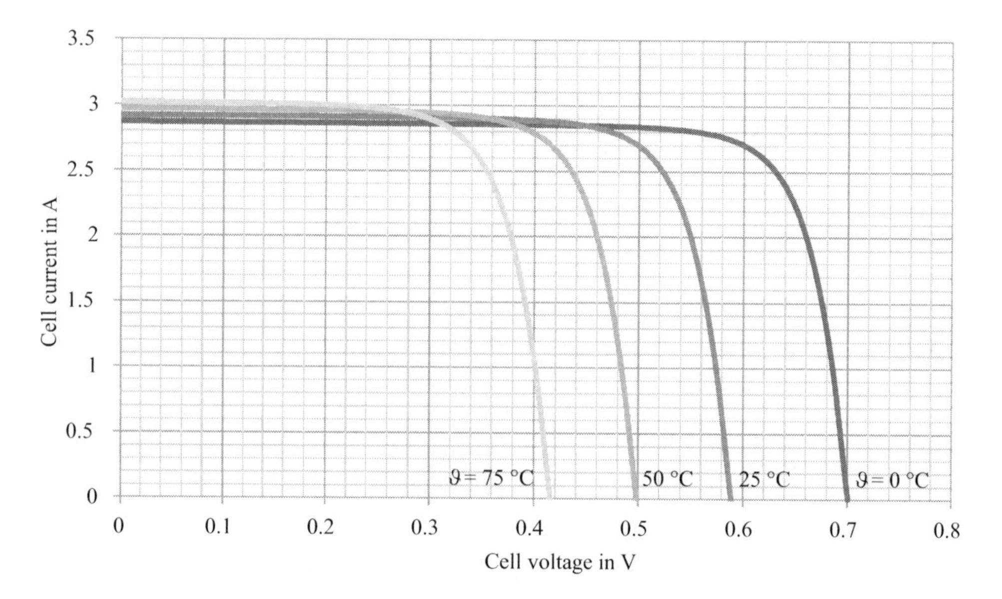

Figure 5.26 How temperatures affect a solar cell's characteristic curve

Table 5.8 Typical temperature coefficients for current, voltage, and output for some common types of solar modules

Cell/module type	α_{UOC} in %/°C	α_{ISC} *in %/°C*	α_P *in %/°C*
Mono-Si	–0.21 – –0.48	+0.02 – +0.08	–0.32 – –0.51
Poly-Si	–0.29 – –0.42	+0.03 – +0.07	–0.32 – –0.51
SR-Si	–0.34 – –0.41	+0.05 – +0.06	–0.42 – –0.51
EFG-Si	–0.38 – –0.50	+0.1	–0.45 – –0.47
HIT	–0.25 – –0.26	+0.03	–0.3
μc-Si / a-Si	–0.30	+0.07	–0.24
a-Si	–0.27 – –0.38	+0.1	–0.18 – –0.23
CdS / CdTe	–0.25	+0.04	–0.18 – –0.25
CIS / CIGS	–0.26 – –0.29	+0.04	–0.35 – –0.5

across different temperatures. In other words, they have a relatively low temperature coefficient. In hot regions and wherever the modules are poorly ventilated, these materials ensure greater power output. Table 5.8 shows some typical temperature coefficients for various cell materials.

In practice, the cell temperature relevant for power generation is far above the ambient temperature. The part of the solar energy that the cell does not convert into electricity heats up the module. The module's temperature ϑ_M, which is roughly equivalent to the cell's temperature, can be calculated from the ambient temperature ϑ_U and the irradiance E incident on the solar module:

$$\vartheta_M = \vartheta_U + c \cdot \frac{E}{1000 \frac{W}{m^2}} \tag{5.54}$$

The factor c indicates the proportionality constant, which is relative to how the solar module is installed. Table 5.9 shows various values for module installation. The exact module temperature can therefore deviate considerably from the value calculated. Wind can also cool modules, thereby reducing temperatures. Cell efficiency also influences temperature. At high efficiencies, less solar energy is converted into heat. In contrast, the module's temperature can increase considerably in spots if it is shaded.

Determining parameters

To simulate a special solar cell, such as with the simplified equivalent circuit, the cell parameters (here I_{Ph} and I_S) are generally determined from measured values in the cell's characteristic curve because a theoretical calculation is extremely complex. To simplify matters, the photocurrent I_{Ph} can be assumed to be the same as the measured short-circuit current I_{SC}. For an ideal diode, the diode factor m is 1. We thus already know two parameters ($I_{Ph} = I_{SC}$ and $m = 1$). The simplified equivalent circuit also includes the diode's saturation current I_S, which can be determined by including open-circuit voltage U_{OC} to find I_S:

$$I_S = \frac{I_{SC}}{\exp\left(\frac{U_{OC}}{U_T}\right) - 1} \approx I_{SC} \cdot \exp\left(-\frac{U_{OC}}{U_T}\right) \tag{5.55}$$

Table 5.9 Proportionality constant c for a calculation of module temperature with various types of installation [DGS08]

Type of installation	*c in °C*
Ground-mounted	22
On roof, large gap	28
On/in roof, good rear ventilation	29
On/in roof, poor rear ventilation	32
On or integrated in façade, good rear ventilation	35
On or integrated in façade, poor rear ventilation	39
Roof integration, no rear ventilation	43
Roof integration, no rear ventilation	55

Then, all of the parameters have been determined for the simplified equivalent circuit with an ideal diode ($m = 1$). However, this model only provides a rough overlapping of the calculated characteristic curve and the values measured. For a non-ideal diode, another diode factor ($m > 1$) is used. The two interrelated parameters m and I_S can be calculated in a software program such as Mathematica™ from the solar cell's characteristic curve for the power generation range in a relatively straightforward manner.

Comparing the simplified equivalent circuit and a real diode produces a very good correspondence between measurements and simulations.

But while the parameters m and I_S are not that difficult to determine, the additional parameters R_S and R_P in the expanded equivalent circuit for the single-diode model are. As the number of free parameters increases, even sophisticated mathematics programs quickly reach their limits. To ensure proper convergence in determining parameters, the starting values have to be chosen close to the final values. It is relatively easy to select the starting values for R_P and R_S.

Parallel resistance R_P can be brought closer to zero increasing the I–V curve for voltage. **Series resistance** R_S can be approximated by increasing the curve beyond open-circuit voltage:

$$R_P \approx \left.\frac{\partial U}{\partial I}\right|_{U=0} \quad (5.56) \qquad R_S \approx \left.\frac{\partial U}{\partial I}\right|_{U>>U_1} \quad (5.57)$$

For the negative diode breakdown U_{Br}, the parameters b and n can be determined like the other parameters – by using the values measured for the negative breakdown.

Electrical description of solar modules

Solar cells in series circuits

Because of the low voltage, solar cells are not used individually, but rather in a series circuit within a module. In return, the modules are either in parallel circuits, series circuits, or a combination of the two.

Because a number of modules are designed for use with 12 V lead batteries, the optimal number of cells for this purpose has become the standard: 32 to 40. There are, however, modules with far more or fewer cells in series circuits for other purposes.

When n cells are in series circuits as shown in Figure 5.27, the current I_i through all the cells i is identical, but the cell voltages U_i add up to produce the module's voltage U:

$$I = I_1 = I_2 = \ldots = I_n \quad (5.58) \qquad U = \sum_{i=1}^{n} U_i \quad (5.59)$$

If all of the cells are identical and have the same conditions (irradiance and the temperature), the following applies for the total voltage:

$$U = n \cdot U_i \quad (5.60)$$

The I–V curve for series circuits is easy to derive from an individual cell's characteristic curve in this case.

The data sheets provided by manufacturers generally only include a few parameters, such as open-circuit voltage U_{OC0}, short-circuit current I_{SC0}, voltage U_{MPP0} and current I_{MPP0} at the maximum power point with an irradiance of E_{1000} = 1000 W/m² and temperature ϑ_{25} = 25 °C, and temperature coefficients α_U and α_I for voltage and current.

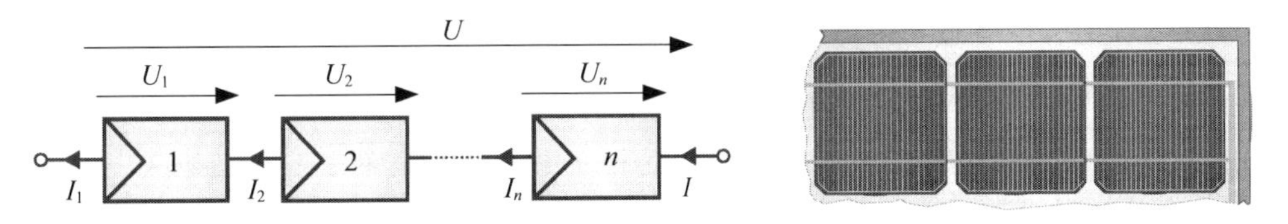

Figure 5.27 Solar cells in a series circuit (left: electrotechnical symbols, current, and voltage; right: view of a section of a module with crystalline cells)

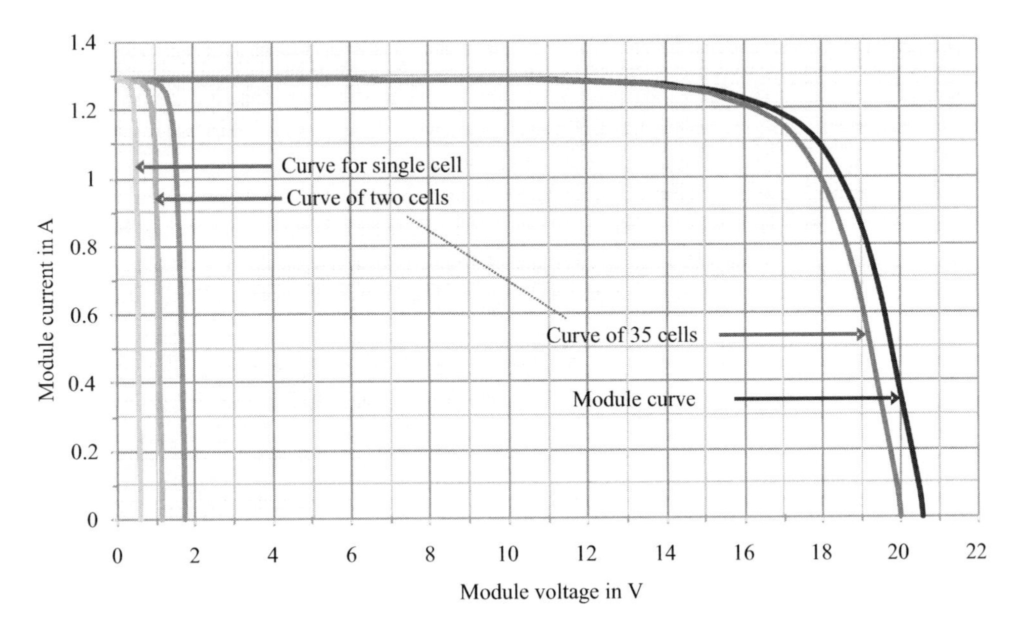

Figure 5.28 A module's characteristic curve based on curves from 36 cells

The formulae

$$U_{OC} = U_{OC0} \cdot \ln(E) / \ln(E_{1000}) \cdot (1 + \alpha_U (\vartheta - \vartheta_{25})) \quad (5.61)$$

$$U_{MPP} = U_{MPP0} \cdot \ln(E) / \ln(E_{1000}) \cdot (1 + \alpha_U (\vartheta - \vartheta_{25})) \quad (5.62)$$

$$I_{MPP} = I_{MPP0} \cdot E / E_{1000} \cdot (1 + \alpha_I (\vartheta - \vartheta_{25})) \quad (5.63)$$

$$I_{SC} = I_{SC0} \cdot E / E_{1000} \cdot (1 + \alpha_I (\vartheta - \vartheta_{25})) \quad (5.64)$$

allow us to roughly determine the module's parameters at different temperatures ϑ and levels of irradiance E. The parameters

$$c_1 = I_{SC} \cdot \exp(-c_2 \cdot U_{OC}) \text{ (5.65) and } c_2 = \ln(1 - I_{MPP} / I_{SC}) / (U_{MPP} - U_{OC}) \quad (5.66)$$

can also be used to show a rough relation between the module's current I and voltage U:

$$I = I_{SC} - c_1 \cdot \exp(c_2 \cdot U) \quad (5.67)$$

For other considerations, see the formulae for rough estimates in [Wag06].

Series circuits under inhomogeneous conditions

In practice, cells do not always operate under identical conditions. Leaves, bird droppings, snow, and other elements – including nearby objects – can shade individual cells. The impact on the module's characteristic curve can be considerable.

If not all of the I–V curves of individual cells are identical, the overall curve is harder to determine. Here, it is assumed that a module with 36 cells in a series circuit has one cell with 75 % less irradiance than the other 35. In this case, the current through all of the cells is identical. The entire characteristic curve can be drawn starting with zero if the current through the cells is preset and the various cell voltages for the cells under full sunlight U_b is determined and added to that of the shaded cell U_a:

$$U = U_a(I) + 35 \cdot U_b(I) \quad (5.68)$$

The characteristic curve can then be easily constructed up to the shaded cell's short-circuit current. This curve, however, only covers a small part of the module's voltage range near the open-circuit voltage. The rest of the curve can only be drawn if the partly shaded cell has greater currents than the cell's short-circuit current. This situation can only occur in the cell's negative voltage range. The shaded cell then consumes power, as described in the equivalent circuit shown in Figure 5.23.

Figure 5.29 shows a point on the module's curve (1). At a given current, the voltage is derived by adding the shaded cell's voltages (1a) to 35 times the voltage of an unshaded cell (1b). If the curve is drawn point for point for various currents, the module's curve then reflects the case of shading, as also shown in the figure.

Shading of the cell clearly reduces the module's output drastically in this example. Although only around 2 % of the module's surface is covered, its maximum output drops by around 70 % from $P_1 = 20.3$ W to $P_2 = 6.3$ W. The partly shaded cell consumes power. In this example, the cell's maximum power loss is 12.7 W when the module short-circuits. If a cell is shaded to a different degree, the losses are located elsewhere along the characteristic curve.

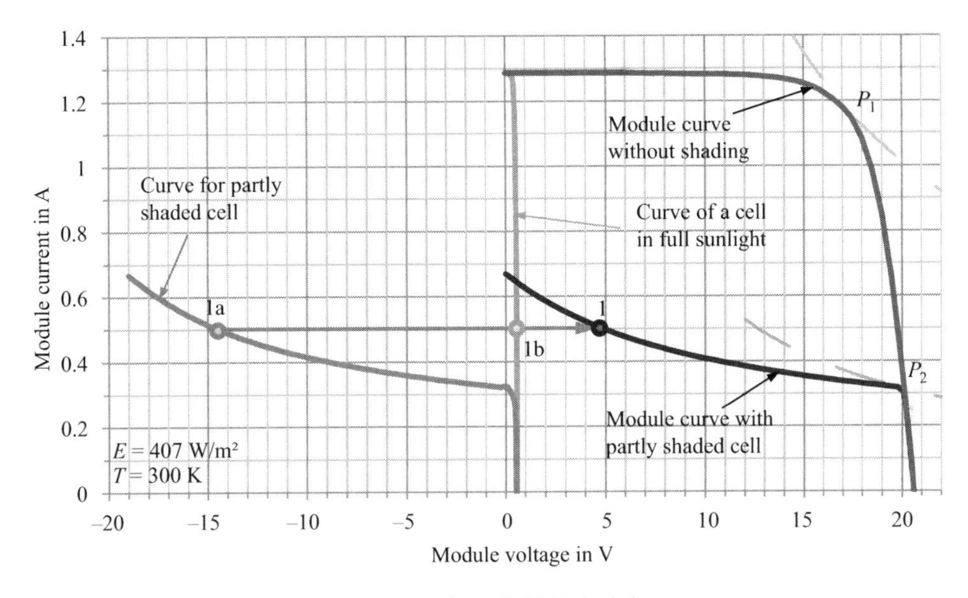

Figure 5.29 A module's characteristic curve with a cell 75 % shaded

In other shading situations and at a higher levels of irradiance, the shaded cell can have a power loss exceeding 30 W. As a result, the cell heats up drastically and can even be destroyed. Individual areas only a few millimetres across – called **hot spots** – heat up and the cell material literally melts away or at least damages the cell's encapsulation [Qua96a].

Bypass diodes are therefore integrated parallel to the solar cells to prevent individual cells from being destroyed by hot spots. Normally, these diodes are inactive. Power only flows over the diodes when the cells are shaded. The bypass diodes switch on when there is a small negative voltage, say, of around -0.7 V, depending on the type of diode. That threshold is reached when the voltage of the shaded cell is equivalent to the sum of the voltages of all other cells along with the voltage of the bypass diode. Bypass diodes are put in to limit the negative voltage at the solar cells, power losses, and the cell's temperature. Figure 5.30 shows how to bypass diodes that are integrated to serve 18 cells each in a module under normal conditions and with a shaded cell.

Generally, bypass diodes are connected to strings of 18 to 24 cells. According to manufacturers, the main reason for this approach is economics. Diodes can affordably be installed in a module's frame and in the connection socket. However, keep in mind that modern high-performance cells have currents up to and exceeding 8 A, so bypass diodes can get very hot during operation and need to be cooled.

Bypass diodes attached to strings with a relatively large number of cells cannot, however, completely prevent the cells from being damaged. Optimal protection is only provided when each cell has its own bypass diode. **Shade-tolerant modules** have bypass diodes for each cell. They were first serially manufactured by Sharp. When the cells in such modules are exposed to varying levels of irradiance, the losses are far lower [Qua96b]. These diodes can be integrated directly into the semiconductor of the solar cells [Has86]. At present, however, commercial module manufacturers are not pursuing this concept further.

Figure 5.31 shows the I–V curves of bypass diodes across a different number of cells. One cell is 75 % shaded. Clearly, the I–V curve shifts towards greater voltages as the number of

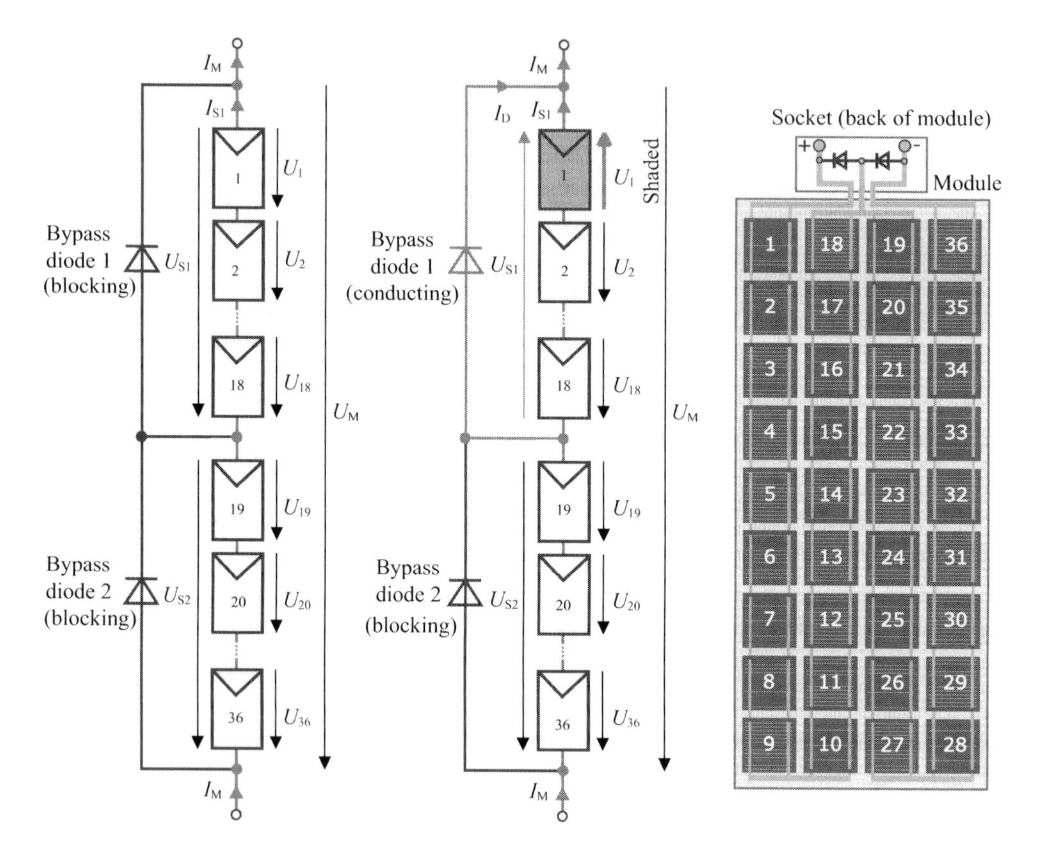

Figure 5.30 Integration of bypass diodes in a solar module with 36 cells. Left: power flow under normal operation; middle: power flow under shading; right: a typical arrangement of cells in diodes in a module

cells in a string decreases; in other words, the bypass diode switches on even at low voltages. The power losses and the loads on the individual cell decrease accordingly.

Figures 5.32 and 5.33 show the current-voltage and power-voltage curves for a module with two bypass diodes, each connected to 18 cells. Here, a single cell has different shading. When a single cell is 50% shaded, the module's output is also cut nearly in half. If the shading is greater, the bypass diode keeps the non-shaded half of the module active. In the process, however, the MPP moves to a much lower voltage. In practice, this switch poses a major challenge for downstream equipment, such as the inverter, which is designed to operate at the maximum power point under all operating conditions. Because power losses are generally far greater than the degree of shading with solar modules commonly sold, shading should be avoided to the extent possible during the planning and installation of systems.

Solar cells in parallel circuits

In addition to series circuits, multiple solar cells can also be arranged in parallel circuits. Usually, they are not because currents are then especially great, leading to greater line losses. This option is therefore only briefly dealt with below.

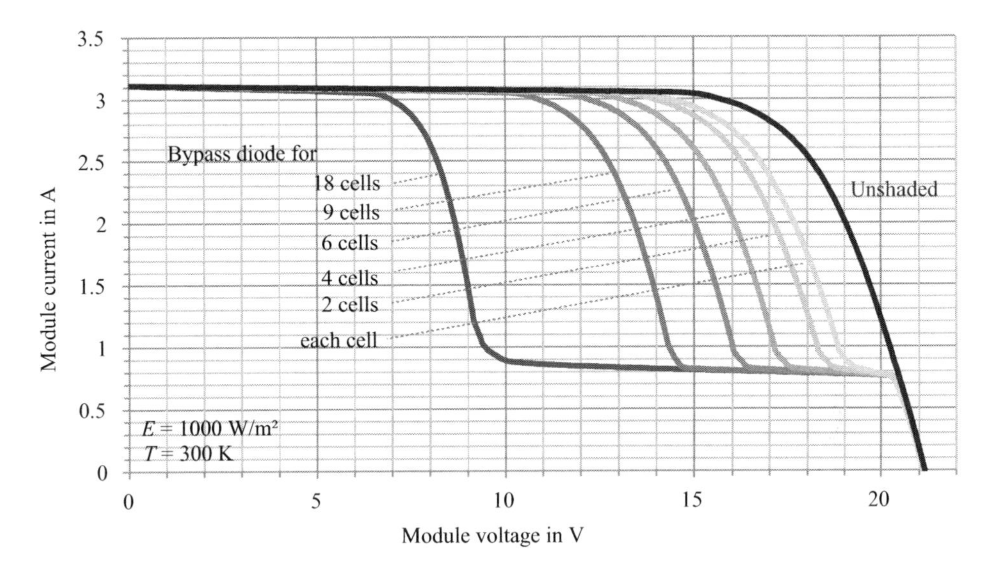

Figure 5.31 A module's characteristic curve with one cell shaded 75 % relative to the number of cells a single bypass diode serves

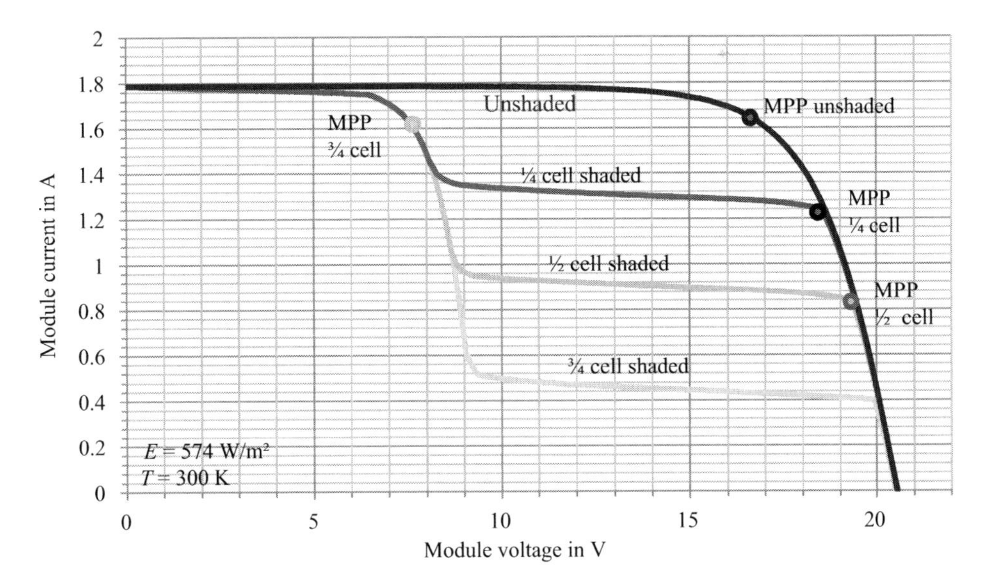

Figure 5.32 I–V curves of a module with 36 cells and two bypass diodes, each covering 18 cells, with one cell shaded differently

When solar cells are in a parallel circuit, all of the cells have the same voltage U. The current of the cells I_i then adds up to produce the total current I.

$$U = U_1 = U_2 = \ldots = U_n \quad (5.69) \qquad I = \sum_{i=1}^{n} I_i \quad (5.70)$$

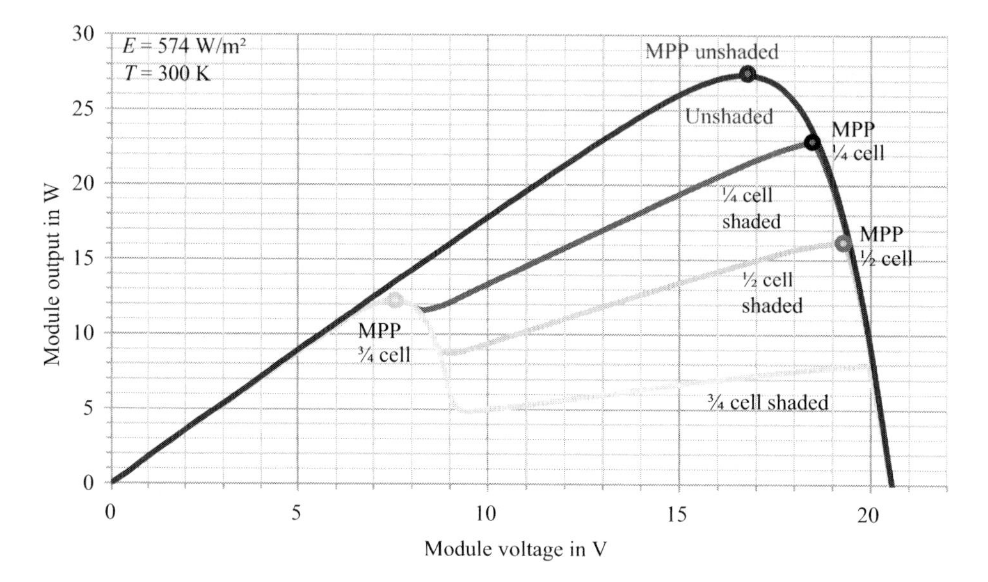

Figure 5.33 P–U curves of a module with 36 cells and two bypass diodes, each covering 18 cells, with one cell shaded differently

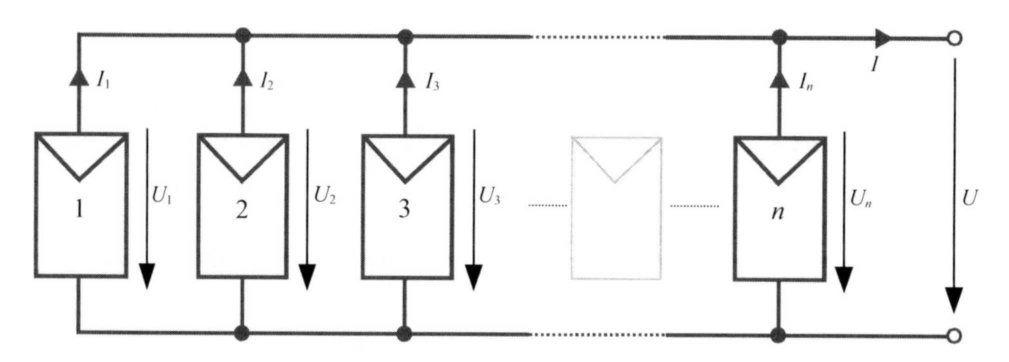

Figure 5.34 n solar cells in a parallel circuit

Solar cells in a parallel circuit are much less sensitive to partial shading. The cells are not expected to break then.

In practice, large solar generators have a large number of solar cells in a serial circuit along with several parallel circuits. In such cases, bypass diodes are only needed to protect the cells in series circuits. The series circuits can be protected with **string diodes**. Because the losses are great and little protection is provided, this option is rarely exercised.

Technical data for solar modules

Table 5.10 provides an overview of the various key data for solar modules. The table includes data for modules using different cell technologies. In addition to modules with polycrystalline and monocrystalline solar cells, various thin-film modules consisting of amorphous silicon, CdTe, and CIS are also listed.

Table 5.10 Technical data for selected solar modules

Designation		*X3–140*	*P230/07*	*SW250*	*E20*	*HIT240*	*FS-387*	*Sm-135*
Manufacturer		Inventux	Solon	SolarWorld	Sunpower	Panasonic	First Solar	Avancis
Number of cells		125	60(6 · 10)	60 (6 · 10)	96 (8 · 12)	72 (6 · 12)	154	n/a
Cell type		aSi/µSi	Poly-Si	Mono-Si	Mono-Si	HIT	CdTe	CIS
MPP output P_{MPP}	Wp	140	260	250	333	240	87.5	135
Nominal current I_{MPP}	A	1.08	8.57	8.05	6.09	5.51	1.78	2.84
Nominal voltage U_{MPP}	V	130	30.45	31.1	54.7	43.7	49.2	47.4
Short-circuit current I_{SC}	A	1.32	8.92	8.28	6.46	5.85	1.98	3.14
Open-circuit voltage U_{OC}	V	166	37.8	37.8	65.3	52.4	61.0	61.5
Temperature coefficient α_{ISC}	%/°C	+0.07	+0.04	+0.043	+0.057	+0.03	+0.04	+0.03
Temperature coefficient α_{UOC}	%/°C	–0.3	–/.33	–3.33	–3.27	–7.25	–5.20	–0.28
Temperature coefficient α_{PMPP}	%/°C	–0.25	–5.43	–5.45	–5.38	–8.30	–0.25	–0.39
Module efficiency	%	9.8	15.9	14.9	20.4	19.1	12.2	12.8
Fill factor		0.64	0.77	0.80	0.79	0.78	0.73	0.70
Length	mm	1,100	1,640	1,675	1,559	1,580	1,200	1,587
Width	mm	1,300	1,000	1,001	1,046	798	600	664
Mass	kg	26	23.5	21.2	18.6	15	12	16
Bypass diodes		0	3	3	3	3	0	1

In the past few years, the efficiency of solar modules has increased considerably and now reaches peak values of more than 20 %. At the beginning of the 1990s, the standard was closer to 10 %. The thickness of cells has decreased so that significantly less cell material is needed, which reduces costs.

The HIT cell – a combination of crystalline and amorphous cell material – is especially efficient; Sanyo was the first manufacturer to produce it commercially. Modules from Sunpower are even more efficient, with rear-contact solar cells made of monocrystalline silicon. Because the contacts are no longer on the front, the entire front side is now available as an active surface; there are no contact fingers that lead to losses.

Shade-tolerant crystalline modules with bypass diodes integrated in the cells are no longer widely available. Manufacturers reduce the number of bypass diodes to the absolute minimum. At the time of writing, Unisolar, however, has sold a module made of amorphous silicon with a bypass diode for each cell.

The MPP output and efficiency are each determined under standard test conditions (1,000 W/m², 25 °C, AM1.5g).

Solar generator and load

Resistive load

Above, we only dealt with characteristic curves for solar cells, modules, and generators. In practice, however, solar modules are not the purpose themselves, but rather a means of producing electricity to serve a load.

The simplest kind of load is an **electrical resistance** R (Figure 5.35). A straight line represents resistance, and it can be used to show the relation between current and voltage:

$$I = \frac{1}{R} \cdot U \tag{5.71}$$

If the current I through the resistance is the same as the solar cell's current, the voltage U can be used to determine the common voltage and hence the maximum power point. Generally, numeric calculations are needed again here.

When **graphically determining the MPP**, the resistance line and the solar cell's I–V curve are drawn into the same diagram. The straight line and the curve overlap at the MPP.

Figure 5.36 shows that the MPP varies greatly depending on the solar module's operating mode. In this example, the module is exposed to an irradiance of 400 W/m² at a temperature of 25 °C near the MPP. Maximum power is output to the resistance here. Under other levels

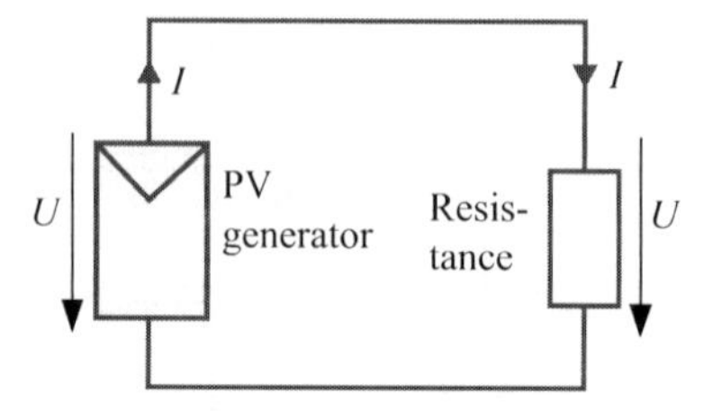

Figure 5.35 Solar generator with resistance

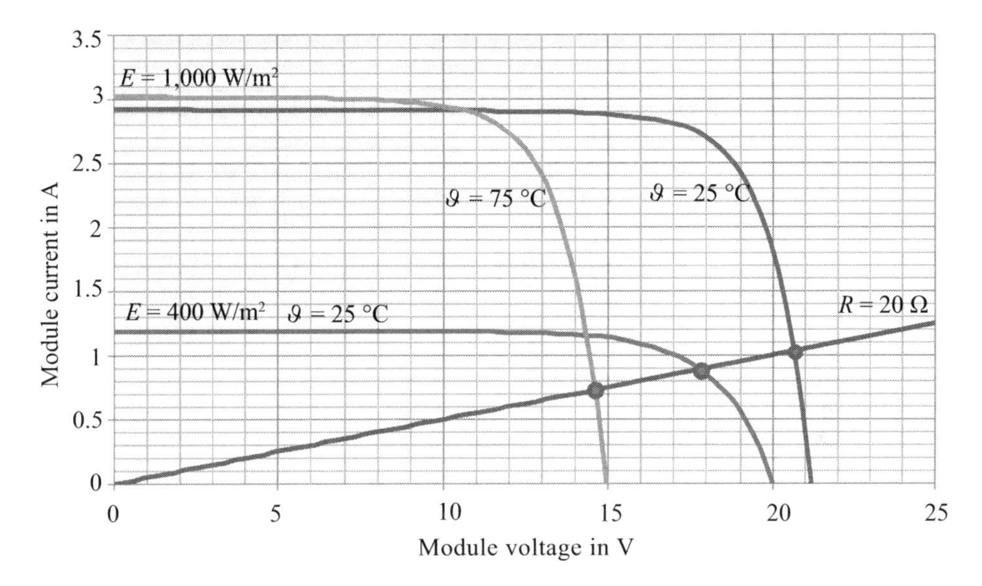

Figure 5.36 A solar module under various operating conditions with electrical resistance

of irradiance and at different temperatures, the module runs at much worse MPPs. Then, the module's actual output is far below its rated output. The voltage and output that reach the resistance vary considerably.

DC–DC converter

The solar generator's output can be greatly improved if a DC–DC converter is installed between the device that consumes power and the solar generator, as shown in Figure 5.37.

The **converter** allows the solar generator to have a different voltage from the appliance. Figure 5.38 shows that constant voltage for the solar generator allows far more power to be produced at greater levels of irradiance than in the example above with a resistance. Power can be increased even further if the solar generator's voltage is adjusted to account for temperature – specifically, if the voltage is increased as temperatures drop.

Good DC–DC converters have efficiencies exceeding 90 %. Only a small part of the power gets lost as heat in the converter. In an ideal converter with an efficiency of 100 %, the input power P_1 and output power P_2 are identical:

$$P_1 = U_1 \cdot I_1 = U_2 \cdot I_2 = P_2 \qquad (5.72)$$

Because of the different voltages, however, the currents I_1 and I_2 are different.

Step-down converter

A step-down converter or buck converter, as shown in Figure 5.39, is used when the load is to have a lower voltage than the solar generator at all times.

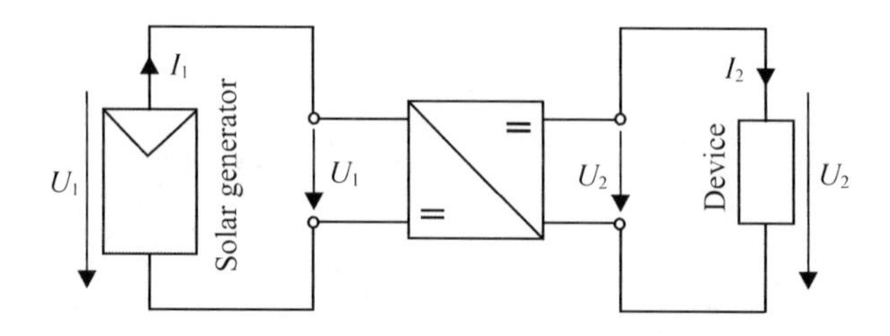

Figure 5.37 Solar generator connected to an appliance via a DC–DC converter

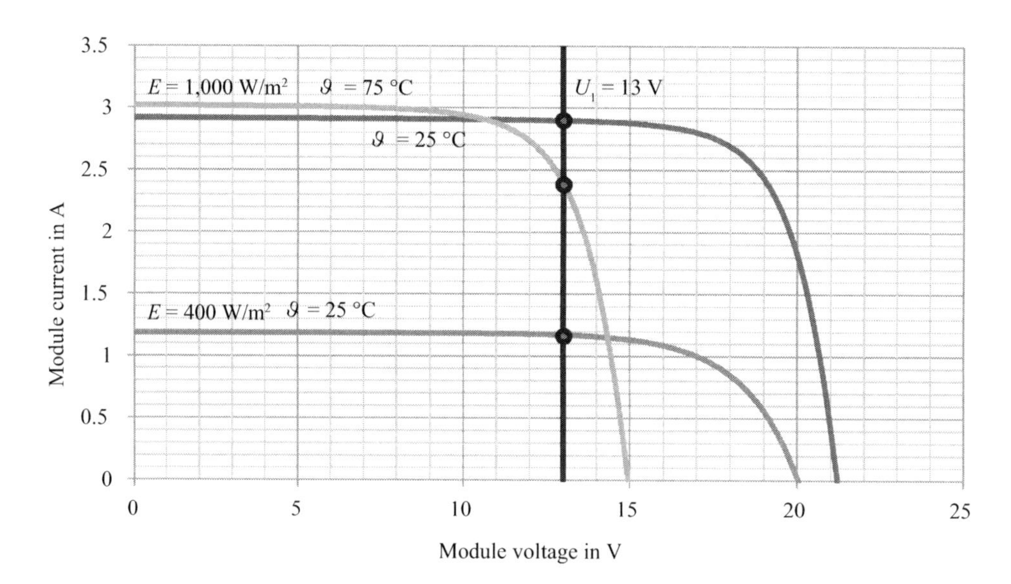

Figure 5.38 A solar module under various operating conditions with a constant voltage load

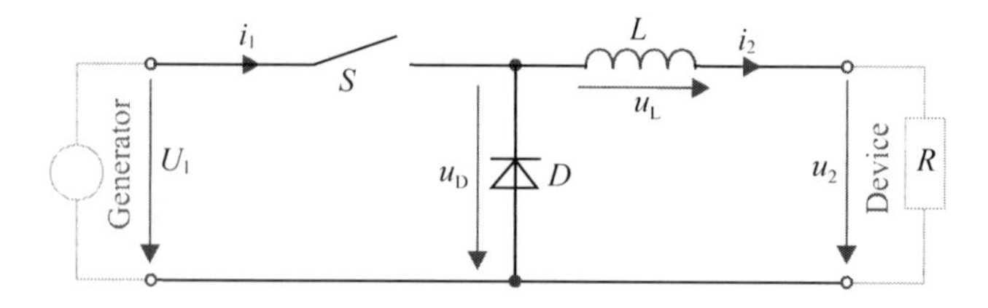

Figure 5.39 A step-down converter with a resistance load

In the following calculations, circuits and diodes are assumed to be ideal elements. The circuit S is closed for the time frame T_E, and the current builds up a magnetic field in which energy is stored because of inductance L. For the voltage u_L of inductance:

$$u_L = L \cdot \frac{di_2}{dt} \tag{5.73}$$

The circuit is then opened for time frame T_A; the inductor's magnetic field weakens, thereby pushing the current through resistance R and diode D.

If we disregard the drop in voltage at the diode in the direction of passage, output voltage u_2 at the appliance is as follows:

$$u_2 = \begin{cases} u_D - u_L = U_1 - u_L & \text{with } u_L > 0 \quad \text{for } 0 \leq t \leq T_E \\ u_D - u_L \approx -u_L & \text{with } u_L < 0 \quad \text{for } T_E \leq t \leq T_E + T_A \end{cases} \tag{5.74}$$

After the duration $T_S = T_E + T_A$, the process is repeated. For the average voltage u_D with **duty cycle** $\delta = T_E / T_S$:

$$\overline{u_D} = U_1 \cdot \frac{T_E}{T_S} = U_1 \cdot \delta \tag{5.75}$$

With $I_N = U_1 / R$ and $\tau = L / R$, the following applies for current i_2 with relation to inductance and the device consuming power:

$$i_2(t) = \begin{cases} I_N - (I_N - I_{min}) \cdot \exp(-t/\tau) & \text{for } 0 \leq t \leq T_E \\ I_{max} \cdot \exp(-(t - T_E)/\tau) & \text{for } T_E \leq t \leq T_S \end{cases} \tag{5.76}$$

The current i_2 ranges between a maximum

$$I_{max} = I_N - (I_N - I_{min}) \cdot \exp(-T_E/\tau) = I_N \cdot \frac{1 - \exp(-T_E/\tau)}{1 - \exp(-T_S/\tau)} \tag{5.77}$$

and minimum

$$I_{min} = I_{max} \cdot \exp(-T_A/\tau) = I_N \cdot \frac{\exp(-T_A/\tau) - \exp(-T_S/\tau)}{1 - \exp(-T_S/\tau)} \tag{5.78}$$

When we then insert I_{min} and $I_{max,}$ we have the following formula for current [Mic92]:

$$i_2(t) = \begin{cases} I_N + I_N \cdot \dfrac{\exp(-T_A/\tau) - 1}{1 - \exp(-T_S/\tau)} \cdot \exp(-t/\tau) & \text{for } 0 \leq t \leq T_E \\ I_N \cdot \dfrac{1 - \exp(-T_E/\tau)}{1 - \exp(-T_S/\tau)} \cdot \exp(-(t - T_E)/\tau) & \text{for } T_E \leq t \leq T_S \end{cases} \tag{5.79}$$

For the average current i_2:

$$\overline{i_2} = I_N \cdot \frac{T_E}{T_S} = I_N \cdot \delta \tag{5.80}$$

Figure 5.40 shows the current i_2 and voltage u_D. In practice, voltage should be relatively constant at the output. For this reason, capacitors C_1 and C_2 are used, as shown in Figure 5.41. When the circuit is open, capacitor C_1 temporarily stores the solar generator's energy.

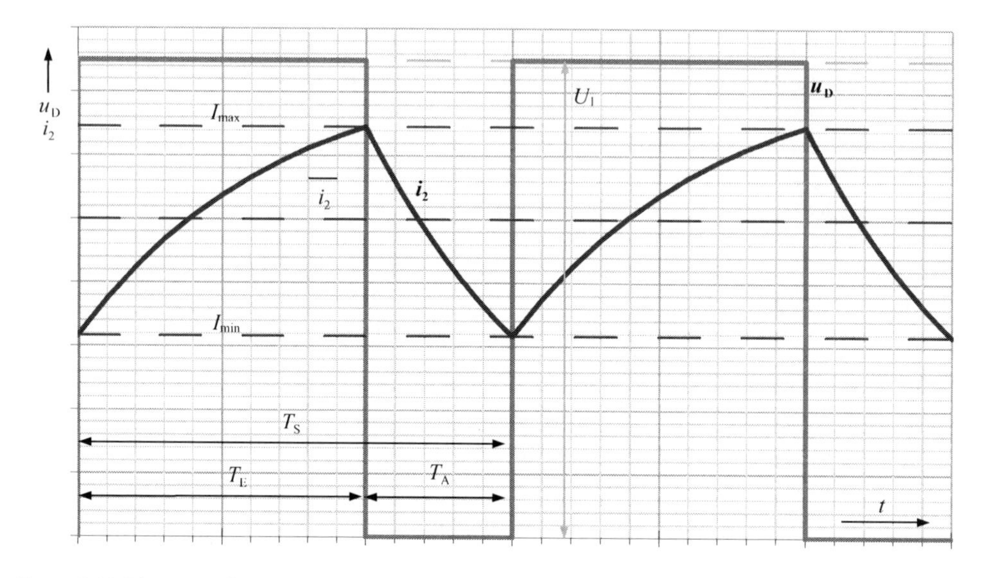

Figure 5.40 The curve for current i_2 and voltage u_D with a step-down converter

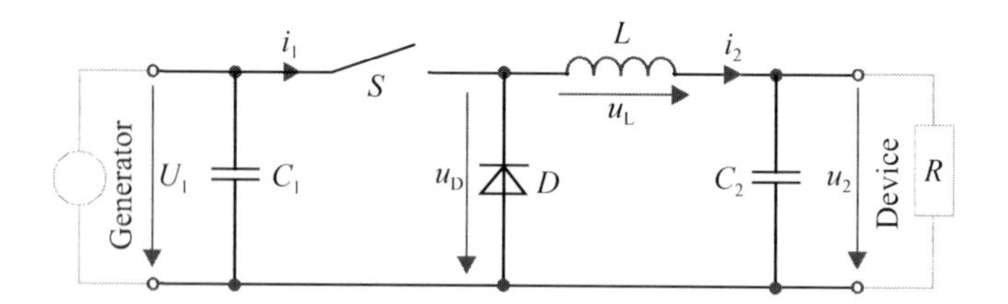

Figure 5.41 Step-down converter with capacitors

For ideal inductance, L the average output voltage

$$U_2 = \overline{u_2} = U_1 \cdot \frac{T_E}{T_S} = U_1 \cdot \delta \tag{5.81}$$

is calculated from the input voltage U_1 and duty cycle δ. For ideal components, the average output voltage

$$I_2 = \overline{i_2} = \overline{i_1} \cdot \frac{T_S}{T_E} = \overline{i_1} \cdot \frac{1}{\delta} \tag{5.82}$$

is calculated based on the average input current and the inverse of the duty cycle. If the average output current I_2 is less than

$$I_{2,\text{limit}} = \frac{T_S}{2 \cdot L} \cdot U_2 \cdot (1 - \frac{U_2}{U_1}) \tag{5.83}$$

the current decreases to zero as a result of inductance during the circuit's lock phase, the diode is blocked, and voltage at inductance falls to zero. Current and voltage thus begin to 'gap'. Proper dimensioning can prevent this problem [Tie02]. If the lower limit current $I_{2,\text{limit}}$ is known, inductance can be calculated thus:

$$L = T_S \cdot \left(1 - \frac{U_2}{U_1}\right) \cdot \frac{U_2}{2 \cdot I_{2,\text{limit}}} \tag{5.84}$$

Clock rates $f = 1 / T$ from 20 kHz to 200 kHz have proven to be a good compromise. The output capacity C_2 is calculated as follows:

$$C_2 = \frac{T_S \cdot I_{2,\text{limit}}}{4 \cdot \Delta U_2} \tag{5.85}$$

from the output voltage's ripple U_2, producing the maximum desired fluctuation ΔU_2.

The circuits S are generally power semiconductors, such as field-effect transistors, bipolar transistors, and thyristors for larger outputs. A number of integrated circuits (ICs) are already available to set the duty cycle for transistors. Circuit breakers are already integrated in a number of ICs for small outputs.

Step-up converters

If the output voltage is to be greater than the input voltage, a step-up or boost converter can be used. A step-up converter is basically the same as a step-down converter with the diode, circuit, and inductance reversed. Figure 5.42 shows a step-up converter.

If circuit S is closed, a magnetic field builds up within inductance L. Voltage $u_L = U_1$ ($u_L > 0$) is produced in the process. If the circuit is open, the voltage at the device consuming power is $u_2 = U_1 - u_L$ ($u_L < 0$), which is greater than the input voltage U_1. Here, no consideration is taken of the voltage decrease at the diode. When the circuit is closed again, capacitor C_2 supports the device's voltage. Diode D prevents the capacitor from being discharged via circuit S. For output voltage U_2, the following applies:

$$U_2 = U_1 \cdot \frac{T_S}{T_A} \tag{5.86}$$

L and C_2 can be dimensioned with $I_{2,\text{limit}} = \frac{1}{2} \cdot U_1 \cdot (1 - U_1 / U_2) \cdot T_S / L$ via

$$L = U_1 \cdot (1 - \frac{U_1}{U_2}) \cdot \frac{T_S}{2 \cdot I_{2,\text{limit}}} \quad (5.87) \text{ and } C_2 = \frac{T \cdot I_{2,\text{limit}}}{\Delta U_2} \tag{5.88}$$

Additional DC–DC converters

Step-up and step-down converters are not the only options for DC–DC. The **buck-boost converter** shown in Figure 5.43 has an output voltage of

$$U_2 = -U_1 \cdot \frac{T_E}{T_A} \tag{5.89}$$

The single-cycle **flyback converter** shown in Figure 5.44 has a transformer instead of inductance. Its output voltages calculated the same as with the buck-boost converter, but

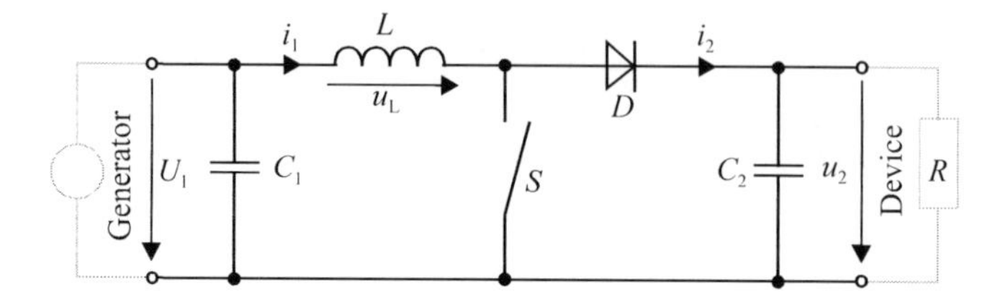

Figure 5.42 How a step-up converter works

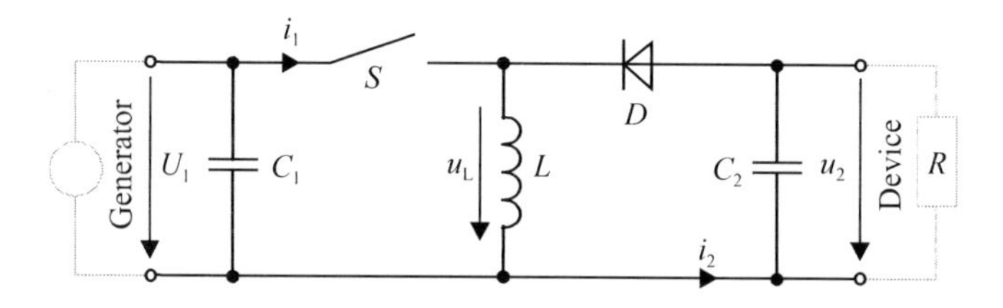

Figure 5.43 A buck-boost converter

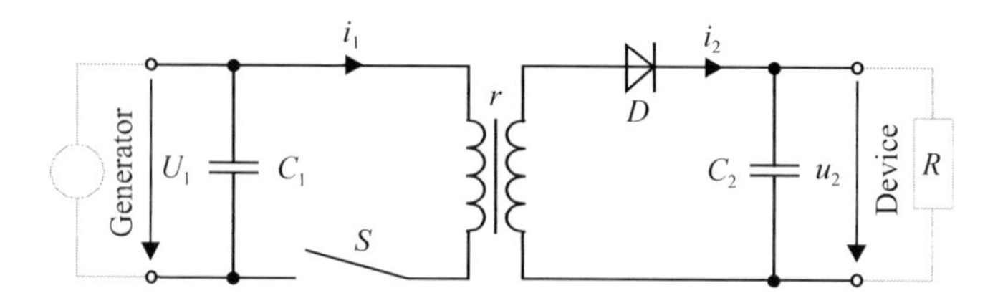

Figure 5.44 How a single-cycle flyback converter works

the transformer's ratio also has to be taken into account and it is relative to the relationship between the winding numbers on the two sides of the transformer.

The following applies for output voltage:

$$U_2 = U_1 \cdot \frac{T_E}{T_A} \cdot \frac{1}{r} \tag{5.90}$$

Push–pull converters are also used for greater outputs, but they require multiple power switches. If capacitors are used instead of inductance, voltage converters can run based on the principle of a charge pump. Examples of circuit converters and control systems for them are given in [Köt96].

MPP trackers

The voltage converters discussed above provide a different voltage level for the solar generator than for the consumption device. If the duty cycle is set to provide a fixed voltage for the solar generator (Figure 5.38), far more energy is produced than would be with power consumption based on resistance, but there would be great losses because of fluctuations in temperature and irradiance. The voltage converter's duty cycle can be changed so that the solar generator's voltage varies.

During normal operation, temperature fluctuations ϑ have the greatest impact on the solar generator's optimal voltage. A temperature sensor can be used to measure the temperature at the solar generator. The temperature coefficients for open-circuit voltage (such as $\alpha_{UOC} = -2.10^{-3}/°C \ldots -5.10^{-3}/°C$ for silicon cells) allow the duty cycle

$$\delta = \frac{U_2}{U_1} = \frac{U_2}{U_{MPP}(\vartheta)} = \frac{U_2}{U_{MPP(\vartheta=25°C)} \cdot (1 + \alpha_{UOC} \cdot (\vartheta - 25\,°C))} \quad (5.91)$$

to be calculated from a step-down converter's MPP voltage U_{MPP} at a reference temperature and output voltage U_2. If the duty cycle is changed again to take account of fluctuations in irradiance, the solar generator will then generally run at its maximum power point. If the voltage converter ensures that the solar generator runs at its MPP, the converter and its control system is called an MPP tracker.

For MPP controls, there are numerous options, a few of which are explained below:

Sensor-based controls: As explained above, MPP voltage is calculated based on measurements taken by temperature and irradiance sensors.

Setting controls based on a reference cell: The curve / open-circuit voltage U_{OC} and the short-circuit current I_{SC} of a reference cell installed near the solar generator have their performance measured. The solar generator's MPP voltage U_{MPP} can be taken from the measurement values. The MPP current I_{MPP} is calculated from the formula for the simple equivalent circuit:

$$I_{MPP} = I(U_{MPP}) = I_{SC} - I_S \cdot \left(\exp\left(\frac{U_{MPP}}{m \cdot U_T}\right) - 1\right) \quad (5.92)$$

For maximum output, the derivation of power from voltage is zero:

$$\frac{d\,P(U_{MPP})}{dU} = \frac{d(U_{MPP} \cdot I(U_{MPP}))}{dU} = I(U_{MPP}) + U_{MPP} \cdot \frac{d\,I(U_{MPP})}{dU} = 0 \quad (5.93)$$

If the formula is used for I_{MPP}, the following applies with respect to MPP voltage U_{MPP}:

$$\begin{aligned} U_{MPP} &= m \cdot U_T \cdot \ln\left(\frac{I_{SC} + I_S}{I_S}\right) - m \cdot U_T \cdot \ln\left(1 + \frac{U_{MPP}}{m \cdot U_T}\right) \\ &= U_{OC} - m \cdot U_T \cdot \ln\left(1 + \frac{U_{MPP}}{m \cdot U_T}\right) \end{aligned} \quad (5.94)$$

This formula can be used with rough estimates or numeric methods. The MPP voltage can therefore be determined from the open-circuit voltage. For a more exact calculation, use the two-diode model.

Perturb and observe: The voltage and current are measured at the voltage converter's input or output, and the power value is calculated and saved, as shown in Figure 5.45.

The duty cycle is changed slightly, thereby changing the voltage, and power output is measured again. If the output has increased, the duty cycle is once again changed in the same direction. If it has worsened, the duty cycle is changed in the opposite direction. At a constant output voltage, the controls simply need to focus on the maximum output current. Power output does not need to be measured.

Incremental conductance: Power and voltage are measured at the generator and multiplied. The derivative dP/dU is created in the process. If it is positive, the generator's voltage must be increased; if it is negative, decreased.

Differential changes: Current and voltage are measured, and their differential changes are determined. The formula

$$\frac{dP}{dU} = \frac{d(U \cdot I)}{dU} = I + U \cdot \frac{dI}{dU} = 0$$

produces

$$I \cdot dU = -U \cdot dI \tag{5.95}$$

meaning that the electronics have to set both of the parameters to the same level.

Curve tracking: Here again, current and voltage are first measured. Depending on the open-circuit voltage $U_A = U_L$, voltage and current are changed as follows alternately:

$$U_B = k \cdot U_A \text{ and } I_C = k \cdot I_B \quad (k < 1)$$

Eventually, two points are located to the left and right of the MPP, and the tracker swings between them.

When part of the solar generator is shaded, a lot of MPP trackers have trouble finding the MPP. Wind shading occurs frequently; power losses can therefore be considerable. A good MPP tracker should therefore provide good results not only for normal operation. In such cases, there are several points of maximum power (Figure 5.33), so if a lower generator output is likely to stem from shading, the entire curve needs to be swept in order to find the maximum power point.

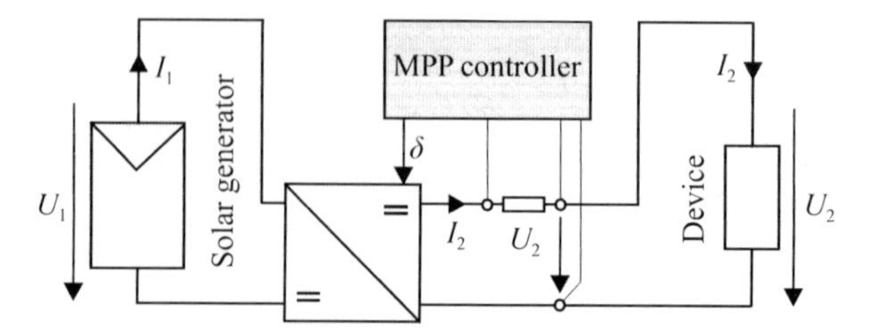

Figure 5.45 How an MPP tracker works

Battery storage

Battery storage types

It is rare for solar generators to be hooked up directly to devices. Generally, more complex systems are used. If there is no grid connection, storage is usually needed. Then energy is available when the sun is not shining, such as at night.

A distinction is made between two types of storage: one to cover a few hours or days during bad weather; the other, to bridge several months, such as to compensate for fluctuations in seasonal irradiation between the summer and the winter. Because long-term storage is generally very complex and expensive, power systems that run all year generally have large PV generators so that enough energy is provided in the winter; alternatively, additional generators are used, such as wind or diesel generators.

For short and midterm storage, batteries have become the standard in terms of cost. Lead batteries are the most widespread because they are the cheapest. Whenever greater energy density is needed to keep the weight down, other battery types are used, such as nickel-metal hydride (NiMH). Over the past few years, very high-performance lithium-ion batteries have become commonplace in high-quality mobile applications, such as cell phones, digital cameras, and laptops. Because they are expensive, however, they are not yet commonly used with solar equipment. Table 5.11 shows some comparative data for a number of battery types.

Lead batteries

Lead batteries are the most common batteries for storing large amounts of energy today. They are the least expensive option. The automotive sector produces large numbers of them as starter batteries. A slightly modified **structure** to produce a longer service life than in the automotive sector is used for lead batteries for solar applications.

In both the automotive and the solar sector, lead batteries have two electrodes. When charged, the positive electrode is made of PbO_2; the negative, of pure lead (Pb). The two electrodes are separated by a membrane, all of which is within a plastic housing. They are

Table 5.11 Data for different types of battery

	Lead	*NiCd*	*NiMH*	*NaS*	*Lithium-ion*
Positive electrode	PbO_2	NiOOH	NiOOH	Na (fluid)	Graphite (nC)
Negative electrode	PbO	Cd	Metals	S (fluid)	$LiMn_2O_4$
Electrolyte	$H_2SO_4 + H_2O$	$KOH + H_2O$	$KOH + H_2O$	β-Al_2O_3	Polymer/salt
Energy density in Wh/l	50–110	80–200	100–350	260–390	250–500
Energy density in Wh/kg	25–50	30–70	60–120	120–220	95–200
Cell voltage in V	2	1.2	1.2	2.1	3.6
Charge/discharge cycles	500–1,500	1,500–3,000	ca. 1,000	1,500–3,000	500–10,000
Operating temperature in °C	0–55	–40– +55	–20– +45	270–350	–20– +55
Self-discharge rate in %/month	5–15	20–30	15–50	0[1)]	<5
Wh efficiency in %	70–85	60–70	60–85	70–85	70–90

Note 1) without energy demand to maintain a high operating temperature

immersed in an electrolyte of diluted sulphuric acid (H_2SO_4). When nearly fully charged at a temperature of 25 °C, the acid has a density of around 1.24 kg/l. The density changes relative to the temperature and the state of charge and can be measured with a hydrometer or via the battery's voltage. Because a single cell of a lead battery has a nominal voltage of 2 V, 6 cells are usually switched in series to produce an operating voltage of 12 V. The number of cells can be changed accordingly to produce a different voltage level.

The following **chemical reactions** take place in a lead battery, as shown in Figure 5.46.

Negative electrode:

$$Pb + SO_4^{2-} \underset{\leftarrow \text{Charge}}{\overset{\text{Discharge}\rightarrow}{\longleftrightarrow}} PbSO_4 + 2e^- \tag{5.96}$$

Positive electrode:

$$PbO_2 + SO_4^{2-} + 4H^+ + 2e^- \underset{\leftarrow \text{Charge}}{\overset{\text{Discharge}\rightarrow}{\longleftrightarrow}} PbSO_4 + 2H_2O \tag{5.97}$$

Net reaction:

$$PbO_2 + Pb + 2H_2SO_4 \underset{\leftarrow \text{Charge}}{\overset{\text{Discharge}\rightarrow}{\longleftrightarrow}} 2PbSO_4 + 2H_2O \tag{5.98}$$

$PbSO_4$ forms during discharging because of the electrolyte, thereby releasing electrons that produce electricity for consumption. When the battery is charged, energy has to be added to it. Pb and PbO_2 are created in the respective electrodes, and more energy has to be added than was taken out. The ratio of the discharge to the charge reflects the charge efficiency. For the charge efficiency, a distinction is made between **Ah efficiency** η_{Ah} and **Wh efficiency** η_{Wh}. Ah efficiency is calculated based on the integrated currents, while Wh efficiency takes account of both current and voltage over the discharge and charge time for the same state of charge, respectively:

$$\eta_{Ah} = \frac{Q_{out}}{Q_{in}} = -\frac{\int_0^{t_{discharge}} I \cdot dt}{\int_0^{t_{discharge}} I \cdot dt} \tag{5.99}$$

$$\eta_{Wh} = \frac{Q_{out} \cdot U_{out}}{Q_{in} \cdot U_{in}} = -\frac{\int_0^{t_{discharge}} U \cdot I \cdot dt}{\int_0^{t_{discharge}} U \cdot I \cdot dt} \tag{5.100}$$

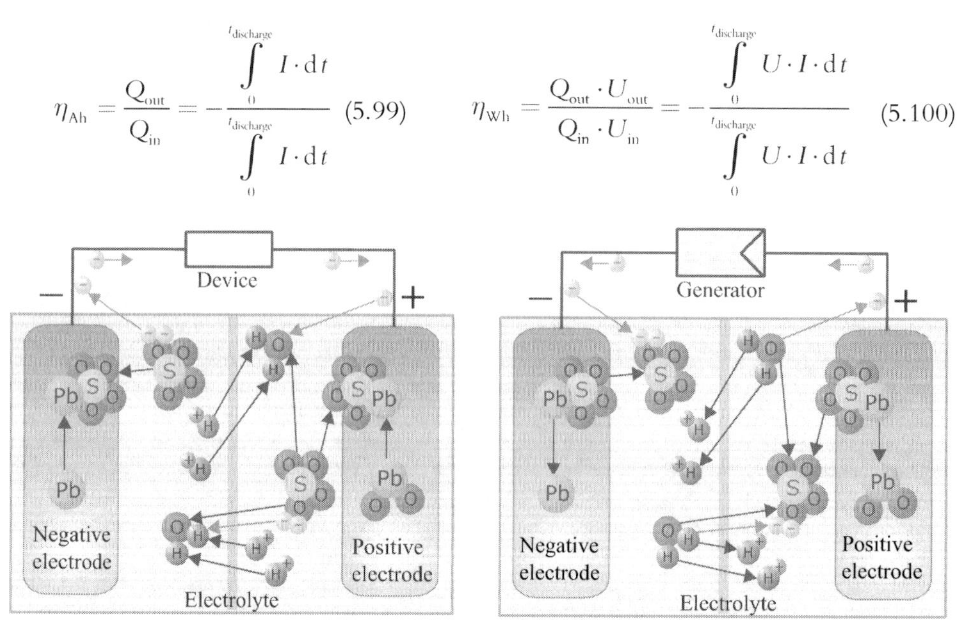

Figure 5.46 Discharging and charging processes within a lead battery

Because voltage is greater during charging than during discharging, Wh efficiency is always lower than Ah efficiency. Depending on the type of battery, the Ah efficiency of a lead battery ranges from 80 to 95 %, with Wh being roughly 10 % lower.

Additional losses that further reduce the overall system's efficiency stem from **self-discharging**. Self-discharge increases along with the temperature; at 25 °C, it is around 0.3 % per day or 10 % per month. Depending on the type of battery, the self-discharge rate can be only half as great.

The **available capacity** of a battery depends on the discharge current, as shown in Figure 5.47. The greater the discharge current, the lower the capacity, and hence the available charge. The discharge voltage is then reached sooner.

Capacity is usually listed along with discharge time when comparing batteries. Here, C_{100} means that this capacity is available when the discharge current is selected so that the battery reaches its end-of-charge voltage in 100 h. If the battery has a capacity of C_{100} = 100 A h, the discharge current is $I_{100} = C_{100}/100$ h = 1 A. If the discharge takes 10 h with a current of 8 A, capacity C_{10} drops below 80 % of C_{100}. At a temperature of 0 °C and a 5 h discharge, capacity even drops to around 50 %. The service life – the battery's number of cycles – decreases along with temperature and discharge depth. The recommended discharge depth is generally around 80 %; levels below 50 % (a less than 50 % charge) should be avoided.

Table 5.12 shows the open-circuit voltage and acid density for a 12 V battery. Depending on the specific model and purpose, the nominal acid density can vary between 1.22 kg/l and 1.28 kg/l. At low temperatures, a greater acid density has a positive effect on operating properties, while lower acid density reduces the impact of corrosion. The voltage values only apply for open circuits – i.e. a battery that has not been charged or discharged for a long time. If the temperature differs from the nominal temperature, a temperature compensation of – 25 mV/°C can be included in the calculation. Charging and discharging increases/decreases, respectively, voltage relative to the strength of current far above/

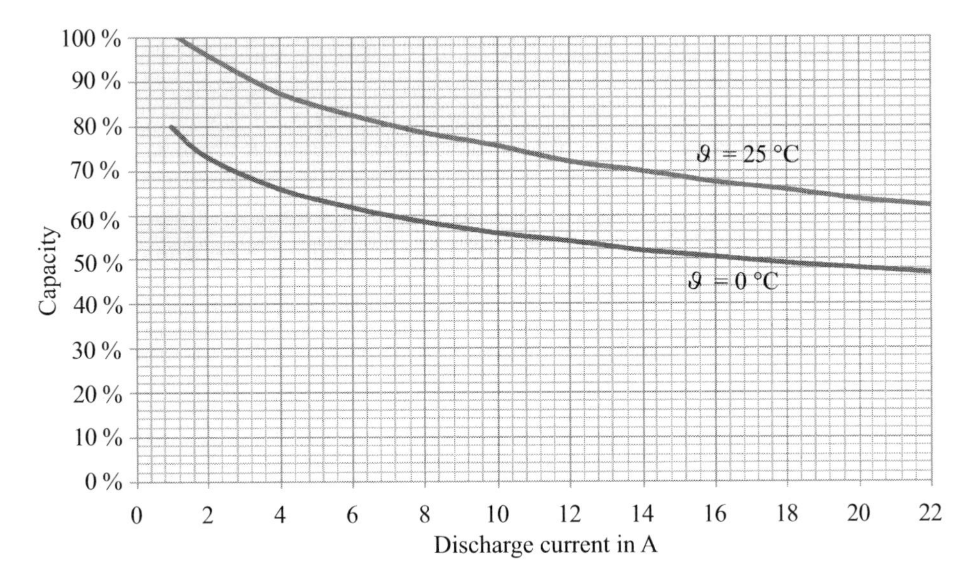

Figure 5.47 A lead battery's available capacity with C_{100} = 100 Ah relative to discharge current and temperature

Table 5.12 Open-circuit voltage and acid density relative to the state of charge of a 12 V lead battery

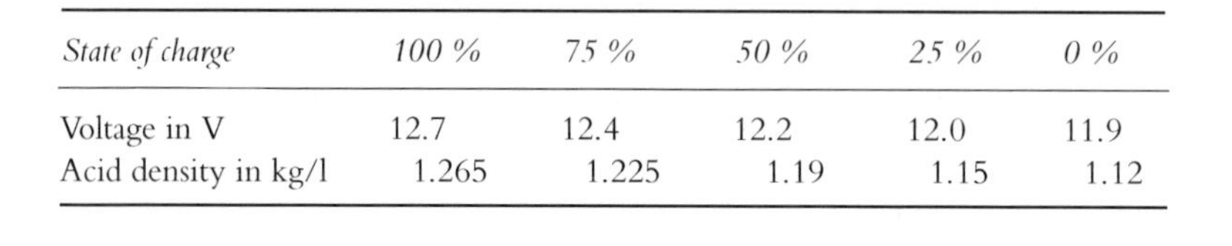

State of charge	*100 %*	*75 %*	*50 %*	*25 %*	*0 %*
Voltage in V	12.7	12.4	12.2	12.0	11.9
Acid density in kg/l	1.265	1.225	1.19	1.15	1.12

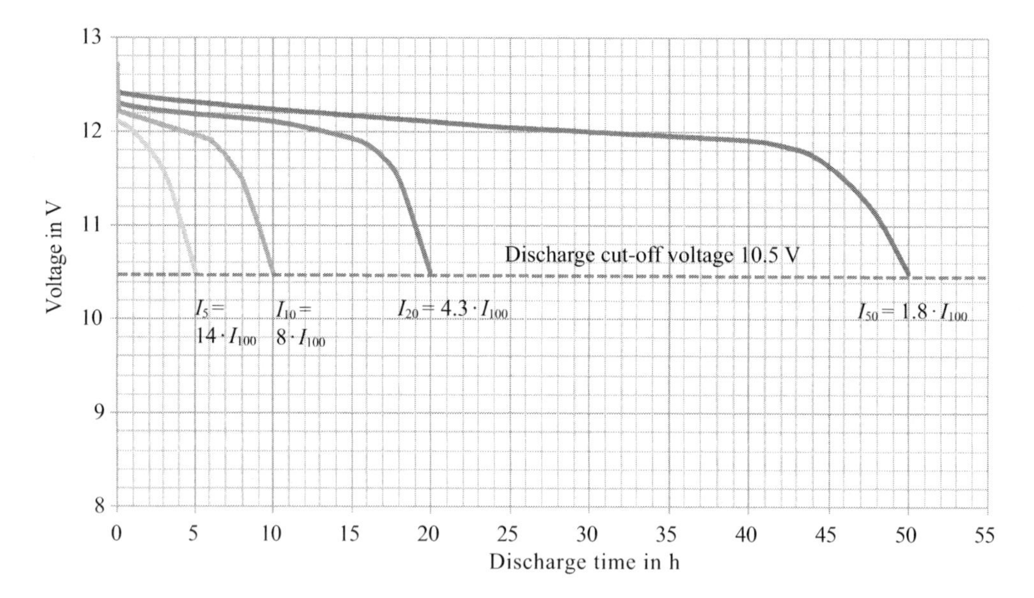

Figure 5.48 Voltage curves relative to discharge time and discharge current

below-circuit voltage. Figure 5.48 shows the voltage curve across the duration of discharge. Starting with a fully charged battery's open-circuit voltage of 12.7 V, voltage quickly drops relative to the charge current. If the initial charge is low, voltage drops more quickly.

The battery's age also has a minor influence on its voltage. Table 5.13 shows how voltage can be used to determine a battery's state of charge.

The battery should not be excessively charged or discharged. If the battery is completely discharged, lead sulphate crystallizes, making it hard to recharge and reconvert completely. The battery is then permanently damaged. Deep discharges should therefore be avoided at all cost. A residual capacity of around 30 % – equivalent to a battery voltage of around 11.4 V – suffices to prevent this damage. At high discharge currents exceeding I_{10}, a low end-of-charge voltage may be recommendable. When batteries are stored for a long time, self-discharge can also cause damage, so the batteries should be recharged frequently.

When a battery is charged, it begins to 'gas' at a voltage of 14.4 V. In other words, the water in the electrolyte is broken up into hydrogen and oxygen. These gases escape from the battery. The battery therefore has to be refilled with water in regular intervals. If this **gassing** process continues for a long time, it can damage the battery. Charging should therefore be restricted to 13.8 V to 14.4 V. In contrast, charging the battery into the gassing phase occasionally is a good idea in order to mix up the electrolyte.

Table 5.13 What a 12 V battery's voltage tells us about its state of charge

Voltage range	*State of charge*
Greater than 14.4 V	Charging discontinued, battery fully charged
13.5 V–14.1 V	Normal voltage range when charging without consumption load
12.0 V–14.1 V	Normal voltage range when charging with consumption load
11.5 V–12.7 V	Normal voltage during discharging
11.4 V	Consumption stops, charging begins

The battery should be stored in a dry room in which temperatures do not drop too low. The gassing process can create explosive oxyhydrogen, so make sure that rooms in which batteries are stored are always **well ventilated**.

If exact simulations are needed, the battery may need to be described by using an electrical equivalent circuit, which has turned out to be a very difficult task. Today, the best model to represent a battery is still a black bucket with a hole in it, as technicians joke. Nonetheless, various attempts have been made to model batteries, such as by Gretsch [Gre78].

Another type of description is the calculation of state of charge. It only requires consideration of a few properties. If the battery is completely charged, it cannot be charged further. To provide protection against deep discharges, the battery should not be discharged below half of its capacity. During charging and discharging, the charge efficiency must be taken into account along with charge losses from self-discharging. Most simulation programs are based on this principle. Figure 5.49 shows a flowchart of this calculation, which is very easy to implement in a computer program. A battery system can be optimized by varying the photovoltaic output and battery capacity.

Other battery types

As mentioned above, other more expensive systems, such as NiMH and Li-ion, are used along with lead batteries because of their greater energy density, faster charge, and longer service life.

Nickel-cadmium batteries were very widespread for a long time. Compared with lead batteries, they have a slightly greater energy density, a greater number of cycles, and a wider temperature range. However, this battery is increasingly unpopular because cadmium is a dangerous substance. The material used within the battery might enter the environment after the batteries are disposed of. Cadmium builds up in the food chain and is absorbed by the human body, which has a hard time completely eliminating it. At sufficient concentrations, cadmium can damage organs and cause cancer. NiCd batteries are therefore now banned for most applications.

Nickel-metal hydride (NiMH) contain substances that are much less problematic, though small amounts of toxic substances are still added to such batteries in practice. Alloys consisting of nickel, titanium, vanadium, zirconium, and chrome are used. The electrolyte in these batteries and in NiCd batteries is diluted caustic potash. NiMH batteries have other benefits compared with NiCd batteries, such as greater density and the lack of a memory effect. In return, they have a low operating temperature range and a very high self-discharge rate (around 1 % per day). As with NiCd batteries, the cell voltage is 1.2 V, so NiMH batteries can easily replace NiCd batteries.

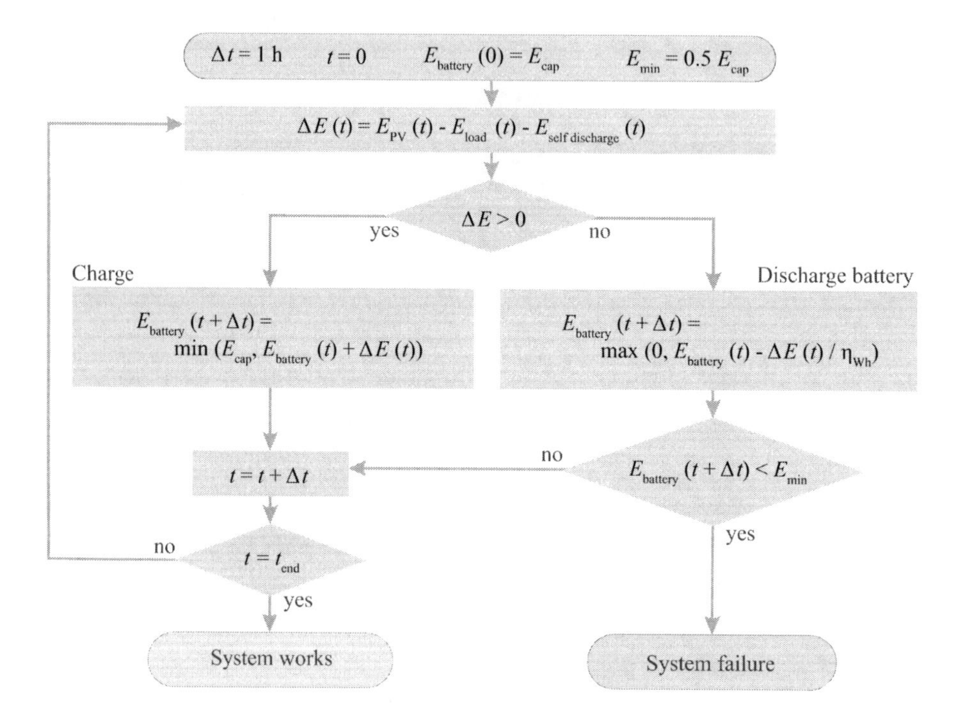

Figure 5.49 Flowchart of the calculation for the state of charge of PV battery systems

A NiMH battery's chemical reactions are as follows:
Negative electrode:

$$MH + OH^- \underset{\leftarrow Charge}{\overset{Discharge \rightarrow}{\longleftrightarrow}} M + H_2O + e^- \quad (5.101)$$

Positive electrode:

$$NiOOH + H_2O + e^- \underset{\leftarrow Charge}{\overset{Discharge \rightarrow}{\longleftrightarrow}} Ni(OH)_2 + OH^- \quad (5.102)$$

Net reaction:

$$NiOOH + MH \underset{\leftarrow Charge}{\overset{Discharge \rightarrow}{\longleftrightarrow}} Ni(OH)_2 + M \quad (5.103)$$

It is much harder to monitor the state of charge with NiCd and NiMH band with lead batteries based on battery voltage because the former are more sensitive to temperature fluctuations, and their voltage even slightly decreases when they are fully charged.

Lithium-ion batteries also have no memory effect and charge quickly. In addition to having great energy density, this battery type also has a low rate of self-discharge. Lower discharge currents are also possible than with batteries that have water-based electrolytes.

The first generation of lithium-ion batteries fell into disrepute because short-circuits and excessive charging caused fires and explosions. Modern cell safety measures now largely rule out such problems. Because of their energy density, lithium-ion batteries are mainly used in mobile applications. New developments in the automotive sector and in stationery PV systems could lead to a boom in the use of these batteries in the next few years. The chemical reactions of a lithium-ion battery – of which there are different versions – are as follows:

Negative electrode:

$$\mathrm{Li}_x\mathrm{C}_n \underset{\leftarrow \text{Charge}}{\overset{\text{Discharge}\rightarrow}{\longleftrightarrow}} n\ \mathrm{C} + x\ \mathrm{Li}^+ + x\ \mathrm{e}^- \tag{5.104}$$

Positive electrode:

$$\mathrm{Li}_{1-x}\mathrm{Mn}_2\mathrm{O}_4 + x\ \mathrm{Li}^+ + x\ \mathrm{e}^- \underset{\leftarrow \text{Charge}}{\overset{\text{Discharge}\rightarrow}{\longleftrightarrow}} \mathrm{LiMn}_2\mathrm{O}_4 \tag{5.105}$$

Net reaction:

$$\mathrm{Li}_{1-x}\mathrm{Mn}_2\mathrm{O}_4 + \mathrm{Li}_x\mathrm{C}_n \underset{\leftarrow \text{Charge}}{\overset{\text{Discharge}\rightarrow}{\longleftrightarrow}} \mathrm{LiMn}_2\mathrm{O}_4 + n\ \mathrm{C} \tag{5.106}$$

In addition to lithium-ion batteries, **sodium-sulphur** (NaS) and **sodium-nickel-chloride** (NaNiCl) batteries are alternatives. NaNiCl batteries are also called **ZEBRA batteries**. A South African developer coined the acronym, which stands for Zeolite Battery Research Africa Project. Both battery types have very great energy density and a very high Ah efficiency.

A sodium-sulphur battery's chemical reactions are as follows:

Negative electrode:

$$2\ \mathrm{Na} \underset{\leftarrow \text{Charge}}{\overset{\text{Discharge}\rightarrow}{\longleftrightarrow}} 2\ \mathrm{Na}^+ + 2\ \mathrm{e}^- \tag{5.107}$$

Positive electrode:

$$x\mathrm{S} + 2\ \mathrm{e}^- \underset{\leftarrow \text{Charge}}{\overset{\text{Discharge}\rightarrow}{\longleftrightarrow}} \mathrm{S}_x^{-2} \tag{5.108}$$

Net reaction:

$$2\ \mathrm{Na} + x\mathrm{S} \underset{\leftarrow \text{Charge}}{\overset{\text{Discharge}\rightarrow}{\longleftrightarrow}} \mathrm{Na}_2\mathrm{S}_x \tag{5.109}$$

NaS batteries have a drawback: an operating temperature of around 300 °C. Lithium-ion batteries are thus considered to have a better chance on the market because they do not constantly consume energy for heat.

Battery systems

The simplest battery systems consist of a photovoltaic generator, a battery, and a device that consumes power. Because the photovoltaic generator has relatively small interior resistance, the battery discharges at low levels of irradiance in the evening and overnight via the generator. A **blocking diode** is used between the battery and the photovoltaic generator to prevent such power losses, as shown in Figure 5.50. At the diode, the power losses constantly occur:

$$P_{\text{loss,diode}} = I_{\text{PV}} \cdot U_{\text{D}} \tag{5.110}$$

so diodes with a low forward voltage drop are used, such as Schottky diodes ($U_{\text{D}} \approx 0.55$ V).

Losses also occur in the **feed lines** with cross-section A, specific resistance ρ and lengths l_1 and l_2 leading to and back from the battery:

$$P_{\text{loss,line}} = I_{\text{PV}} \cdot (U_{\text{L1}} + U_{\text{L2}}) = I_{\text{PV}}^2 \cdot (R_{\text{L1}} + R_{\text{L2}}) = I_{\text{PV}}^2 \cdot \frac{\rho}{A} \cdot (l_1 + l_2) \tag{5.111}$$

Copper cables ($\rho_{\text{Cu}} = 0.0175\ \Omega\ \text{mm}^2/\text{m}$) with a length of ($l_1 = l_2 = 10$ m) and a cross-section of $A = 1.5\ \text{mm}^2$ at a current of $I_{\text{PV}} = 6$ A have significant line losses of $P_{\text{loss, line}} = 8.4$ W. With a 100 W photovoltaic generator, these losses make up almost 12 % of the output power if the diode pass-through losses of 3.3 W are included. To keep these losses down, feed lines should be as short as possible and have large cross-sections. In 12 V systems, the voltage should not drop by more than 3 % or 0.35 V in the lines from the photovoltaic generator to the battery or by more than 7 %, equivalent to 0.85 V, in the lines from the battery to the device consuming power. In the example above, the cables should therefore have a cross-section of 6 mm².

When the lines are longer, losses can be reduced if the system voltage is increased, such as when multiple batteries are in a series circuit, which increases battery voltage U_{Bat}.

The battery at the photovoltaics generator is:

$$U_{\text{PV}} = U_{\text{bat}} + U_{\text{D}} + U_{\text{L1}} + U_{\text{L2}} \tag{5.112}$$

Here, diode voltage U_{D} is nearly constant, while the voltage drops U_{L1} and U_{L2} within the lines proportional to the photovoltaic current I_{PV}. The battery voltage U_{Bat} is relative to the charge current and the state of charge.

Overall, the voltage at the photovoltaics generator slightly increases along with the current – in other words, as irradiation increases – and it varies slightly with the battery's state of charge. When a solar generator is directly connected to a battery, the power point is generally properly set even if irradiation incident on the solar generator changes, as shown in Figure 5.51. For this reason, voltage converters and MPP trackers in combination with battery systems are not often used because the power consumed by the additional electronics may be higher than the additional energy gains. An MPP tracker may, however, be useful if the solar generator is run under fluctuating irradiation for a long time.

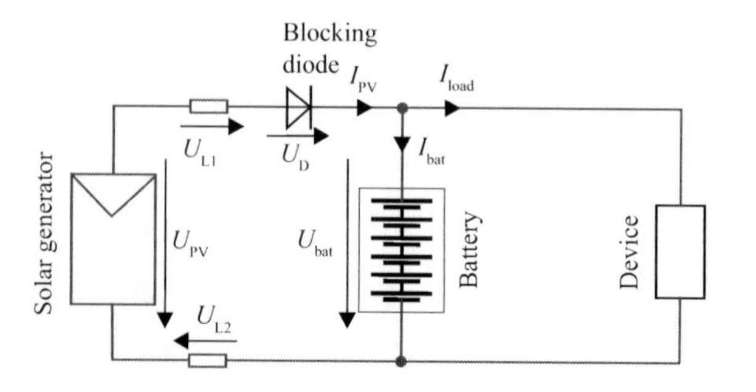

Figure 5.50 A simple photovoltaic system with battery storage

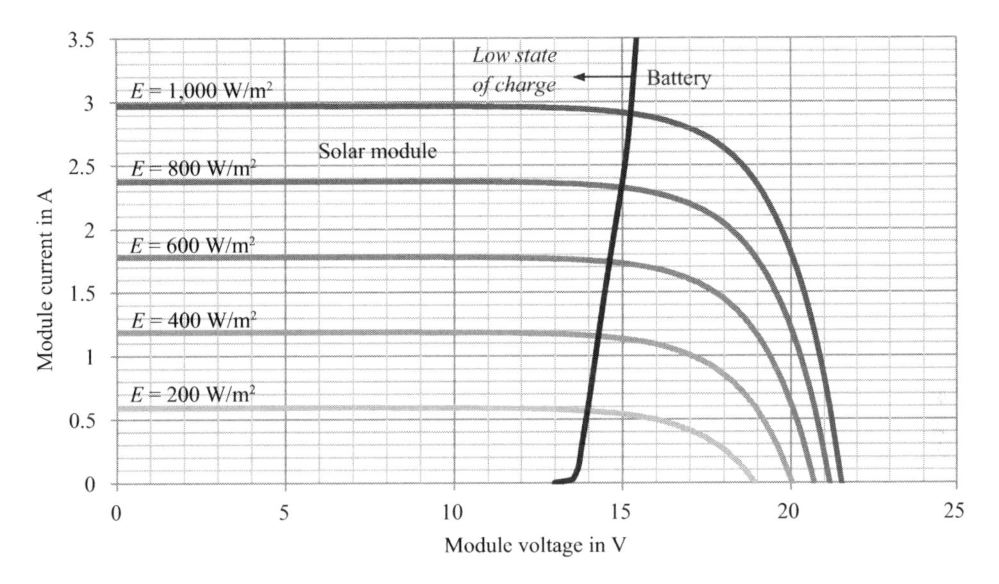

Figure 5.51 A solar generator's power points with a lead battery for storage with a blocking diode and 0.1 Ω line resistance without load

In a simple system in which the photovoltaic generator and power consumers are directly connected to the battery, the battery is not protected from excessive charging or deep discharging. Such systems should therefore only be used when states of charge detrimental to the battery can be prevented. Otherwise, the battery would quickly be damaged and become worthless.

Most battery systems therefore have a **charge controller** (Figure 5.52). With lead batteries, charge controllers generally monitor voltage. Specifically, they measure the battery's voltage U_{Bat}.

If it drops below the lower limit for voltage of deep discharge (around 11.4 V for a 12 V lead battery), switch S_2 separates power consumers from the battery. Once the battery has recovered – that is, once the battery's voltage has reached a minimum level – devices are reconnected. Switch S_1 stops the battery from being charged further if its voltage exceeds its cutoff level (around 14.4 V for a 12 V lead battery). Here, a distinction is made between **series controllers** (Figure 5.53) and **shunt controllers** (Figure 5.54).

Power semiconductors and field effect transistors (Power MOSFETS) are used as the switches here. The drawback of serial controllers is that pass-through losses always occur at switch S_1 when the battery is being charged. In contrast, good MOSFETs have a resistance smaller than 0.1 Ω but the losses that field effect transistor BUZ11 entails with a pass-through resistance of 0.04 Ω at 6 A still at up to 1.44 W. If the solar generator's voltage is monitored along with the battery's, no blocking diode is needed as long as the charge controller opens switch S_1 if the solar generator's voltage drops below the battery's.

Shunt controllers are more widely used. When the battery is fully charged, this controller short-circuits the solar generator via switch S_1. The solar generator's voltage drops to the voltage loss at the switch (< 1 V), and the blocking diode prevents current from flowing

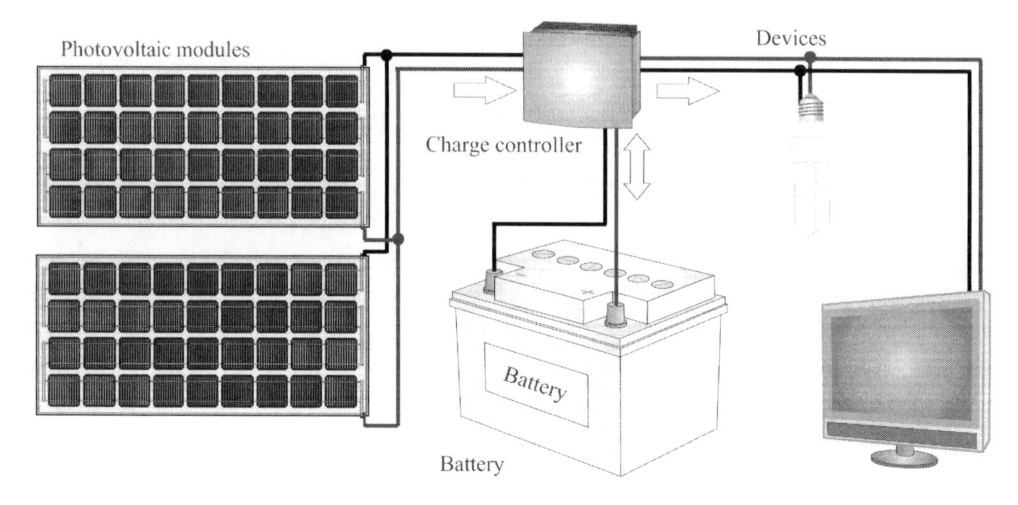

Figure 5.52 A photovoltaic battery system with a charge controller [Qua09]

from the battery via the switch. In normal operation, the short circuit does not pose any problems for the solar generator. If, however, the solar module is shaded unevenly, the load on shaded cells can be tremendous.

Individual battery cells are sometimes used to differing extents, especially in large battery systems (such as those used in electric vehicles), so these cells age more quickly. If a cell fails, it detrimentally affects other cells as well. It is therefore a good idea to use battery management systems that not only measure overall battery voltage, but also provide optimal protection for individual cells.

In addition to simple systems consisting only of a photovoltaic generator, battery, charge controller, and consumption device, there are also larger, more complex systems for off-grid power supply, such as to mountain cottages. Usually, the photovoltaic generator is combined here with another source of energy, such as a wind or diesel generator. The combination reduces costs and increases supply security. Such a combination, however, requires a more complex energy management system. Usually, some appliances that consume power will not be available in DC versions. If AC devices are used, an inverter is required (described further down below).

Other storage options

In addition to electrochemical storage in batteries, there are other ways to store electrical energy. The storage options include:

- capacitors
- superconducting coils
- flywheels
- pumped hydropower
- compressed air
- synthetic gas, methane, and hydrogen.

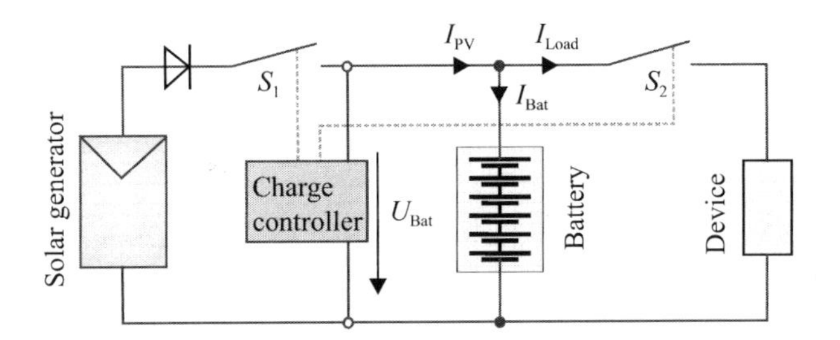

Figure 5.53 A photovoltaic battery system with a serial charge controller

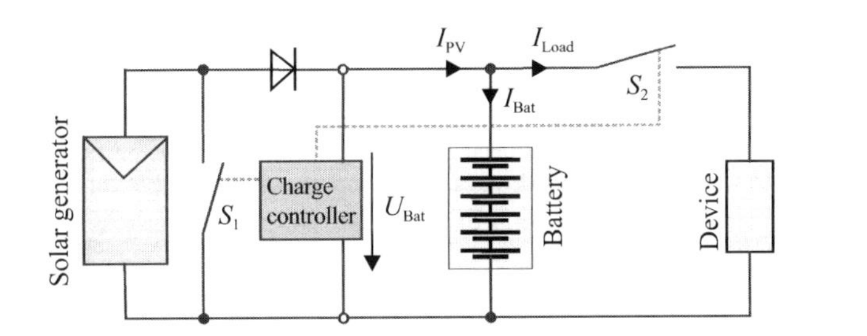

Figure 5.54 Photovoltaic battery system with a shunt controller

For off-grid solar energy systems, these storage options play only a minor role. Pumped hydro storage is discussed in greater detail in the chapter on hydropower; likewise, hydrogen production and storage are dealt with in that chapter.

Inverters

Inverter technology

Up to now, the discussion has assumed that all devices consuming power run on direct current. Power from the grid, however, is alternating current. An off-grid inverter is required to run common appliances connected to an off-grid PV system. If the power from a photovoltaic system is to be exported to the grid, a grid inverter is used. These two types of inverters use similar technology, but there are some decisive differences in terms of design and requirements.

Switchable valves are needed when using power electronics to turn direct-current into alternating current. The following power semiconductor elements can be used for this purpose; depending on the design, the voltages can clearly exceed 1,000 V, or the currents can exceed 1,000 A:

- power MOSFETs (power field effect transistors)
- bipolar power transistors
- IGBTs (insulated gate bipolar transistors)
- thyristors (controllable diodes)
- triacs (dual-direction thyristors)
- GTO thyristors (gate turn-off thyristors).

Below, we take a closer look at only **MOSFETs** (metal-oxide semiconductor field effect transistors). Figure 5.55 shows the circuit of a self-blocking n-channel MOSFET with three connections: G (Gate), S (Source) and D (Drain).

If there is voltage at input G and voltage U_{GS} is greater than threshold voltage U_{th}, the transistor switches. It then becomes conductive between the inputs S and D. If voltage U_{GS} falls below the threshold voltage, the transistor blocks. A body diode is integrated in the field effect transistors between inputs S and D. It facilitates switching because it can derive excess voltage from inductance within the circuit. Because body diodes are usually relatively slow, an additional parallel, fast Schottky diode can be useful.

In an inverter, current periodically has to be transferred from one branch to another. This process is also called **commuting**. To this end, various valves have to be opened or closed. Inverters used to require external controllers with thyristors, but today most inverters have their own controllers with transistors or MOSFETs as switches.

Rectangle and trapezoid inverters are very straightforward and are therefore described below in greater detail first. Because the quality of the output current from these inverters is not very great, pulse-width modulation inverters and resonant inverters described below are more commonly used in photovoltaics.

Rectangle inverter

A **B2 or H bridge** is a very simple way to create an inverter circuit, as shown in Figure 5.56.

It consists of four valves connected to the alternating current grid via a transformer. Below, inverters connected to a fixed grid are discussed further.

At regular intervals, valves one and three are opened, followed by two and four, to produce a roughly rectangular alternating current in the transformer. Valves 1 and 2 can also be replaced with non-controllable diodes so that only half of the components need to be controlled. In this case, the bridge is half-controlled. The valves are switched with a delay around control angle α for the voltage's zero-crossing. Figure 5.52 shows the power curve for a B2 circuit.

The curve looks very different from the desired sine curve. A Fourier analysis – also called a **harmonic analysis** – assesses the power curve. A 2π periodic function f(ω*t*) can be broken down into a convergent series in the interval $-\pi \leq \omega t \leq +\pi$:

$$f(\omega t) = \frac{a_0}{2} + \sum_{n=1}^{\infty} \left(a_n \cdot \cos(n \cdot \omega t) + b_n \cdot \sin(n \cdot \omega t) \right) \tag{5.113}$$

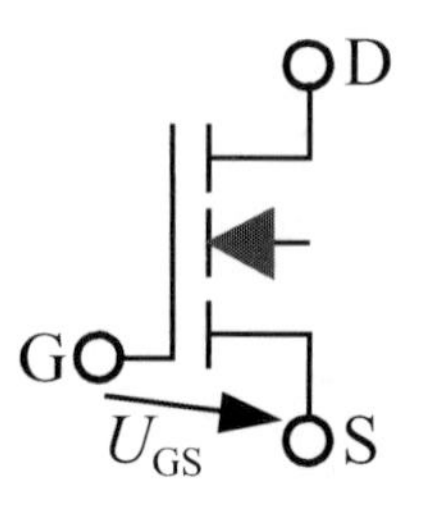

Figure 5.55 Circuit of a self-blocking n-channel MOSFET

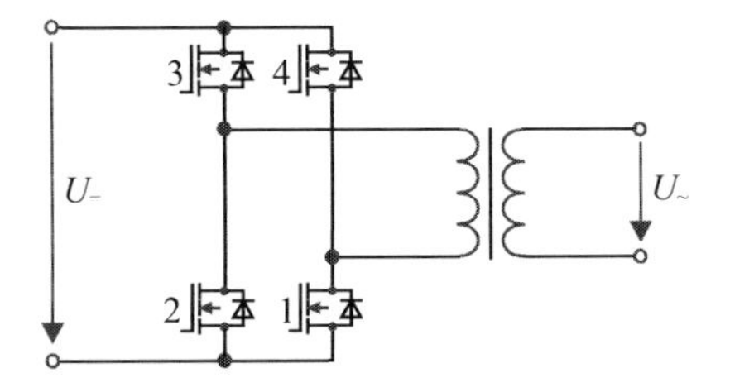

Figure 5.56 H-bridge (B2)

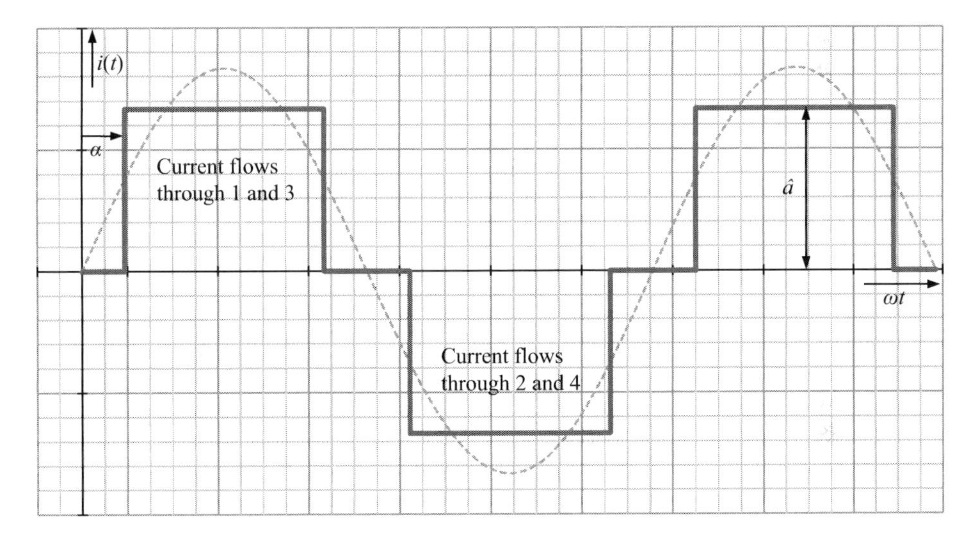

Figure 5.57 An idealized power curve for a semi-controlled B2 bridge

Coefficients a_n and b_n can be determined by means of:

$$a_n = \frac{1}{\pi} \cdot \int_{-\pi}^{\pi} f(\omega t) \cdot \cos(n \cdot \omega t) \, d\omega t \quad (n = 0, 1, 2, \ldots) \tag{5.114}$$

$$b_n = \frac{1}{\pi} \cdot \int_{-\pi}^{\pi} f(\omega t) \cdot \sin(n \cdot \omega t) \, d\omega t \quad (n = 1, 2, \ldots) \tag{5.115}$$

The following harmonic analysis can be performed for the current's rectangular shape with the amplitude $\hat{a}$ and control angle α of the B2 bridge:

$$f(\omega t) = \frac{4 \cdot \hat{a}}{\pi} \cdot \left[\cos\alpha \cdot \sin\omega t + \tfrac{1}{3} \cdot \cos 3\alpha \cdot \sin 3\omega t + \tfrac{1}{5} \cdot \cos 5\alpha \cdot \sin 5\omega t + \cdots\right] \tag{5.116}$$

In addition to the desired sinusoidal fundamental frequency (atomic number 1), there is also the function of oscillations of other periodic durations (atomic number ≥ 2), undesired

harmonics. Figure 5.58 shows the fundamental frequency and harmonics up to ordinal number 7. If harmonics are added up to ordinal number 7, the rectangle is already visibly mapped, as the thick black curve shows. Additional harmonics would have to be included for a more precise mapping. In practice, an oscilloscope and subsequent computer analysis can be used to perform a harmonic analysis. More complex devices can even determine the share of harmonics independently.

In general, the current can be represented based on the fundamental frequency with amplitude $\hat{i}_1$ and the harmonics with amplitudes $\hat{i}_2$, $\hat{i}_3$ etc.:

$$i(t) = \hat{i}_1 \cdot \sin(\omega t) + \hat{i}_2 \cdot \sin(2 \cdot \omega t) + \hat{i}_3 \cdot \sin(3 \cdot \omega t) + \quad \ldots \tag{5.117}$$

Class A devices with an input current ≤ 16 A can cause harmonic currents in the low-voltage level of the grid, so VDE 0838 part 2 and DIN EN 61000–3–2 specifies maximum values for harmonic currents [DIN05].

In addition to harmonic currents, the inverter can also cause harmonic voltages. VDE 0839 part 2–2 and DIN EN 61000–2–2 specify the admissible harmonic voltages relative to the fundamental frequency's voltage [DIN03]. If multiple harmonics occur simultaneously, the **total harmonic distortion** (THD)

$$D = \sqrt{\sum_{n=2}^{40} \left(\frac{\hat{u}_n}{\hat{u}_1} \right)^2} \tag{5.118}$$

expresses the effect. It is calculated from the voltage amplitudes $\hat{u}_n$ of the harmonics and the fundamental frequency's voltage amplitude $\hat{u}_1$. The maximum admissible total harmonic distortion is $D = 0.8$. In addition to amplitudes, the calculations often use root mean squares.

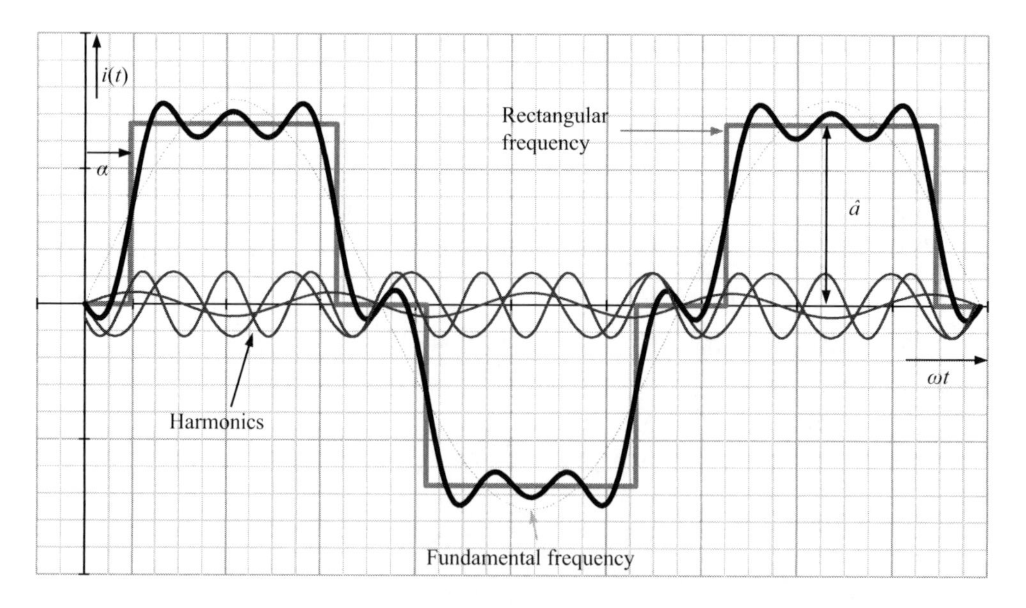

Figure 5.58 Making a rectangular curve out of various sinusoidal waves

Harmonic distortion is another quality factor:

$$k = \sqrt{\frac{\sum_{n=2}^{\infty} U_n^2}{\sum_{n=1}^{\infty} U_n^2}} \tag{5.119}$$

It is derived from the ratio of the root mean squares for harmonics to the root mean squares for fundamental frequencies and harmonics. The harmonic distortion of good inverters is below 3 %. If the harmonics are too great, filters must be used to reduce them.

Often, transformers are integrated in inverters. Transformers separate the inverter from the grid and can adjust voltage between the inverter's and the grid's voltage. However, transformers always lead to losses and can be dispensed with in principle. But then, greater safety measures must be taken because the grid and the solar generator are no longer galvanically isolated.

The B2 bridge only supplies alternating current into only one phase conductor. On the grid, which has three phases, the result is asymmetries. Therefore, lines larger than 4.6 kVA have to have three-phase current in the three-phase conductors in Germany. Chapter 6 (wind power) discusses the basic properties of three-phase power in greater detail. Figure 5.59 shows the six-pulse bridge **(B6 bridge)** used in the inverter circuit to generate three-phase power. In this circuit, the valves are opened cyclically so that alternating current and voltage is produced in three phases shifted by a third each.

In addition to B2 and B6 bridges, others such as M2 and M3 midpoint circuits and other bridge circuits are used. These options work the same, however, as those described above.

Modern inverter topologies

In inverters that use **pulse width modulation (PWM)** one of the aforementioned circuits – such as B2 and B6 – is used. Here, the valves are not only opened and closed once every half-wave, but several times to produce pulses of different widths, as shown in Figure 5.60. After filtering, the fundamental wave is sinusoidal. The quality of a sine wave is much better than that of rectangular waves from inverters; specifically, there are fewer undesirable harmonics. For this reason, most inverters now used have PWM.

To produce especially high inverter efficiencies, other circuit concepts are used in addition to the bridges described above. Single-phase inverters sometimes have an **H5 or HERIC circuit** (Highly Efficient and Reliable Inverter Concept). They are based on the concept of H bridges. Other transistors ensure that choke current can be drawn off during switching in the freewheeling phase, which reduces switching losses.

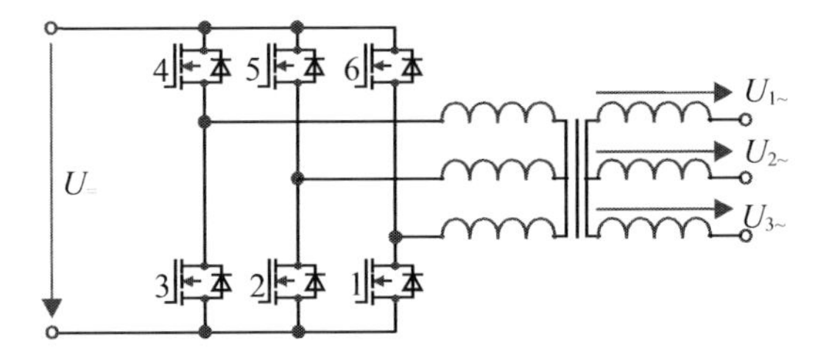

Figure 5.59 Six pulse bridge (B6)

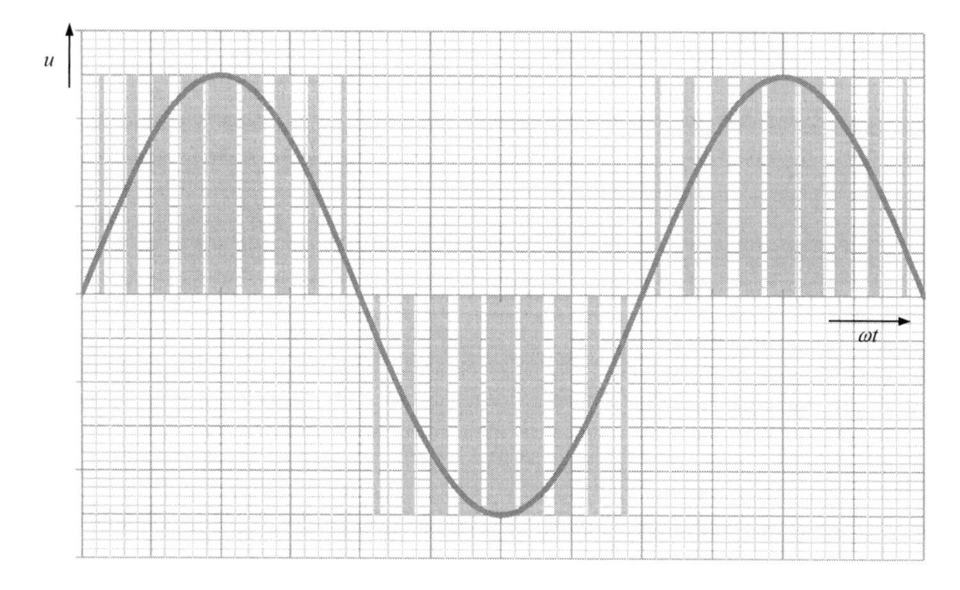

Figure 5.60 A curve from pulse width modulation

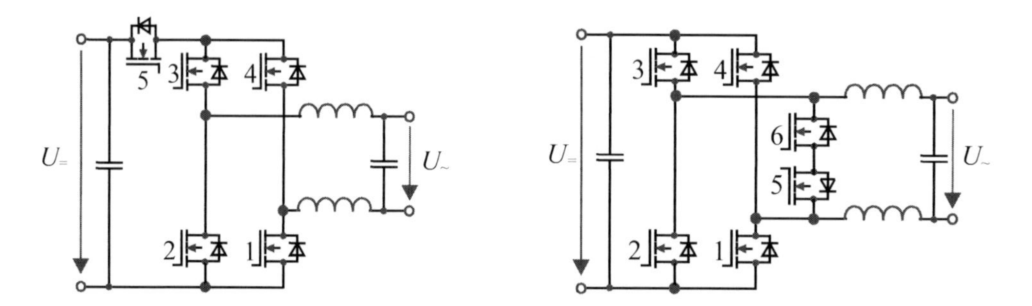

Figure 5.61 Highly efficient inverter circuits. Left: H5 topology, right: HERIC topology

Inverters for photovoltaics

An inverter's functions and tasks

Photovoltaic inverters perform far more tasks than merely converting direct into alternating current, as Figure 5.62 shows. The grid always determines the output voltage. The inverter has to adjust the DC voltage for the photovoltaic generator so that it always runs at its MPP. Good inverters have very high MPP tracking speeds. The **MPP tracking efficiency** can exceed 99 % considerably even under changing irradiance conditions. In other words, the losses due to deviations from the actual MPP are smaller than 1 %. A step-up converter between the photovoltaic generator and the bridge adjusts DC voltage. The bridge in a pulse width modulation inverter can also adjust DC voltage, however. The step-up converter is not needed then, and leaving it out improves the inverter's efficiency. Then, however, the solar generator's voltage needs to be high. The minimum MPP voltage for a single-phase inverter without a transformer is then around 350 V.

A **transformer** can adjust the alternating current voltage and provide galvanic isolation between the grid and the photovoltaic generator. Inverters without transformers are now commonly used in photovoltaics. By doing without a transformer, losses are reduced, thereby enabling especially great inverter efficiencies. To ensure safety, however, residual current has to be monitored universally and the grid connection reliably if there is a flaw. Before using an inverter that has no transformer, make sure that the photovoltaic modules you are using have been approved by the manufacturer for this combination. Certain moduels undergo potential-induced degradation (PID) when used with inverters that have no transformer. TCO corrosion in particular destroys the front contacts of thin-film modules and can destroy them in the process. Grounding one of the module's poles – generally the minus pole – considerably reduces the risk of degradation.

Inverters for off-grid applications differ fundamentally from those designed for grid connections. Off the grid, inverters often have a battery charge controller integrated in them. In return, grid-connected inverters have to fulfil specific criteria and perform certain tasks in line with grid quality, while such requirements are less strict for off-grid inverters. VDE-AR-N 4105 [VDE11] specifies these rules for low-voltage grid connections in Germany. The rules include maximum values for voltage changes and harmonic currents.

Comprehensive grid and equipment protection must perform the following functions and disconnect the photovoltaic system from the grid reliably:

- protection against voltage drops, $U < 184$ V
- protection against voltage increases, $U > 253$ V
- protection against frequency drops, $f < 47.5$ Hz
- protection against frequency increases, $f > 51.5$ Hz
- detection of grid connection loss (unplanned micro-grid).

The system can only be switched on again when the grid voltage lies within 195.5 and 253 V, with the grid frequency being between 47.5 and 50.05 Hz. At a grid frequency above 50.2 Hz, the power output has to be reduced linearly relative to frequency down to 48 % at 51.5 Hz.

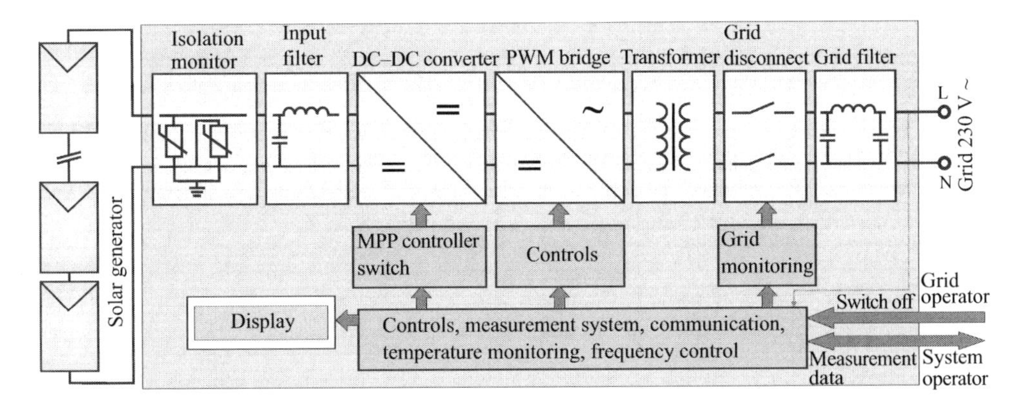

Figure 5.62 The components in a photovoltaic inverter

The unbalanced load must not exceed 4.6 kVA. Three unconnected, single-phase inverters distributed across three different phases could then produce a maximum AC power of 13.8 kVA. If the output needs to be greater, either three-phase inverters need to be used, or single-phase inverters need to be connected, though the unbalanced load still must not be allowed to exceed 4.6 kVA.

Starting at 3.68 kVA, the inverter also has to be able to provide reactive power as defined by the grid operator. Then, more photovoltaic systems can be integrated on the grid at a single connection point.

Inverter efficiencies

Inverters used in photovoltaics rarely work in nominal operation at nominal output P_N. Because the level of irradiance changes, the inverter runs at partial load over long periods of time. It is therefore important for the inverter to be efficient even when running at low output. The inverter should not be over-dimensioned. To take a simple example, if a 1 kW inverter is connected to a 500 W photovoltaic system, the inverter will only run at a maximum of 50 % of its nominal output. Greater losses can be expected if the inverter runs constantly at partial load. The inverter itself should consume as little power as possible. Shutting it off at night is one way to reduce its power consumption.

One way of comparing inverters is in terms of their European efficiency:

$$\begin{aligned}\eta_{\mathrm{Euro}} &= 0.03\cdot\eta_{5\%} + 0.06\cdot\eta_{10\%} + 0.13\cdot\eta_{20\%} \\ &\quad +0.1\cdot\eta_{30\%} + 0.48\cdot\eta_{50\%} + 0.2\cdot\eta_{100\%}\end{aligned} \tag{5.120}$$

It represents the average weighted efficiency. Account is taken of the inverter's response at partial loads that commonly occur under average irradiation levels in Western Europe.

Figure 5.63 shows the **efficiency curve** relative to input power. Modern inverters reach maximum efficiencies of up to 98 %. Even under partial loads, efficiencies are still high, and they only collapse under extreme partial loads. The figure also shows the energy output for this particular DC system output relative to the overall yield for Berlin. The share of very high outputs is small because such levels only occur a few hours per year. The middle range makes up the biggest share, which is why efficiency at 50% partial load has such a heavy weighting in European efficiency. The calculations described here are based on hourly averages of irradiance. If more realistic, shorter intervals are used for irradiance, the energy shares shift towards greater outputs.

In other parts of the world, however, levels of irradiation are very different, so the energy shares look a lot different. Figure 5.64 shows the shares of energy relative to hourly DC system output, but this time for Los Angeles, not Berlin. Clearly, greater system output is more frequent. The California Energy Commission (CEC) has also defined an average efficiency with a much greater weighting of high power outputs than is used in the calculation of European efficiency. **CEC efficiency** is defined as follows:

$$\eta_{\mathrm{CEC}} = \frac{1}{3}\sum_{i=1}^{3}\left.\begin{pmatrix}0.04\cdot\eta_{10\%} + 0.05\cdot\eta_{20\%} + 0.12\cdot\eta_{30\%} \\ +0.21\cdot\eta_{50\%} + 0.53\cdot\eta_{75\%} + 0.05\cdot\eta_{100\%}\end{pmatrix}\right|_{U_i} \tag{5.121}$$

In addition to operation under partial load, the inverter's input voltage influences its efficiency. Figure 5.64 shows the efficiency curve for two different voltages. CEC efficiency

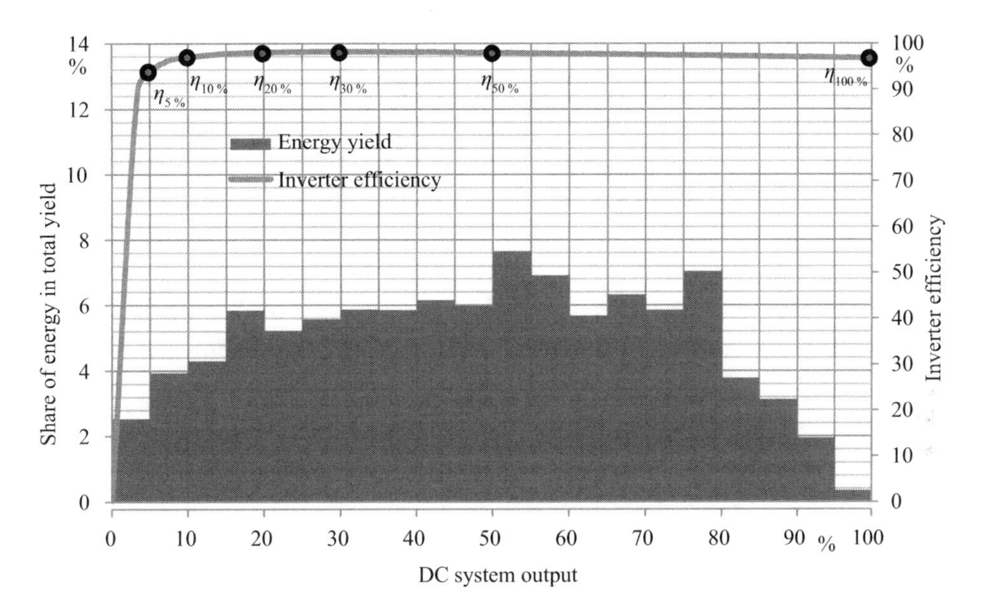

Figure 5.63 Inverter efficiency relative to DC system output and typical energy shares for the average hourly system output in overall yield for Berlin

also takes account of these factors by determining the average efficiencies for three different DC voltages ($U_1 = U_{MPP,min}$, $U_2 = U_{MPP,nom}$, and $U_3 = U_{MPP,max}$) at the lower end of the inverter's MPP range, under nominal voltage, and at the top end of the range.

Table 5.14 shows the **technical data** for a number of inverters. The inverters shown cover the entire output range from module inverters up to central megawatt inverters.

Modern inverters have very high efficiencies. The maximum efficiency can be 98 % at 6 kVA. The use of new silicon-carbide (SiC) semiconductor transistors might even increase efficiency to 99 %.

System concepts

There are three basic concepts for grid-connected inverters in photovoltaic systems:

- central inverters
- string inverters
- module inverters (micro-inverters).

Figure 5.65 shows solar modules connected to a central inverter. Here, modules are switched in series until the desired DC voltage is attained. For greater outputs, multiple strings of the same size are connected in parallel. The string diodes shown in the graphic often do not need to be used because they provide little protection; leaving them out can be a good idea because they always cause pass-through losses.

The **master–slave principle** is increasingly applied to improve performance under partial load. Here, multiple inverters are interconnected to produce one big inverter. Figure 5.66

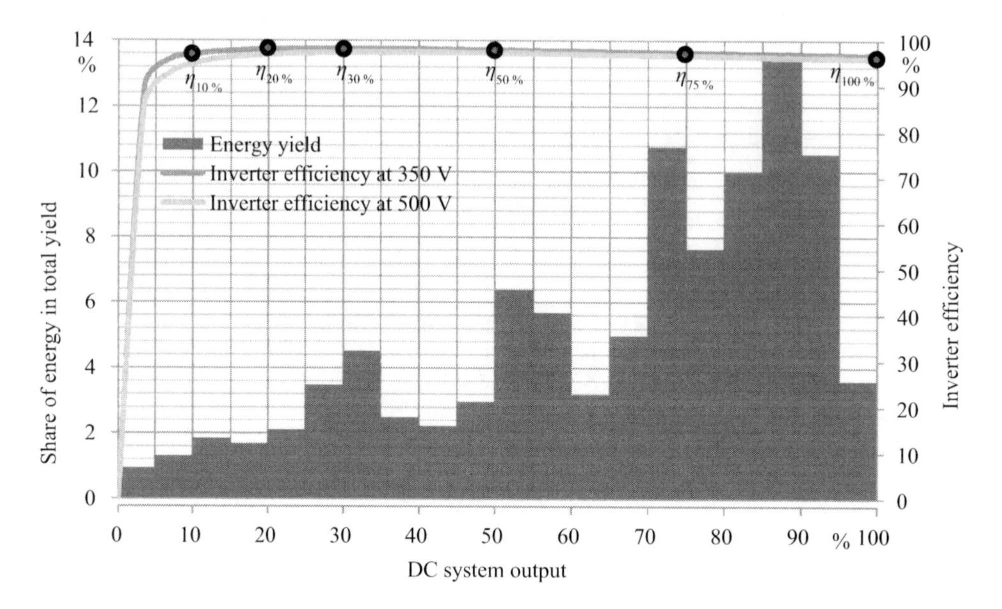

Figure 5.64 Inverter efficiency relative to DC system output at two different DC voltages and typical energy shares for the average hourly system output in overall yield for Los Angeles

Table 5.14 Selected technical data for a number of photovoltaic inverters

Device		*Enecsys SMI-D360*	*Sunways NT2500*	*SMA 6000TL*	*REFUSOL 15 K*	*SINVERT PVS2400*
Topology		Enecsys	HERIC	H5	UtraEta	PVS
Phases		1	1	1, connectable	3	3
Nominal output AC	kVA	0.34	2.5	6.0	15.0	2,400
Max. AC power	kVA	0.34	2.5	6.0	16.5	2,400
Max. DC power	kW	0.38	2.65	6.2	17.5	2,452
Max. DC voltage	V	54	900	700	900	1,000
MPP range DC	V	30–42	340–750	333–500	460–800	570–750
Max. DC current	A	13.4	7.5	19	36	4416
Power exported starting at	W	n/a	4	10	20	n/a
Overnight consumption	W	<0.03	<0.1	0.25	<0.2	n/a
Harmonic distortion, AC current	%	<5	<3	<4	<2.5	n/a
Max. efficiency	%	95.4	97.8	98.0	>98.0	98.7
European efficiency	%	93.5	97.4	97.7	97.7	98.6
Mass	kg	1.8	26	31	38	8520

shows the principle of a central inverter consisting of three units. Initially, only one inverter outputs power. At a third of overall output, it is running at maximum capacity. When the generator's output increases, additional units are switched on one by one.

Under partial shading and when individual modules have lower output, the central inverter concept is particularly unproductive because the losses are above average in such cases. If partial shading is expected or if individual parts of the photovoltaic system have a different orientation, it makes sense to separate strings or use smaller units or individual string inverters. Another benefit is that the length of the direct current lines needed is also shorter.

Micro-inverters are the optimal solution for partial shading. They allow a different voltage to be set for each module. Cabling is also much more straightforward because no direct current lines are needed at all. Another benefit is that the photovoltaic system is easy to expand. The lower inverter efficiency and higher cost largely outweigh these benefits, however, so that micro-inverters have not become the standard. Figure 5.67 shows the two options.

Direct consumption of photovoltaics

The following sections discuss first a simple off-grid combination of a photovoltaic generator and a battery and, second, a simple grid-connected system. These two options have dominated photovoltaics up to now. But the trend is towards more complex systems optimized for the direct consumption of solar power.

In numerous regions, the cost of solar power is now lower than the retail electricity rate. Germany reached this level of 'grid parity' in 2012. Straightforward grid-connected systems will remain attractive as long as Germany has its current feed-in tariffs. It is also clear that photovoltaic systems will have to make do without high feed-in tariffs in the future.

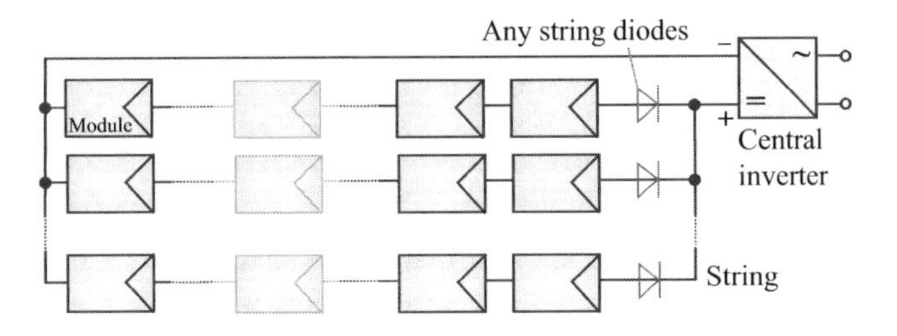

Figure 5.65 A photovoltaic system consisting of multiple strings connected to a central inverter

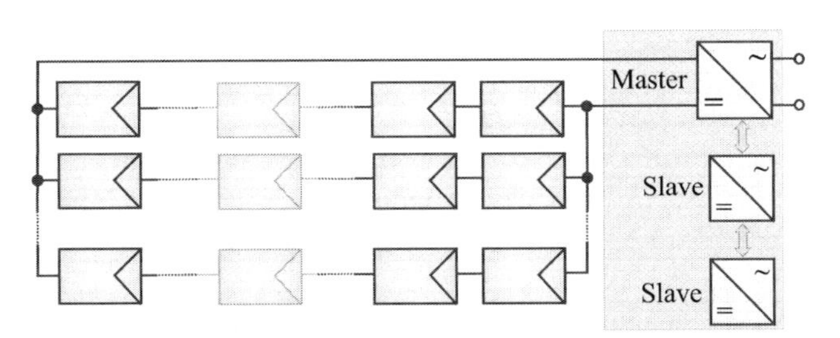

Figure 5.66 Central master–slave inverter for photovoltaic system

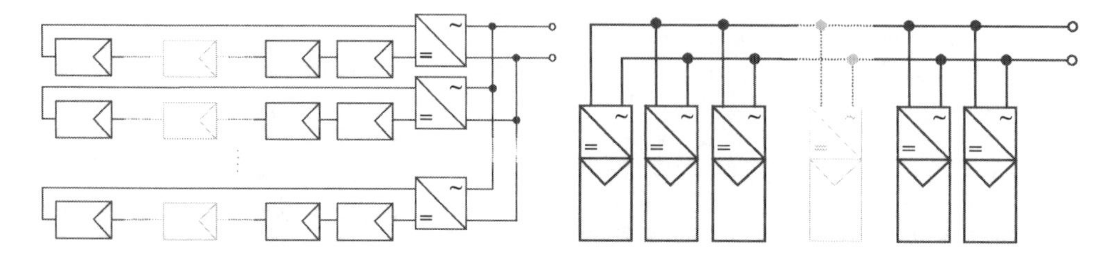

Figure 5.67 Photovoltaic generator with string inverters (left) and micro-inverters (right)

If the owner of the photovoltaic system consumes most of its power directly, photovoltaic systems still pay for themselves. But this option generally only works with extremely small systems. The larger the system, the greater the amount of solar power that cannot be consumed immediately. There is always the option of exporting excess power to the grid. But as the share of solar power on the grid increases, the potential revenue decreases. In a few years, peak solar power production can be expected to begin exceeding total power demand in Germany on the grid. When there is a lot of sunshine, solar power sold to the grid will be relatively worthless.

In the future, systems that can store excess solar power for later use without having to be exported to the grid or use it to create heat will increasingly become important. At present, it seems that batteries will be the main type of storage for grid-connected systems. A control system sends the excess power to the battery. Power is only exported to the grid when the battery is full; the system also shuts down the solar **energy system** in extreme cases. If the solar output cannot meet demand, power can be taken from the battery to cover the gap. When the battery is empty, power can be taken from the grid. The system can also react to the grid's demand with smart charging management; then, excess power would be exported to the grid when the grid needs it the most. If devices are properly timed, the share of solar power consumed directly can be increased further, thereby optimizing affordability. In principle, the battery could then also be used to stabilize the grid.

Theoretically, such a system could work as a micro-grid during a power outage, thereby increasing individual supply security. This option would require, however, an inverter capable of creating a micro-grid. A disconnect also has to separate the system from the grid so that a micro-grid can be produced. Figures 5.68 and 5.69 show **grid-connected photovoltaic systems with battery storage**.

Battery storage can in principle be connected either to the AC or DC lines. If it is connected to AC lines, the battery storage can be integrated to an already completed photovoltaic system. In this case, however, a second inverter is needed for the battery, which leads to additional costs and power losses. If the storage is connected to DC lines, only one inverter is needed for both the battery and the photovoltaic system.

Instead of battery storage, hydrogen storage can also be integrated in grid-connected photovoltaic systems. Batteries generally are only designed for short-term storage, while hydrogen storage can be seasonal. Hydrogen can be produced during the summer from excess power to provide energy during the winter. When properly dimensioned, the components in such a system can provide complete **independence** from the grid. Photovoltaic systems with hydrogen storage have, however, much greater losses during charging and discharging that battery systems do. Furthermore, hydrogen storage is still very expensive, so such systems are still only used in very small numbers.

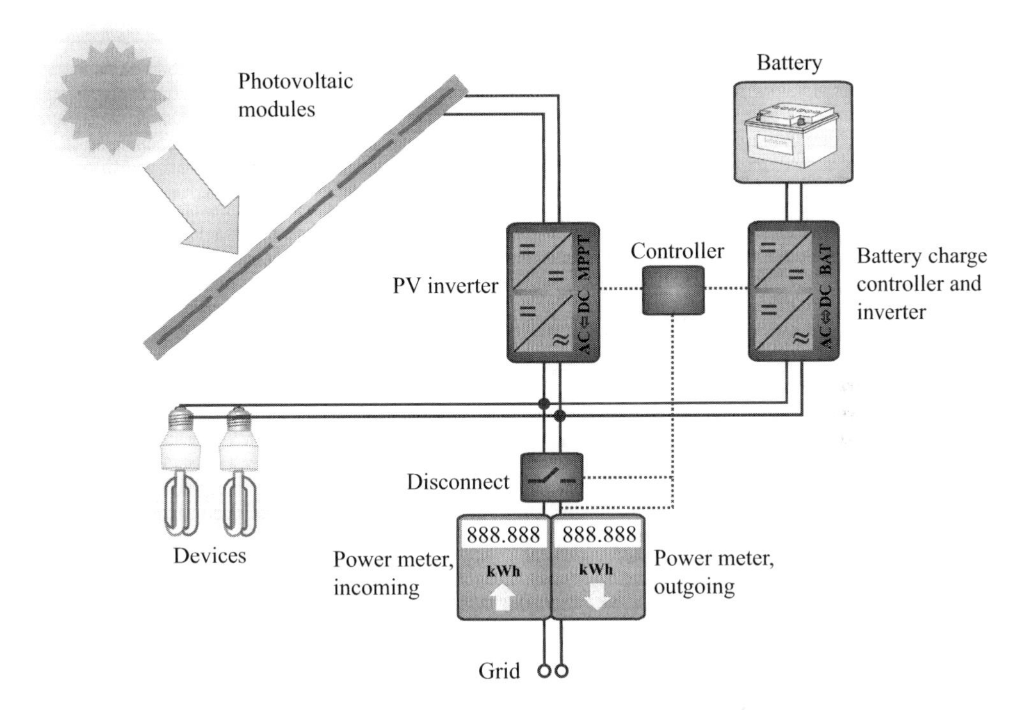

Figure 5.68 Grid-connected photovoltaic system with AC-coupled battery storage

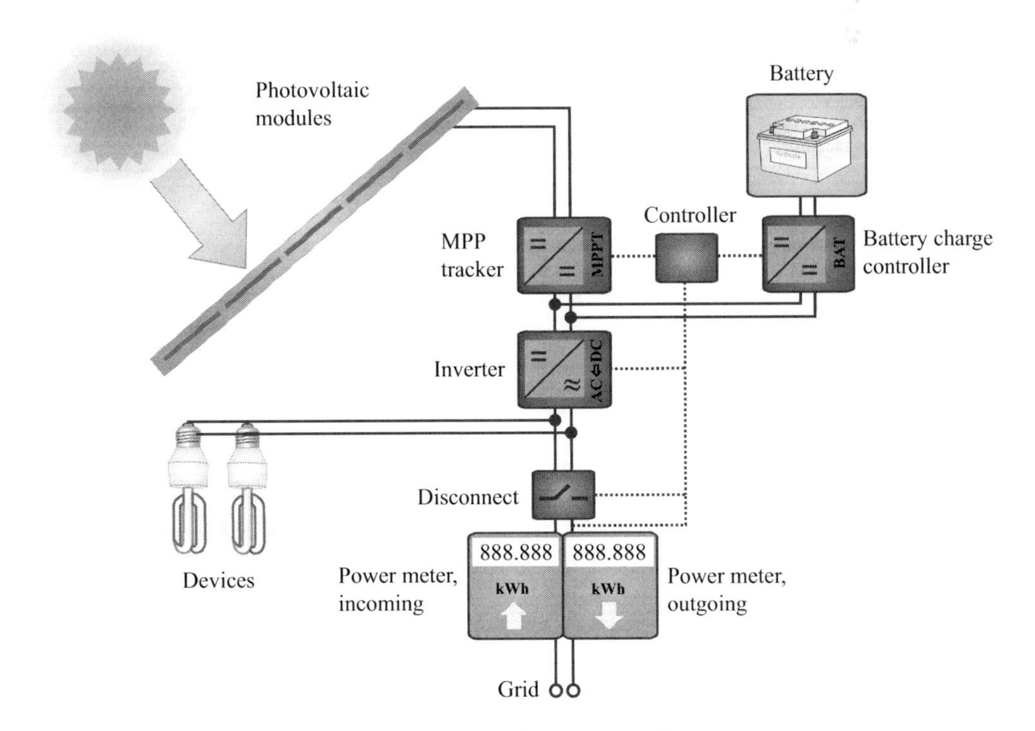

Figure 5.69 Grid-connected photovoltaic system with DC-coupled battery storage

Another way of increasing direct consumption is thermal storage. A simple cartridge heater uses excess solar power to store heat. This approach even saves money if the offset fuel costs are higher than the possible revenue from the sale of the solar power. If the heat storage unit is full or if there is not enough solar power to cover power consumption, power can be purchased from the grid. If a heat pump is already in place or planned, it can directly replace the cartridge heater. Heat pumps generally do not pay for themselves as additions to existing heating systems if the heat pump is merely to serve as an efficient way of using excess solar power.

Demand for cooling will grow in the future. Because cooling loads closely correlate to levels of irradiance, inexpensive vapour-compression refrigerators are a good way of exploiting excess solar power. Figure 5.71 depicts such a system, which can be combined with battery storage if need be.

Another way of storing excess solar power is the methanation of hydrogen so that this synthetic hydrogen can be exported to natural gas networks. Chapter 10 discusses this option in detail.

Planning and design

Off-grid systems

The planning and design of photovoltaic systems is very different if the goal is to be connected to the grid or off the grid. This section only discusses the simple combination of photovoltaics and batteries for off-grid power supply. Hybrid systems with other power generators, such as wind generators, can generally only be properly dimensioned with the aid of simulation programs.

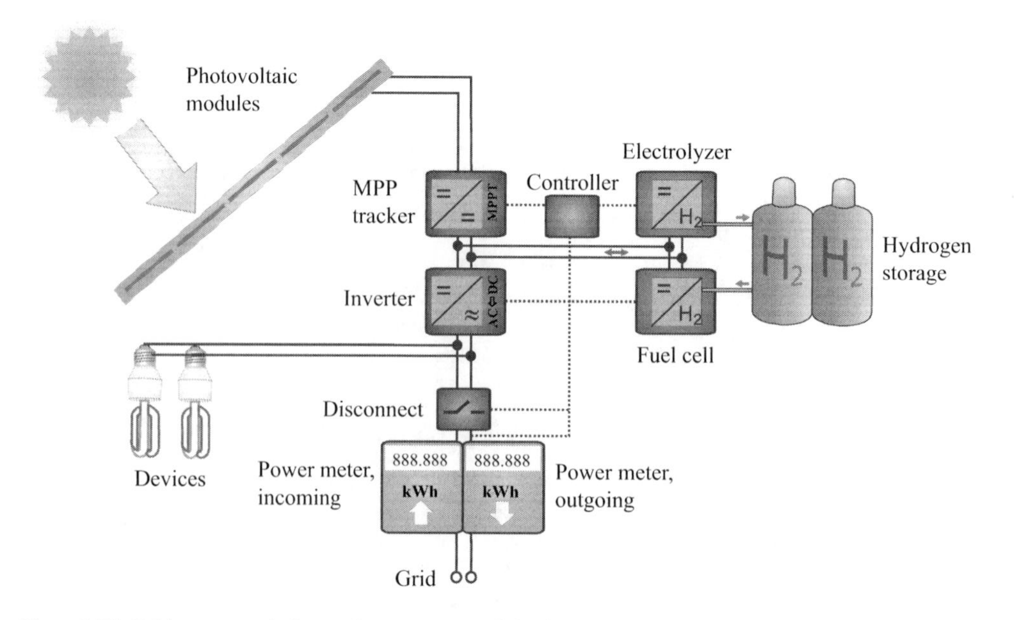

Figure 5.70 Grid-connected photovoltaic system with hydrogen storage

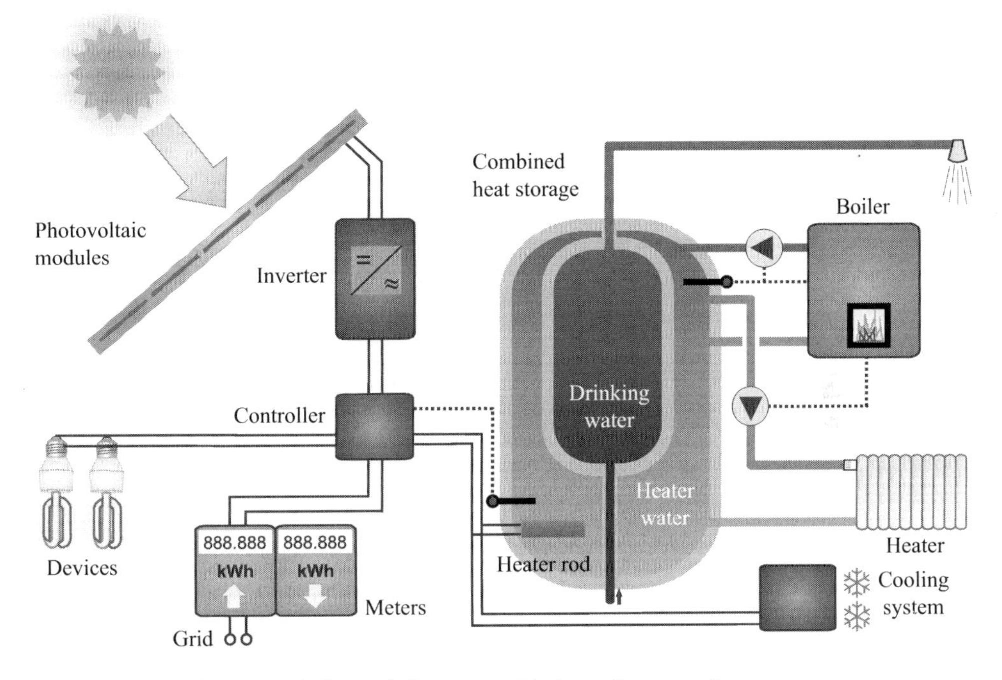

Figure 5.71 A grid-connected photovoltaic system with thermal storage of excess power

Off-grid systems should generally only have enough storage to cover a few days in order to keep costs down. While seasonal storage from the summer for the winter is technically possible, it is expensive in countries like Germany, where the differences over the year are great.

An off-grid photovoltaic system should therefore be designed for the worst month of the year, which is typically December in Germany. Here, the orientation should have a much greater angle than would be common for grid-connected systems. A 60 to 70° angle facing the south would optimally capture the sun in December. Because supply security is not a problem in the summer, the losses from this suboptimal orientation would be acceptable then.

Figure 5.72 shows the daily and monthly averages for irradiation in Berlin on a horizontal plane and on a surface tilted 60° towards the south. The 60° angle provides far greater yield, especially on days with greater daily averages for irradiation. On cloudy days, the yield even drops slightly from this orientation. Table 5.15 shows the monthly and yearly sums for various locations and orientations. In the winter, irradiation in Cairo can be up to six times greater than in Berlin. Photovoltaic off-grid systems can then be accordingly smaller and less expensive.

The photovoltaic generator now needs to be dimensioned so that it can cover monthly power demand with the monthly supply of solar energy. The battery ensures that supply is smooth. Figure 5.72 shows that only a few days need to be bridged.

Table 5.15 Monthly and annual sum of solar irradiation $H_{G,gen,M}$ / $H_{G,gen,a}$ in kWh/m² for various locations and orientations

Location	*Berlin*		*Freiburg*		*Madrid*		*Cairo*
Orientation	30° South	60° South	30° South	60° South	30° South	60° South	30° South
January	28	31	44	51	95	108	157
February	49	52	61	64	105	111	163
March	87	86	103	102	169	164	200
April	117	106	131	117	171	149	204
May	159	136	156	133	148	122	214
June	151	127	162	133	206	162	214
July	161	135	175	145	223	175	221
August	153	135	162	141	213	180	220
September	108	104	128	120	179	165	210
October	68	70	87	91	134	135	201
November	37	42	52	59	95	104	160
December	21	25	38	45	81	92	146
Year	1,139	1,042	1,296	1,198	1,819	1,667	2,310

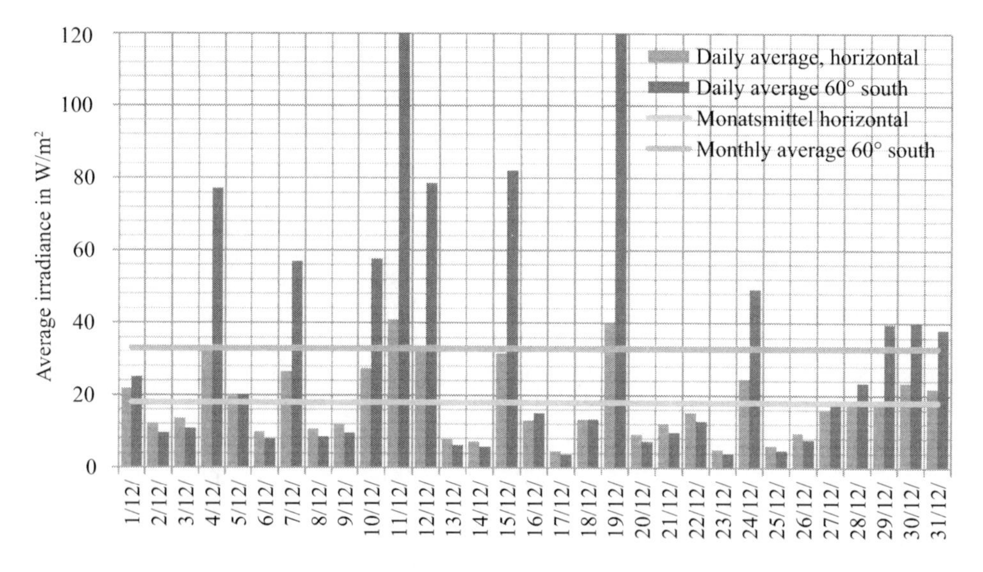

Figure 5.72 Daily and monthly average levels of irradiance on a horizontal plane and a surface facing 60° to the south in December in Berlin

The required MPP output P_{MPP} from the solar modules can be roughly estimated from the irradiation $H_{G,gen,M}$ in the worst month in kWh/m² on the slanted module plane, power demand $E_{demand,M}$ in that month, a safety bonus f_S of at least 50 % and performance ratio *PR*:

$$P_{MPP} = \frac{(1+f_S)\cdot E_{demand,M}}{PR} \cdot \frac{1\,\frac{kW}{m^2}}{H_{G,gen,M}} \tag{5.122}$$

The performance ratio *PR* describes typical system losses from dirt on the modules, shading, partial load, heat losses, and losses from lines and the battery. The values are generally around 0.7 for unshaded off-grid systems.

A lead battery should be dimensioned so that it is only discharged to 50 % under normal conditions and can completely meet demand for a certain number of days. For safe operation in the winter, around 4 to 5 reserve days d_R are needed in Germany, while only 2 to 3 would be required in countries with greater levels of irradiation. The battery voltage U_{bat} (such as 12 V) is used to calculate the required battery capacity:

$$C = \frac{2 \cdot E_{demand,M}}{U_{bat}} \cdot \frac{d_R}{31} \tag{5.123}$$

For example, a photovoltaic system is to power an 11 watt light bulb for three hours a day even in the winter in a garden cottage. The monthly power demand is then $E_{demand,M} = 31 \cdot 11\ W \cdot 3\ h = 1{,}023\ Wh$. With a safety bonus of 50 % = 0.5 and a performance ratio of 0.7, the required MPP output of a module facing the south at a 60° angle in Berlin would be:

$$P_{MPP} = \frac{(1+0.5) \cdot 1023\ Wh}{0.7} \cdot \frac{1\ \frac{kW}{m^2}}{25\ \frac{kWh}{m^2}} = 87.7\ W$$

With four reserve days and a battery voltage of 12 V, the battery capacity would be:

$$C = \frac{2 \cdot 1023\ Wh}{12\ V} \cdot \frac{4}{31} = 22\ Ah$$

Grid-connected systems

While battery systems are almost always designed to meet the demand load, this load hardly plays a role when designing grid-connected systems. Usually, it is assumed that the grid will be able to cover all loads. The focus is therefore usually on the available space and budget.

For the area available A_{PV} for the installation of a photovoltaic generator, the capacity that can be installed is relative to module efficiency η_{PV} (cf. Table 5.6):

$$P_{PV} = \eta_{PV} \cdot A_{PV} \cdot 1000\ \frac{W}{m^2} \tag{5.124}$$

For instance, a roof area of 50 m² and an efficiency of 14 % for crystalline silicon solar modules add up to an output of 7 kW_p.

To prevent frequent operation under partial loads, the inverter's nominal capacity should be the same as the total capacity of the solar modules if the system is optimally oriented and unshaded or if the system is in a location with a lot of irradiation. Otherwise, it might make sense to have the inverter's output 10 to 30 % smaller. Inverter manufacturers generally provide design recommendations.

Once you have chosen an inverter, you have to decide what the optimal number of modules is for the desired photovoltaic module type. Here, various boundary conditions need to be kept in mind. First, photovoltaic generator must not exceed the inverter's maximum DC voltage $U_{max,inverter}$ lest the inverter be damaged. The maximum possible voltage is the sum of the open-circuit voltages U_L of all of the photovoltaic modules connected in series under great irradiance and at low module temperatures. In the German climate, irradiance is generally assumed to be 1,000 W/m²; and the module's temperature, –10 °C. Use the formulae in the section 'Solar cells in series circuit' earlier in this chapter to convert STC data from a photovoltaic module's datasheet.

The maximum number n_{max} of photovoltaic modules that can be connected in series is then:

$$n_{max} = \frac{U_{max,inverter}}{U_{OC}(1000\ \frac{W}{m^2}, -10\ °C)} \tag{5.125}$$

Next, make sure that the photovoltaic generator's MPP never leaves the inverter's MPP range under any operating condition. Maximum MPP voltage occurs at high irradiance at low temperatures; minimum MPP voltage, at low irradiance. Figure 5.73 shows different voltage ranges.

Finally, the number of parallel module strings needs to be determined from the admissible current and output.

If a photovoltaic generator is planned and its output and surface area have been defined, the next step is generally a yield forecast. The ideal energy yield E_{ideal} of a photovoltaic system is calculated from the photovoltaic surface area A_{PV}, the photovoltaic efficiency η_{PV}, and solar irradiation $H_{G,gen}$ incident on the modules:

$$E_{ideal} = A_{PV} \cdot \eta_{PV} \cdot H_{G,gen} = \frac{P_{PV} \cdot H_{G,gen}}{1000\ \frac{W}{m^2}} \tag{5.126}$$

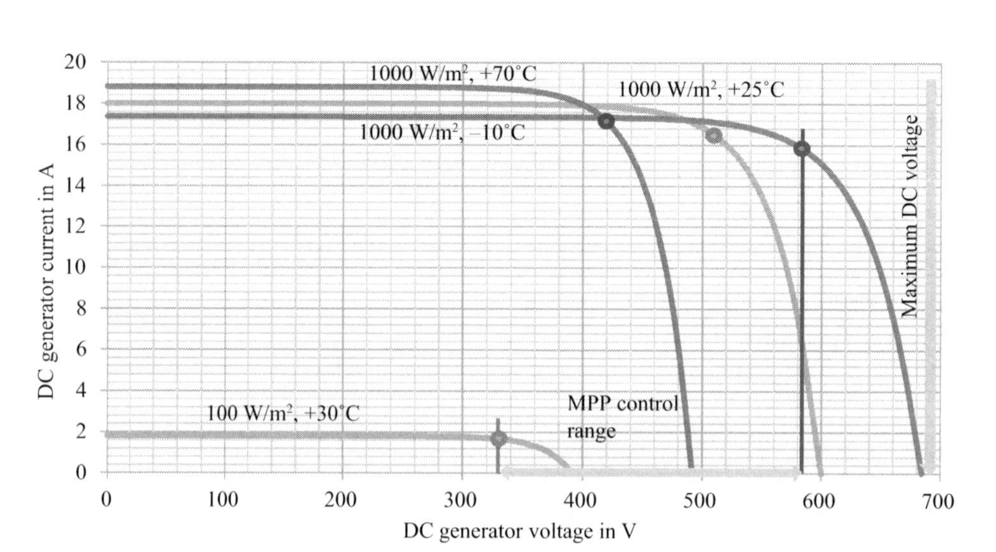

Figure 5.73 Voltage ranges for a photovoltaic generator and an inverter

If the module efficiency η_{PV} is 10 % and irradiation $H_{G,gen}$ is 1,100 kWh/(m² a) on a 50 m² area facing 30° to the south, the photovoltaic system can have 5 kW_p, enough to produce an annual ideal energy yield of 5,500 kWh/a.

In reality, a photovoltaic system's energy yield is much lower because the following effects cause power losses in practice:

- lower efficiency from hotter modules
- lower efficiency under partial load
- lower actual output than indicated on data sheets
- mismatch losses when modules and cells of different output are connected
- reflection losses from sunlight incident at an angle
- efficiency changes from different spectral compositions
- losses from dirt and snow
- losses from shading
- line and diode losses
- the MPP tracker's tracking flaws
- the inverter's voltage conversion losses and own power consumption
- module and inverter failure.

The **performance ratio** *PR* describes the ratio between actual and ideal energy yield:

$$E_{real} = PR \cdot E_{ideal} \tag{5.127}$$

The average system has a performance ratio of 0.75 %. This value can be used as an average when designing a new system. Very good systems have values exceeding 0.8. The performance ratio of inefficient systems can drop below 0.6. In such cases, inverter failures and long-term shading play a crucial role.

The change in annual irradiation has a relatively small impact on the performance ratio. The PR value therefore is a good way of assessing the quality of completed photovoltaic systems. The real yield E_{real} comes from the photovoltaic system's meter. To determine the ideal yield E_{ideal}, either the annual irradiation on the photovoltaic generator must be measured or horizontal measurements close to the photovoltaic system have to be converted to the plane of the modules.

If the 5 kW_p photovoltaic system from the example above produces a real annual yield of 4,500 kWh/a, the performance ratio is 81.1 %. The system thus produces nearly optimal operating results.

In addition to the performance ratio *PR*, the specific yield Y_F provides additional information on the system. It is calculated from the ratio of the real yield E_{real} to the photovoltaic nominal output P_{PV}:

$$Y_F = \frac{E_{real}}{P_{PV}} \tag{5.128}$$

Specific yield depends considerably upon location but can be used to compare similar systems within a location. It is similar to the full load hours given for other technologies. For the example above, the specific yield is:

$$Y_F = \frac{4500\ \frac{\text{kWh}}{\text{a}}}{5\ \text{kW}_\text{p}} = 900\ \frac{\frac{\text{kWh}}{\text{a}}}{\text{kW}_\text{p}} = 900\ \frac{\text{h}}{\text{a}}$$

If the same system were installed in northern Africa at an irradiation of 2,200 kWh/(m² a), the yield could be 9,000 kWh/a, but the performance ratio would remain unchanged at 81.8 %. In contrast, the specific yield would increase to 1,800 h/a.

The designs and yield calculations provided here only allow for rough estimates. For a more detailed calculation, use professional design and simulation software.

Direct-consumption systems

Up to now, countries like Germany, Spain, and Italy have mainly installed grid-connected photovoltaic systems, but systems tailored to direct consumption will dominate the market there soon. This option is financially attractive when the cost of solar power from the system is below the retail rate, a situation called **grid parity**. German households reached grid parity in 2012 (see Chapter 11). Other countries are soon to follow. Direct-consumption systems will then be attractive without any compensation for power sold to the grid – provided that the solar power generated is largely consumed without producing excesses.

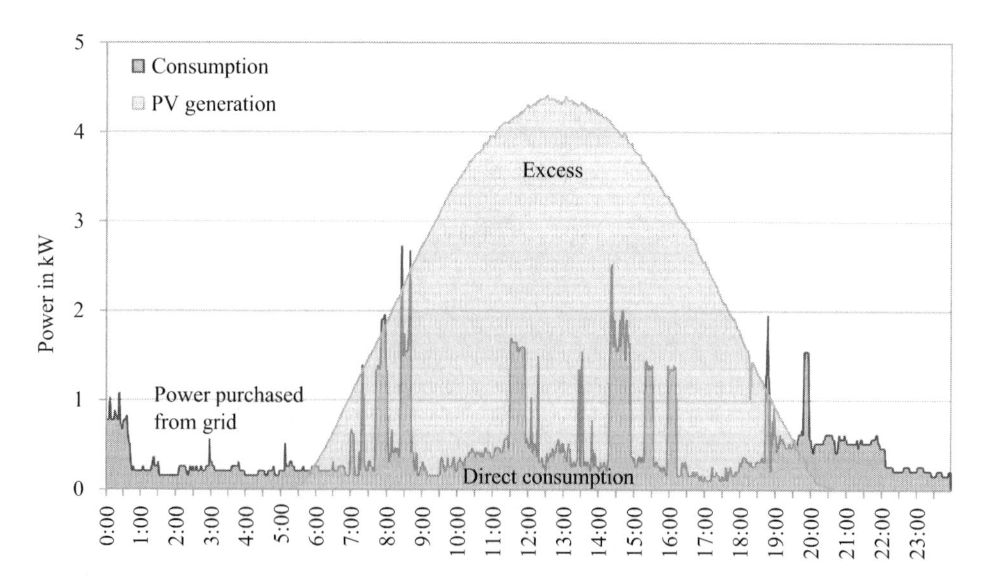

Figure 5.74 Typical load for a single-family home in Germany on a sunny spring weekend with a resolution of one minute compared with production from a 5 kW photovoltaic system

Towards that end, immediate complete consumption of all solar power generated is practically impossible without additional storage because demand in households fluctuates greatly (Figure 5.74).

The **self-consumption rate** e can be determined from the PV system's power produced $\overline{P}_{\mathrm{PV}}$ over a certain interval of time Δt and the power $\overline{P}_{\text{feed-in}}$ generally for a year.

$$e = \frac{\sum \overline{P}_{\mathrm{PV}} \cdot \Delta t - \sum \overline{P}_{\text{feed-in}} \cdot \Delta t}{\sum \overline{P}_{\mathrm{PV}} \cdot \Delta t} = \frac{\sum \min\left(\overline{P}_{\mathrm{PV}}; \overline{P}_{\text{consumption}}\right) \cdot \Delta t}{\sum \overline{P}_{\mathrm{PV}} \cdot \Delta t}$$
$$= \frac{\sum \overline{P}_{\text{self-consumption}} \cdot \Delta t}{\sum \overline{P}_{\mathrm{PV}} \cdot \Delta t} \tag{5.129}$$

If the time interval Δt is too large, peak generation could be flattened, thereby making the share directly consumed seem larger than it is. For the most exact calculation of the share of power consumed directly, the time interval should not be longer than 15 minutes, though an interval of one minute is preferable. The larger the photovoltaic system, the lower the possible self-consumption rate.

Without storage, high shares of directly consumed solar power are only possible with relatively small photovoltaic systems, as Figure 5.75 shows. In a typical five-person household in Germany, a 90 % rate is only possible with a 1 kW system. Then, only 10 % of the power generated would not be directly consumed within the household immediately. Such a system would pay for itself without any compensation for power exported to the grid as soon as solar power from the system costs 10 % less than the retail power rate. As the PV system gets bigger or the household and its consumption smaller, the difference between the retail rate and the cost of solar power must be larger.

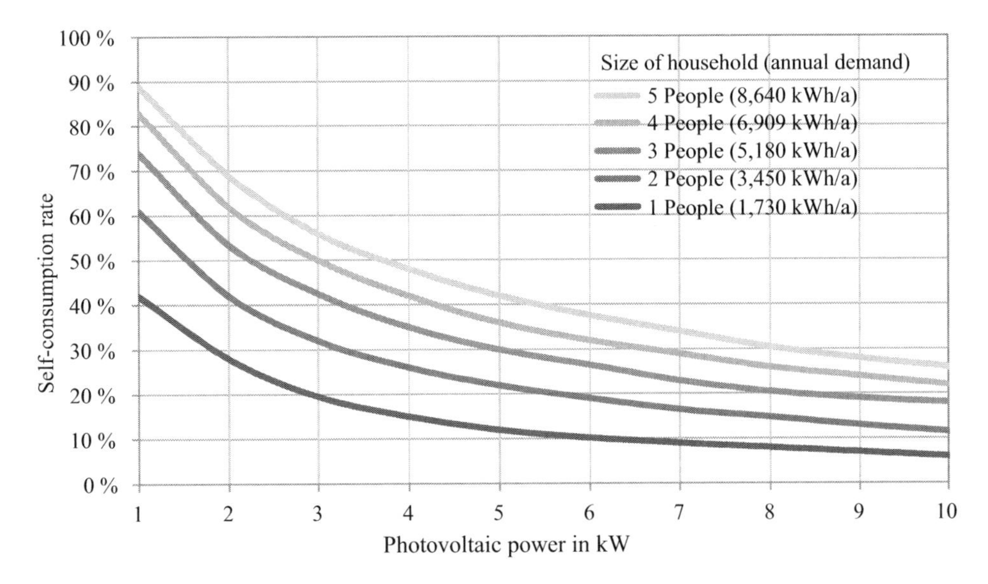

Figure 5.75 Typical annual average self-consumption rate in a German single-family home relative to the size of the PV system and household

(Data: [Wen12])

The self-consumption rate is often greater for industrial and commercial consumers, whose power consumption is more regular. Standard load profiles for various types of businesses can be used to estimate the potential self-consumption rate.

In Germany, the following **standard load profiles** are used by distribution grid operators with a 15-minute resolution:

- G0: General commercial
- G1: Commercial, workdays 8 am to 6 pm
- G2: Commercial with great or primary consumption in the evening
- G3: Commercial, around-the-clock
- G4: Stores/hairdressers
- G5: Bakeries
- G6: Weekend operation
- L0: Farms
- L1: Farms with dairy production or other ancillary facilities
- L2: Other special farms
- H0: Households

Figure 5.76 shows solar power generation on a sunny spring workday compared with the power demand at a typical commercial enterprise with a G0 standard load profile. If the photovoltaic system is properly designed, the self-consumption rate can be very high because the generation curve for solar power will remain completely within the area of demand on cloudy days.

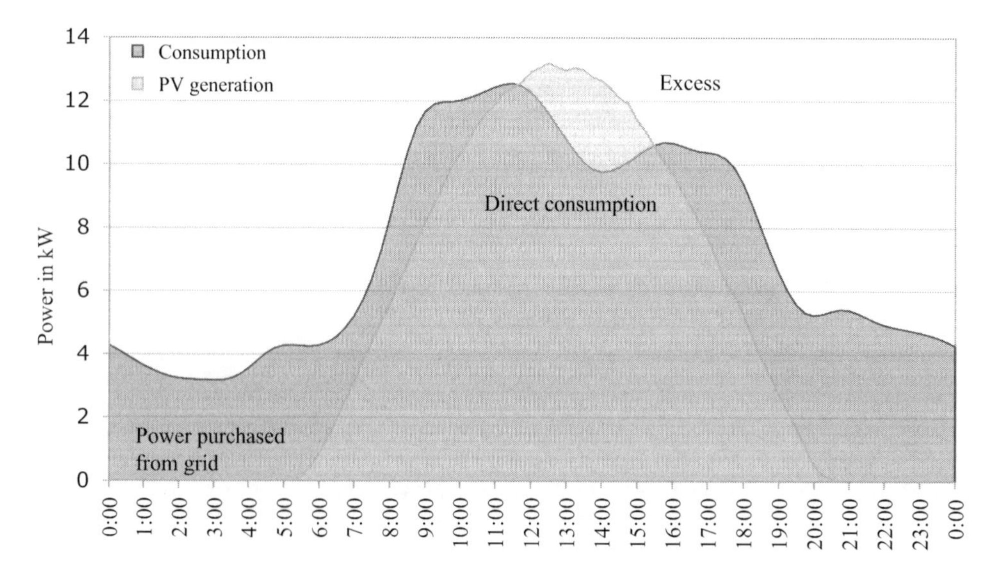

Figure 5.76 A typical load curve for a commercial enterprise (standard load profile G0) with an annual power consumption of 60,000 kWh on a spring workday with a 15-minute resolution compared with power generated by a 15 kW photovoltaic system in Germany

It is, however, also possible for a photovoltaic system to pay for itself even at low self-consumption rates. Here, you simply have to decide how the excess solar power is to be used, and there are various options. In general, the option that pays the most will be chosen.

As long as there is attractive payment for power sold to the grid, such as German feed-in tariffs, grid exports of excess power makes sense. In numerous countries, the conditions

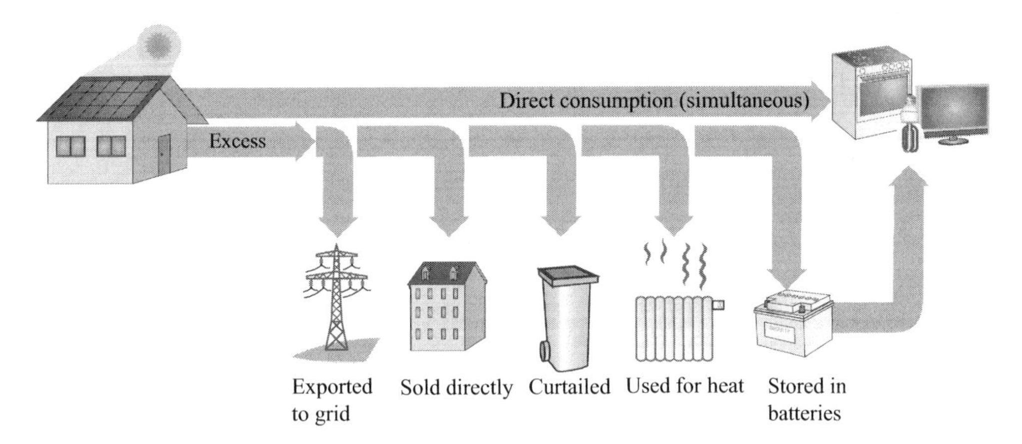

Figure 5.77 Ways of using excess power from a solar energy system not consumed directly

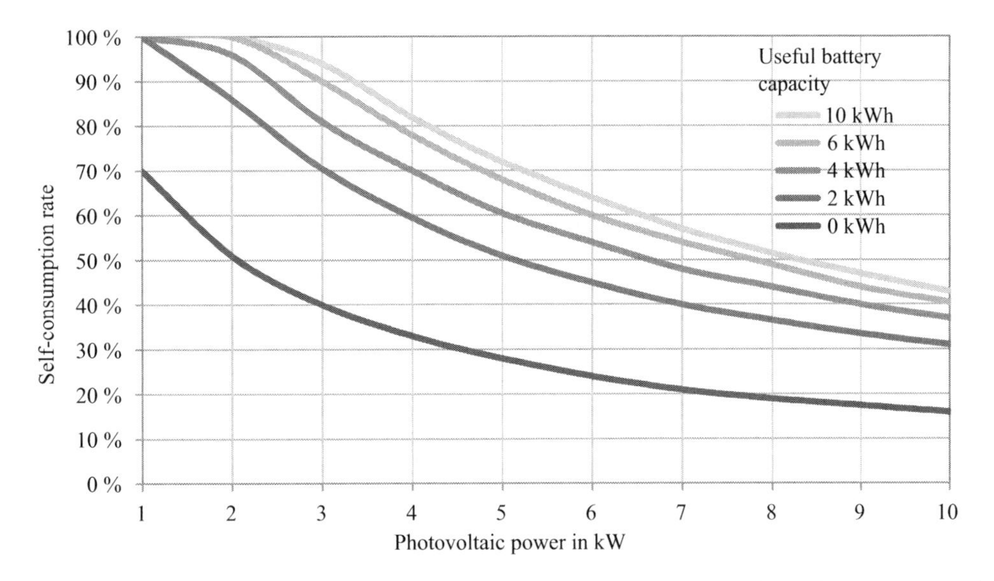

Figure 5.78 Typical annual averages of the self-consumption rate relative to photovoltaic capacity and battery storage capacity in a German single-family household with an annual consumption of 4,700 kWh

(Data: [Wen12])

are either worse than in Germany, or power cannot be sold to the grid at all. Production of power from the photovoltaic system may then need to be curtailed. There is no technical obstacle here; the MPP tracker simply needs to move voltage into an unfavourable range so that power production is reduced to the desired level. But before a photovoltaic system is curtailed and solar power production artificially reduced, other buyers can be looked for (Figure 5.77).

Owners of small photovoltaic systems will soon be able to serve as power providers by selling their excess electricity to their own customers or resellers at low prices. In the midterm, the revenue will probably not be great because the amount of excess solar power will be relatively low. Then, technical solutions that increase the self-consumption rate will become more important. Battery systems can temporarily store a lot of the excess and make it available at times of low irradiation or overnight, thereby increasing the share of direct consumption noticeably. Typical households will need a battery storage capacity of 4 to 6 kWh for this purpose (Figure 5.78).

Another way of increasing direct consumption is to use excess solar power to generate heat (Figure 5.71). If a photovoltaic system with a battery that has a useful storage capacity of 5 kWh is combined with an 800-litre storage tank for drinking water, a 4 to 5 kW photovoltaic system can reach 90 to 100 % direct consumption. In this case, some 30 % of the solar energy would be used to generate heat.

The share can vary considerably depending on consumption and system configuration. Estimating the self-consumption rate is even more complex for industrial and commercial firms. A detailed design is only possible if you use a simulation program

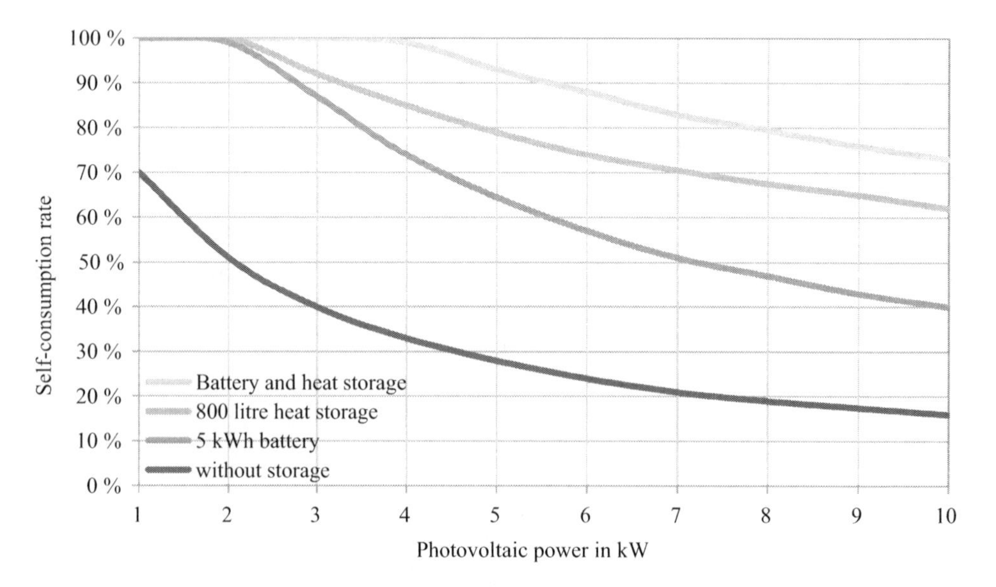

Figure 5.79 Typical annual averages of the self-consumption rate for different self-consumption systems relative to photovoltaic and battery storage capacity in a German single-family household with an annual consumption of 4,700 kWh

(Data: [Qua12])

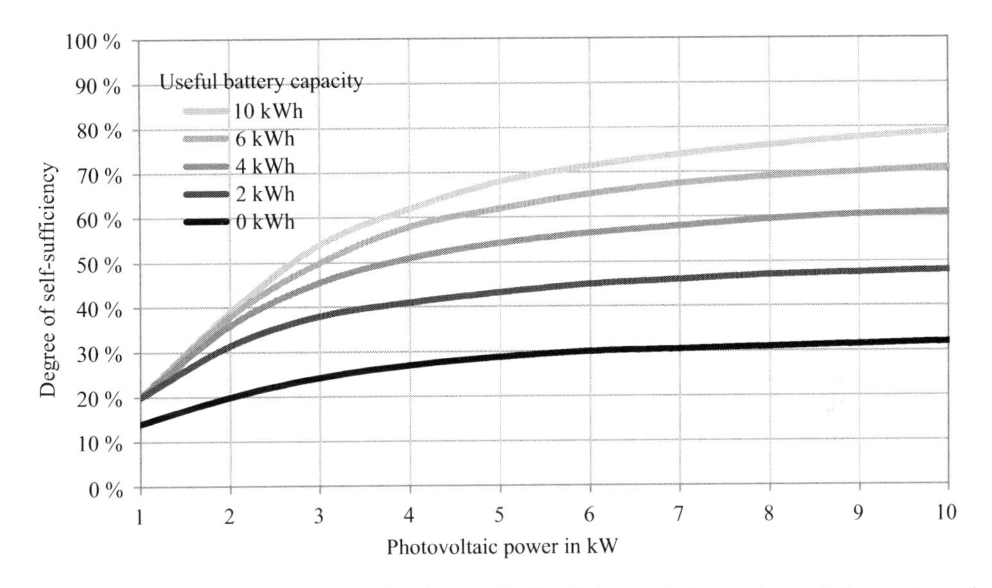

Figure 5.80 Typical annual averages of the degree of self sufficiency relative to photovoltaic capacity and battery storage capacity in a German single-family household with an annual consumption of 4,700 kWh

(Data: [Wen12])

based on good consumption profiles. The resolution of the simulation should be by the minute if possible or at least in increments of 15 minutes in line with standard load profiles in order to represent the simultaneity of power generation and consumption. Even without simulations, the values from this chapter can at least be used as starting points for system design.

A lot of consumers increasingly want to consume the largest possible share of their own solar power in order to become less dependent on the grid. Here, one speaks of the degree of self-sufficiency. It is largely equivalent to the solar coverage rate for solar thermal systems.

The **degree of self-sufficiency** a is calculated by adding up the total consumption $\overline{P}_{\text{consumption}}$ and amount of power purchased from the grid $\overline{P}_{\text{grid-supply}}$ over a given timeframe Δt:

$$a = \frac{\sum \overline{P}_{\text{consumption}} \cdot \Delta t - \sum \overline{P}_{\text{grid-supply}} \cdot \Delta t}{\sum \overline{P}_{\text{consumption}} \cdot \Delta t} = \frac{\sum \min\left(\overline{P}_{\text{PV}}; \overline{P}_{\text{consumption}}\right) \cdot \Delta t}{\sum \overline{P}_{\text{consumption}} \cdot \Delta t} = \frac{\sum \overline{P}_{\text{self-consumption}} \cdot \Delta t}{\sum \overline{P}_{\text{consumption}} \cdot \Delta t} \tag{5.130}$$

At a 100 % degree of self-sufficiency, no power is purchased from the grid; the photovoltaic system directly covers all power consumption. In practice, however, complete autonomy in a country like Germany is difficult to attain. Even with a 10 kW photovoltaic system, a typical single-family household will only obtain a roughly 35 % degree of self-sufficiency without battery storage. With battery storage of an affordable size, that level can, however, be increased to up to 80 % (Figure 5.80).

6 Wind energy

Introduction

Unlike the ways of using solar energy directly as described in previous chapters, wind power is a way of using it indirectly. The sun creates temperature differences on the earth, thereby creating wind. The wind can then be used to drive power generators. It attains much higher energy densities than incident sunlight. While the maximum irradiance on earth is around 1 kW/m^2, the energy density of wind during a heavy storm is around 10 kW/m^2, and it can even exceed 25 kW/m^2 in strong ones. Indeed, tornadoes and hurricanes even exceed 100 kW/m^2. In contrast, a common wind at a velocity of 5 m/s only has a low energy density of 0.075 kW/m^2.

The **history of wind energy** goes back centuries. More than 3,000 years ago, the wind is said to have been used for irrigation. Historic documents show that it was used to grind grain in Afghanistan in the seventh century BC. These windmills applied the principle of resistance and were very simple designs of low efficiency according to current standards. Wind energy began to conquer Europe in the twelfth century. Over the centuries, windmills continued to be further developed. In the seventeenth and eighteenth centuries, they were used to drain land in the Netherlands. By that time, they had become technically sophisticated and were even able to independently track the wind. In North America, countless 'western mills' used to pump water dotted the landscape in the nineteenth century. At the beginning of the twentieth century, steam engines and combustion engines proved to be serious competition for wind energy, which became even more marginal during electrification. It was not until the oil crises of the 1970s that wind energy experienced a renaissance. These days, the wind is almost exclusively used to generate electricity, in contrast to previous centuries. Germany now has a flourishing wind power sector. The latest turbines are high-tech marvels with capacities up to and exceeding 5 MW. By 2011, the German wind power sector had already created more than 100,000 jobs, with annual sales coming in at more than €10 billion.

Wind power will meet a large part of Germany's future energy demand. The country's **Feed-in Act** of 1991 and the **Renewable Energy Act** (EEG) of 2000 were adopted to give wind power the legal framework it needs to become competitive with conventional energy sources. Since 2009, Germany has also offered very attractive financial terms for offshore wind turbines. After photovoltaics, wind power has the second greatest potential for power generation among all renewable sources in Germany. It alone could meet more than half of German power demand. The potential is much greater in other countries, such as the UK.

In Germany, wind power is now also an example of the development of a future energy policy. Some energy supply firms are concerned about the competition and are calling for the EEG to be scrapped. Other arguments are also made against wind power, such as blight, noise, and health effects. While some of these points of criticism have some merit, we should be careful when environmental protection concerns are the main arguments brought against wind power. There is no denying that wind turbines change the appearance of a pristine coastal landscape. But if the alternative is fossil energy, these coastal landscapes that we want to keep wind turbines from interfering with may no longer exist in a few centuries as a result of the greenhouse effect.

Supply of wind energy

How wind is created

The sun causes wind. Gigantic amounts of solar energy constantly reach the earth. If the earth did not constantly reflect this incident solar energy back into space, it would constantly heat up and eventually burn up. At the equator, more solar energy reaches the earth than is reflected back into space. At the poles, the path that sunlight takes is longer than at the equator, however. The situation is therefore the reverse at the poles. There, there is much less incident sunlight, and more energy is reflected into space than is incident. As a result, a giant amount of energy is transported from the equator to the poles.

The main way this heat is transported is through a global exchange of air mass. Gigantic global circulations of air pump heat from the equator to the poles. These gigantic air circulations are called **Hadley cells** (Figure 6.1). The earth's rotation diverts these currents. The result is relatively constant winds, which were crucial for oceangoing sailboats for

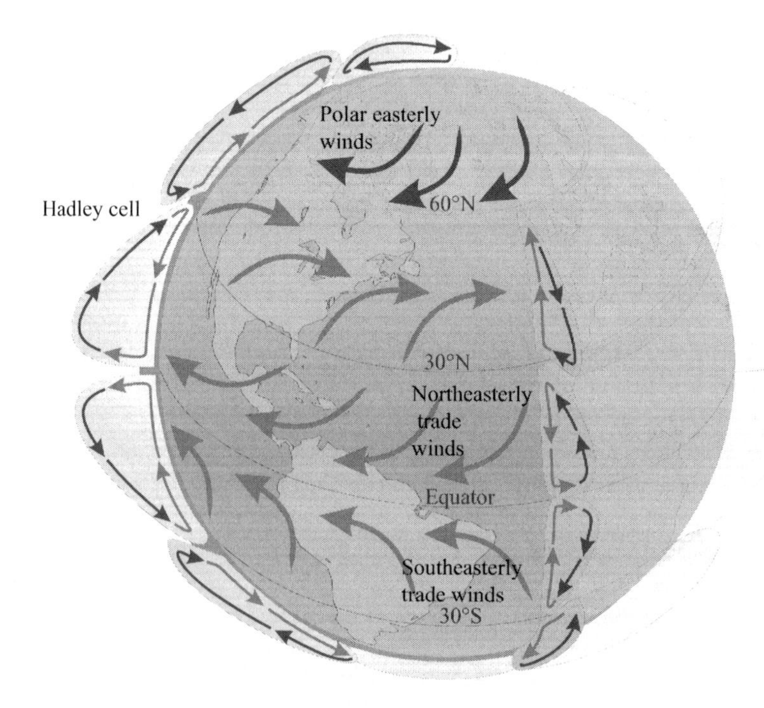

Figure 6.1 Global circulation and creation of wind

a long time. In the tropics north of the equator, there is a relatively constant wind from the Northeast. This wind is therefore called the northeasterly trade wind. South of the equator, we find the southeasterly trade wind.

In addition to these **large balancing currents**, there are smaller currents caused by high-pressure and low-pressure areas. Coriolis forces change the direction of the straight streams of high-pressure and low-pressure areas. The earth's rotation causes air masses in the northern hemisphere to turn to the right; in the southern hemisphere, they turn to the left. The wind then rotates around low-pressure areas.

In **coastal areas**, there is a lot of wind. On the one hand, there are fewer obstacles to the wind in such areas; on the other, there are also local compensation currents in coastal regions. The sun generally heats up the land more than it does the water during the day. The result is local pressure differences that draw air from the water onto the land. These currents can reach up to 50 kilometres onto land. At night, the land cools down faster than the water does, so compensation currents occur in the opposite direction.

Categories of wind velocities

In meteorology, wind velocities are often indicated according to the **Beaufort scale**, as shown in Table 6.1. This scale allows wind velocities to be roughly measured without complicated devices. For technical purposes, the scale is not very useful, however. For more accurate assessments, velocities v are indicated in SI unit m/s. Table 6.1 shows categories of wind velocities in the Beaufort scale along with actual wind velocities.

Wind velocity distribution

To determine the annual amount of wind energy for a specific location over a year, the wind velocity distribution is often calculated either from comparative measurements or statistical

Table 6.1 Classifications of wind velocities in the Beaufort scale

Bg	*v in m/s*	*Designation*	*Effect*
0	0–0.2	No wind	Smoke rises straight up
1	0.3–1.5	Light air	Wind direction only detectable from smoke
2	1.6–3.3	Light breeze	Palpable wind, leaves rustle
3	3.4–5.4	Gentle breeze	Leaves and thin twigs move
4	5.5–7.9	Moderate breeze	Wind moves twigs and thin branches, carries dust
5	8.0–10.7	Fresh breeze	Small trees begin to sway
6	10.8–13.8	Strong breeze	Thick branches move, wind begins to whistle
7	13.9–17.1	Near gale	Trees in motion, hard to walk
8	17.2–20.7	Fresh gale	Twigs broken off trees
9	20.8–24.4	Strong gale	Minor damage to buildings and roofs
10	24.5–28.4	Whole gale	Trees uprooted
11	28.5–32.6	Violent storm	Heavy damage
12	≥ 32.7	Hurricane force	Severe damage

Bg: Beaufort scale *v*: wind velocity in m/s

parameters based on tables and wind maps. The distribution can either be given in a table or described in statistical functions.

Figure 6.2 shows the **relative frequency distribution** $h(v)$ of wind velocities v for a location in northern Germany. This distribution shows how often a particular wind velocity occurs. The image also shows two different frequency distribution functions explained below.

The measurement interval can be a problem in determining frequency distributions. For technical reasons, an average is often taken across several minutes or hours. As we will see below, the energy in wind does not increase linearly as wind velocity picks up, so the result can be distortions in subsequent calculations when strong, but brief wind velocities are not properly taken into account. It is therefore a good idea to keep the time intervals as short as possible for measurements and records of wind velocity.

The **average wind velocity** is calculated thus:

$$\bar{v} = \sum h(v) \cdot v \tag{6.1}$$

The average wind velocity itself does not tell us much about a particular location because it does not tell us whether there are long periods of calm followed by storms or a constant, moderate wind. Nonetheless, they are often used to classify locations.

There are sophisticated wind maps for average wind velocities based on several years of serial measurements taken in Germany and Europe (such as [Tro89]). While the average wind velocity on the German coast can exceed 6 m/s at a height of 10 m, it drops below 3 m/s on the continent. The only good wind conditions far from the German coast lines are

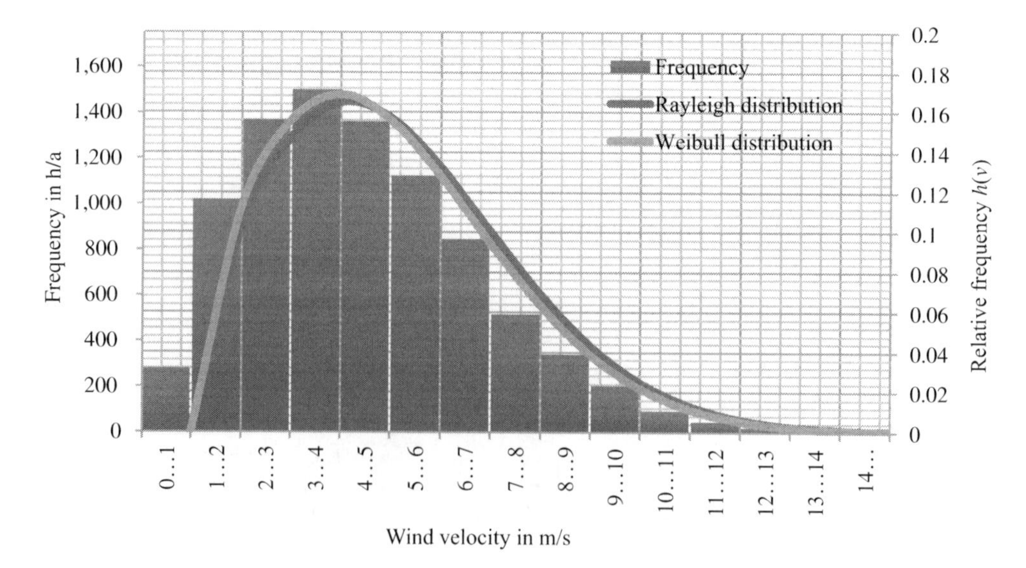

Figure 6.2 Frequency distribution of wind velocities at a location on the German North Sea coast at a height of 10 m

found in mountain regions. Digital atlases are now also available, and computer programs (such as [Ris09]) can be used to interpolate wind velocities for specific locations if no local measurements are available.

A better way of assessing a location's wind velocity is a frequency distribution of wind velocity either as a frequency distribution of wind velocity intervals described above or as a constant statistical function. The Weibull and Rayleigh distributions are used as statistical functions.

The **Weibull distribution** of wind velocity v is based on a shape or form parameter k and scale parameter a:

$$f_{\mathrm{Weibull}}(v) = \frac{k}{a} \cdot \left(\frac{v}{a}\right)^{k-1} \cdot \exp\left(-\left(\frac{v}{a}\right)^{k}\right) \tag{6.2}$$

The shape and scale parameters vary from one location to another. Table 6.2 shows a number of parameters for various locations in Germany.

The average wind velocity

$$\overline{v} = a \cdot \left(0.568 + \frac{0.434}{k}\right)^{\frac{1}{k}} \tag{6.3}$$

can be roughly estimated from the Weibull parameters [Mol90]. The parameter a for $k = 2$ stems from the average wind velocity:

$$a_{k=2} = \frac{\overline{v}}{0.886} \approx \frac{2}{\sqrt{\pi}} \cdot \overline{v} \tag{6.4}$$

Enter a and $k = 2$ in the Weibull distribution to produce the **Rayleigh distribution**:

$$f_{\mathrm{Rayleigh}}(v) = \frac{\pi}{2} \cdot \frac{v}{\overline{v}^2} \cdot \exp\left(-\frac{\pi}{4} \cdot \frac{v^2}{\overline{v}^2}\right) \tag{6.5}$$

Average wind velocity is used in the Rayleigh distribution to produce a distribution of wind velocities. Figure 6.3 shows Rayleigh distributions for various average wind velocities.

Table 6.2 Weibull parameter and average wind velocities at a height of 10 m at various locations in Germany [Chr89]

Location	*k*	*a*	*v in m/s*	*Location*	*k*	*a*	*v in m/s*
Berlin	1.85	4.4	3.9	Munich	1.32	3.2	2.9
Hamburg	1.87	4.6	4.1	Nuremberg	1.36	2.9	2.7
Hanover	1.78	4.1	3.7	Saarbrücken	1.76	3.7	3.3
Heligoland	2.13	8.0	7.1	Stuttgart	1.23	2.6	2.4
Cologne	1.77	3.6	3.2	Wasserkuppe	1.98	6.8	6.0

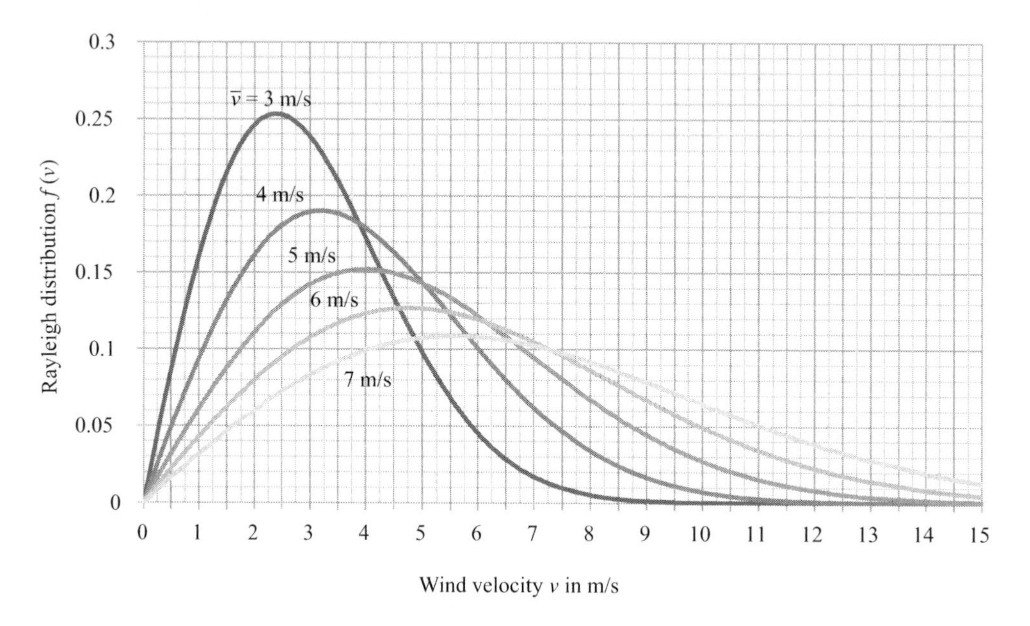

Figure 6.3 Rayleigh distributions for various average wind velocities $\overline{v}$

Influence of environment and height

Average wind velocity is usually measured at a height of 10 m. Objects on the ground can cause wind velocities to fluctuate considerably over just a few hundred metres. Hills and mountains influence wind velocity. On the side of a mountain top or rim facing the wind vertically, the wind velocity can speed up to twice its value in an area without obstacles. On the side facing away from the wind, in contrast, wind velocities are much lower.

Objects, plants, and uneven ground near a location can also slow down wind velocity considerably. Individual large obstacles do not affect a wind turbine if its entire swept area is three times higher than the obstacle or if the wind turbine is sufficiently far from the obstacle. In extreme cases, the distance has to be 35 times the height of the obstacle. If the distance is lower, strong wind turbulence can reduce the amount of wind energy that can be exploited.

Wind turbines usually have a hub height much higher than 10 m. As they increase in height, wind velocity also increases because the wind on the ground is blocked by unevenness and objects.

Leeward: The side facing away from the wind
Windward: The side that is facing the wind
Wind direction: The direction from which the wind is coming
90°: East wind; 180°: South wind; 270°: West wind; 360°: North wind

N
NW NO
W O
SW SO
S

Figure 6.4 Terms used to describe wind direction

The wind velocity $v(h_2)$ at height h_2 can be calculated by using the **logarithmic wind profile** and the roughness length z_0 at a wind velocity $v(h_1)$ at a height of h_1:

$$v(h_2) = v(h_1) \cdot \frac{\ln\left(\dfrac{h_2 - d}{z_0}\right)}{\ln\left(\dfrac{h_1 - d}{z_0}\right)} \tag{6.6}$$

Obstacles can shift the boundary layer off the ground. Parameter d can be included for this offset. If there is a lot of space between obstacles, $d = 0$. Otherwise, d can be assumed to be equal to 70 % of the obstacle's height.

The **roughness length** z_0 indicates the height at which the wind is completely blocked. The greater the roughness length, the greater the impact on the wind. Table 6.3 shows a breakdown of various terrain classes according to roughness length.

The rougher a terrain, the more wind velocities increase with height, as shown in Figure 6.5. Therefore, taller towers are generally a good idea for wind turbines far inland, while they may not be needed as much on the coast.

The greater the height, the lower the influence of the ground on wind velocity. Wind that is completely unaffected by height is called **geostrophic wind**, and it is generally only found at heights far above 200 m.

Another factor that plays a role for the log wind profile is the Hellmann potential.

It follows from $z = \sqrt{h_1 \cdot h_2}$ and $a = \dfrac{1}{\ln \dfrac{z}{z_0}}$ that $\dfrac{v(h_2)}{v(h_1)} = \left(\dfrac{h_2}{h_1}\right)^a$ (6.7)

For $z = 10$ m and $z_0 = 0.01$ m, a is roughly $1/7$, so one also speaks of the $1/7$ potential law. Also called the wind profile power law, this potential is only valid if the offset d of the border layer to the ground is zero.

Table 6.3 Roughness lengths z_0 for various terrain classes in Davenport [Chr89]

Terrain classes in Davenport	*Roughness length z_0 in m*	*Description of surface*
1 – sea	0.0002	Open water
2 – smooth	0.005	Marsh
3 – open	0.03	Open flat land, meadows
4 – open to rough	0.1	Agricultural land with short crops
5 – rough	0.25	Agricultural land with tall crops
6 – very rough	0.5	Park landscapes with bushes and trees
7 – closed	1	Regular obstacles (forests, villages, suburbs)
8 – inner cities	2	Centres of big cities with tall and low buildings

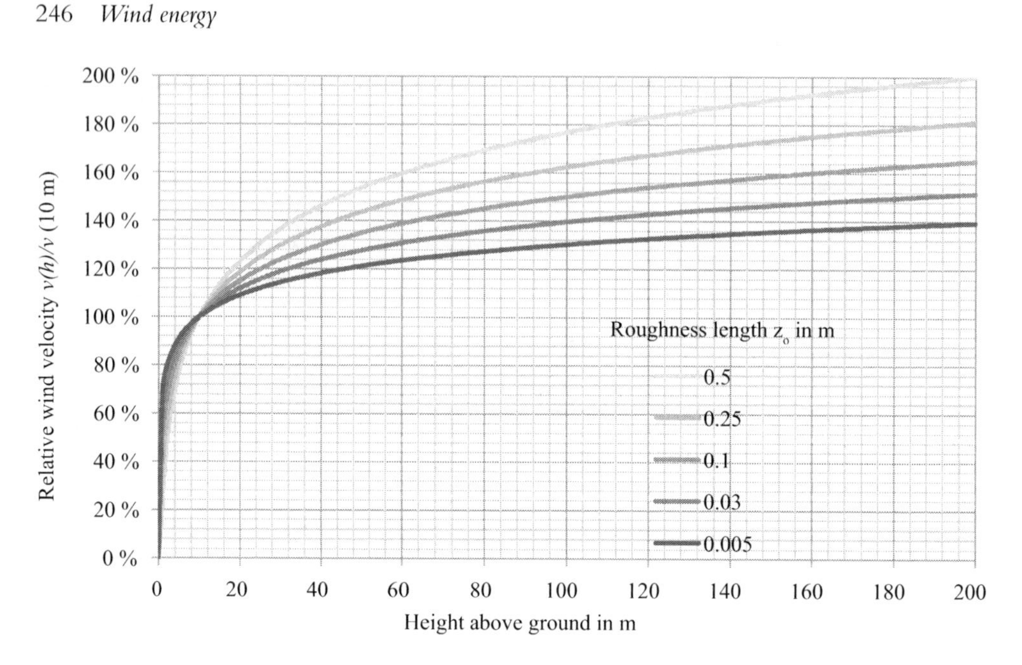

Figure 6.5 Increase in wind velocity at greater heights relative to roughness length compared with a wind velocity at a height of 10 m

Use of wind energy

The wind's energy content

At wind velocity v, the **kinetic energy** E in the wind can be calculated using the following general formula:

$$E = \frac{1}{2} \cdot m \cdot v^2 \tag{6.8}$$

The power P contained in the wind is calculated by differentiating energy across time. At a constant wind velocity v, the following applies:

$$P = \dot{E} = \frac{1}{2} \cdot \dot{m} \cdot v^2 \tag{6.9}$$

From the mass

$$m = \rho \cdot V \tag{6.10}$$

which is the product of density ρ and volume V, the **mass air flow rate**

$$\dot{m} = \rho \cdot \dot{V} = \rho \cdot A \cdot \dot{s} = \rho \cdot A \cdot v \tag{6.11}$$

of air with a density ρ passing across area A at velocity v is calculated. We can then calculate **wind power**:

$$P = \frac{1}{2} \cdot \rho \cdot A \cdot v^3 \tag{6.12}$$

Table 6.4 Air density relative to temperature, $p = 1$ bar = 1000 hPa [VDI94]

Temperature ϑ in °C	*−20*	*−10*	*0*	*10*	*20*	*30*	*40*
Density ρ in kg/m³	1.377	1.324	1.275	1.230	1.188	1.149	1.112

Air density changes according to air pressure p and temperature ϑ. Table 6.4 shows how the value changes relative to temperatures; density changes proportionally to air pressure.

During gale force winds (wind category 11), at a wind velocity of 30 m/s and a temperature of 10 °C, the wind contains 16.6 kW/m². At such high power densities, the devastation that a storm can cause is easy to imagine. In contrast, winds at 1 m/s do not even contain 1 W/m². Clearly, high average wind velocities are essential for a wind turbine to have good output.

Wind turbines **take energy out** of the wind. The turbine reduces wind velocity v_1 to wind velocity v_2 and makes use of the power differential. If this event took place in an enclosed tunnel with rigid walls under constant air pressure, the change of velocity v_2 would also affect velocity v_1, because the same amount of air that enters the tunnel has to leave it. In other words, the mass of air flow $\dot{m}$ is the same in front of and behind the turbine.

In the field, when a wind turbine slows down the wind, the mass airflow in front of and behind the turbine also remains constant. But as Figure 6.6 shows, the wind behind the turbine goes through a larger area than the wind does in front of the turbine. At constant air pressure and density ρ, the following applies:

$$\dot{m} = \rho \cdot \dot{V} = \rho \cdot A_1 \cdot v_1 = \rho \cdot A \cdot v = \rho \cdot A_2 \cdot v_2 = const \tag{6.13}$$

The wind velocity

$$v = \tfrac{1}{2} \cdot (v_1 + v_2) \tag{6.14}$$

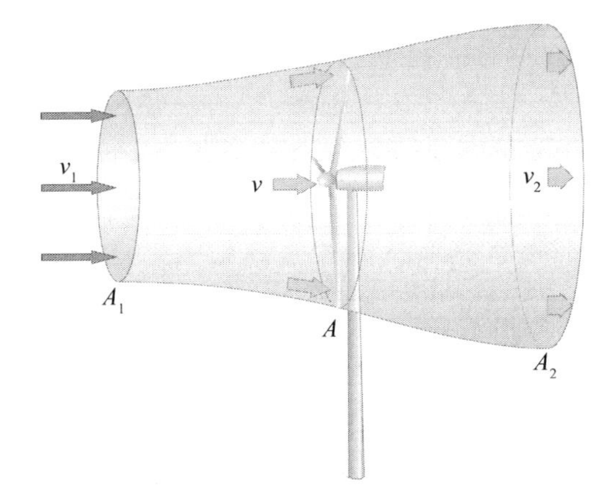

Figure 6.6 Airflow around a wind turbine in the field

at the height of a wind turbine is the average of wind velocities v_1 and v_2. The power taken out of the wind P_N can be calculated from the difference between these two wind velocities:

$$P_N = \tfrac{1}{2} \cdot \dot{m} \cdot (v_1^2 - v_2^2) \tag{6.15}$$

$\dot{m} = \rho \cdot A \cdot v = \rho \cdot A \cdot \tfrac{1}{2} \cdot (v_1 + v_2)$ produces:

$$P_N = \tfrac{1}{4} \cdot \rho \cdot A \cdot (v_1 + v_2) \cdot (v_1^2 - v_2^2) \tag{6.16}$$

Without the influence of the wind turbine, the wind's power over area A is:

$$P_0 = \tfrac{1}{2} \cdot \rho \cdot A \cdot v_1^3 \tag{6.17}$$

The ratio of the power taken from the wind P_N to the power contained in the wind P_0 is called the **coefficient of power** c_P and is calculated as follows:

$$c_P = \frac{P_N}{P_0} = \frac{(v_1 + v_2) \cdot (v_1^2 - v_2^2)}{2 \cdot v_1^3} = \frac{1}{2} \cdot (1 + \frac{v_2}{v_1}) \cdot (1 - \frac{v_2^2}{v_1^2}) \tag{6.18}$$

The maximum coefficient of power was calculated by Betz and is called the ideal COP or Betz's law $c_{P,Betz}$ [Bet26].

With $\zeta = \frac{v_2}{v_1}$ we use $\frac{d c_P}{d\zeta} = \frac{d\left(\frac{1}{2} \cdot (1+\zeta) \cdot (1-\zeta^2)\right)}{d\zeta} = -\frac{3}{2} \cdot \zeta^2 - \zeta + \frac{1}{2} = 0$

to calculate the ideal velocity ratio $\zeta_{id} = \frac{v_2}{v_1} = \frac{1}{3}$.

When inserted in the formula for the coefficient of power, we get:

$$c_{P,Betz} = \frac{16}{27} \approx 0.593 \tag{6.19}$$

If wind with an original wind velocity of v_1 is slowed down to a third of its velocity behind a wind turbine ($v_2 = 1/3 \cdot v_1$), the theoretical maximum power that can be taken from the wind is equivalent to around 60 % of the power contained in the wind.

Under real conditions, this optimum is not reached. Good turbines have a power coefficient of around 0.5. The **efficiency** η for the power taken from the wind can be defined using the ratio of the power harvested P_N to the maximum harvestable power, the ideal power P_{id}:

$$\eta = \frac{P_N}{P_{id}} = \frac{P_N}{P_0 \cdot c_{P,Betz}} = \frac{P_N}{\frac{1}{2} \cdot \rho \cdot A \cdot v_1^3 \cdot c_{P,Betz}} = \frac{c_P}{c_{P,Betz}} \tag{6.20}$$

Drag turbine

If an object is set up perpendicular to the wind, the wind exerts a force F_W on the object. This resistance is calculated from the wind velocity v, the area the wind passes over A and a general drag coefficient c_W:

$$F_W = c_W \cdot \frac{1}{2} \cdot \rho \cdot A \cdot v^2 \tag{6.21}$$

Figure 6.7 shows the resistance coefficients of various bodies. The power required to resist these forces is calculated using $P_W = F_W \cdot v$ in:

$$P_W = c_W \cdot \frac{1}{2} \cdot \rho \cdot A \cdot v^3 \tag{6.22}$$

If the wind moves an object at velocity u in the direction of the wind, the drag is

$$F_W = c_W \cdot \frac{1}{2} \cdot \rho \cdot A \cdot (v - u)^2 \tag{6.23}$$

and the power used is

$$P_N = c_W \cdot \frac{1}{2} \cdot \rho \cdot A \cdot (v - u)^2 \cdot u \tag{6.24}$$

For instance, let us roughly calculate the **power of an anemometer** used to measure wind velocity v. Here, two open hemispherical cups spin around a common rotating axis. The wind pushes one of the cups from the front; the other, from the back (Figure 6.8).

Here, the resulting force F consists of one component that pushes forwards while the other pushes back [Gas07]:

$$F = c_{W1} \cdot \frac{1}{2} \cdot \rho \cdot A \cdot (v - u)^2 - c_{W2} \cdot \frac{1}{2} \cdot \rho \cdot A \cdot (v + u)^2 \tag{6.25}$$

The following applies for the power used:

$$P_N = \frac{1}{2} \cdot \rho \cdot A \cdot \left(c_{W1} \cdot (v - u)^2 - c_{W2} \cdot (v + u)^2\right) \cdot u \tag{6.26}$$

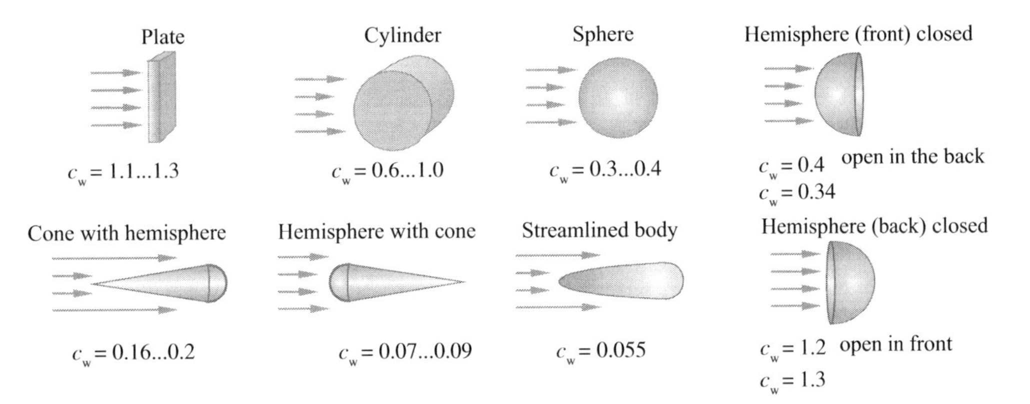

Figure 6.7 Resistance coefficients of various bodies [Her12]

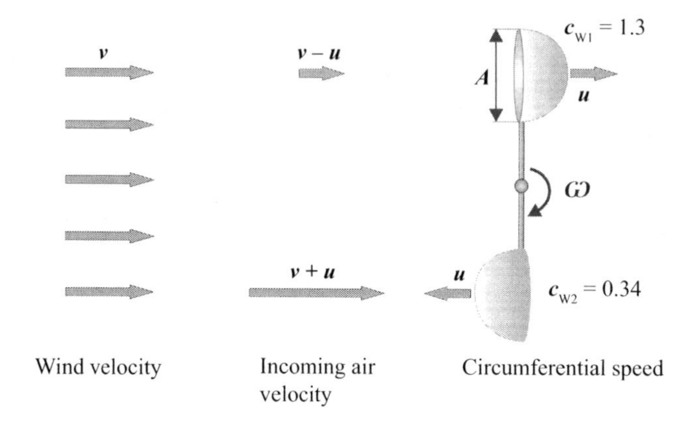

Figure 6.8 Model of an anemometer used to calculate power

The ratio of the **circumferential speed** u to wind velocity v is called the **tip-speed ratio** λ:

$$\lambda = \frac{u}{v} \tag{6.27}$$

The tip-speed ratio is always less than one for drag turbines. Including the tip-speed ratio, the following applies for power:

$$P_N = \tfrac{1}{2} \cdot \rho \cdot A \cdot v^3 \cdot \left(\lambda \cdot \left(c_{W1} \cdot (1-\lambda)^2 - c_{W2} \cdot (1+\lambda)^2\right)\right) \tag{6.28}$$

The **power coefficient** for the anemometer is calculated as follows:

$$c_P = \frac{P_N}{P_0} = \frac{P_N}{\frac{1}{2} \cdot \rho \cdot A \cdot v^3} = \lambda \cdot \left(c_{W1} \cdot (1-\lambda)^2 - c_{W2} \cdot (1+\lambda)^2\right) \tag{6.29}$$

The anemometer's maximum power coefficient is around 0.073, far below the ideal power coefficient of 0.593 calculated above. This power coefficient is reached at a tip-speed ratio of around 0.16, in other words when the wind velocity V is roughly 6 times greater than the circumferential velocity u.

The optimal power coefficient $c_{P,opt,W}$ for drag turbines is calculated with

$$c_P = \frac{P_N}{P_0} = \frac{\frac{1}{2} \cdot \rho \cdot A \cdot c_W \cdot (v-u)^2 \cdot u}{\frac{1}{2} \cdot \rho \cdot A \cdot v^3} = c_W \cdot (1 - \frac{u}{v})^2 \cdot \frac{u}{v} \tag{6.30}$$

and $u/v = 1/3$, with the maximum drag coefficient being $c_{W,max} = 1.3$, producing

$$c_{P,opt,W} = 0.193 \tag{6.31}$$

This value is also far below the ideal of 0.593. Therefore, modern turbines rarely use the drag principle, employing instead the lift principle, which provides much greater power coefficients, meaning that the energy in the wind can be exploited much better.

Lift turbines

When the airflow across the top of a body is faster than at the bottom, pressure builds up at the bottom, creating a **lift effect** described by Bernoulli:

$$F_A = c_A \cdot \tfrac{1}{2} \cdot \rho \cdot A_P \cdot v_A^2 \tag{6.32}$$

It can be calculated with the **lift coefficient** c_A, air density ρ, volume flow v_A and projection area A_P. The rotors of most modern turbines use the lift effect. The projection area

$$A_P = t \cdot r \tag{6.33}$$

of a rotor blade is calculated from its span t and width, which is roughly equivalent to the rotor's radius r.

As with drag turbines, lift turbines also have **drag**:

$$F_W = c_W \cdot \tfrac{1}{2} \cdot \rho \cdot A_P \cdot v_A^2 \tag{6.34}$$

However, the lift is generally much greater than the drag for a lift turbine. The ratio of the two forces is called the **lift-to-drag ratio**:

$$\varepsilon = \frac{F_A}{F_W} = \frac{c_A}{c_W} \tag{6.35}$$

Sometimes, the reverse drag-to-lift ratio is also used. Good profiles have lift-to-drag ratios of up to 400.

The **volume flow** of air in the two formulae above

$$v_A = \sqrt{v_W^2 + u^2} \tag{6.36}$$

is calculated from the wind velocity v_W and circumferential velocity u (Figure 6.9). The tip-speed ratio

$$\lambda = \frac{u}{v_W}$$

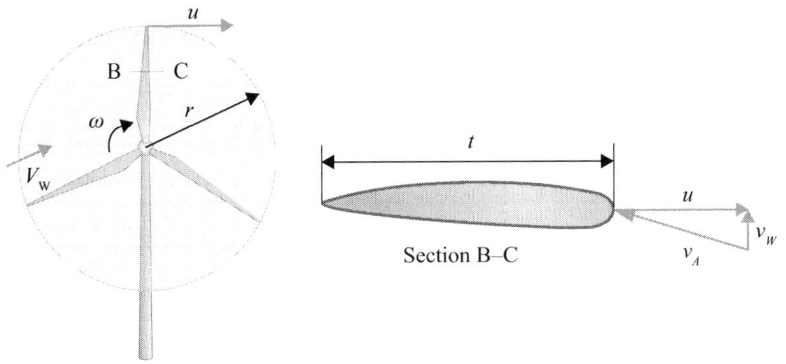

Figure 6.9 Wind velocity v_W and rotor rotation are used to calculate the flow rate v_A

then produces

$$v_A = v_W \cdot \sqrt{1+\lambda^2} \tag{6.37}$$

Figure 6.10 shows the ratio of drag $\mathbf{F_W}$ and lift $\mathbf{F_A}$. The resulting force is calculated by vector addition:

$$\mathbf{F_R} = \mathbf{F_W} + \mathbf{F_A} \tag{6.38}$$

It can be divided into an axial component $\mathbf{F_{RA}}$ and a tangent component $\mathbf{F_{RT}}$. A tangent component $\mathbf{F_{RT}}$ causes the rotor to rotate. The axial component is also called thrust (Figure 6.11).

The lift coefficient c_A and the drag coefficient c_W vary considerably according to the rotor blade's angle α. For $\alpha < 10°$, the following roughly applies [Gas07]:

$$c_A \approx 5{,}5 \cdot \alpha \cdot \pi / 180° \tag{6.39}$$

If the rotor blade's pitch ϑ is changed as shown in Figure 6.10, its angle α also changes, as does the coefficient of power c_P. If the rotor's angle is large, the maximum coefficient of power is much smaller, and the coefficient of power shifts towards smaller tip-speed ratios. This effect is called **pitch control.** To start up a wind turbine, the rotor angle is increased. At high wind velocities, the rotor blades are turned out of the wind to limit power. The rotor's angle ϑ is also called the pitch angle. The drag coefficient c_W is negligible at angles smaller than 15°.

The resulting force causes torque M. The power P_N harvested from the wind can be calculated from the coefficient of power c_P for the power P_0 originally contained in the wind, as with drag turbines:

$$P_N = c_P \cdot P_0 = c_P \cdot \tfrac{1}{2} \cdot \rho \cdot A \cdot v_W^3 \tag{6.40}$$

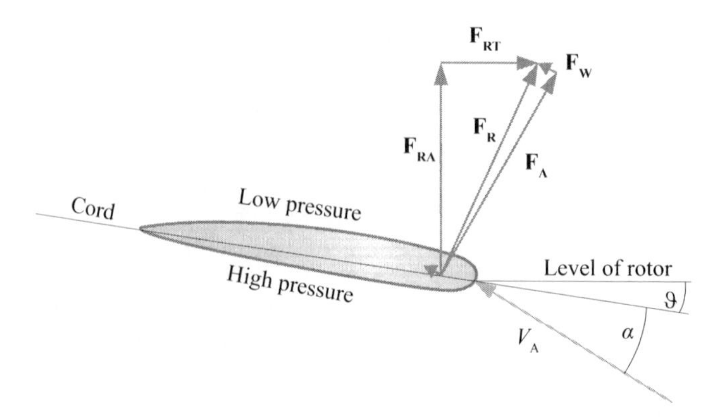

Figure 6.10 Forces at play for a lift turbine

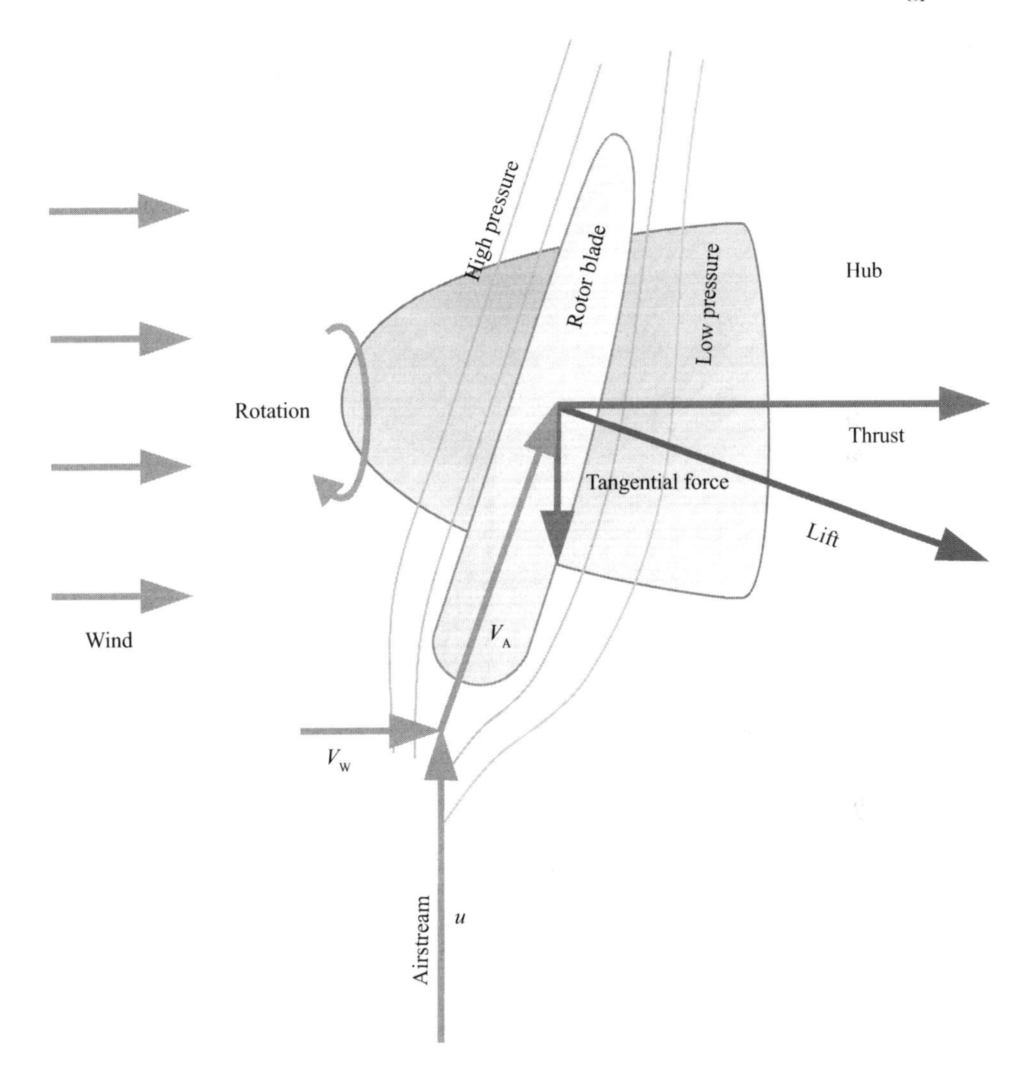

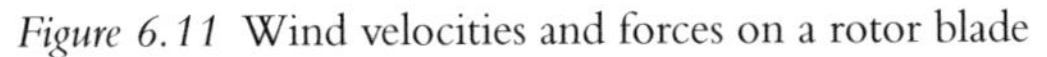

Figure 6.11 Wind velocities and forces on a rotor blade

With

$$M = \frac{P_N}{\omega} = \frac{P_N \cdot r}{u} \tag{6.41}$$

the following applies for **torque**:

$$M = c_P \cdot \frac{v_W}{u} \cdot \tfrac{1}{2} \cdot \rho \cdot A \cdot r \cdot v_W^2 \tag{6.42}$$

Torque M can be represented with the **torque coefficient**

$$c_M = c_P \cdot \frac{v_W}{u} = \frac{c_P}{\lambda} \tag{6.43}$$

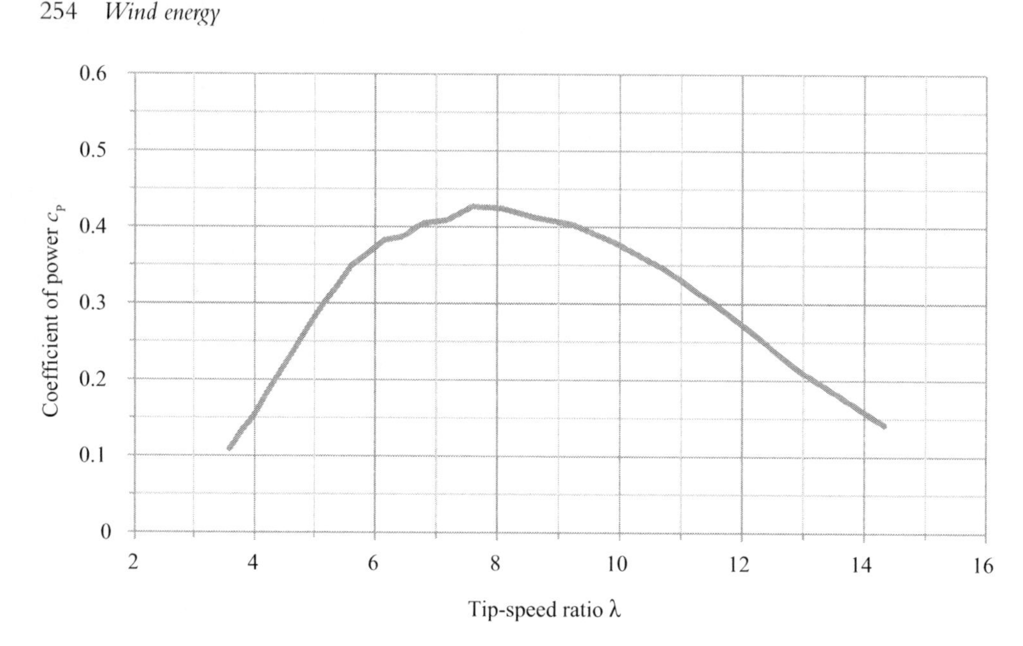

Figure 6.12 Coefficient of power c_p relative to the tip-speed ratio λ of a Vestas V44 600 kW turbine (Data: [Ves97])

and $A = \pi r^2$ as follows:

$$M = c_M \cdot \frac{1}{2} \cdot \rho \cdot A \cdot r \cdot v_W^2 = c_M \cdot \frac{1}{2} \cdot \rho \cdot \pi \cdot r^3 \cdot v_W^2 \tag{6.44}$$

If the torque M or power P of a wind turbine is known relative to wind velocity v_W, the coefficient of power c_p can be calculated at constant speed. Figure 6.12 shows the curve for the coefficient of power relative to the tip-speed ratio of the 600 kW wind turbine built at the end of the 1990s. The maximum coefficient of power is 0.427, much closer to the Betz optimum than a drag turbine is. Modern turbines even have a maximum power coefficient of around 0.5.

The curve for the power coefficient now has to be calculated in a compensated process along the rotor blade to account for aerodynamic fluctuations. Usually, the power coefficient is instead measured relative to the tip-speed ratio. The curve for the power coefficient can be roughly described as a cubic polynomial:

$$c_p = a_3 \cdot \lambda^3 + a_2 \cdot \lambda^2 + a_1 \cdot \lambda + a_0 \tag{6.45}$$

The coefficients a_3 to a_0 can be determined from measurements in program such as Matlab™ and MS Excel™. Figure 6.13 shows the two power coefficient curves and an approximation of a cubic polynomial.

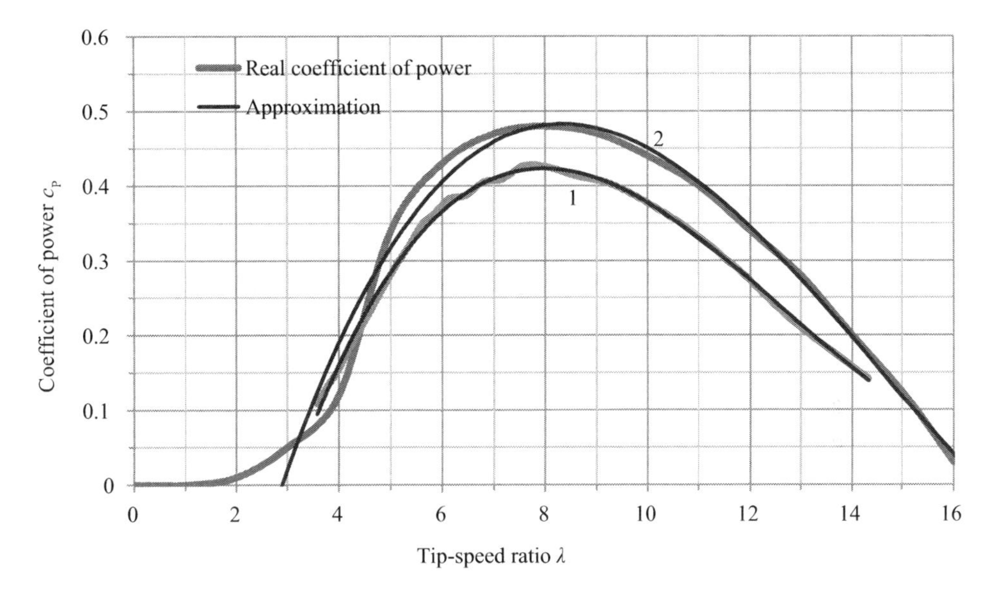

Figure 6.13 Power coefficients and approximation of a cubic polynomial

Wind turbine designs

Above, we talked about how power can be taken out of the wind using the drag and lift effect. In practice, this energy is used to perform mechanical work, such as in old windmills. Modern wind pumps also use wind power mechanically.

But the production of electricity is much more interesting and more common today. Here, a wind rotor drives an electrical generator. There are numerous concepts for rotors, which are described below.

Vertical-axis wind turbines

Some of the oldest designs are windmills with a vertical axis. The first ones were built more than 1,000 years ago and exploited the drag effect. Today, there are various types of wind power generators with a vertical axis, described in Figure 6.14.

A distinction is made between:

- Savonius rotors
- Darrieus rotors
- H rotors.

The **Savonius rotor** mainly exploits the drag principle, like the previously described cup anemometer. It consists of two scoops, each opening in the opposite direction. Near the axis, the scoops overlap a bit, so that the wind moves from one to the other after each semi-rotation. To a small extent, this design thus also uses the lift principle, so that the efficiency of a Savonius rotor is somewhat better than that of a simple drag turbine, though it is still much worse than a lift turbine. If optimally designed, a Savonius rotor has a maximum

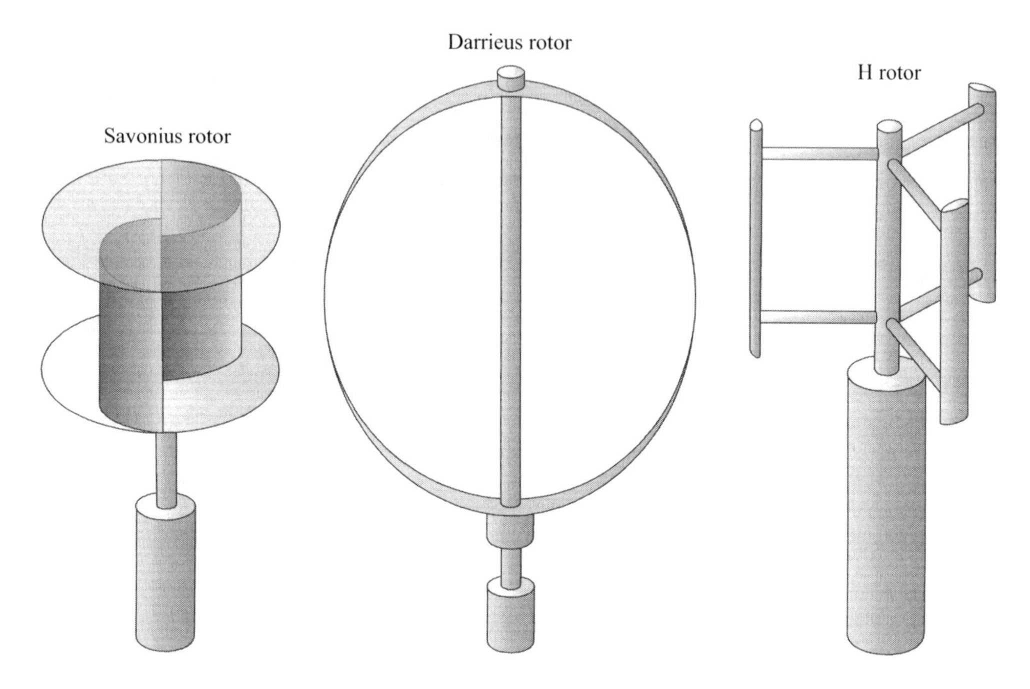

Figure 6.14 Vertical-axis rotors

power coefficient of 0.25 [Hau96]. One strong point of Savonius rotors is that they start up at low wind velocities. They are therefore used on production plants and in utility vehicles for ventilation purposes and to start up Darrieus rotors. In addition to their poor efficiency, Savonius rotors are also material-intensive. Large units are therefore not built.

The **Darrieus rotor** was patented by Frenchman Georges Darrieus in 1929. It consists of two or three rotor blades in the shape of a parabola. The profile of these rotorblades is similar to that of a lift turbine. The Darrieus rotor thus also exploits the lift principle. But unlike a rotor with a horizontal axis, the vertical axis constantly leads to a changing blade angle in the wind. Darrieus rotors are far more efficient than Savonius rotors, but still only around 75% as efficient as a rotor with a horizontal axis. A major drawback of Darrieus rotors is that they are not self-starting. They require a starter, which can be either a separate generator or a Savonius rotor.

The **H rotor,** also called the H-Darrieus rotor, is a further development of the concept. The H stands for Heidelberg-Motor, the manufacturer, so the turbine is also called the Heidelberg rotor. This rotor has a synchronous electric motor directly integrated in the rotor's structure; it has no gearbox. Like the Darrieus rotor, it employs the lift principle. The three rotor blades of an H rotor are perpendicular to the axis and kept in position by struts connected to the vertical axis. The H rotor was designed for extreme weather conditions in the Antarctic and the Alps; it is especially robust.

Wind turbines with a vertical axis have a number of benefits. Their structure is relatively simple. The electric generator, the gearbox (if required), and electrical controls can be stored in the ground station. It is therefore relatively easy to perform maintenance on the vertical-axis rotors. In addition, they do not have to track the wind. They are therefore especially suited to regions where the wind direction changes quickly.

Despite these benefits, vertical-axis turbines have not become the standard and are only used for special applications. Their low efficiency, material fatigue due to frequently changing loads, and generally greater amount of material means that they have not been able to compete with horizontal-axis rotors economically.

Horizontal-axis wind turbines

Structure

Most wind turbines currently used to generate electricity have a horizontal axis. It was mainly midsize firms that pushed this development. Today, wind turbines have reached a high technical level. At the end of the 1980s, a few newly installed wind turbines were larger than 100 kilowatts, but current turbines can have more than 5,000 kilowatts, equivalent to five megawatts.

The following **components** are the main ones in the horizontal-axis turbines for power generation:

- rotor blades, rotor hub, rotor brake, and possibly rotor pitch mechanism
- electrical generator, possibly with gearbox
- wind measurement system and wind tracking system (azimuth adjustment)
- nacelle, tower, and foundations
- electrical cabinet, controls, and grid connection.

Figure 6.15 shows the cross-section of a horizontal-axis wind turbine.

Rotor blades

A distinction is made between modern horizontal-axis wind turbines that have one, two, and three rotor blades. Modern wind turbines rarely have more than three rotor blades. The smaller the number of roller blades, the less material is needed.

Single-blade rotors have to have a counterweight on the rotor blade on the other side of the hub. Single-blade rotors do not spin evenly, and there is a great load on the material. At present, few wind turbine prototypes have only a single blade. Such designs will not become the standard in general.

The optimal power coefficient of three-blade rotors is slightly lower than that of two-blade rotors. But three-blade rotors are visually calmer and fit the landscape better. The mechanical load is also lower when three blades are used instead of one or two. The benefits of three-blade rotors outweigh the drawbacks of the extra material needed so that **three-blade rotors** are the most common today.

The shape of a rotor blade decisively affects the feasible power coefficient. Rotor blades should taper off from the hub to the tip. [Kle93] explains that the depth of a rotor blade t relative to its distance r from the hub for a rotor with z blades, radius R, rated tip-speed ratio λ, and lift coefficient c_A can be calculated as follows:

$$t = \frac{1}{z} \cdot \frac{8}{9} \cdot \frac{2\pi}{c_A} \cdot \frac{R}{\lambda \sqrt{\lambda^2 \left(\frac{r}{R}\right)^2 + \frac{4}{9}}} \approx \frac{1}{z} \cdot \frac{8}{9} \cdot \frac{2\pi}{c_A} \cdot \frac{R}{\lambda^2 \left(\frac{r}{R}\right)} \qquad (6.46)$$

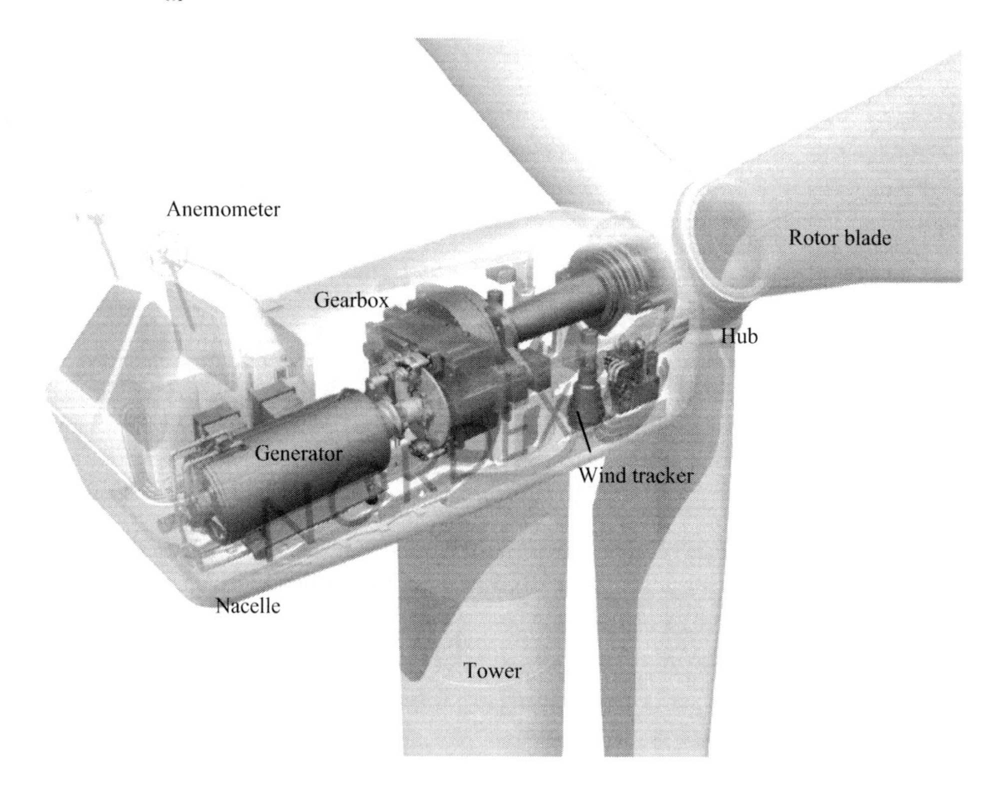

Figure 6.15 Design and components of a wind turbine
(Graphic: Nordex AG)

Near the rotor hub, the optimum shape is often deviated from for design reasons. In addition to aerodynamic advantages, the tapering of the rotor blade towards the tip also saves material and cost. Timber and metal used to be the main materials used, but these days a plastic reinforced with fibre optics is used almost exclusively because it is easier to shape and weighs less; likewise, carbon fibres are increasingly becoming commonplace.

The direction of incoming wind v_A differs considerably between the hub and the blade tip. While the incoming wind velocity at the blade tip largely comes from the circumferential direction, the real wind velocity component determines the speed at the hub. The rotor blade profile should take account of these parameters in order to produce an optimal power coefficient. Figure 6.16 shows the change in rotor blade depth and its face.

Wind velocity areas

The rated tip-speed ratio is closely linked to the number of rotor blades. For three-blade rotors, the maximum power coefficient is reached at a tip-speed ratio of around 7 to 8, compared with 10 for two-blade rotors and 15 for single-blade rotors, though there are great deviations from one turbine type to another. In other words, single-blade rotors turn much faster than comparable rotors with two or three blades.

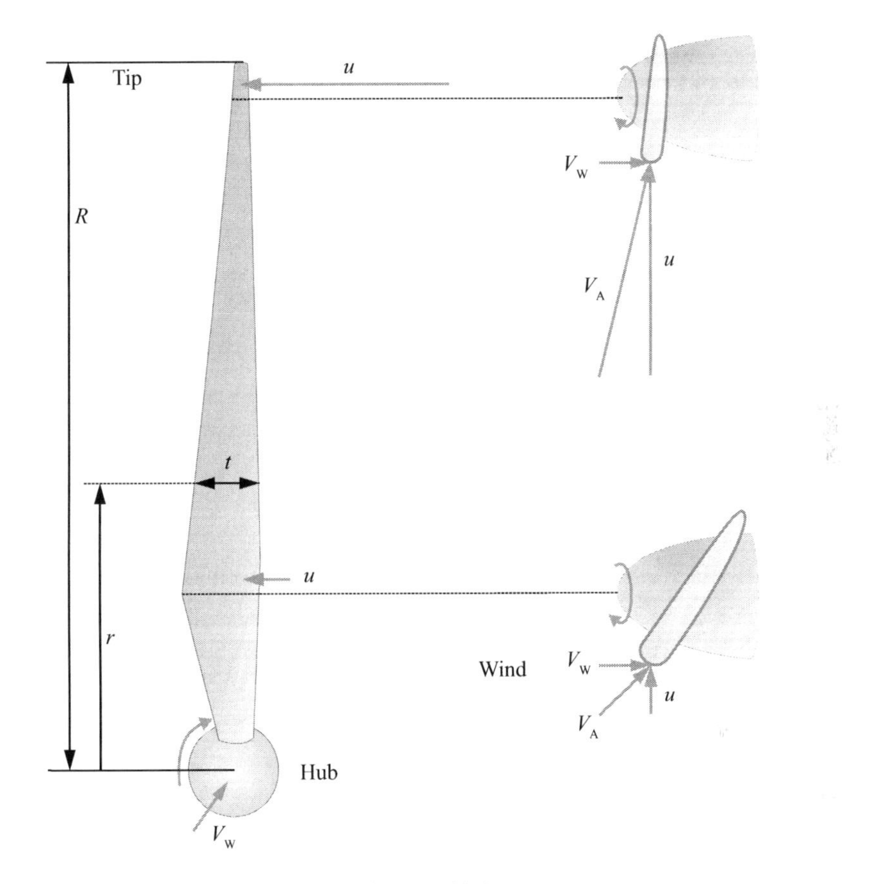

Figure 6.16 Changes in depth and angle across the rotor blade

The optimal tip-speed ratio λ_{opt} is used to specify the **rated wind speed**

$$v_{design} = \frac{u}{\lambda_{opt}} = \frac{2 \cdot \pi \cdot r \cdot n}{\lambda_{opt}} \tag{6.47}$$

for the rotor's diameter r and speed n. For instance, a wind turbine with three blades, a rotor radius of $r = 22$ m, a speed of $n = 28\ \text{min}^{-1} = 0.467\ \text{s}^{-1}$ and an optimal tip-speed ratio of $\lambda_{opt} = 7.5$ has a design wind velocity of $v_{design} = 8.6$ m/s. At this wind velocity, the turbine reaches maximum efficiency. The design wind velocity is mainly important for turbines that run at constant speed. Variable-speed turbines can reach their optimal levels at other wind velocities by changing the rotor speed. For them, indicating a rated wind speed makes little sense.

Running a wind turbine at low wind speeds is not recommended. Little or no power can then be harvested from the wind, and the turbine might even consume more than it produces. A rotor brake therefore keeps a wind turbine from starting up below a defined **minimum wind velocity.**

The design wind velocity was described above. A wind turbine's **nominal wind velocity** is the wind velocity at which the turbine produces its nominal output. The nominal wind velocity is greater than the design wind velocity. Above the nominal wind velocity, the turbine's output is maintained at a constant level. At a high wind velocity, the turbine could overload and be damaged. It is therefore switched off when wind velocities are too high. The rotor brake then holds the turbine still, and the rotor is turned out of the wind if possible. The following values are typical for various wind velocities:

- Startup velocity v_{start} = 2.5 m/s – 4.5 m/s²
- Design wind velocity v_{design} = 6 m/s – 10 m/s²
- Nominal wind velocity v_N = 10 m/s – 16 m/s²
- Switch-off wind velocity v_{stop} = 20 m/s – 34 m/s²
- Survival wind velocity v_{life} = 50 m/s – 70 m/s

Figure 6.17 shows the generator's real power and coefficient of power for a variable-speed 2,300 kW wind turbine relative to wind velocity. The ability to run at variable speed allows the maximum power coefficient of 0.5 to be attained over a wider range of wind velocities.

Limiting power output and switch-off during storms

A varying amount of power can be harvested from the wind depending on its velocity. Above the nominal wind velocity, output must be maintained at a constant level lest the power generator be overloaded. There are two basic approaches to limiting the power output of a wind turbine:

1. stall control
2. pitch control.

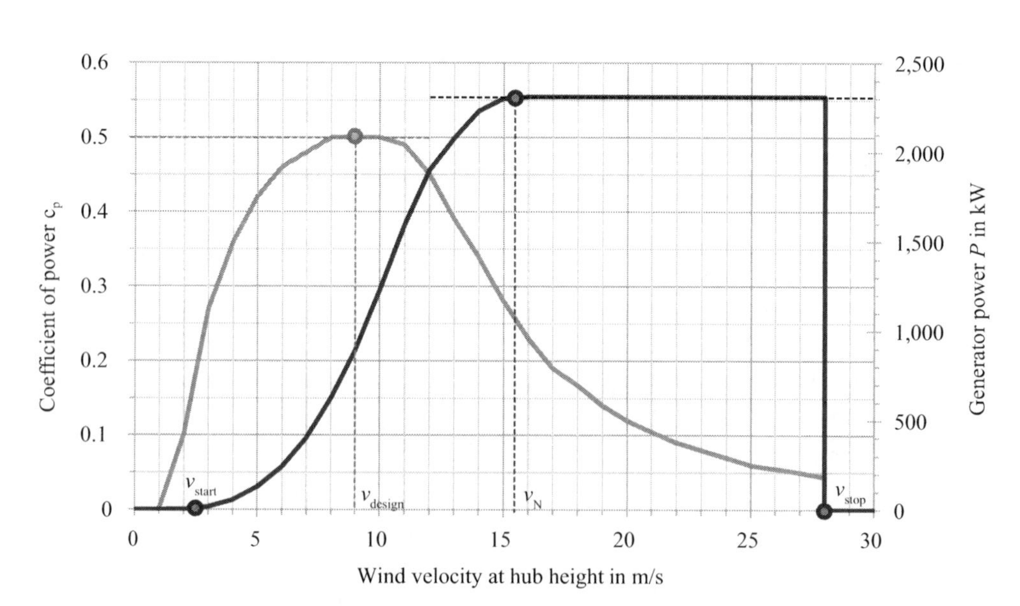

Figure 6.17 The generator's real power and power coefficient relative to wind velocity for a 2.3 MW Enercon E-70

(Data: [Ene06])

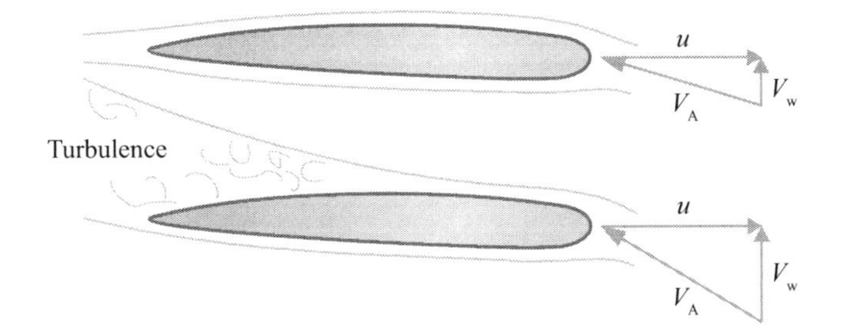

Figure 6.18 Interrupted airflow from the stall effect at large wind velocities

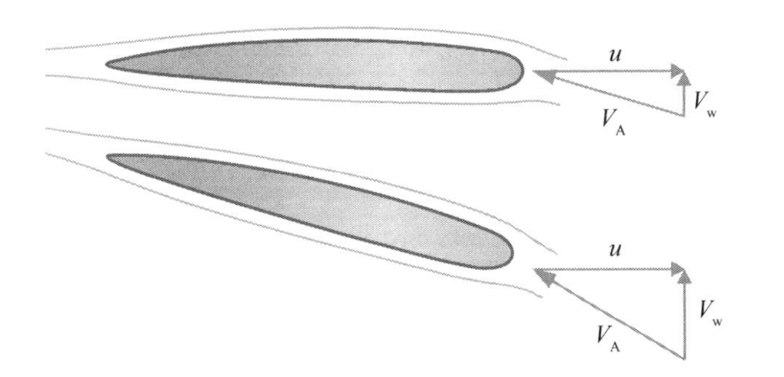

Figure 6.19 Blade pitch adjustments at various wind velocities

Stall control is based on the interruption (stall) of airflow at a certain blade angle (Figure 6.18). As a result, the lift is almost completely lost. In this way, the power passed on from the wind to the rotor can be limited. In stall-controlled turbines, the rotor speed n and the circumferential speed u are kept constant. As wind velocities v_W increase, the rotor blade's angle also increases. But the rotor blade itself is not turned with stall control; the blade's pitch therefore remains constant. Stall control is provided with little technical effort simply through design changes. The drawback of stall control is that you have little influence on the process, which occurs passively. The maximum output of a newly designed rotor blade is hard to predict with stall control because the stall process cannot be sufficiently described mathematically. Once the maximum output has been reached, stall-controlled turbines have their power reduced, and it does not remain constant, as shown in Figure 6.20.

Because of these drawbacks, current modern grid-connected wind turbines preferably use **pitch control**, even though this option is more complicated. Stall control is now only used with small generators and micro-grids.

In contrast to stall control, the pitch of the blade is actively changed with pitch control. When the turbine is starting up and at great wind velocities, the rotor blade is turned into the wind (Figure 6.19). The blade's angle is then reduced, thereby actively changing the rotor blade's power harvest. A pitch-controlled turbine requires a more complicated design because the rotor blade has to pivot at the hub. The blades of small systems can be turned mechanically by means of centrifugal force. Larger turbines have an electric motor that puts the rotor blade in the desired position. Electricity is consumed in the process, however.

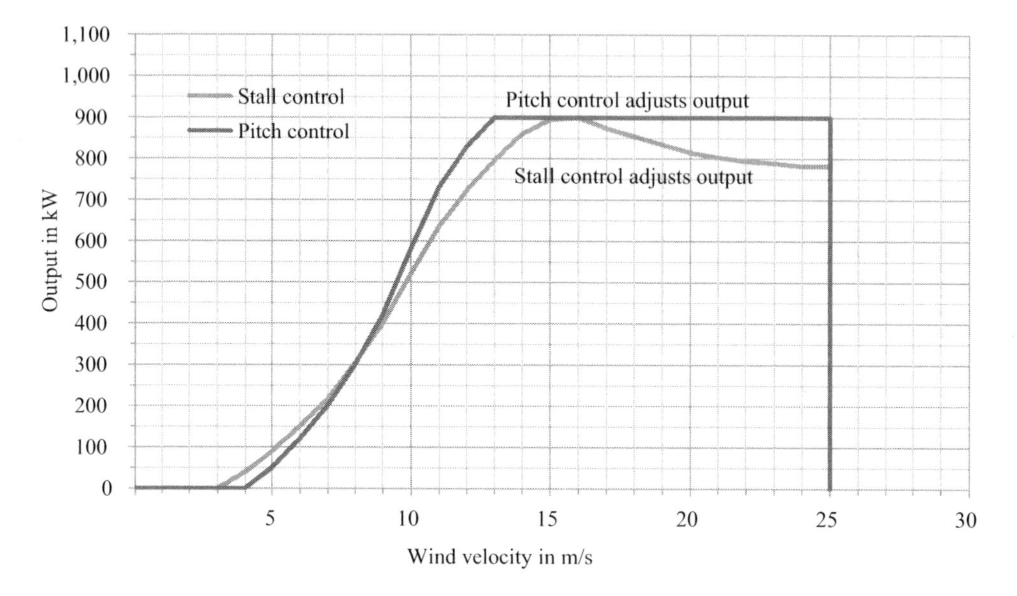

Figure 6.20 The power curve for stall-controlled and pitch-controlled wind turbines

When a wind turbine is switched off under high winds, pitch control can turn the rotor blade into the flag position. Power harvesting is then minimized, and the turbine is prevented from being damaged. Stall-controlled turbines used to have integrated aerodynamic breaks. For instance, the tip of the blade would open up to slow down the rotation. When the tip is turned 90°, the rotor cannot spin as fast.

Tracking the wind

Unlike vertical-axis wind turbines, those with a horizontal axis need to track the wind. The rotor blades are turned into the wind so that the wind always flows past the rotor blades at the desired angle. If the wind direction changes quickly and there are gusts, pitch-controlled turbines in particular have a problem. The power fluctuations can be great, though the speed can be varied to compensate for these fluctuations.

A distinction is made between upwind turbines and downwind turbines in terms of the rotor's position. **Upwind turbines** have the rotor on the side of the tower facing the wind, while **downwind turbines** have the rotor on the other side. Downwind turbines have one drawback: the rotor blades constantly move behind the tower's slipstream. The results are strong mechanical loads and additional noise from turbulence around the turbine and nacelle. For this reason, most large turbines are upwind. Downwind turbines have the advantage of having the rotor blade optimally positioned in the wind by wind pressure. Small wind generators often attract the wind with a wind vane. The wind vane keeps the upwind rotor perpendicular to the wind.

Horizontal-axis turbines generally track the wind with the entire nacelle – including the rotor, gearbox, and generator – turning on the tower. A wind measurement system on the nacelle determines the wind velocity and direction. Electric or hydraulic drives move the nacelle and rotor if need be. Once the nacelle has reached the desired position, azimuth brakes hold it in position. In practice, there are always minor deviations between the wind direction and the optimal rotor position. This deviation is called the yaw angle and is around 5° on average.

Tower, foundation, gearbox, and generator

The **tower** is one of the most important parts of a wind turbine. Its job is to hold up the nacelle and rotor blades. Because wind velocity increases considerably with height, higher towers can also increase a turbine's yield. In the pioneer years of wind power, most towers were open pylons. They have the advantage of using little material and therefore being inexpensive. Today, such pylons are rarely used with wind turbines for visual reasons. Small generators have simple towers supported with guy-wires. Large wind turbines have hollow towers made of steel or concrete with a round cross-section. The towers can be 100 m or taller. A standard 120 m tower with a diameter of 6 m at the base tapering off to 5.5 m at the top is a massive 750 t. As wind turbines get bigger, transporting and assembling towers and rotor blades become more difficult. Such a wind turbine needs a heavy concrete foundation to anchor such a mass.

A wind turbine's **nacelle** houses the rotor bearings, the gearbox, and the electrical generator. Because the generator is generally designed to run at high speed, a gearbox is needed to convert rotor speed. The gearbox adjusts the rotor's slower speed to the generator's faster speed. But the gearbox also comes at a price. It not only has a price tag itself, but also reduces power production by increasing friction; it also increases noise and maintenance. Some turbines do without a gearbox (called 'direct-drive turbines'), but they require a special generator design. It must have a large number of electrical poles to ensure that the rotor and the grid are coordinated even at low rotor speeds. The larger number of poles increases the cross-section, however, and hence the generator's dimensions and mass.

Offshore wind turbines

There are great hopes at present for offshore wind turbines. After all, wind farms can cover a lot of area at sea without conflicting with other usages. At the same time, there is far more wind at sea than on land, and it also blows more regularly. The yield at offshore wind turbines is thus greater – as much as 50 % or more higher than under optimal conditions on land.

Offshore wind turbines have a lot in common with the technology used on land. Systems used at sea generally must be robust and require little maintenance. After all, when the weather is bad and the seas are rough, offshore turbines are not accessible. Special ships are required for major maintenance tasks, but even they only allow work to be done when the sea is relatively calm. Aggressive salt water is another problem for offshore systems. All of the components therefore require special protection from corrosion.

Special ships with cranes are used for construction. Special foundations anchor the wind turbines on the seafloor (Figure 6.22). For economic reasons, the water cannot be too deep. At present, the maximum water depth is around 50 m, but the load-bearing capacity of the seafloor also plays a role. If the ground is too soft, the foundations become more complicated. These complications can make projects unaffordable even in shallow water.

Grid connections are also more complicated for offshore wind farms than for onshore ones. Underwater cables connect individual wind turbines to the transformer station. It

Figure 6.21 Construction of a wind turbine. Top left: foundations. Right: tower. Bottom left: rotor and nacelle

(photos: Bundesverband WindEnergie e.V. and ABO Wind AG)

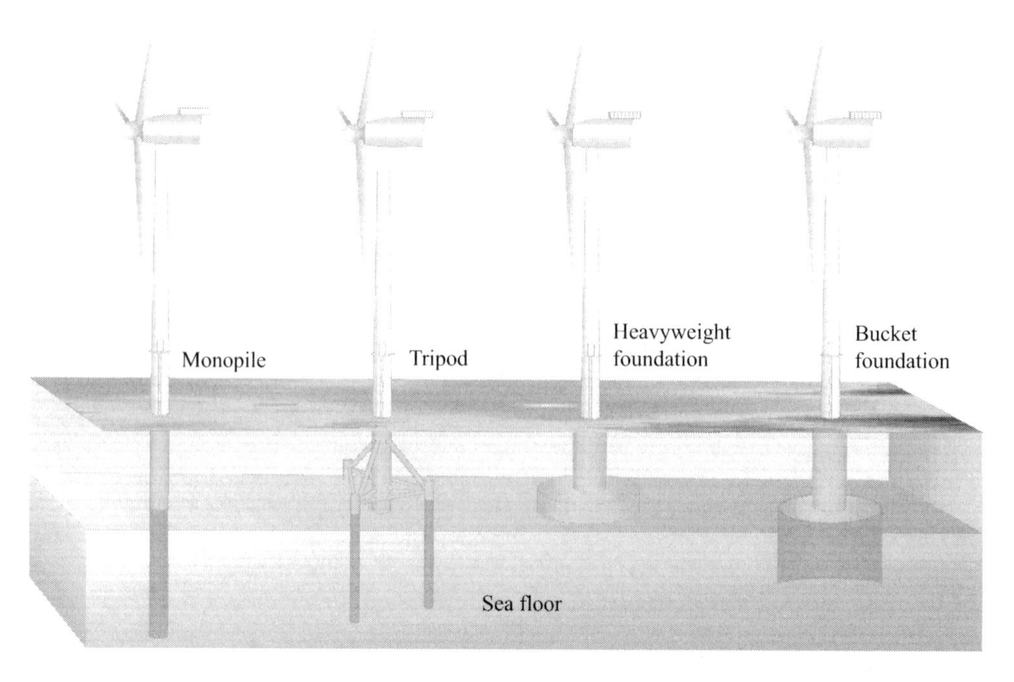

Figure 6.22 Foundations for offshore wind turbines

is also located within the wind farm at sea and looks a bit like a small oil rig. The **transformer station** converts the voltage from the wind turbines into high voltage so that transmission losses are reduced. If the distance to the coast is too great, a direct-current cable might be needed because the losses from alternating current in undersea cables are relatively great. A special inverter converts alternating current into direct current. On land, it is converted back into alternating current.

In Germany, the sea can be divided into two legal zones. First, there is the zone of territorial waters within 12 nautical miles (22.2 kilometres) of the coast. After that comes the **exclusive economic zone** (EEZ), which reaches out up to another 200 nautical miles (370.4 kilometres). In the Baltic, Germany's EEZ is much smaller than in the North Sea because of all of the neighbouring countries. Within German territorial waters, the adjacent German state handles permits for wind farms. Because of the visual impact on the coast, very few wind farms are planned within this zone. In the EEZ, however, wind farms are so far from the coast that they practically can no longer be seen. The German Maritime and Hydrography Agency (BSH) handles permits in this zone.

Electrical machines

A crucial part of wind turbines is the generator, which converts the kinetic energy from the wind into electrical energy. A large part of this chapter is therefore dedicated to electrical machines. Before the role they play in wind power can be explained, however, a few basic principles need to be explained.

Electrical machines can be categorized by the type of electricity:

- direct current
- alternating current
- pulse current
- three-phase electric power.

Direct-current machines are among the oldest ones because they are so straightforward; they are used in numerous applications today, such as in the motors that drive windscreen wipers in cars. The universal motor that runs on alternating current is mainly used in household appliances. Step motors that position printers, for instance, run on pulse current.

For wind turbines, however, **three-phase current machines** are the best option, and they can be divided into **induction motors** and **synchronous motors**.

Electrical alternating current calculation

Three-phase motors run on three-phase alternating current. Before we discuss this further, we first need to review what alternating current is.

The time-dependent curve of alternating voltage can be described

$$u(t) = \hat{u} \cdot \cos(\omega \cdot t + \varphi_u) \qquad (6.48)$$

with the amplitude $\hat{u}$, reference point phase angle φ_u and angular frequency

$$\omega = 2\pi \cdot f \qquad (6.49)$$

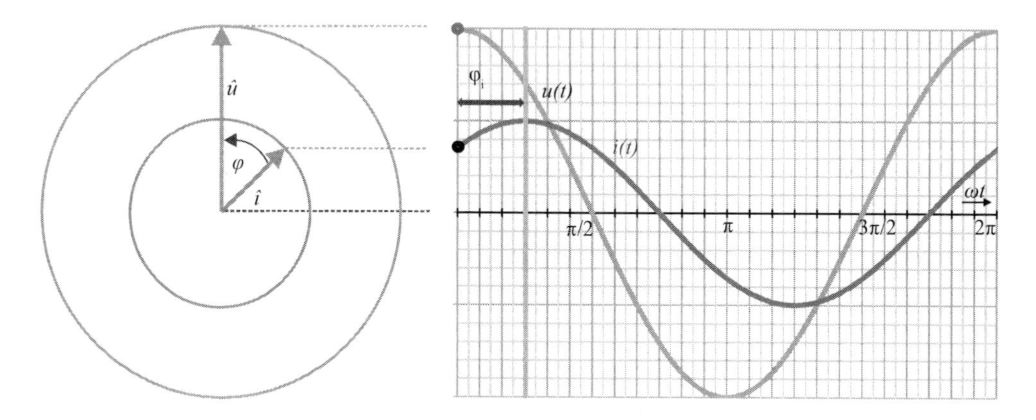

Figure 6.23 Temporal curve of current and voltage along with amplitudes ($\varphi = \pi/4$)

The **current** i with phase angle φ_i and amplitude $\hat{i}$ combines to produce:

$$i(t) = \hat{i} \cdot \cos(\omega \cdot t + \varphi_i) \tag{6.50}$$

For the **phase angle** φ between current and voltage, the following applies:

$$\varphi = \varphi_u - \varphi_i \tag{6.51}$$

If the phase angle is positive, the voltage is ahead of the current; if it is negative, behind it. Figure 6.23 shows the temporal curve of current and voltage. With the reference point phase angles $\varphi_u = 0$ and $\varphi_i = -\pi/4$, the phase angle is $\varphi = +\pi/4$, meaning that voltage is ahead of current here.

Another common description adds amplitude – here, the amplitude of voltage – as a vertical reference in a vector diagram. The other amplitudes are added relative to that at a phase angle. Figure 6.23 shows such a vector diagram of amplitudes.

An indication of an average is interesting in describing changing parameters. The **root mean square** is usually used as a temporal average because the sine curve used to create an arithmetic average has both positive and negative values. The root mean square of a function $u(t)$ with a periodic duration of $T = 1/f$ is defined as follows:

$$u_{\mathrm{eff}} = U = \sqrt{\frac{1}{T} \cdot \int_0^T u^2(t)\,\mathrm{d}t} \tag{6.52}$$

When the voltage and current are sinusoidal, the root mean squares are:

$$U = u_{\mathrm{eff}} = \frac{\hat{u}}{\sqrt{2}} \approx 0,707 \cdot \hat{u} \quad (6.53) \quad \text{and} \quad I = i_{\mathrm{eff}} = \frac{\hat{i}}{\sqrt{2}} \approx 0,707 \cdot \hat{i} \tag{6.54}$$

The vector diagram for root mean squares would look like Figure 6.23 except that the vectors are shorter for root mean squares. We can imagine these vectors mathematically as a complex plane. The real axis (Re) is often vertical, contrary to the normal horizontal axis commonly used in maths. **Complex parameters** are represented by underlined symbols below.

With

$$\underline{U} = U \cdot e^{j\varphi_u} = U \cdot e^{j0} = U$$

we get the following complex representation for the example above for the vector of the current's root mean square with phase angle φ_i and imaginary unit j ($j^2 = -1$):

$$\underline{I} = I \cdot e^{j\varphi_i} \tag{6.55}$$

Such components as inductors and capacitors cause a phase shift between current and voltage. To describe complex planes, imaginary resistance – called **electrical reactance –** is introduced.

Figure 6.24 shows a serial circuit with resistance and inductance along with the vector for current and voltage. Here, voltage $\underline{U}_1$ is put into the real axis as a reference ($\varphi_u = 0$). In this example, the current $\underline{I}$ turns around the reference point's phase angle $\varphi_i = 3\pi/4$ so the phase angle between current and voltage is also $\varphi = -3\pi/4$. This value was arbitrarily chosen in this **example**.

The current $\underline{I} = I \cdot e^{j\varphi_i}$ is then used to calculate voltage at the resistance R:

$$\underline{U}_R = R \cdot \underline{I} = R \cdot I \cdot e^{j\varphi_i} \tag{6.56}$$

The voltage vector $\underline{U}_R$ points in the same direction as the current vector $\underline{I}$. Voltage at the inductor is calculated from its induction L / electrical reactance $X = \omega L$ along with $j = e^{j\frac{\pi}{2}}$ as follows:

$$\underline{U}_L = j\omega L \cdot \underline{I} = jX \cdot \underline{I} = e^{j\frac{\pi}{2}} \cdot X \cdot I \cdot e^{j\varphi_i} = X \cdot I \cdot e^{j(\frac{\pi}{2}+\varphi_i)} \tag{6.57}$$

Voltage $\underline{U}_2$ can be determined from:

$$\underline{U}_2 = \underline{U}_1 - \underline{U}_L - \underline{U}_R \tag{6.58}$$

When the voltage vectors $\underline{U}_R$ and $\underline{U}_L$ are shifted in parallel, the vector $\underline{U}_2$ is the one that closes the diagram.

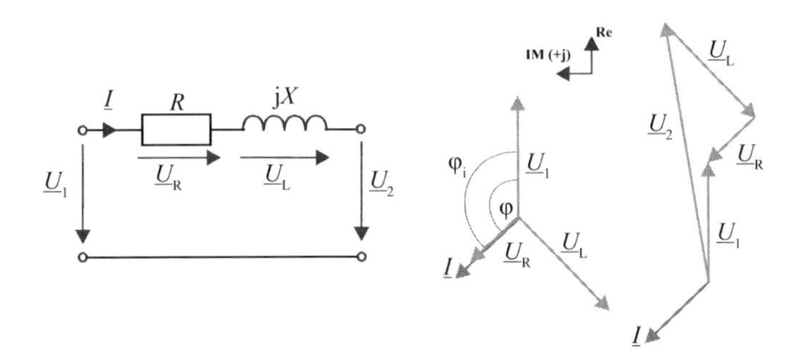

Figure 6.24 Serial circuit of resistance and inductor with vector diagrams

The **momentary power** $p(t)$ is calculated as in the calculation of direct current:

$$p(t) = u(t) \cdot i(t) \tag{6.59}$$

The average power or **real power** P is calculated from the difference of positive and negative areas in the $p(t)$ curve and the time axis. The formula

$$P = \frac{1}{T} \int_{0}^{T} u(t) \cdot i(t) \cdot \mathrm{d}t \tag{6.60}$$

determines the real power P in harmonic voltage and current curves with the phase angle φ:

$$P = \tfrac{1}{2} \cdot \hat{u} \cdot \hat{i} \cdot \cos\varphi = U \cdot I \cdot \cos\varphi \tag{6.61}$$

If there is a phase shift between current and voltage by $\pm\pi/2$, the positive area becomes as big as the negative area, and the real power is zero. This does not mean that no power is available. Rather, all of the energy simply swings back and forth between the generator and the consumer. This share of the energy that moves back and forth also exists at other phase angles and is called reactive power Q:

$$Q = U \cdot I \cdot \sin\varphi \tag{6.62}$$

When the phase angle is positive, the reactive power also is, and we speak of inductive reactive power. When the phase angle is negative, the reactive power also is, and we speak of capacitive reactive power. Sometimes, other vector diagrams are used for current and voltage, and the phase angle is turned 180°, which also confusingly changes the plus and minus signs.

The **apparent power** S is calculated as follows:

$$\underline{S} = P + \mathrm{j} \cdot Q \text{ and } |\underline{S}| = S = \sqrt{P^2 + Q^2} = U \cdot I \tag{6.63}$$

Real power P is indicated in the unit W (watt). Reactive, apparent, and real power are the products of current and voltage in terms of these units. But to make a clearer distinction, other units are used. Reactive power Q is indicated in the unit var (volt amp reactive), while apparent power S is given in the unit VA (volt amp).

The **power factor**

$$\cos\varphi = \frac{P}{S} \tag{6.64}$$

describes the ratio of real power P to apparent power S. Because the cosine for the negative and positive phase angles produces the same value, the power factor is often described as being inductive or capacitive. Then you know whether the phase angle is positive or negative.

Rotating magnetic field

When current flows through an electrical conductor, a magnetic field is created, as shown in Figure 6.25 for a conductor and an inductor.

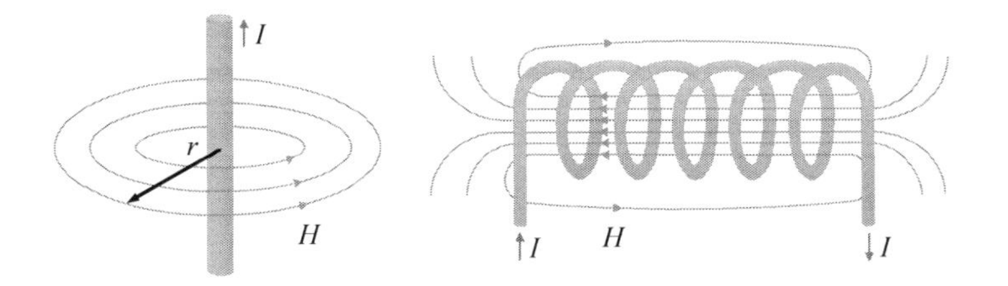

Figure 6.25 Magnetic fields around a conductor and inductor with current flowing through them

The conductor with electric current I has at a distance r from the conductor a magnetic field strength:

$$H = \frac{I}{2 \cdot \pi \cdot r} \tag{6.65}$$

In addition to magnetic field strength H, **magnetic flux** B can be determined from the magnetic field constant

$$\mu_0 = 4 \cdot \pi \cdot 10^{-7} \frac{\text{Vs}}{\text{Am}} \approx 1{,}257 \cdot 10^{-6} \frac{\text{Vs}}{\text{Am}} \tag{6.66}$$

and the material-dependent permeability number μ_r as follows:

$$B = \mu_0 \cdot \mu_r \cdot H \tag{6.67}$$

To create a rotating magnetic field from the magnetic fields around current conductors, inductors through which three-phase alternating current is flowing are installed at an offset of 120°. The three inductors U, V, and W with connections U1, U2, V1, V2, W1, and W2 are arranged at an offset of 120° here as shown in Figure 6.26.

If the **three-phase windings** offset temporally by 120° are arranged at a 120° special offset, a rotating magnetic field results, as shown in Figure 6.27. Here, the currents from

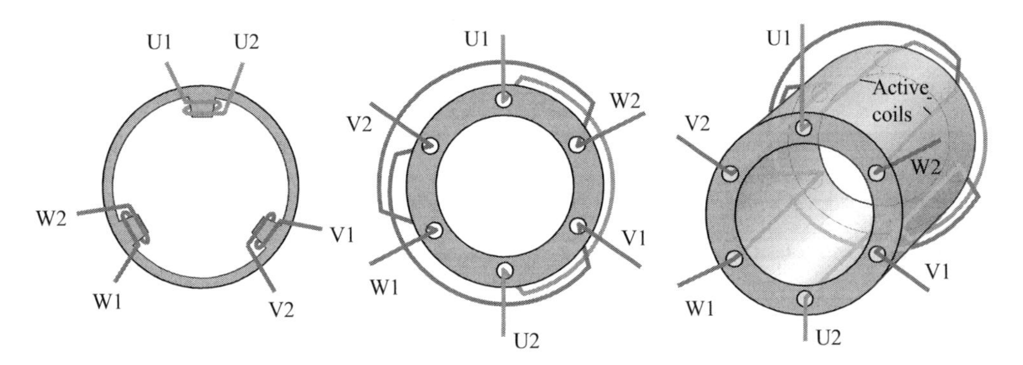

Figure 6.26 Left: cross-section through a stator to produce a rotating magnetic field with three inductors offset by 120° (concentrated winding); middle: cross-section; right: three-dimensional view of an integrated three-phase power winding (distributed winding)

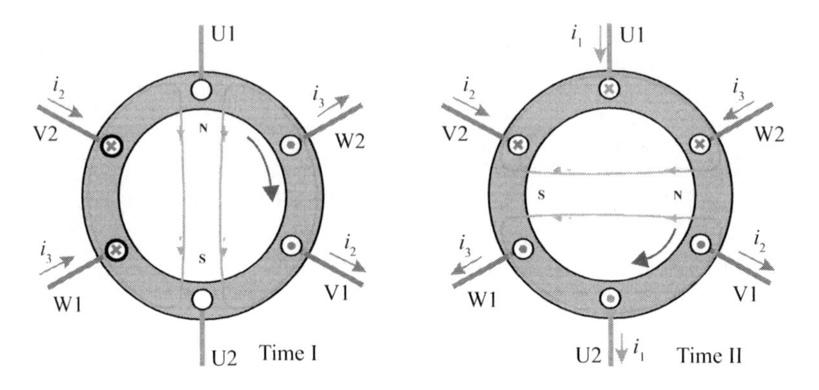

Figure 6.27 A change in the magnetic field at two different points in time when sinusoidal current with three phases offset by 120° is applied

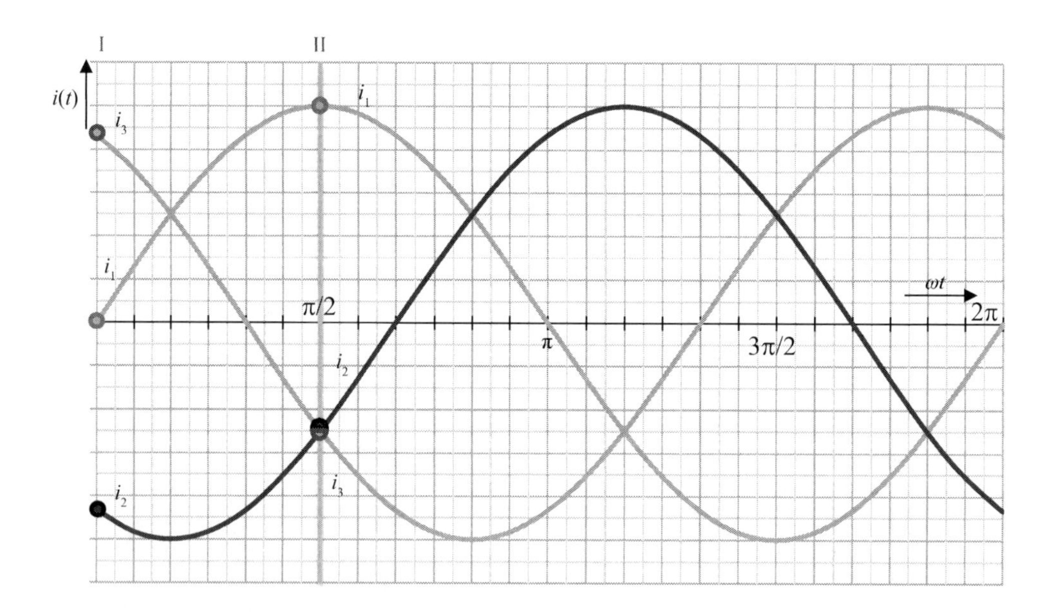

Figure 6.28 Three-phase current to generate a rotating magnetic field

Figure 6.28 is shown at two different points in time: I ($\omega t = 0$) and II ($\omega t = \pi/2$). Clearly, the magnetic field's north pole has moved 90° clockwise at time II. If we trace the curve at multiple different points in time, we get a magnetic field that constantly changes its direction – a **rotating magnetic field**.

If a magnetic pin is put onto the stator inside the three-phase winding, it would constantly turn. The speed would be directly equal to the frequency of the current, with one rotation being completed with every repetition of the sine curve. In Europe, the grid frequency is $f = 50$ Hz, producing the synchronous speed of n_S, so the speed of a circumferential magnetic field is $n_S = 50\ \text{s}^{-1} = 3{,}000\ \text{min}^{-1}$.

In the arrangement above, the magnetic field only has two poles or a single pair of poles N-S ($p = 1$), but the stator could have more windings to create more pairs of poles. If the grid frequency remains the same, the speed is cut in half every time the number of pairs of poles doubles. The **synchronous speed** n_S is thus determined by the grid frequency f and the number of pole pairs p:

$$n_S = \frac{f}{p} \tag{6.68}$$

The division of poles

$$\tau_p = \frac{d \cdot \pi}{2 \cdot p} \tag{6.69}$$

is determined by the number of pole pairs p and the diameter d of the holes in the stator. The following applies for the magnetic induction B with amplitude $\hat{B}$ distributed spatially and temporally across the circumference along route x:

$$B(x,t) = \hat{B} \cdot \sin(\frac{\pi \cdot x}{\tau_p} - \omega \cdot t) \tag{6.70}$$

In practice, the six connections U1, U2, V1, V2, W1, and W2 of the inductors are connected in a star or delta connection to create a rotating magnetic field, as shown in Figure 6.29. This approach reduces the number of line conductors to three: L1, L2, and L3. Often, these conductors are labeled as R, S, and T. If need be, a neutral line N can be added, but it is not necessarily needed for three-phase machines.

Between the root mean squares of the conductor–conductor voltages and the conductor–neutral line voltages, the following applies:

$$U = U_{12} = U_{23} = U_{31} = \sqrt{3}U_{1N} = \sqrt{3}U_{2N} = \sqrt{3}U_{3N} \tag{6.71}$$

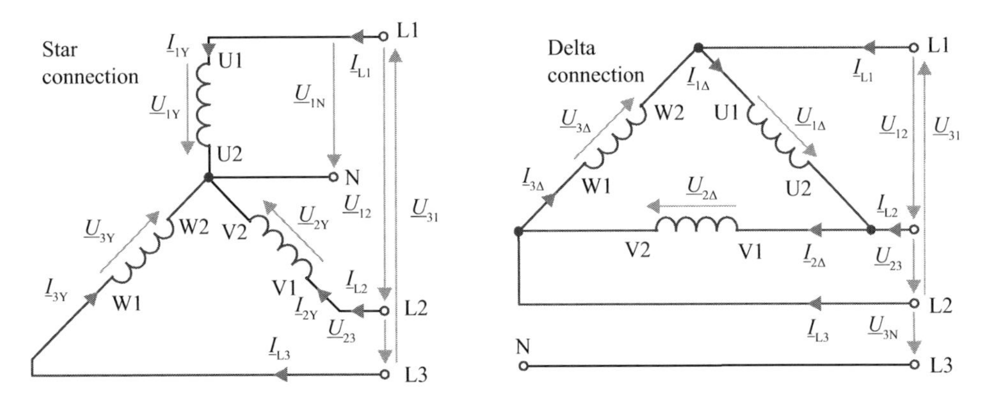

Figure 6.29 Star and delta connections

At 230 V and between two conductors $\sqrt{3} \cdot 230$ V = 400 V. The individual voltages are phase-shifted by $2\pi/3$.

In a **star connection**, the voltages U_Y at the inductors (also called string voltage) are the same as the conductor–neutral line voltages. In the **delta connection**, the string voltages U_Δ corresponded to the conductor–conductor voltages U, producing the following root mean squares:

$$U = U_\Delta = \sqrt{3} \cdot U_Y \tag{6.72}$$

In the star connection, the currents I_Y through the inductors are the same as the conductor currents I:

$$I = I_L = I_Y \tag{6.73}$$

In the delta connection, the root mean squares of the currents I_Δ through the inductors are reduced by the factor $\sqrt{3}$ when the conductor currents are distributed. The result is:

$$I = \sqrt{3} \cdot I_\Delta \tag{6.74}$$

Because the string voltages U_Y at the inductors are reduced by the factor $\sqrt{3}$ in a star connection, however, the string currents I_Y are also reduced by the factor $\sqrt{3}$ relative to a delta connection:

$$I_\Delta = \sqrt{3} \cdot I_Y \tag{6.75}$$

The conductor currents I in a star connection with the same inductor system as a delta connection are therefore not identical with the conductor–conductor voltages.

The overall real power of asymmetrical three-phase system with a star connection is calculated from the sum of the real power of the three strings:

$$P = 3 \cdot U_Y \cdot I_Y \cdot \cos\varphi = \sqrt{3} \cdot U \cdot I \cdot \cos\varphi \tag{6.76}$$

Likewise, the formula for the delta connection is:

$$P = 3 \cdot U_\Delta \cdot I_\Delta \cdot \cos\varphi = \sqrt{3} \cdot U \cdot I \cdot \cos\varphi \tag{6.77}$$

The power uptake of an asymmetrical overall circuit can be determined from measurements of a conductor's current and voltage. It follows from $U_\Delta = \sqrt{3} \cdot U_Y$ and $I_\Delta = \sqrt{3} \cdot I_Y$ that

$$P_\Delta = 3 \cdot P_Y \tag{6.78}$$

In other words, the power uptake of a delta connection is three times greater than that of a comparable star connection.

The **reactive power** Q and the **apparent power** S of a delta star connection are calculated similarly to real power using the following formulae:

$$Q = \sqrt{3} \cdot U \cdot I \cdot \sin\varphi \quad \text{and} \tag{6.79}$$

$$S = \sqrt{3} \cdot U \cdot I \tag{6.80}$$

Synchronous motors

Design

A synchronous motor consists of a **stator** and a rotor. The stator is the stationary part that creates a rotating magnetic field, as explained in the previous section. In a housing made of layered metal sheets, the stator's coils are usually open grooves along the interior boreholes (distributed winding).

As explained above, a magnetic compass needle within the stator would begin to turn at the frequency of the rotating magnetic field. But instead of a magnetic needle, the synchronous motor's stator has a **rotor**, which can be driven, for instance, by the rotor blades of a wind turbine. Such a rotor also has to be magnetic to follow the rotating magnetic field at the synchronized speed. In a synchronous motor, permanent magnets or **excitation coils** with direct current flowing through them create the rotor's magnetic field. The current running through the rotor coils generally has to come from slip rings.

There are two types of rotor designs as shown in Figure 6.30. A **turbo rotor** consists of a solid cylinder. Grooves along the sides are provided for the excitation coils. Because of its weight, a turbo rotor can take up centrifugal forces better. In return, more material is needed.

As the name indicates, **salient-pole rotors** have clearly visible poles. The rotors have two, four, or more poles (Figure 6.30). The electro-technical description of salient-pole

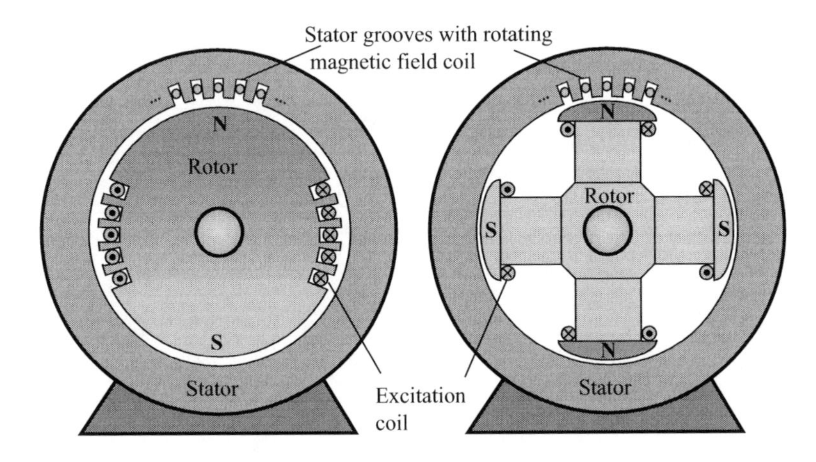

Figure 6.30 A cross-section of a synchronous motor. Left: turbo rotor; right: salient-pole rotor

rotors is much more complicated than for turbo rotors because the former has asymmetries. Below, we only talk about turbo rotors. For more details about salient-pole rotors, see the appropriate literature (such as [Mül94] and [Fis06]).

The speed $n_S = f_1 / p$ of the stator's magnetic field is determined by the magnetic current's frequency f_1 and the stator's number of pole pairs p. At a frequency of 50 Hz and two poles ($p = 1$), the speed is $n_S = 3{,}000\ \text{min}^{-1}$.

The rotor in a synchronous motor has the same speed as the stator. The north pole of the rotor thus always follows the stator's south pole. But when the synchronous motor is under a load, the rotor's north pole and the stator's south pole are not directly opposite each other. Rather, the load on the motor shifts the poles, producing pole angle ϑ. As the load increases, so does the pole angle.

If the synchronous motor is a generator in a wind turbine, the rotating magnetic field's synchronous speed also sets the rotor speed. The poles of the rotor and stator are shifted here as well, only in the other direction, with the rotor's pole ahead of the stator's in the direction of rotation. The pole angle is thus no longer negative, but positive. The stronger the generator is driven, the greater the pole angle ϑ. The load therefore also changes the pole angle. The speed remains constant, however; in other words, the rotor is synchronized with the stator frequency. Hence the name 'synchronous motor'. The rotor's speed can only be changed by changing the rotating magnetic field's frequency f or number of pole pairs p.

Electrical description

Each string in a three-string stator winding creates a main field linked to the rotor. This is taken into account in the main inductance L_h and main electrical reactance $X_h = 2 \cdot \pi \cdot f_1 \cdot L_h$. In addition to the main field, however, stray fields not linked to the rotor are also created. The stray electrical reactance X_σ accounts for these fields. The rotor's field brings about voltage in the stator when it turns. This voltage is called pole wheel voltage $\underline{U}_p$. The root mean square of the pole wheel voltage is proportional to that of the rotor's excitation current I_E when linear

$$U_p \sim I_E \tag{6.81}$$

meaning that the excitation current can be used to set the rotor voltage.

In addition to the drop in voltage at the electrical reactance, there is an ohmic voltage drop at stator resistance R_1. The following formula then describes the circuit between stator current $\underline{I}_1$ and stator voltage $\underline{U}_1$ on a winding:

$$\underline{U}_1 = \underline{U}_p + \underline{I}_1 \cdot R_1 + \underline{I}_1 \cdot \mathrm{j}(X_h + X_\sigma) \tag{6.82}$$

The impact of stator resistance R_1 is so small in large machines that it can generally be disregarded. The main reactance X_h and scattered reactance X_σ can be combined to produce synchronous reactance:

$$X_d = X_h + X_d \tag{6.83}$$

The result is the following equation for a simplified **equivalent circuit** with a single string (Figure 6.31):

$$\underline{U}_1 = \underline{U}_p + \underline{I}_1 \cdot jX_d \tag{6.84}$$

The big advantage of synchronous motors over induction motors, which will be discussed below, is that they can also generate reactive power as needed. A synchronous motor's

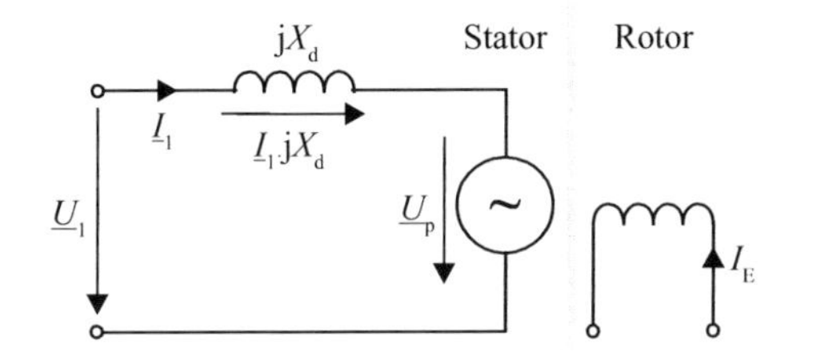

Figure 6.31 Simplified equivalent circuit ($R_1 = 0$) of a turbo rotor for one string

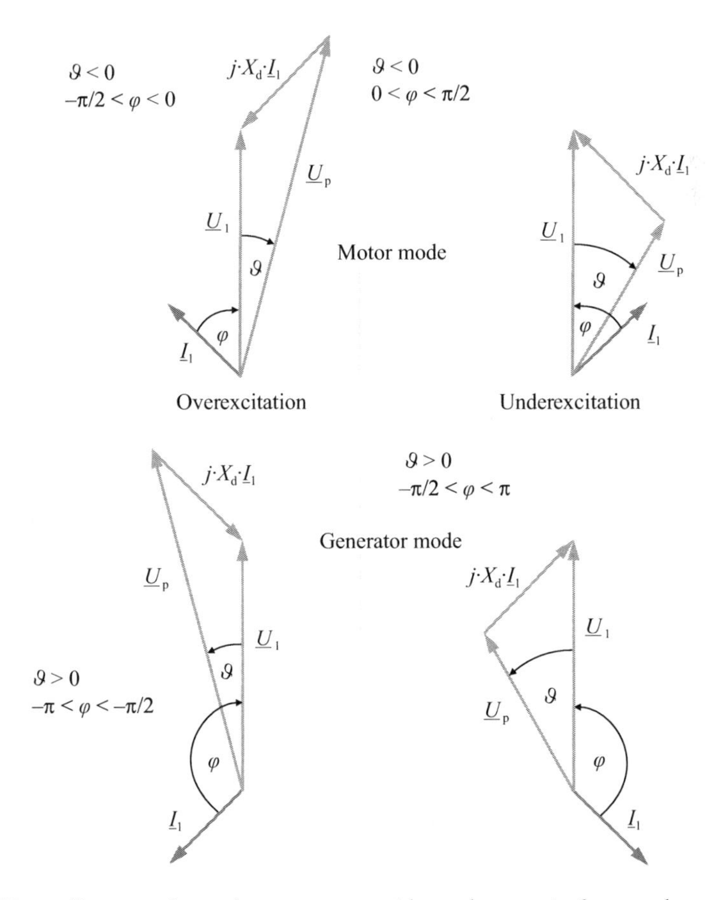

Figure 6.32 Vector diagrams of a synchronous motor with a turbo rotor in four-quadrant mode

stator current can go through all phase angles. This is also called **four-quadrant** operation. Figure 6.32 shows **vector diagrams of a turbo rotor** for four different operating modes.

Depending on the phase angle, a distinction is made between underexcitation and **over-excitation.** With underexcitation, the synchronous motor acts as a winding that consumes inductive reactive power. In contrast, the motor acts as a capacitor when it is overexcited, thereby supplying inductive reactive power.

The angles in the vector diagram can be used to produce the following set of relations:

$$I_1 \cdot \cos\varphi = -\frac{U_p}{X_d} \cdot \sin\vartheta \tag{6.85}$$

In other words, stator current I_1 and phase angle φ are relative to the load and excitation. As the load – and hence the torque M – increases, so does pole wheel angle ϑ. The rotor voltage U_p is relative to excitation/excitation current.

The stator's real power

$$P_1 = 3 \cdot U_1 \cdot I_1 \cdot \cos\varphi \tag{6.86}$$

can be calculated from the real power of the stator's three individual winding strings. The rotating field's synchronous speed n_S is calculated as follows:

$$M = \frac{P_1}{2 \cdot \pi \cdot n_S} \tag{6.87}$$

to produce the **torque** relative to real power:

$$M = -\frac{3 \cdot U_1}{2 \cdot \pi \cdot n_S} \cdot \frac{U_p}{X_d} \cdot \sin\vartheta \tag{6.88}$$

When used as a generator, the pole wheel angle ϑ is positive, whereas power P and torque M are, by definition, negative, so the motor produces power. The rotor's torque is greater than the torque from the real power because losses from friction and losses from the gearbox and generator still have to be taken into account.

Figure 6.33 shows the curve of torque M over pole wheel angle ϑ. The synchronous motor reaches its maximum admissible torque M_K at a pole wheel angle of $\pm\pi/2$. A change in the pole wheel voltage and hence the excitation current thus changes the amount of torque. If the machine's load torque exceeds the maximum admissible torque, the machine is no longer synchronous. If used as a motor, the motor trips up and stalls. If used as a generator, the rotor turns faster than the stator's rotating field, and the generator spins loose. The tremendous centrifugal forces of a wind turbine would eventually mean that the rotor is damaged in the process. Sufficient safety measures must be taken to prevent this outcome, such as with aerodynamic breaks.

Synchronisation

Before a synchronous generator can be connected to the grid, it first has to be synchronized with it. Up to now, it is been assumed that the stator's frequency – the frequency

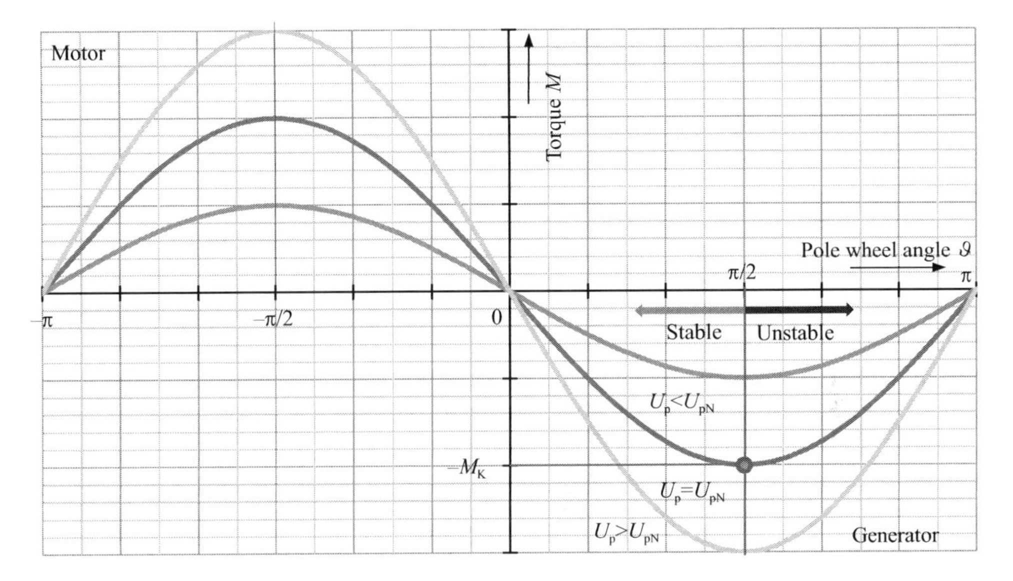

Figure 6.33 Torque curve of a synchronous motor with a turbo rotor relative to pole wheel angle ϑ and pole wheel voltage U_p

of its voltage – corresponded with the grid frequency. When a synchronous generator is ramped up, the rotor's speed – and hence the stator's frequency – differs from the grid frequency, however. If the synchronous generator were immediately connected to the grid, the result would be severe compensation events, such as strong compensation currents. The result would be an unacceptable load on the grid, and the generator could be damaged in the process. For this reason, the following **synchronization conditions** must be fulfilled before a synchronous generator is connected to the grid:

- The grid and the stator must have the same phase sequence (direction of rotation).
- The stator and grid voltage have to have the same frequency.
- The stator and grid voltage have to have the same voltage amplitude.
- The stator and the grid voltage have to have the same phase position.

Modern synchronous generators are automatically synchronized using electronic connection devices. The following minimum synchronization conditions must be fulfilled [VDE94]:

- Voltage difference $\Delta U < \pm 10\,\% \cdot U_N$
- Frequency difference $\Delta f < \pm 0.5$ Hz
- Phase angle difference $\Delta\varphi < \pm 10°$

The induction motors described below do not require synchronization.

Induction motor

Design and modes of operation

The **stator** – the part of the induction motor that does not move – essentially has the same design as in a synchronous motor. In both, a rotating field is created across a delta of three coils.

The **rotor**, in contrast, is fundamentally different. The rotor neither has permanent magnets, nor direct-current coils to create a magnetic field, but instead three-phase windings. A distinction is made between two types of rotors for induction motors: the squirrel-cage rotor and the slip-ring rotor.

The **squirrel-cage rotor** has three-phase windings in the rotor's grooves, and the lines are connected at both ends by shorting rings. In contrast, **slip-ring rotors** have their windings connected internally at one end and via slip-rings and graphite brushed to the outside at the other. This approach improves the response of the induction motor during the ramp-up phase, among other things.

The major benefit of induction motors is they have a simple, robust design; squirrel-cage rotors do not have slip rings subject to wear and tear.

The rotating field's synchronous speed n_S in the stator is calculated as above from the grid frequency f_1 and the number of pole pairs p:

$$n_S = \frac{f_1}{p} \tag{6.89}$$

When the motor is not moving, the rotating field crosses over the standing rotor, creating voltage in the rotor coil lines. The result is current in the conductor bars, which have a tangential force on the rotor, causing it to move. The rotor then turns at speed n, which is always lower than synchronous speed n_S, because a differential speed is needed to produce voltage in the rotor. The relative difference between the rotor speed n and synchronous speed n_S is called **slip**:

$$s = \frac{n_S - n}{n_S} \tag{6.90}$$

If the induction motor is used as a generator, the rotor runs faster than the stator's rotating field. The machine then feeds real power to the grid via the stator. Table 6.5 shows the various operating conditions of an induction motor.

Table 6.5 Speed and slip under various operating conditions

Operating condition	*Speed*	*Slip*
Standstill	$n = 0, n_S > 0$	$s = 1$
Motor mode	$0 < n < n_S$	$0 < s < 1$
Synchronous mode (neutral)	$n = n_S$	$s = 0$
Generator mode	$n > n_S$	$s < 0$
Braking	$n < 0, n_S > 0$	$1 < s < \infty$

Unlike synchronous generators, inductive generators always need inductive reactive power, as we will see below. It can come from the grid via an overexcited synchronous motor or be provided by power electronics. When an inductive motor is used in a microgrid, the **reactive power** can come from capacitors providing inductive current.

Equivalent circuits and current locus curves

In terms of its electrical design, an induction motor resembles a **transformer** consisting of two coupled windings. In an ideal transformer, the windings are connected without any stray fields. In the real world, however, transformers have stray fields and ohmic losses, which can be taken into account through the addition of real and reactive resistance in the equivalent circuit (Figure 6.34).

The following voltage equations then apply for the transformer:

$$\underline{U}_1 = R_1 \cdot \underline{I}_1 + jX_{1\sigma} \cdot \underline{I}_1 + jX_{h1} \cdot \underline{I}_1 + jX_{12} \cdot \underline{I}_2 \tag{6.91}$$

$$\underline{U}_2 = R_2 \cdot \underline{I}_2 + jX_{2\sigma} \cdot \underline{I}_2 + jX_{h2} \cdot \underline{I}_2 + jX_{12} \cdot \underline{I}_1 \tag{6.92}$$

Windings w_1 and w_2 can be used along with $\underline{I}_2 = \frac{w_1}{w_2} \cdot \underline{I}'_2 = tr \cdot \underline{I}'_2$, $\underline{U}_2 = \underline{U}'_2 \cdot tr^{-1}$, $R_2 = R'_2 \cdot tr^{-2}$, $X_{2\sigma} = X'_{2\sigma} \cdot tr^{-2}$, $X_{h2} = X_h \cdot tr^{-2}$, $X_{12} = X_h \cdot tr^{-1}$ and $X_{h1} = X_h$ to produce primary voltage equations:

$$\underline{U}_1 = R_1 \cdot \underline{I}_1 + jX_{1\sigma} \cdot \underline{I}_1 + jX_h \cdot (\underline{I}_1 + \underline{I}_2') \tag{6.93}$$

$$\underline{U}'_2 = R'_2 \cdot \underline{I}'_2 + jX'_{2\sigma} \cdot \underline{I}'_2 + jX_h \cdot (\underline{I}_1 + \underline{I}'_2) \tag{6.94}$$

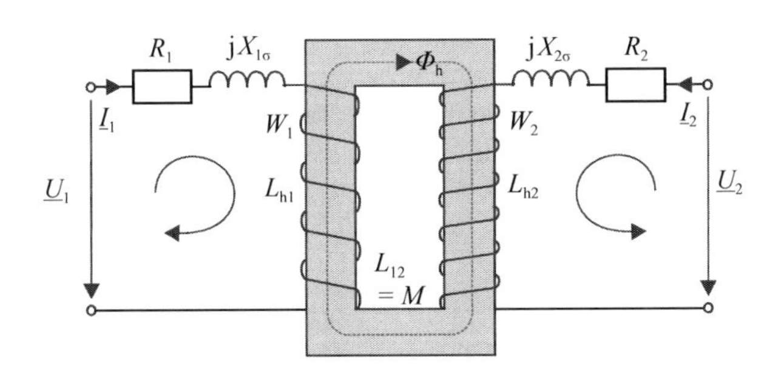

Figure 6.34 An ideal transformer with real and reactive resistances

Unlike a transformer, however, an inductive motor has a short-circuited rotor coil relative to the transformer's secondary winding ($\underline{U}'_2 = 0$). The induction motor's rotor frequency f_2 differs from the stator's frequency f_1. Both frequencies are related to the slip s:

$$\omega_2 = 2 \cdot \pi \cdot f_2 = s \cdot 2 \cdot \pi \cdot f_1 = s \cdot \omega_1 \tag{6.95}$$

The following applies for reactances in the rotor circuit:

$$X = \omega_2 \cdot L = s \cdot \omega_1 \cdot L \tag{6.96}$$

If $X'_{2\sigma}$ and X_h are replaced in the secondary voltage equation by $s \cdot X'_{2\sigma}$ and $s \cdot X_h$, $\underline{U}'_2 = 0$ for the rotor:

$$0 = \frac{R'_2}{s} \cdot \underline{I}'_2 + \mathrm{j}X'_{2\sigma} \cdot \underline{I}'_2 + \mathrm{j}X_h \cdot (\underline{I}_1 + \underline{I}'_2) \tag{6.97}$$

The equation

$$\frac{R'_2}{s} = R'_2 + R'_2 \cdot \frac{1-s}{s} \tag{6.98}$$

produces the single-string **equivalent circuit** for an induction motor (Figure 6.35).

$X'_2 = X'_{2\sigma} + X_h$ produces the following for the rotor's voltage equation:

$$\underline{I}'_2 = -\frac{\mathrm{j}X_h}{\frac{R'_2}{s} + \mathrm{j}X'_2} \cdot \underline{I}_1 \tag{6.99}$$

If the stator's voltage equation is added and resolved for $\underline{I}_1$, $X_1 = X_{1\sigma} + X_h$ produces:

$$\underline{I}_1 = \frac{\frac{R'_2}{s} + \mathrm{j}X'_2}{(R_1 + \mathrm{j}X_1) \cdot \left(\frac{R'_2}{s} + \mathrm{j}X'_2\right) + X_h^2} \cdot \underline{U}_1 \tag{6.100}$$

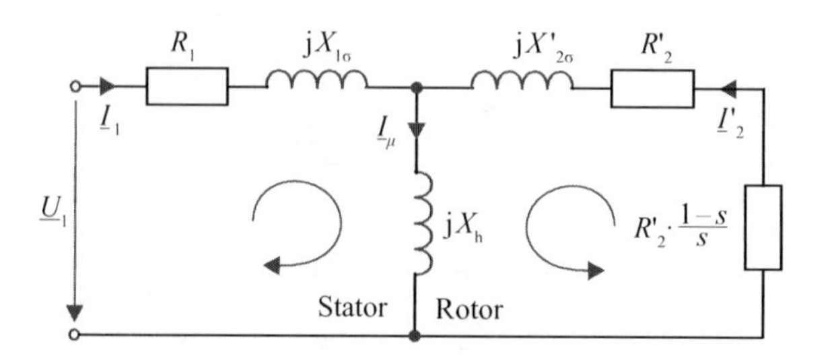

Figure 6.35 Single-string equivalent circuit for an induction motor

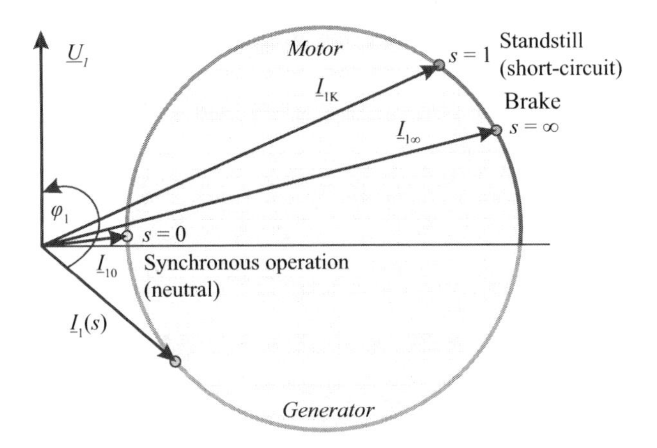

Figure 6.36 Diagram based on Heyland and Ossanna, determination of current locus curves for stator current

This equation is also called the equation for the stator current's **locus curve**. At constant grid voltage $\underline{U}_1$, grid frequency f_1, the stator current merely depends upon the slip s if resistances and reactances remain unchanged.

Figure 6.36 shows the curve for stator current $\underline{I}_1$ relative to slip. The stator voltage $\underline{U}_1$ is a reference parameter in the real axis. The current travels in a circle relative to the machine's mode of operation. The circle is named after **Heyland and Ossanna**; in the former, stator resistance R_1 is not taken into account. The circle shows that current is far greater when the machine is ramping up from standstill than when it is close to neutral.

Under the greatly simplified assumption $X_h \rightarrow \infty$, X_h brings the current to zero. For $R_1 \approx 0$ and $X_\sigma = X_{1\sigma} + X'_{2\sigma}$, the simplified equivalent circuit is shown in Figure 6.37. It is used to derive the Kloss formula discussed in the section 'Torque-speed curves' below.

For power, the simplified expression is:

$$\underline{I}'_2 = -\underline{I}_1 = -\frac{\underline{U}_1}{\frac{R'_2}{s} + \mathrm{j}X_\sigma} \tag{6.101}$$

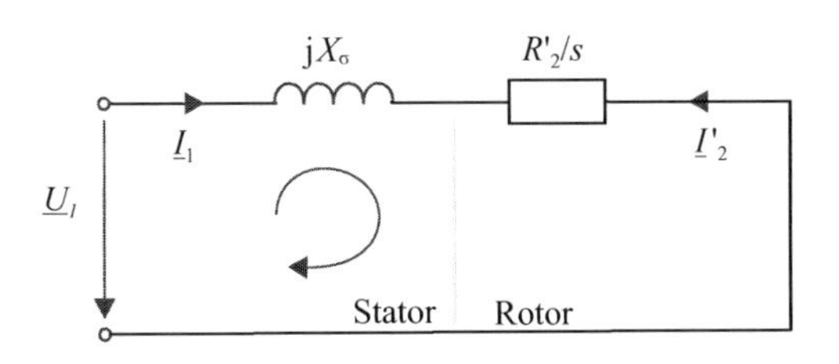

Figure 6.37 Simplified single-string equivalent circuit of an inductive motor

Power balance

For the owner of a wind turbine, the generator efficiency is decisive; after all, you want as much of the power absorbed by the generator to be exported to the grid. The ratio of the power that the generator exports to the grid P_1 to the power absorbed by the generator P_2 represents its efficiency η. Inductive generators used in wind turbines currently reach efficiencies up to and exceeding 97 %. Although this efficiency is very high, the losses that occur do make a difference. When the power exceeds 1,000 kW, even 3 % losses can build up a considerable amount of heat. Figure 6.38 shows the power balance and the various losses from an inductive machine.

The mechanical power available at the machine's clutch P_2 is reduced by friction losses P_R from contact with materials and the air. In the rotor windings, heat losses occur: the copper losses P_{Cu2}. The iron losses P_{Fe2} in the rotor are negligible here. The reduced power is passed on from the rotor to the stator as airgap power P_L. The following relation applies:

$$P_L = P_2 - P_R - P_{Cu2} \tag{6.102}$$

The copper losses P_{Cu2} from the rotor coil are converted into resistance R'_2 (see Figure 6.35). The following then applies for the three three-phase windings:

$$P_{Cu2} = 3 \cdot I'^2_2 \cdot R'_2 \tag{6.103}$$

The **air gap power** P_L can be calculated as follows:

$$P_L = 3 \cdot I'^2_2 \cdot \frac{R'_2}{s} = \frac{P_2 + P_R}{1 - s} \tag{6.104}$$

In the stator, there are also heat losses / copper losses

$$P_{Cu1} = 3 \cdot I_1^2 \cdot R_1 \tag{6.105}$$

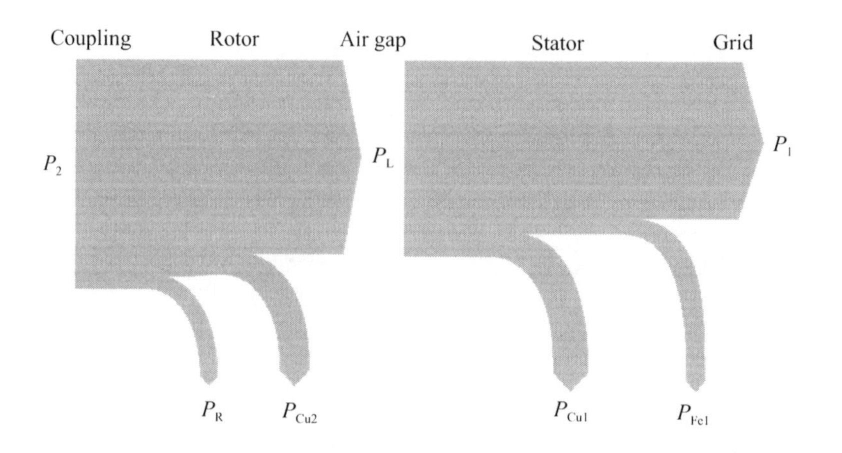

Figure 6.38 Power balance of an inductive generator

and iron losses P_{Fe1}. The **stator power**

$$P_1 = 3 \cdot U_1 \cdot I_1 \cdot \cos\varphi \tag{6.106}$$

is then exported to the grid.

Torque–speed curves and typical generator data

The relation between torque and speed is interesting in induction motors because the speed of the rotor changes relative to the driving torque. Below, the machine's **internal torque** is used; it is calculated from the air gap power P_L and the speed n_S of the stator's rotating field:

$$M_i = \frac{P_L}{2 \cdot \pi \cdot n_S} = \frac{3}{2 \cdot \pi \cdot n_S} \cdot \frac{R'_2}{s} \cdot I'^2_2 \tag{6.107}$$

The internal torque M_i is the product of the mechanical torque M at the shaft (which is negative when the machine is used as a generator) and the friction torque M_R (positive):

$$M_i = M + M_R = \frac{P_2}{2 \cdot \pi \cdot n} + \frac{P_R}{2 \cdot \pi \cdot n} \tag{6.108}$$

The power equation from the simplified equivalent circuit is used with

$$I'^2_2 = |\underline{I}'_2|^2 = \frac{|\underline{U}_1|^2}{\left|\frac{R'_2}{s} + jX_\sigma\right|^2} = \frac{U_1^2}{\frac{R'^2_2}{s^2} + X_\sigma^2} \tag{6.109}$$

for the internal torque

$$M_i = \frac{3}{2 \cdot \pi \cdot n_S} \cdot \frac{R'_2}{s} \cdot \frac{U_1^2}{\frac{R'^2_2}{s^2} + X_\sigma^2} = \frac{3 \cdot U_1^2}{4 \cdot \pi \cdot n_S \cdot X\sigma} \cdot \frac{2}{\frac{R'_2}{s \cdot X\sigma} + \frac{s \cdot X\sigma}{R'_2}} \tag{6.110}$$

The formulae

$$M_{iK} = \frac{3 \cdot U_1^2}{4 \cdot \pi \cdot n_S \cdot X_\sigma} \tag{6.111}$$

and

$$s_K = \frac{R'_2}{X_\sigma} \tag{6.112}$$

produce the **Kloss formula**:

$$M_i = M_{iK} \cdot \frac{2}{\frac{s}{s_K} + \frac{s_K}{s}} \tag{6.113}$$

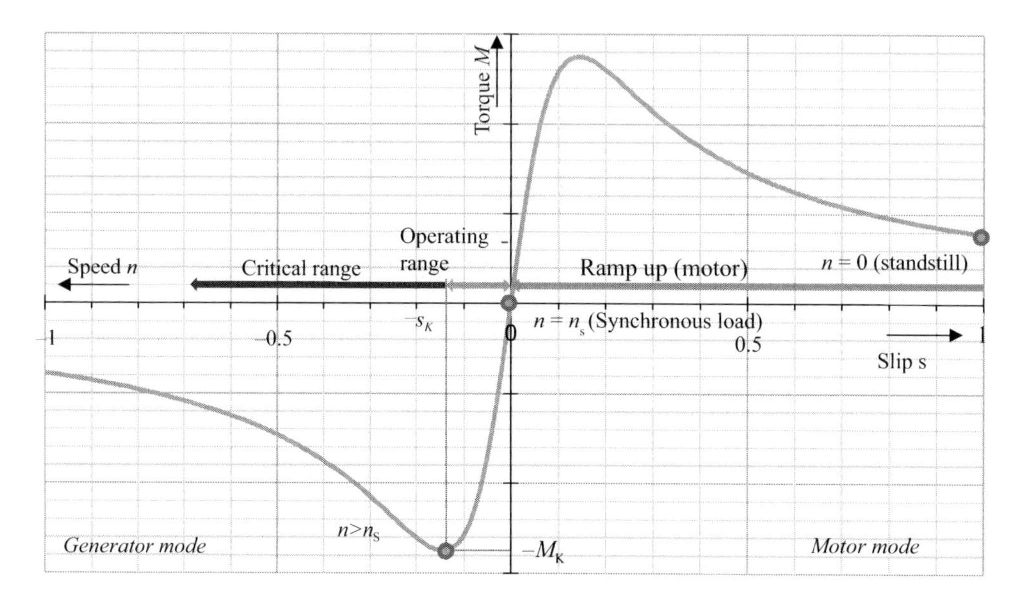

Figure 6.39 The torque-speed curve of an induction motor

If the slip s corresponds to s_K, the second term is 1. The internal torque M_i reaches its maximum. The maximum admissible torque is M_{iK}, whereas the maximum admissible slip is s_K.

Figure 6.39 shows the torque-speed curve for an induction motor. When an induction generator ramps up, it starts off as a motor beginning with slip $s = 1$. When it crosses the synchronous operating point at $s = 0$, the generator reaches its normal operating range. If the generator exceeds its maximum torque M_K, the speed increases quickly and uncontrollably, and the system is destroyed mechanically. An induction motor therefore has an operating range of 0 to $-s_K$, and the speed is slightly greater than synchronous speed n_S.

Table 6.6 shows the data for an induction generator in a wind turbine. The nominal slip – the relative rotation speed in nominal operation – is -0.05 = -5 %: The normal speed ranges from 1,515 min^{-1} to 1,650 min^{-1}. It is calculated from the number of pole pairs $p = 2$ and the grid frequency f_1:

$$n = (1 - s) \cdot \frac{f_1}{p} \cdot \frac{60\,\text{s}}{\text{min}} \tag{6.114}$$

Table 6.6 Technical data for a 600 kW induction generator in a wind turbine [Ves97]

Nominal power P_N	600 kW	Reactive power Q under full load	324 kvar
Nominal voltage U_N	690 V	cos φ under full load	0.88
Grid frequency f_1	50 Hz	Speed range n	1,515–1,650 min^{-1}
Nominal current I_N	571 A	Nominal speed n_N	1,575 min^{-1}
Pole number $2 \cdot p$	4	Slip range s	0 – −0,1
Winding connection	Star	Nominal efficiency η_N	95.2 %

The power factor $\cos\varphi$ of 0.88 is far from the ideal $\cos\varphi = 1$. Under partial load, $\cos\varphi$ worsens, reaching 0.5 at 25 % of load. The result is a large share of reactive power that has to be reduced by compensation circuits. After compensation, such as from capacitors, $\cos\varphi$ reaches values around 0.99.

Electrical system components

Induction generator directly connected to the grid

The **Danish concept** is a very simple design. It is mainly used for small and midsized wind turbines made in Denmark. In this concept, a stall-controlled induction generator is directly connected to the grid (Figure 6.40). A gearbox is used to adjust the rotor speed and the generator speed. The benefit of this design is its simplicity. The induction motor does not need to be synchronized with the grid as with a synchronous motor. The operating speed sets itself. Great, undesired startup currents sometimes occur when large generators are connected to the grid. **Soft starters** can be used to reduce them.

The rotor's stall control limits its speed at high wind velocities. The generator can accommodate fast changes in wind velocity by changing its speed and slip s. Induction generators used in wind turbines can withstand speed changes of around 10 %. However, larger slip values lead to greater losses and lower efficiency.

For this reason, induction generators are used with **variable slip.** These generators do not have a cage rotor that is short-circuited at both ends, but rather one with adjustable resistances R_L within the rotor circuit. Either the rotor coils run externally over the slip rings where resistances are connected or the controllable resistances rotate with the rotor. Figure 6.41 shows how an induction motor's torque-speed curve changes when resistances are connected within the rotor circuit. When the slip values are low, the torque curve flattens out. Because power increases proportionally to torque, great power output is possible at higher speeds, and jumps in power can be better accommodated.

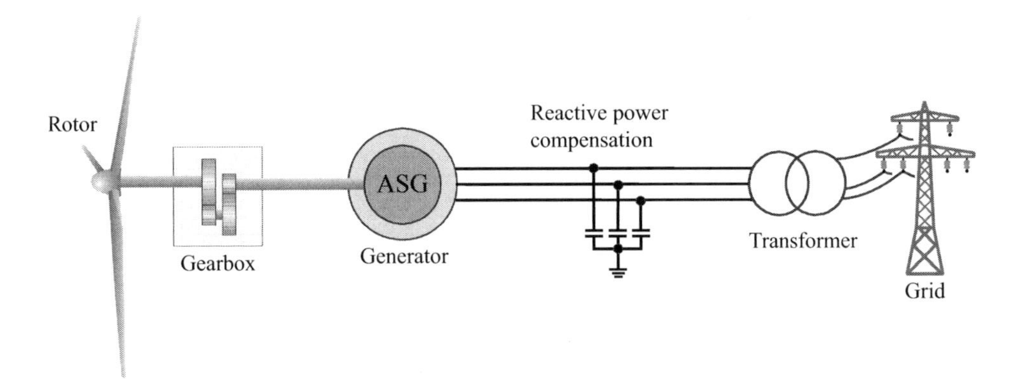

Figure 6.40 Induction generator with a direct grid connection

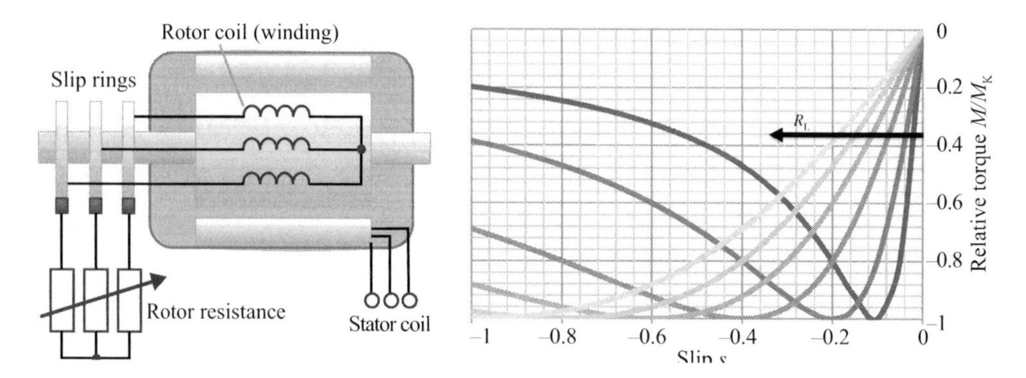

Figure 6.41 The torque curve relative to slip s with changeable rotor resistances R_l

A speed-power diagram shows how well the induction generator is connected to the grid.

$\lambda = \frac{2 \cdot \pi \cdot r}{v} \cdot n,$ the rough estimate of c_p for a specific wind turbine,

$$c_p = 0.00068 \cdot \lambda^3 - 0.0297 \cdot \lambda^2 + 0.3531 \cdot \lambda - 0.7905$$

and $P = c_p \cdot P_0 = c_p \cdot \frac{1}{2} \cdot \rho \cdot A \cdot v^3$

can be used to calculate the rotor power P relative to the speed n at a constant wind velocity v (Figure 6.42). The induction generator, however, can only withstand slight deviations and speed. The grid frequency and the connected gearbox specify the rotor's speed, which can only be varied via the slip, which increases with the load. Induction generators with variable slip allow the increase to be varied at greater power. In this way, great fluctuations in loads – and the resulting loads on the grid – can be remedied somewhat.

Figure 6.42 also shows that the rotor speed has a considerable impact on the amount of wind energy that can be harvested. Because the rotor speed is nearly constant, maximum power cannot be harvested at every wind velocity. In this example, no power is harvested at wind velocities below 4 m/s because the rotor is too fast. The maximum power yield occurs at a wind velocity of around 8 m/s. The percentage of the power that can be harvested drops along with the wind velocity.

System concepts with two different speed settings are therefore used. One option is for the wind turbine to have two different induction generators, one after the other. Another option is **pole-changing generators**. Here, the stator has two windings with a different number of poles, and the circuit switches from one to the other. Because the speed is directly related to the number of poles, two speeds can be set. Figure 6.43 shows that the range in which an optimal amount of power can be harvested from the wind can be expanded into two different speed ranges. The first induction generator in this example is used at wind velocities of 3 m/s to 7 m/s, while the second is used at faster wind velocities above 7 m/s.

The biggest drawback of induction motors is that they always need inductive reactive current. When connected to the grid, this current can be taken from the grid. Power

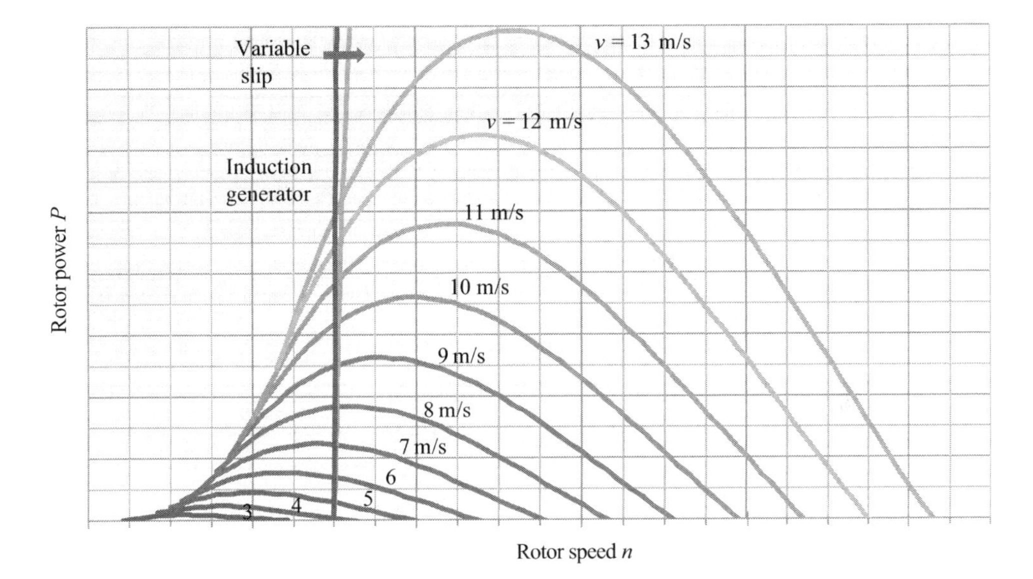

Figure 6.42 Power points of an induction generator directly connected to the grid

providers, however, require large fees for this service, making compensation for reactive power a necessity. As shown in Figure 6.40, capacitors can be used to lower this demand for reactive power from the grid, but it can generally only be designed for a few power points. At other power points, the differential reactive power can be taken from the grid. Modern power electronics also include controls for reactive power.

The situation is different in a micro-grid. Because the demand for reactive power increases along with power, reactive power has to be provided within the micro-grid. It can come either from power electronics or a synchronous machine acting as a phase shifter to meet the demand for reactive power.

Synchronous generator directly connected to the grid

Synchronous generators do not require additional reactive power, which they get from the excitation current; furthermore, synchronous generators can both absorb and produce reactive power. Permanent magnets can also be used for the excitation of synchronous generators. In this case, however, no additional controls are needed for reactive power. On the other hand, permanent magnets are relatively expensive. Therefore, thyristors are generally used for excitation; they take three-phase current from the grid and convert it into the desired direct current (Figure 6.44).

Unlike induction motors, synchronous generators have an absolutely constant speed. A synchronous generator's operating curve in the speed-power diagram is a vertical line without the slight curve at higher power levels, as with the induction generator. Here, large fluctuations in load cannot be accommodated with slip; they are instead passed on almost entirely to the grid. In addition to the burden on the grid, the turbine is exposed to great mechanical loads then. A torque limiter can be used to accommodate strong gusts, but only at the cost of greater wear and tear.

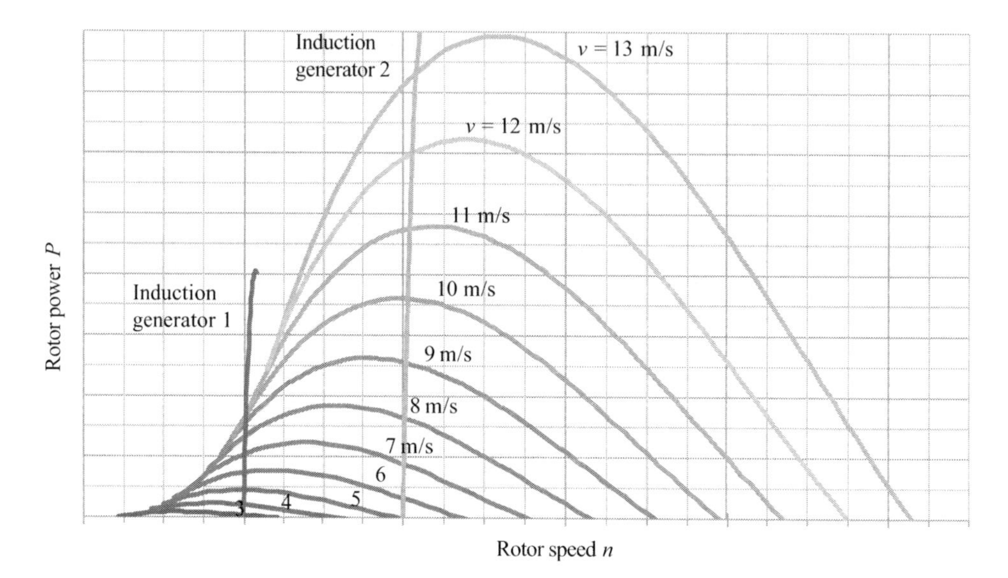

Figure 6.43 A wind turbine's power points with two induction generators at different speeds

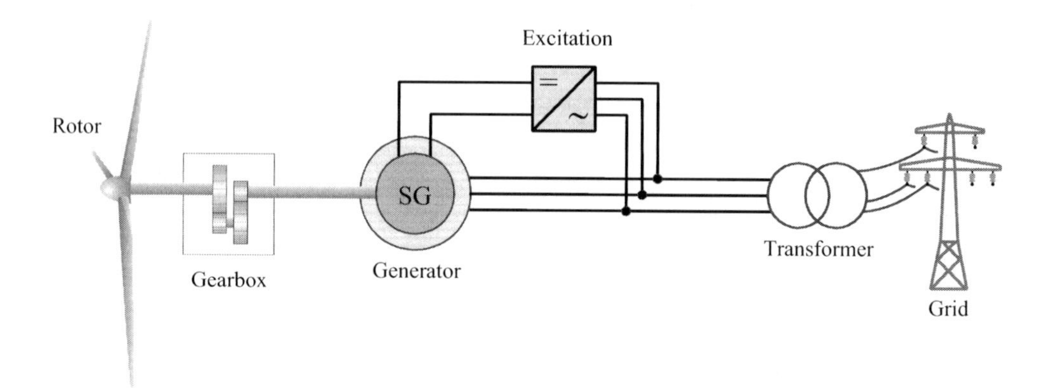

Figure 6.44 Synchronous generator with a direct grid connection

For this reason, synchronous generators are rarely connected directly to the grid (Figure 6.44). They are generally only directly connected in micro-grids. They then power pumps driven by three-phase motors or appliances running on direct current via converters and battery storage. This approach allows for various power points at different speeds for the synchronous generator, thereby reducing loads considerably.

Synchronous generator with converter and intermediate circuit

Modern power electronics can be used to compensate for the drawbacks of synchronous generators connected directly to the grid. Here, the synchronous generator is first connected to an intermediate direct-current circuit and a frequency converter before being connected to the grid (Figure 6.45). The generator can then be set to a different frequency than that of the grid. The generator's speed is directly influenced when its frequency is changed. The speed can vary across a wide range in order to accommodate wind velocity optimally.

Figure 6.46 shows that changes in speed allow the maximum amount of wind energy to be harvested at low and medium wind velocities. At greater wind velocities, power has to be curtailed. Maintaining constant speed is one option, and the stall principle is used at high wind velocities. There is nonetheless a risk of the rotor being overloaded and breaking at very high wind velocities. Therefore, additional controls are needed, such as pitch controls. Then, the coefficient of power is reduced, and the rotor curves shift towards smaller speeds, as shown in Figure 6.46.

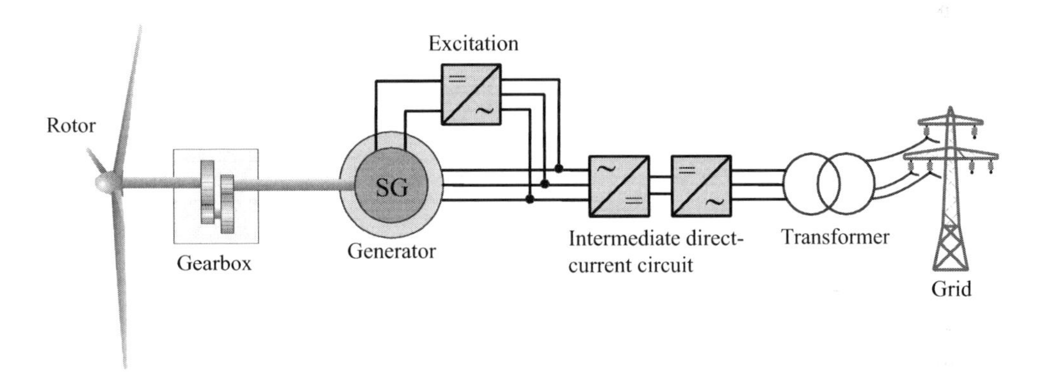

Figure 6.45 Synchronous generator with intermediate direct-current circuit

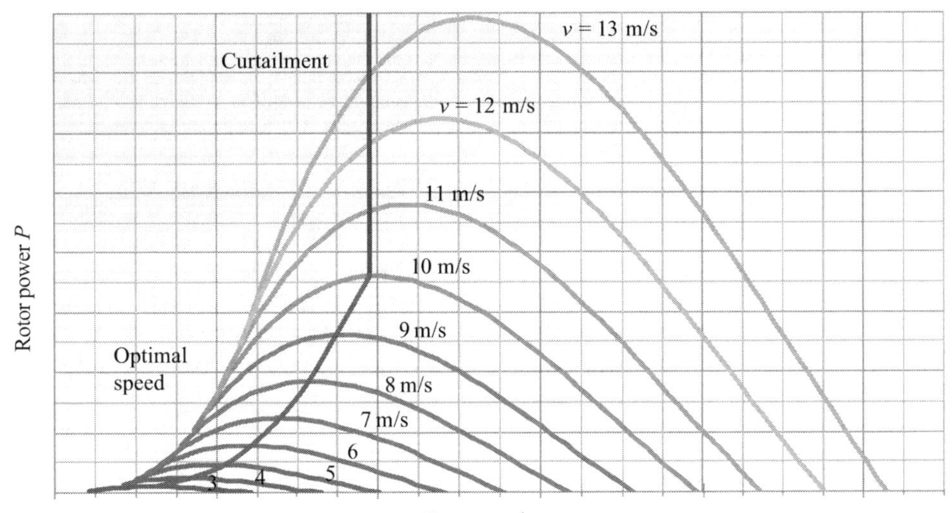

Figure 6.46 Power points of a variable-speed wind turbine

The gearbox is then no longer needed to adapt the rotor and generator speeds because the rotor's frequency converter allows other speeds to be attained. This type of wind turbine is currently produced in large quantities in the megawatt and multi-megawatt range. Because power electronics cannot, however, convert the frequency into an infinite range, the synchronous generators used have 80 and more poles. The drawback of generators with so many poles is their relatively large size and great mass. In return, material is not needed for the gearbox, thereby also saving costs. **Direct-drive turbines** have other benefits as well, such as low noise. If the converter is an inverter with pulse-width modulation, the power electronics can vary the demand for reactive power.

Speed-controlled induction generators

Above, we saw that the directly connected induction generator can use variable slip to change speed. Because slip causes great losses in the generator's rotor circuit when it is too great, the slip was previously limited to below 10 %. But obviously, rotor power can also be used. The generator's rotor power can be exported to the grid via an intermediate

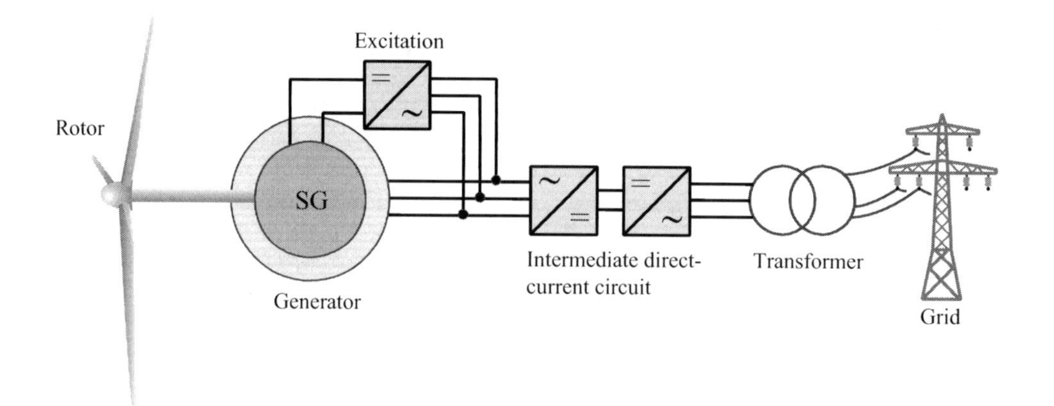

Figure 6.47 Direct-drive synchronous generator with an intermediate direct-current circuit

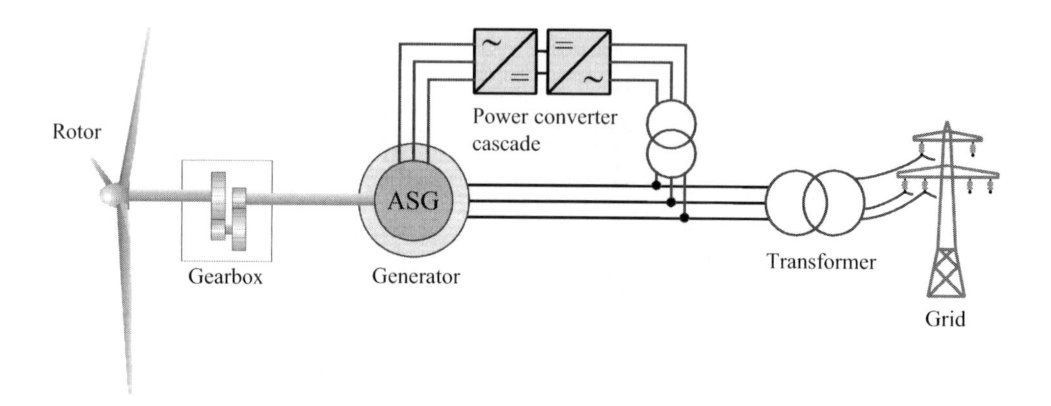

Figure 6.48 Variable-speed induction generator with converter in the rotor circuit

circuit and an inverter (Figure 6.48). If the generator speed can only be varied above the grid frequency, the circuit is called a **supersynchronous converter cascade.** The great demand for reactive power is a drawback in this approach.

While the supersynchronous converter cascade can only export power from the rotor to the grid, a **doubly fed induction generator** can also take power from the grid into the rotor. Here, a converter is installed with either an intermediate circuit or directly.

The doubly fed induction generator can be both subsynchronous and supersynchronous, meaning that the rotor speed can be below and above the grid frequency. In this way, the generator's demand for reactive power can be controlled. Effects on the grid and higher costs partly offset the benefits of this system. Because this system can change speed and adjust demand for reactive power, it offers similar benefits to a direct-drive synchronous generator with a converter and is therefore used frequently.

Micro-grid systems

In micro-grids, either storage or a second type of generator (such as a diesel generator) is needed to ensure supply security when the wind is not blowing. Small systems generally use batteries to store electricity. To charge the battery, a converter converts the wind generator's three-phase current into direct current. Induction generators also need reactive power compensation because the generator otherwise cannot output any power.

When the battery is full, a charge controller must separate the generator from the battery to protect the battery from overcharging. Small wind generators generally do not have active motors or brakes that take them out of the wind or stop them from turning. Unlike photovoltaic modules, generators should not be run in neutral or short-circuited if the wind continues to turn the rotor after it is switched off electrically. If the electric generator is short-circuited, the coils could burn up. If the generator runs in neutral, the rotor will quickly speed up, and the wind turbine could be mechanically destroyed. One option is to connect the generator to a heat resistor that limits the electric current and speed (Figure 6.49).

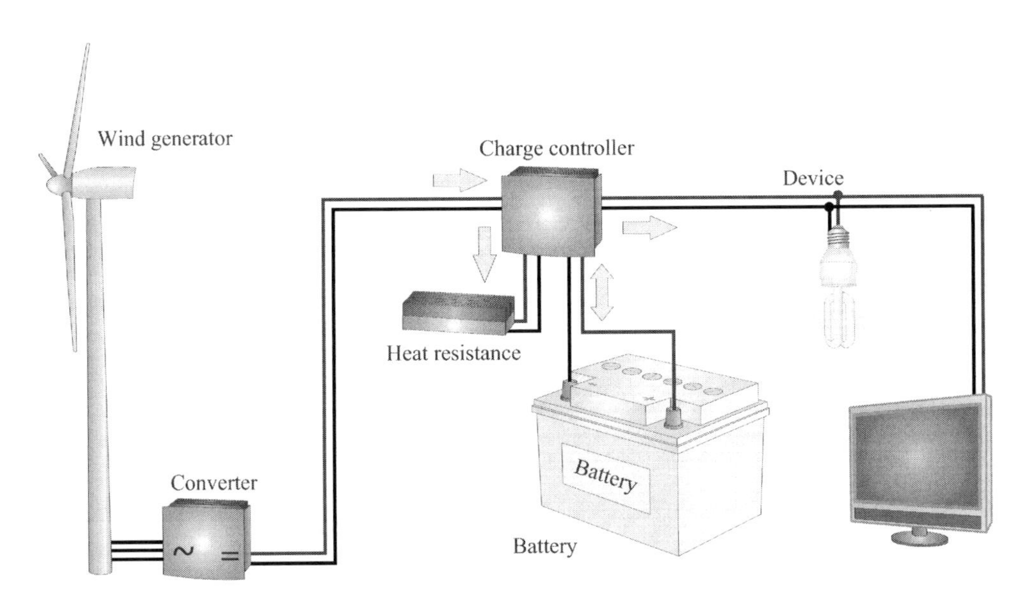

Figure 6.49 The principle of a micro-grid with wind power

A combination with photovoltaic modules generally increases supply security in a micro-grid because wind generators and photovoltaics complement each other well. The battery storage can then be smaller, thereby reducing costs.

Wind turbines connected to the grid

System yield

Before a wind turbine is built, it makes sense to conduct a yield analysis. To do so, assumptions must be made or measurements taken about the wind velocity distribution at the planned site. Often, special expert reports are drawn up for large turbines and wind farms. Manufacturer data for power $P(v)$ relative to wind velocity v (see Figure 6.17) and the distribution of wind velocities $\mathrm{f}(v)$ can be used to calculate the **average power**:

$$\overline{P} = \int_{v=0}^{\infty} \mathrm{f}(v) \cdot P(v) \cdot \mathrm{d}v \qquad (6.115)$$

The Weibull and Rayleigh distributions described at the beginning of this chapter can be used to determine the wind velocity distribution. If measurements have already been taken of wind velocity, usually a frequency distribution $\mathrm{h}(v)$ is provided for various wind velocity intervals i as shown in Figure 6.2. The average power $P(v_i)$ of a wind turbine at interval i produces its average power:

$$\overline{P} = \sum_i \mathrm{h}(v_i) \cdot P(v_i) \qquad (6.116)$$

Finally, the **energy yield** E is calculated over a certain time t (usually one year):

$$E = \overline{P} \cdot t \qquad (6.117)$$

Grid connection

Wind turbines are usually built in relatively sparsely populated areas, which often have only weak grid connections. The capacity of modern wind turbines can greatly exceed 1,000 kW, so it makes sense to check before construction whether the grid would be able to take up the wind power generated at all. To keep line losses down, greater voltage is needed to transport greater amounts of power. The voltage levels are divided up as follows:

- Low voltage 0.1 kV to 1 kV
- Medium voltage 1 kV to 30 kV
- High voltage 30 kV to 230 kV

Generally, a transformer is used to export power to the medium-voltage grid. In Germany, the technical criteria for connections of wind turbines to medium-voltage lines are set forth in the 'Technische Richtlinie Erzeugungsanlagen am Mittelspannungsnetz' published by the BDEW [BDEW08]. This guideline also applies for other generators of renewable power, such as hydropower, biomass power plants, and large photovoltaic systems.

Facility owners must comply with all rules and regulations for grid connections. The grid connection should be coordinated with the grid operator during the planning phase. If a grid connection is technically possible, it must be registered with the grid operator. When construction has been completed, the connection can be made. But first, a declaration of compliance is required to confirm that the renewable energy generator is in line with all of the rules and regulations. Before going into operation, the facility undergoes an inspection in the presence of the grid operator to confirm that it works properly.

When the grid is working properly, renewable energy generators must not deviate from its voltage by more than 2 % at the connection point. The grid rules also put a limit on short-term voltage fluctuations. One aspect is called **flicker**. A series of voltage fluctuations can cause conventional light bulbs to change their illuminance. There are also limits to the admissible harmonics (see Chapter 5).

Because failures are possible both with wind turbines and on the grid, safety devices are required; they used to take wind turbines off the grid immediately when a failure occurred. Back in the 1990s, the capacity of renewable energy generators on the grid was relatively small, but in the first half of 2014, Germany got 27 % of its power supply from renewables. Modern renewable energy generators are therefore no longer disconnected; instead, they remain connected to the grid to provide support to keep up voltage.

In dynamic grid support, the generators try to maintain voltage when it fluctuates at the high-voltage and ultra high-voltage levels. If all wind turbines connected to the grid were disconnected at the same time in reaction to a short-term dip in voltage, the result would be a complete power outage. To prevent this from happening, wind turbines must stay on the grid for $\leq$ 150 ms if there is a considerable drop in voltage.

If too much power is exported to the grid, its frequency increases. In turn, it drops if there is not enough capacity. The grid operator balances supply and demand to keep the frequency stable. Below a grid frequency f_N of 47.5 Hz and above 51.5 Hz, one speaks of a grid failure, and systems are disconnected from the grid.

In general, the frequency should not rise above 50.2 Hz. If it does, all generators must reduce their currently available real power P_M by a differential of ΔP:

$$\Delta P = 20 \cdot P_M \cdot \frac{50{,}2\ \text{Hz} - f_N}{50\ \text{Hz}} \tag{6.118}$$

Real power can only be increased again if the grid frequency returns to a value of $f_N \leq 50.05$ Hz.

Power from large wind farms will continue to increase. When the size of the wind farm exceeds a gigawatt, connections may have to be made at the ultra-high-voltage level. This is especially the case for offshore wind farms, where total capacity is very large. They are connected to the (medium to ultra-high-voltage) grid on land.

As the share of renewables increases, so does the grid structure. Large wind farms are increasingly replacing conventional power plants. However, the grid connection points for wind farms are often far from the nodes built for conventional power plants. The current grid therefore has to be upgraded and expanded in some spots. New ultra high-voltage lines are especially needed for the grid integration of large offshore wind farms. But the link above the lines needed is reasonable and will be affordable. There is thus a good chance that a growing number of very large renewable energy generators will be integrated in the grid to help protect the climate.

7 Hydropower

Introduction

The number of hydropower facilities is currently far lower than during the technology's golden age at the end of the eighteenth century, both worldwide and in Europe. Back then, Europe had 500,000 to 600,000 watermills alone [Kön99], most of them in France, though thousands of watermills were found in other European countries as well. The mills were not only used to grind grain, but also to drive a wide range of tools and machines. Along streams, watermills with water wheels up to 18 metres in diameter were common sights. The average capacity of watermills at the time was rather small at 3 to 5 kW, though the largest ones peaked at above 40 kW.

A distinction is made between undershot, breast shot, and overshot waterwheels for historic watermills. **Overshot waterwheels** use the water's potential energy; this is also called the **reaction principle**. Good waterwheels have efficiencies exceeding 80 %. [Brö00]. In contrast, the breast shot wheel only uses the water's kinetic energy based on the **action principle**.

Steam engines eventually replaced watermills. But unlike windmills, watermills never completely disappeared when fossil energy became commonplace. At the end of the nineteenth century, when electrification began, hydropower played a major role. At the beginning, small turbines were used to drive an electric generator. But the size of these systems grew quickly. The amount of water and storage needed became much greater than what was required for watermills.

Today, hydropower is the largest source of renewable energy worldwide for electricity. But because local potential varies, the share of hydropower in some countries is much greater than in others. For instance, Norway currently gets almost all of its electricity from carbon-neutral hydropower. Countries like Brazil, Austria, Canada, and Switzerland also have shares far above 50 % (Table 7.1). In Europe, hydropower is especially important in the Alpine countries and in the far north.

In Germany, the share is greatly below the global average at less than 4 %. But Germany has not failed to tap its hydropower potential. Indeed, hydropower made up more than 20 % of German power supply as recently as 1950. Although it has been expanded considerably, its share of power supply has nonetheless plummeted since. Quite simply, power consumption outstripped the growth in hydropower production due to a lack of suitable locations. Today, little hydropower potential remains in Germany to be tapped. Although half as much hydropower is produced in Germany as in Austria or Switzerland, its share in German power supply is far lower.

Table 7.1 The share of hydropower in power generation in 2010

Country	*Paraguay*	*Norway*	*Brazil*	*Venezuela*	*Canada*	*Switzerland*
Share	100 %	95.2 %	81.9 %	72.0 %	59.9 %	55.8 %
Country	*Austria*	*Sweden*	*China*	*Russia*	*USA*	*Germany*
Share	53.8 %	49.3 %	18.0 %	16.7 %	6.2 %	3.3 %

(Data: [EIA12])

Hydropower potential

The earth is called a Blue Planet because 71 % of its surface is water. Without the sun, however, our planet would not be blue, but a white desert of ice. It is the sun that makes 98 % of this water fluid. There are some 1.4 billion km³ of water on the planet. 97.4 % of it is saltwater in the oceans, with only 2.6% being freshwater. Almost 3 quarters of this freshwater is locked up in polar ice, ocean ice, and glaciers; the rest is mainly in groundwater and soil humidity. Rivers and lakes only contain 0.02 % of the earth's water. The sun causes 980 l/m² of it to evaporate on average over the year, equivalent to 500,000 km³ annually (Figure 7.1).

Roughly 22 % of the solar energy incident on earth drives this gigantic circulation of water. This amount of energy is 3,000 times greater than primary energy demand on Earth. But only a small fraction of that can be tapped.

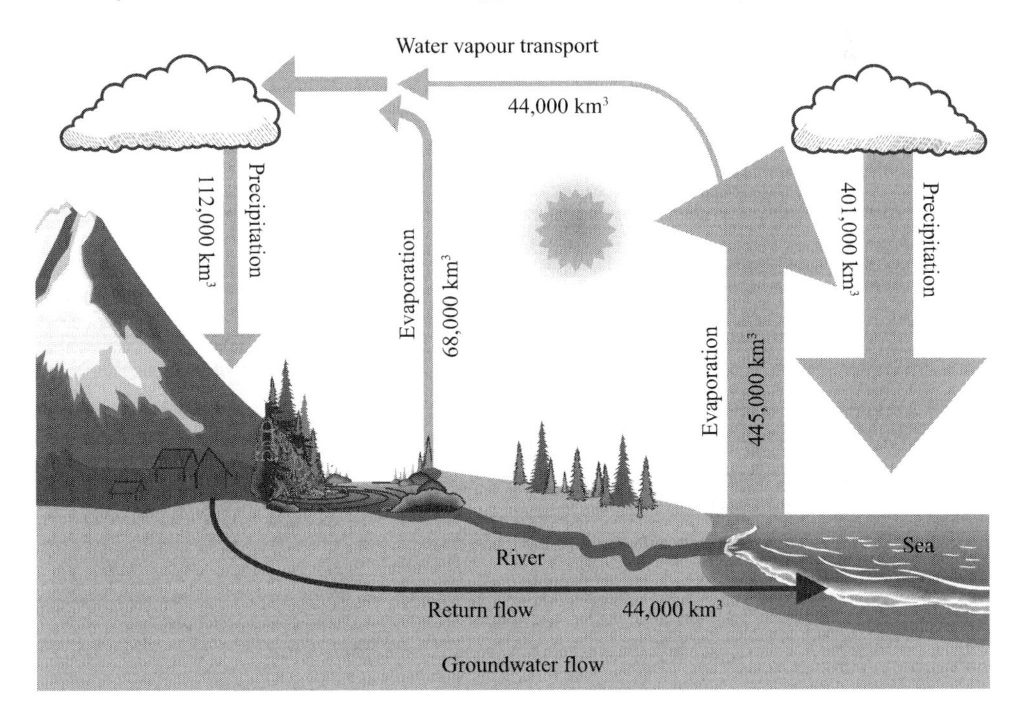

Figure 7.1 The earth's water cycle

Roughly 225,000 km³ is in rivers and lakes across the planet, with a total energy content of 160 EJ. Along with volume, altitude is decisive in terms of tapping water as an energy source. The water that flows down a mountain stream across several hundred metres of altitude may contain more energy than a large river moving just a few metres towards the ocean. Around a quarter of the water in rivers and lakes could be used to produce hydropower, so that theoretically 10 % of our current global primary energy demand could be met with hydropower. In Europe, the potential of hydropower has largely been tapped, though there is considerable untapped potential in other parts of the world.

The amount of energy from hydropower could easily be doubled worldwide. Small hydropower units fit in landscapes and nature especially well. For economic reasons, however, the largest possible structures are still built.

In addition to regional differences, the amount of solar and wind power that can be harvested mainly depends upon the space available, while the potential depends directly on rivers when it comes to hydropower. In addition to flowing water, the amount of precipitation and melting water plays a large role in the potential, as do altitude differences.

Two parameters largely characterize the flow of water. The discharge Q is the water's volumetric flow rate, whereas water level W indicates the height of the water's surface above a certain point over the ground. These two parameters are constantly measured at various points of major rivers and published.

Like photovoltaics and wind power, hydropower fluctuates over the year and from one year to another depending on the weather. Figure 7.6 shows the example of the Rhine's average discharge at Rheinfelden over more than 50 years. Floods can lead to considerable deviations in the overall discharge from one year to another. In contrast, the useful water potential fluctuates less because the large amounts of water during floods cannot be used in practice.

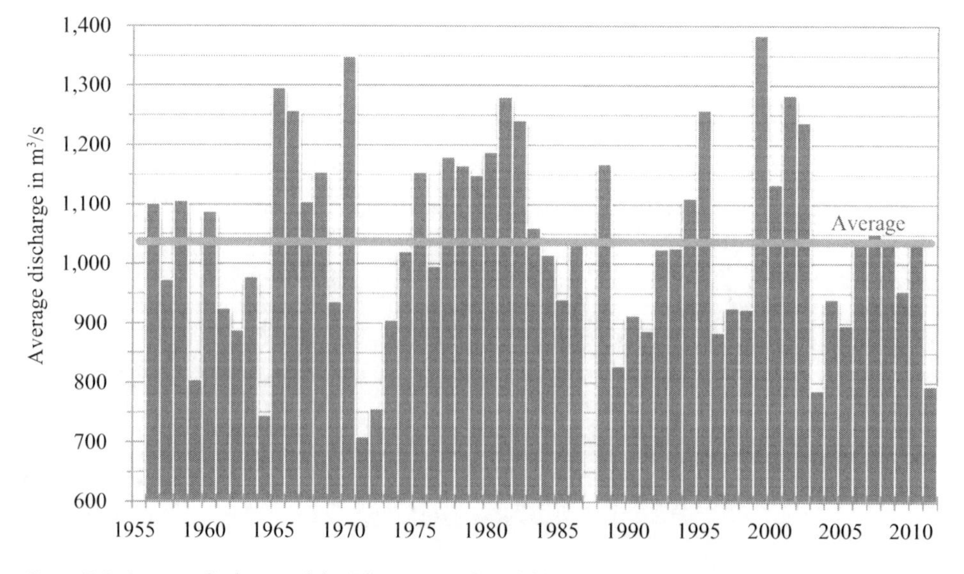

Figure 7.2 Average discharge of the Rhine near Rheinfelden from 1956 to 2011

(Data: [LGRP; LUBW; BAFU12], no data for 1987)

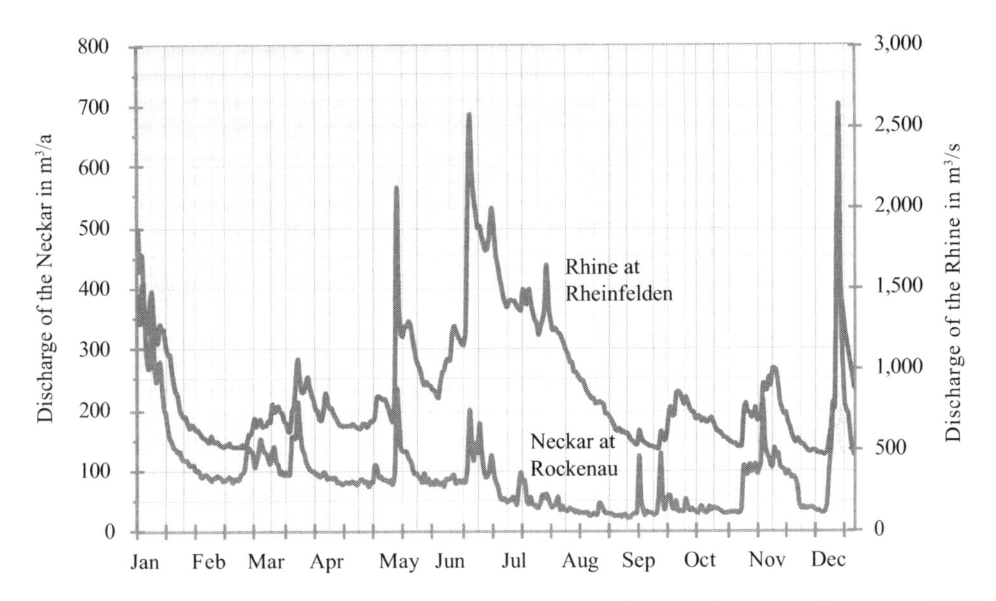

Figure 7.3 Discharge of the Rhine near Rheinfelden and the Neckar at Rockenau over the course of 1991 (Data: [LUBW])

Figure 7.3 shows a curve for the discharge of the Rhine and the Neckar rivers over a year. Because they drain neighbouring areas, the curve is similar over the year. Nonetheless, the two rivers display a few different regional flooding events, and the Rhine also has a great amount of discharge in the spring due to melting water. Aside from the flooding events, which always happen at different times, the discharge remains quite similar from one year to the other. It cannot be ruled out, however, that climate change will rapidly melt glaciers in the Alps, thereby changing the flow of rivers in the long run.

When hydropower facilities are planned, **load duration curves** are often used. Here, a river's discharge values are lined up in order. The result shows what discharge level occurs on how many days in a year. Figure 7.4 shows the data for the load duration curves from 1991. Because one year can deviate significantly from another, it is good to compare load duration curves over longer time frames.

During the design of a hydropower plant, its rated discharge Q_A is specified (Figure 7.5). This factor is the amount of water at which the power plant reaches its capacity. If the river's discharge rises above the plant's rated discharge, the excess water has to be allowed to flow around the facility unused.

If the greatest amount of power is to be generated, the turbines would need to have a high rated discharge. However, if the river's discharge drops below the plant's rated discharge, the plant will not have enough water to run at full capacity. The turbines will then either run at lower efficiency under partial load or individual turbines will be switched off and remain unused. If the focus is on the optimal use of the turbines, a lower rated discharge will be chosen. In practice, a compromise is usually made between maximum power production and optimal utilization of the turbines when determining rated discharge.

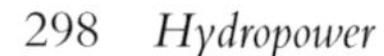

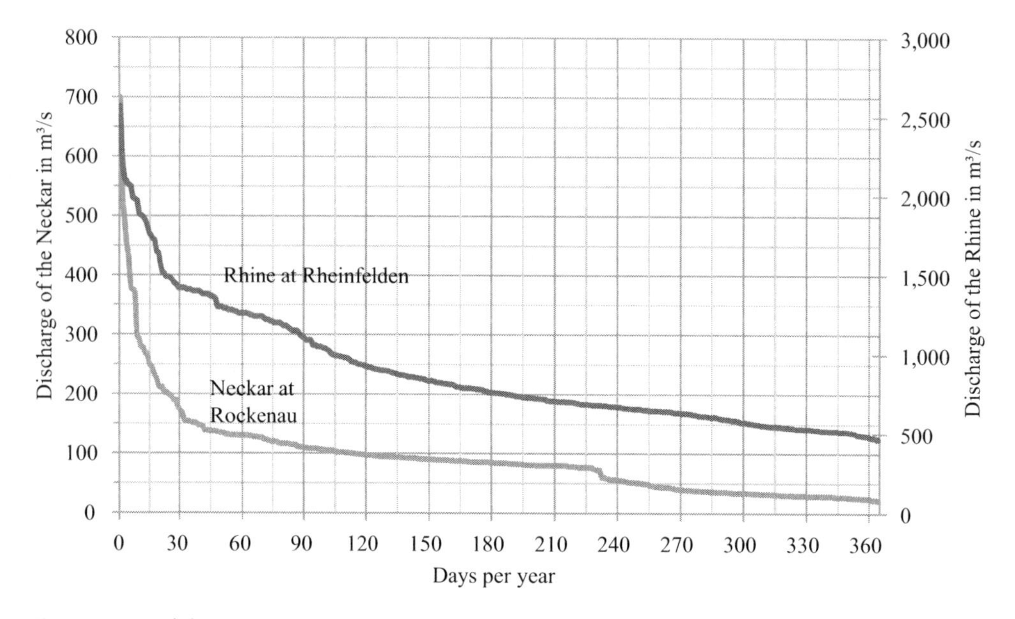

Figure 7.4 Load duration curves for the Rhine at Rheinfelden and the Neckar at Rockenau in 1991

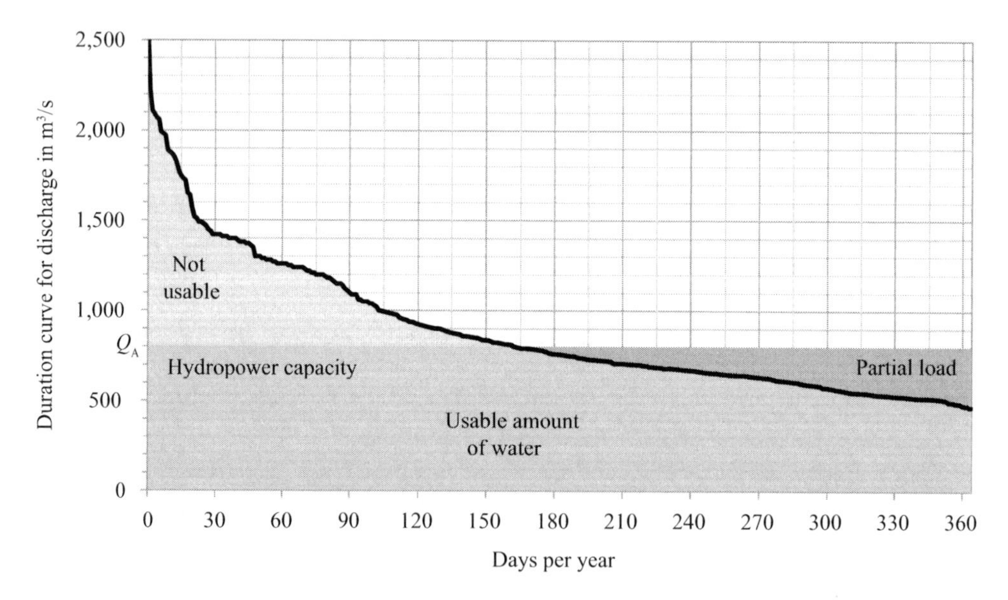

Figure 7.5 Determining rated discharge based on a low duration curve

Hydropower plants

Electricity was generated from water in Germany for the first time in 1891. Because of its low specific power generation costs, hydropower potential has been largely tapped in Germany already. Large plants with capacities exceeding 100 MW are especially interesting financially. Hydropower is not uncontroversial, however, because of the negative impact resulting from the dramatic changes made to the natural world. In the future, the focus will therefore be on increasing the efficiency of existing hydropower plants through modernization and building micro-hydropower units. The latter have largely been overlooked up to now because of their greater specific cost. Outside Germany, new large hydropower plants continued to be built, however, such as the controversial Three Gorges Dam in China, which has a capacity of 18.2 GW.

In general, a distinction is made between the following types of hydropower plants:

- run-of-river plants
- storage dams
- pumped-storage plants.

Run-of-river plants

Run-of-river plants (Figure 7.6) have a dam that backs up the flow of water to create a higher water level behind the dam than in front of it. The water is then directed through a turbine that drives an electric generator. A transformer then converts the generator's voltage into grid voltage.

The **water's power**

$$P_W = \rho_W \cdot g \cdot Q \cdot H \tag{7.1}$$

is calculated from its density ρ_W ($\rho_W \approx 1{,}000\ \mathrm{kg/m^3}$), useful height H (in m), gravitational acceleration g ($g = 9.81\ \mathrm{m/s^2}$) and discharge Q (in $\mathrm{m^3/s}$).

The useful height H can be determined with sufficient accuracy from the geodesic height difference of the water surfaces behind and in front of the dam. The power P_W is therefore roughly proportional to the discharge Q. In contrast, the useful power decreases if the discharge is very great. If the discharge is above the rated discharge Q_A, the turbines cannot handle all of the water themselves, so part of it has to be directed around the dam without being used to generate energy. In addition, the useful height H also increases

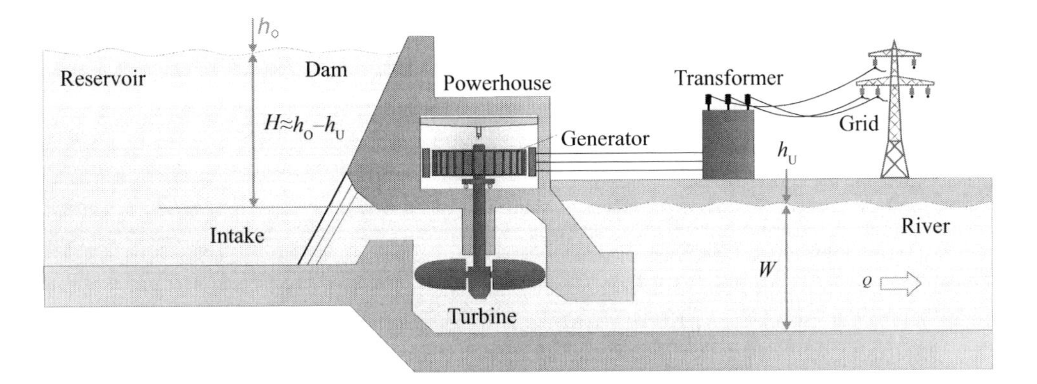

Figure 7.6 Diagram of a run-of-river plant

at high levels of discharge because of the great amount of water behind the dam and the rising water level W.

While data about the temporal curve of discharge Q are available for numerous rivers in Germany, the temporal curve of useful height H is generally not recorded. However, water levels W are available for numerous rivers. It is assumed that run-of-river plants maintain water height h_O behind the dam at a relatively constant level. The lower water height h_U is represented by water level W. Normally, the rated discharge Q_A and design head H_A of **run-of-river plants** are indicated at the levels for which the plant was designed. Water level W and discharge Q over the course of the year can be used to determine the water level W_A for a given rated discharge Q_A. The variable height difference before and after the dam H is then calculated with the following formula:

$$H = (H_A + W_A) - W \tag{7.2}$$

Turbine efficiency η_T, generator efficiency η_G, and other losses (such as gearbox losses, transformer losses and downtime, all of which is expressed by the loss factor f_Z which ranges from 3 % to 10 %) are used to calculate the power plant's electrical output:

$$P_{el} = (1 - f_Z) \cdot \eta_G \cdot \eta_T \cdot \rho_W \cdot g \cdot Q \cdot H \tag{7.3}$$

In addition to the height difference between water before and after the dam H_A and the rated discharge Q_A, the bottleneck capacity $P_{el,N}$ is usually also indicated. Then, we can calculate the overall efficiency in nominal operation:

$$\eta_N = \frac{P_{el,N}}{\rho_W \cdot g \cdot Q_A \cdot H_A} \tag{7.4}$$

Depending on the type of power plant, it can fall below 60 % or exceed 90 %. The overall efficiency η_N, height difference before and after the dam H_A and time series for water levels W and discharge Q can then be used to calculate the hydropower plant's electrical output:

$$P_{el} = \eta_N \cdot \rho_W \cdot g \cdot \min(Q; Q_A) \cdot (H_A + W_A - W) \tag{7.5}$$

If we roughly assume that the water level does not change significantly during the operation of a hydropower plant and that its efficiency remains constant over a certain power range, then power production is directly proportional to the discharge Q.

Storage dams

Run-of-river power plants practically have no storage capacity, and power production greatly depends upon water supply. In contrast, storage dams can compensate for natural fluctuations in the water supply. Here, a dam raises the water level to provide the required pressure but also to ensure a constant flow of water. The water level behind the dam can be only a few metres higher, as with run-of-river plants, or several hundred metres, especially in mountainous areas. In such cases, the water passes through pressurized lines to reach the turbine.

Figure 7.7 Aerial view of the Itaipu power plant
(photo: Itaipu Binacional [Ita04])

Constructed from 1975 to 1991 and expanded in 2004, the **Itaipu power plant** on the border between Brazil and Paraguay is one of the most impressive hydropower plants in the world with a current total capacity of 14 GW. In 2000, its 18 generators set a record by producing 93.428 TWh of electricity, enough to meet 24 % of Brazil's power demand along with 95 % of Paraguay's. Six such power plants would be able to generate more electricity than Germany consumes. Table 7.2 shows the power plant's enormous size. Figure 7.7 shows an aerial photo.

Such giant hydropower plants are, however, controversial because they change the local environment so greatly. A lot of large power plants have a very negative impact on local conditions. The Assuan Dam in Egypt is a well-known example. Since its construction, there has been no flooding downstream on the Nile to supply regions along its banks with nutrients. Artificial irrigation has led to higher salinity in the soil. Agricultural harvests have been reduced, and great changes have been detected in the delta. For instance, the negative impact of erosion is making itself felt.

Table 7.2 Technical data of the Itaipu power plant [Ita04]

Reservoir		*Dam*		*Generator units*	
Surface of the basin	1,350 km²	Max. height	196 m	Units	20 (10 at 50 Hz and 10 at 60 Hz)
Extent	170 km	Total length	7,760 m	Nominal output	715 MW each
Volume	29 trillion m³	Volume of concrete	8.1 million m³	Mass	3,343 / 3,242 t each

Before hydropower plants are built, the benefits and drawbacks should be carefully weighed up. While hydropower is a source of low-carbon electricity, local effects can only be minimized if the project is planned so that the power plant is properly integrated in rivers and nature.

Pumped storage plants

In addition to run-of-river plants and storage dams, pumped-storage power plants are another alternative for hydropower. They not only make use of the natural water supply, but also store energy. When there is excess electricity, water is pumped from a lower basin into a higher one. When power is needed again, the water flows from the top basin through a pressurized pipeline down to a turbine and back into the lower basin to generate electricity (Figure 7.8). The surge tank absorbs pressure changes when the power plant quickly ramps up and down. Figure 7.9 shows an example of a pumped storage plant in Spain. The largest German pump-storage facility is in Goldisthal, Thüringen; completed in 2004, it has a capacity of 1,060 MW.

A distinction is made between pumped-storage power plants with and without natural water inflows. Systems without natural inflows can only store energy; they should not be considered sources of renewable electricity.

Hydropower plants with storage dams or pumped storage store the water's potential energy as electricity. The amount of energy that can be stored

$$E = V \cdot \rho \cdot g \cdot h_{\mathrm{P}} \cdot \eta_{\mathrm{RTG}} \tag{7.6}$$

is called its potential or its rated power

$$P = \dot{E} = \dot{V} \cdot \rho \cdot g \cdot h_{\mathrm{P}} \cdot \eta_{\mathrm{RTG}} = Q \cdot \rho \cdot g \cdot h_{\mathrm{P}} \cdot \eta_{\mathrm{RTG}} \tag{7.7}$$

It is calculated from the storage volume V, the potential energy height h_{P}, gravitational acceleration ($g = 9.81$ m/s^2), water density ($\rho \approx 1{,}000$ kg/m^3), the discharge Q, and efficiency η_{RTG} of pipelines, turbines, and generators in generating electricity.

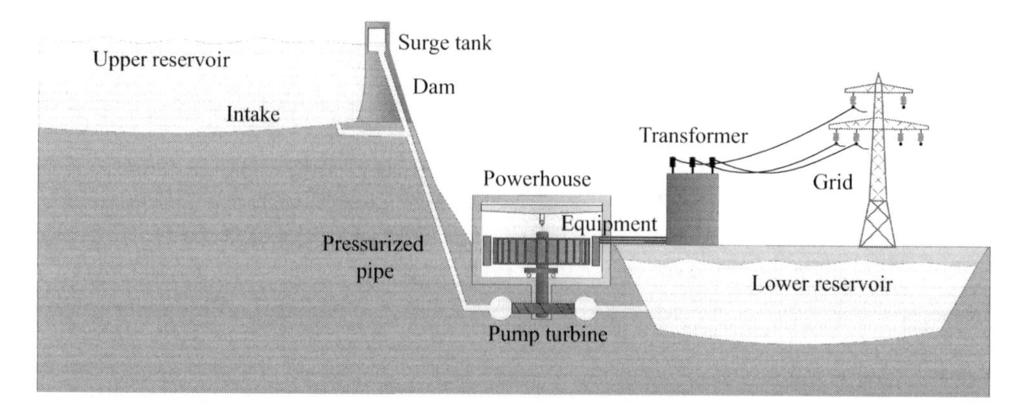

Figure 7.8 How a pumped storage power plant works [Qua09]

Figure 7.9 A pumped storage power plant near Málaga, Spain. Left: upper storage basin; right: lower storage basin with pressurized pipeline and surge tank

Enormous storage volumes are needed for larger amounts of energy. The Goldisthal pumped storage facility has an upper basin with a useful volume V of 12 million m^3. The wall surrounding the basin is 3,370 m long. The elevation difference h_P is 302 m; the electric capacity, 1,060 MW. The pumped storage power plant therefore contains 8,480 MWh, equivalent to eight hours of full power production [Vat03]. The efficiency η_{RTG} is therefore around 85 %.

In addition to storing energy, pumped storage power plants are also used to maintain frequency, provide reactive power, and quickly compensate for extreme power fluctuations (Gie03). The Goldisthal can ramp up its turbines from a complete standstill in only 98 s, a very short time compared with thermal power plants. Back in the 1960s, pumped storage power plants had already reached peak efficiencies exceeding 77 % (Figure 7.10). Modern facilities can have efficiencies exceeding 80 %.

In 1997, public pumped-storage power plants in Germany were used for an average of 1,950 h/a; those with natural water inflow averaged 902 h/a, compared with 780 h/a for those without. German pumped storage power plants had a net generation capacity of 4.042 TWh in 1996, compared with a pump capacity of 5.829 TWh, equivalent to an average efficiency of just under 70 %. The fleet includes, however, a large number of old power plants.

Pumped storage power plants could allow our current fleet of large conventional power plants, which do not ramp up and down quickly, to be better utilized. Likewise, pumped storage power plants would play a major role in a future power supply with a large share of renewable energy.

In the past few decades, pumps have mainly run overnight to improve the capacity utilization of baseload power plants. During the day, pumped storage power plants helped meet peak power demand (Figure 7.11). As the share of renewable power grows, however, pumped storage power plants will be used at different times of day; the effect is already making itself felt in Germany.

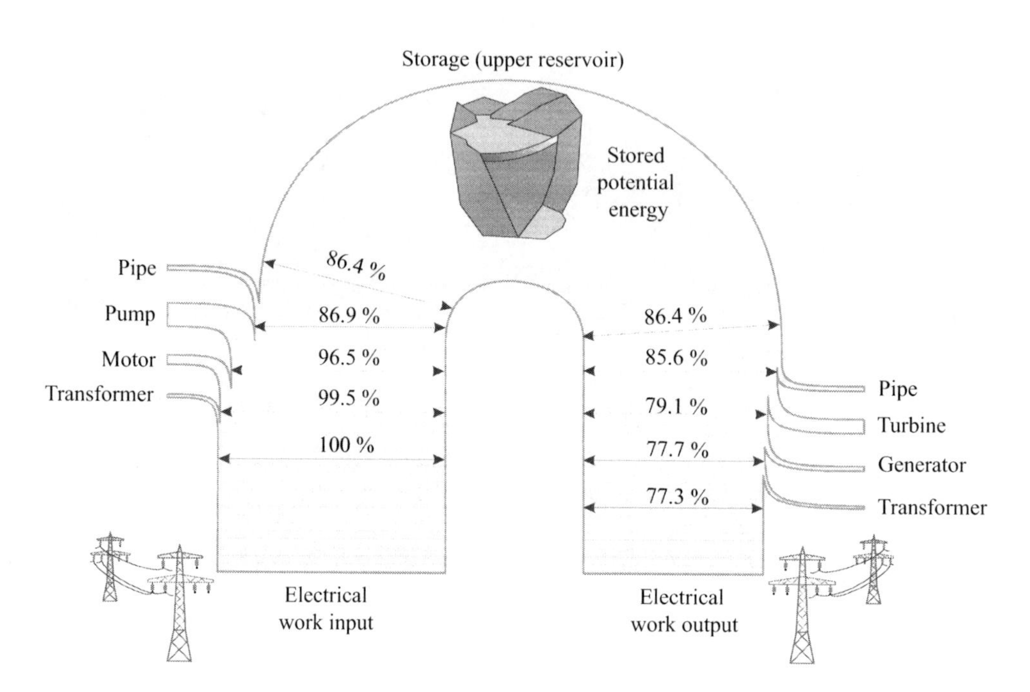

Figure 7.10 Losses and efficiency of a pumped storage power plant (based on [Böh62])

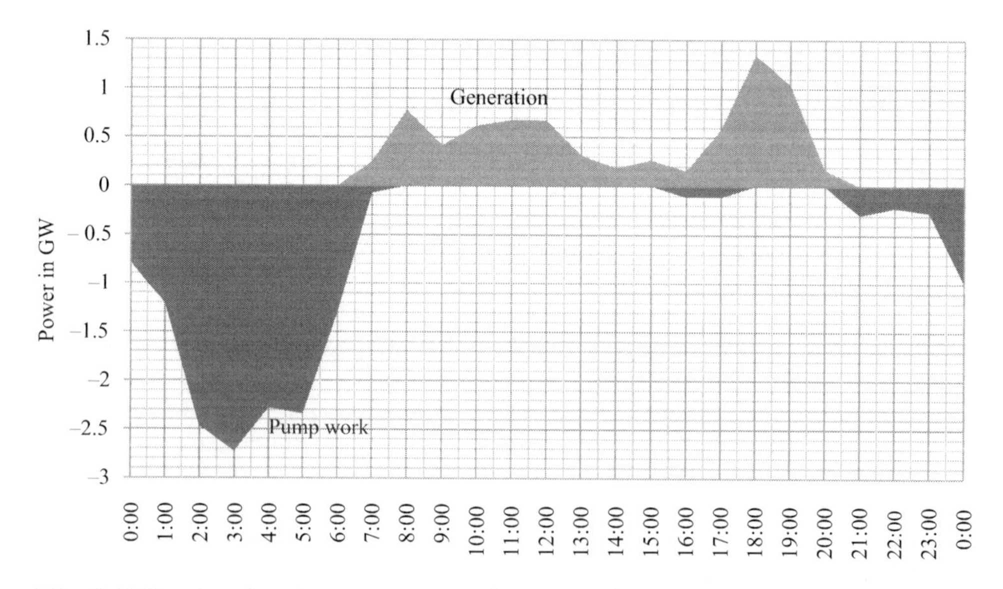

Figure 7.11 Pump work and power generation from pumped storage power plants in Germany on 15.01.1996

(Data: [VIK98])

Water turbines

Turbine types

Modern water turbines have little in common with the waterwheels of previous centuries. Turbines are optimized to suit the water's elevation and flow rate for specific applications (Figure 7.12). In the Itaipu power plant, they have capacities exceeding 700 MW.

In principle, a distinction can be made between reaction turbines and impulse turbines. In the latter, pressure does not change; the water's kinetic energy is used. **Impulse turbines** include:

- Pelton turbines
- Turgo turbines
- Ossberger turbines.

Reaction turbines include:

- Kaplan turbines
- bulb turbines
- propeller turbines
- Francis turbines.

In reaction turbines, water pressure is greater at the turbine's input than at its output. The turbine uses the pressure change – and hence the potential energy – to create kinetic energy.

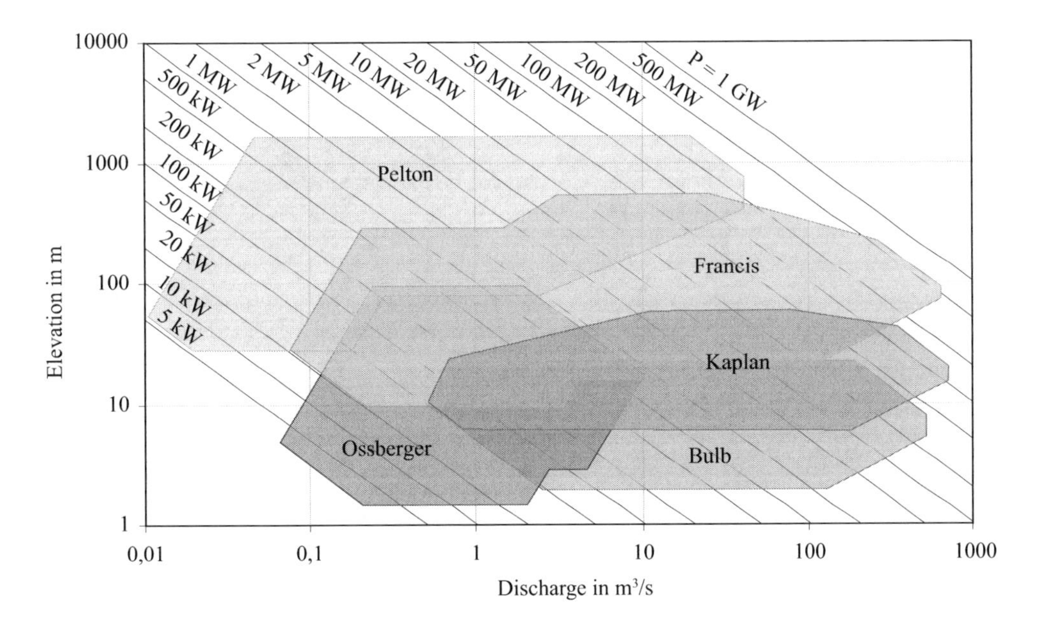

Figure 7.12 Use of water turbines relative to water elevation and discharge

Kaplan turbine and bulb turbine

If the water levels are not that different, the turbine developed by and named after Austrian engineer Viktor Kaplan in 1912 is used. This turbine looks like a large ship's propeller, which the water flows past.

It has three to eight adjustable blades. Unlike **propeller turbines**, which have fixed blades, this one can be adjusted to suit different application conditions, so that it has a better efficiency under partial load. Figure 7.13 shows a photo and a cross-section of a Kaplan turbine. The large synchronous generator with numerous poles is above the turbine in this case.

Kaplan turbines are mainly used when water levels are very close and large amounts of water are available. These conditions are typical for run-of-river power plants. Kaplan turbines reach efficiencies between 80 and 95 %. **Bulb turbines** are similar to Kaplan turbines, but the former have a horizontal axis, making them suitable for even smaller height differences. Figure 7.14 shows a cross-section of a bulb turbine. The generator is behind the turbine here.

Ossberger turbine

The Ossberger turbine has an efficiency of around 80 %. Because the turbine consists of three parts that can be successively and separately exposed to water, it has an especially high efficiency under partial load. It is therefore mainly used for small applications and height differences up to 100 m whenever multiple parallel turbines are not a good option. It is also relatively robust in terms of soiling. In return, this turbine type has a relatively low speed that generally requires a gearbox.

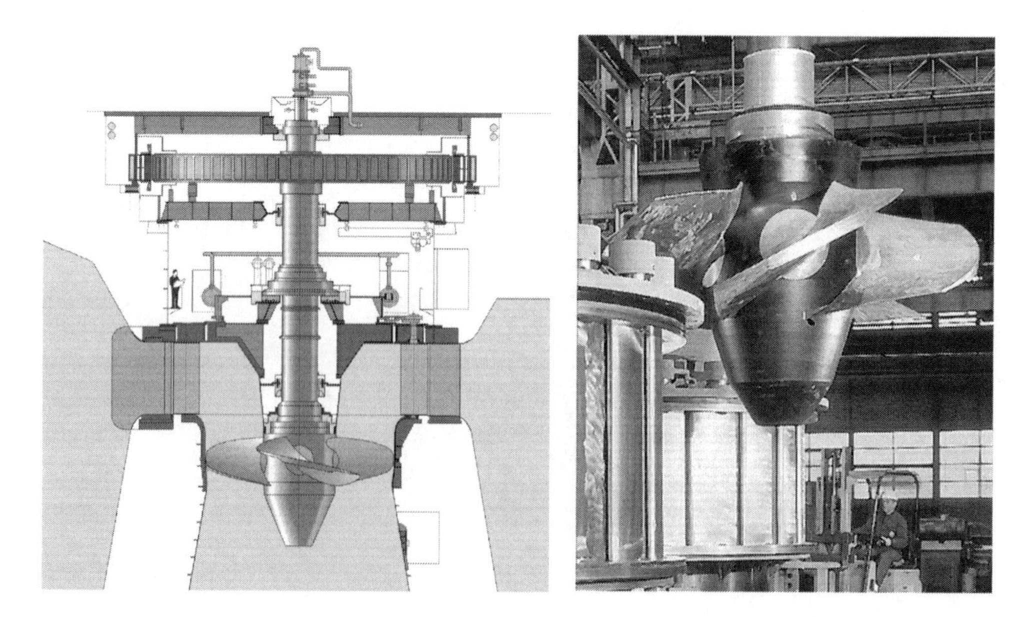

Figure 7.13 Drawing of a Kaplan turbine with a generator (left) and a photo of one (right)
(Source: Voith Siemens Hydro Power Generation [Voi08])

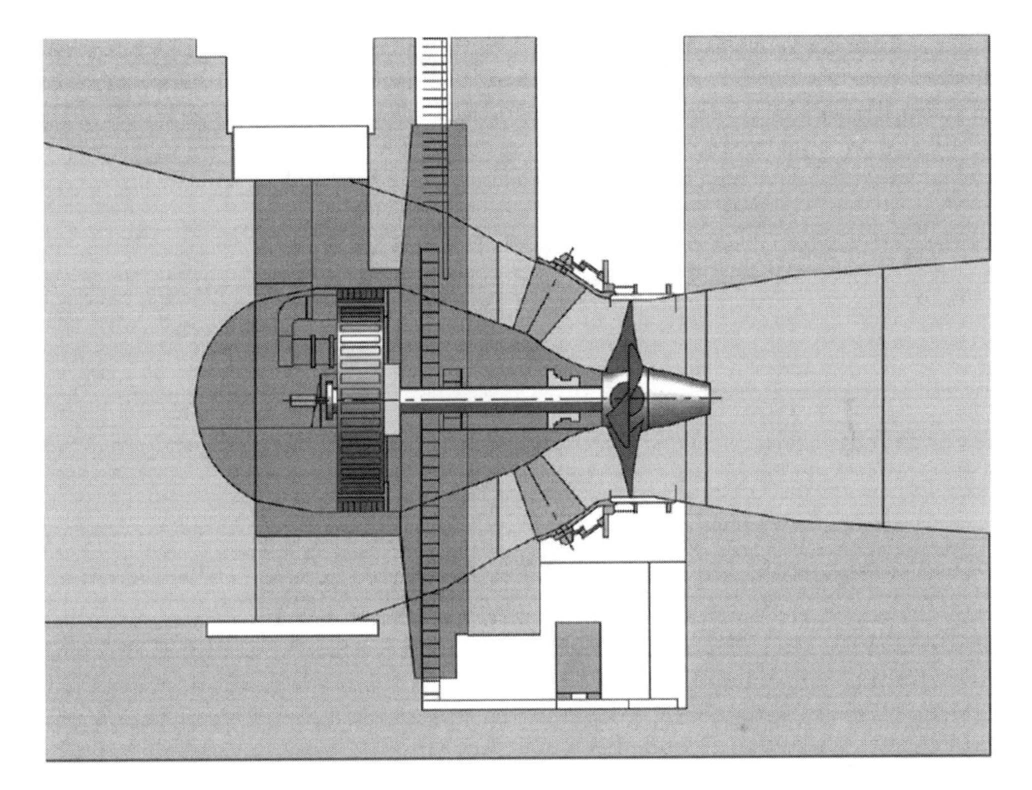

Figure 7.14 Drawing of a bulb turbine with generator

(Source: Voith Siemens Hydro Power Generation [Voi08])

Francis turbine

The Francis turbine was developed in 1849 by British engineer James Bichemo Francis. This turbine is especially suited for large height differences, therefore making it useful particularly in combination with storage dams.

The turbine has an efficiency exceeding 90 %, though its efficiency under partial load is only moderate. This turbine is available in three speeds: slow, normal, and fast. The Francis turbine can also be used as a pump in pumped-storage power plants. Figure 7.15 shows a 715 MW Francis turbine at the Itaipu power plant along with the cross-section of a 265 MW Francis pump turbine at the pumped storage power plant in Goldisthal.

Pelton turbine

In 1880, the American Lester Allen Pelton developed the turbine named after him. It is mainly suited for great height differences, such as in mountain regions. The turbine reaches very high efficiency levels at 90 to 95 %. Water reaches the turbine through pressurized pipes. A jet then sprays the water onto the scoop-shaped blades at high velocity. Figure 7.16 shows a photo and a cross-section of a Pelton turbine. The **turbo turbine** is a special form of the Pelton turbine.

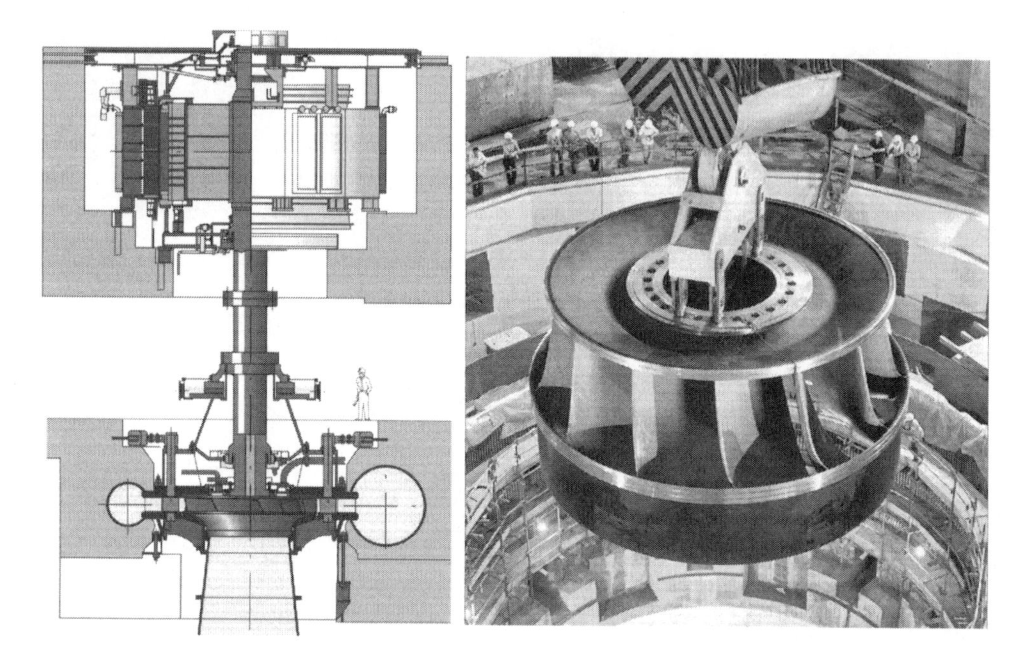

Figure 7.15 Drawing of a Francis pump turbine at the Goldisthal pumped storage power plant (left) and a photo of a Francis turbine at the Itaipu power plant (right)

(Source: Voith Siemens Hydro Power Generation [Voi08])

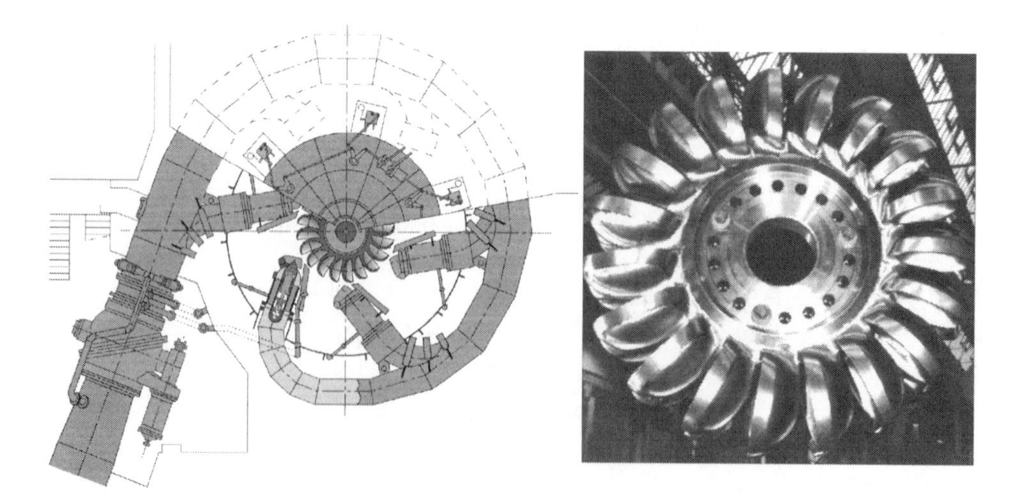

Figure 7.16 Drawing of a Pelton turbine with six jets (left) and a photo of a Pelton turbine (right)

(Source: Voith Siemens Hydro Power Generation [Voi08])

Turbine efficiency

Turbine efficiency η_T mainly depends upon discharge Q (Figure 7.17). As the useful height H decreases, turbine efficiency only drops slightly so that this relationship is not dealt with further below.

Below a minimum discharge Q_{min} the turbine does not produce any power. Discharge Q and rated discharge Q_A can be used to calculate turbine efficiency η_T:

$$q = \frac{Q - Q_{min}}{Q_A} \tag{7.8}$$

The following empirical approach roughly describes these relations:

$$\eta_T = \begin{cases} 0 & \text{for } Q \leq Q_{min} \\ \dfrac{q}{a_1 + a_2 \cdot q + a_3 \cdot q^2} & \text{for } Q_{min} < Q < Q_A \\ \eta_{T,N} & \text{for } Q_A \leq Q \end{cases} \tag{7.9}$$

Table 7.3 shows the parameters used to determine efficiency curves from Figure 7.17. The turbine's efficiency is then used to calculate the turbine's mechanical power:

$$P_{mech} = \eta_T \cdot P_W \tag{7.10}$$

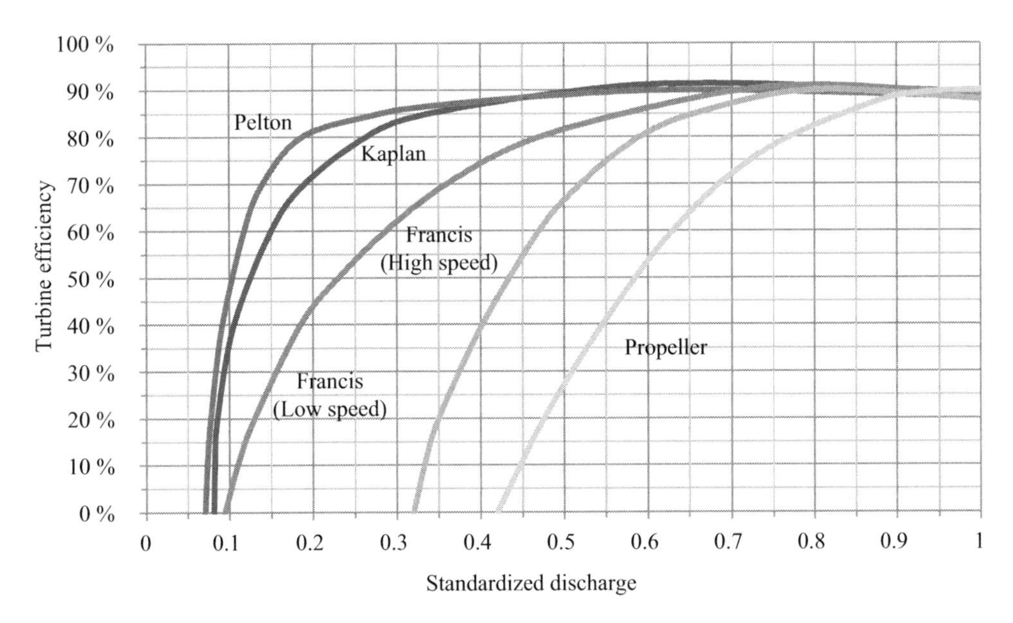

Figure 7.17 The efficiency of individual turbine types relative to the discharge Q standardized for the rated discharge Q_A

(Based on [Raa89])

Table 7.3 Typical parameters used to determine turbine efficiency

	Q_{min} / Q_A	$\eta_{T,N}$	a_1	a_2	a_3
Kaplan	0.081	0.895	0.045	0.965	0.1
Pelton	0.07	0.885	0.03	0.99	0.1
Francis	0.095	0.89	0.18	0.63	0.31
Propeller	0.42	0.9	0.25	0.28	0.69

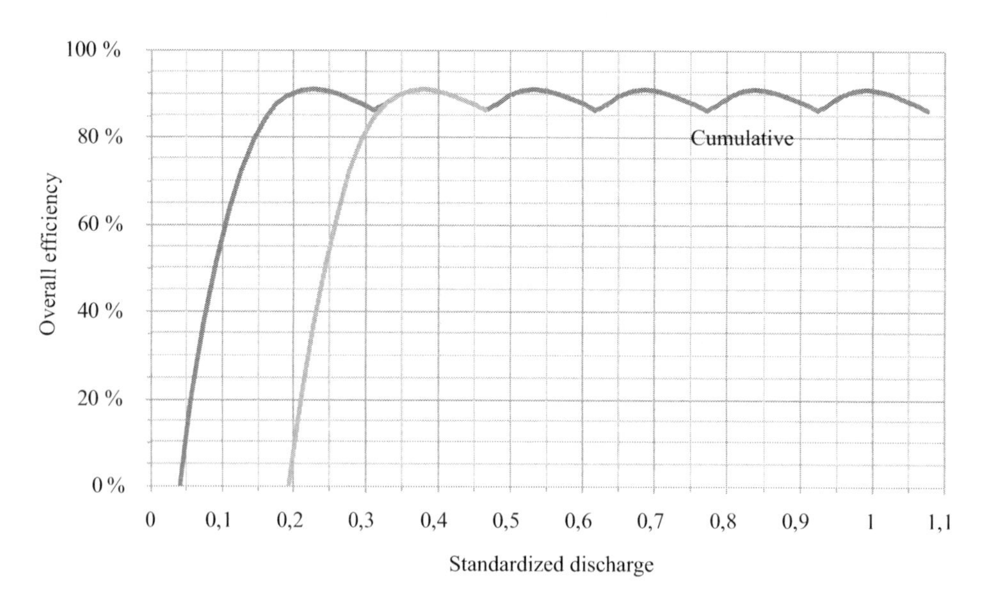

Figure 7.18 The curve of overall efficiency relative to a standard discharge in a power plant with six turbines

Synchronous generators are generally used in hydropower plants. The efficiency η_G of electricity generators is nearly constant over a large spectrum and only drops noticeably under extremely low loads.

Usually, multiple parallel turbines are used in midsize and large hydropower plants. When there is not enough hydropower, individual turbines can then be switched off so that the remaining turbines still run close enough to their nominal output. Figure 7.18 shows the curve for the efficiency of multiple parallel turbines relative to discharge. The curve remains relatively constant over a wide operating range.

Other hydropower systems

Tidal power plants

The tides are the result of gravitational forces between the moon, the sun, and the earth. As the earth turns, these gravitational forces continuously change direction. The masses of water in the oceans follow these forces. The result is the tides, which can change the elevation of water by more than 1 m on the high seas. Roughly every 12 hours, tides caused by the moon occur at a particular point on the planet. The tides build up in coastal regions. In extreme cases, water level can change in such areas by more than 10 m as a result of the tides.

The principle of exploiting the principal energy of the tides is relatively straightforward. A dam separates a bay along the coast in an area with great height differences. When the tide comes in, the water flows through a turbine into the bay; when the tide goes out, it flows the other way. The turbine and the connected generator convert the water's energy into electricity. Power production fluctuates in the process. When the tides reverse, it drops to zero.

Back in the Middle Ages, tidal mills were built. Today, few modern tidal power plants exist to produce power. In addition to the systems listed in Table 7.4, a number of smaller power plants also exist, especially in China.

The largest such power plant is currently the Shiwa-ho in South Korea, which has 10 25.4 MW Kaplan turbines. The dam that surrounds the 56.5 km² basin is 12.7 km long. With a total power production of 550 GWh the plant reaches just over 2,000 full-load hours. The dam was originally built to create new land. The oldest and best-known tidal power plant is located in the Rance delta in France. When tidal power plants are built, the corrosive properties of salty seawater must be taken into account. The greatest obstacle towards the construction of new tidal power plants, however, is the relatively high investment cost and local environmental impacts relative to the modest capacity utilization. While numerous new projects are on the drawing board, it is unclear which ones will actually be developed for financial reasons.

Ocean current power plants

While large dams are needed for tidal power plants, it is far easier for ocean current power plants to be integrated in their environment with a low impact. These power plants are similar to wind turbines (see Chapter 6), only that the rotor turns underwater. Figure 7.19 shows a prototype, whose rotor can be lifted above the water surface for maintenance; it was developed with funding from the German Environmental Ministry in cooperation with German research institute ISET and installed off the coast of North Devon, England [Bar04].

The physical properties of wind turbines can, in principle, be transferred to ocean current power plants. The main difference is that water has a much greater density than air. Ocean current power plants can therefore make far more power from lower current speeds than wind turbines can. The specific power density of water at a current velocity of 2 m/s already reaches 4 kW/m².

Table 7.4 Tidal power plants worldwide [Gra01; Wik12]

Country	*System*	*Year of installation*	*Height difference in m*	*Installed capacity in MW*
South Korea	Shiwa-ho	2011	5.8	254
France	Rance	1967	5.6	240
Canada	Annapolis	1983	5.5	20
China	Ganzhtan	1970	1.3	5
China	Jiangxia	1980	2.5	3.2
Russia	Kislaya	1968	1.4	2

Ocean current power plants can, however, only be used in regions with relatively high current velocities at moderate water depths of around 25 m. These conditions mainly occur around peninsulas, in bays, between islands, and in straits. Shipping often limits options even further, but there is nonetheless still great potential. In Germany, for instance, such systems could be installed on the southern tip of Sylt, where current velocities reach 3 m/s [Gra01]. Further development and serial production would reduce costs very quickly, so that ocean current power plants could become an additional source of climate-friendly electricity over the midterm.

Wave power plants

For decades, great hopes have been placed on the development of wave power plants. The potential of wave energy is tremendous. The oceans have a total surface of 360.8 million km^2. If half of this water mass is raised by 0.5 m, the potential energy of 0.6 EJ is stored. However, only coastal regions with shallow water can be used to harvest wave power. In Germany, the useful potential is equivalent to less than 1 % of power demand.

In terms of design, a distinction is made between:

- floating systems
- chamber systems
- TapChan units.

Figure 7.19 Left: a prototype in the Seaflow project off the British west coast; right: maintenance ship in a planned ocean current power plant

(Photo: ISET)
(Graphic: MCT)

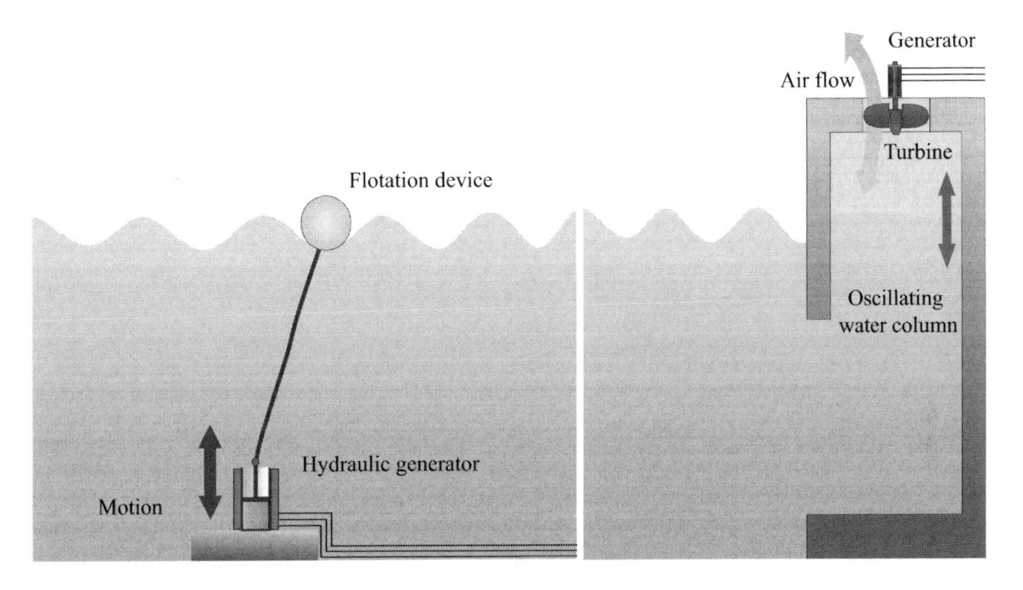

Figure 7.20 How wave power plants work; left: floating system; right: chamber system [Qua09]

Floating systems use the potential energy of waves. A floating device follows the motion of the waves. It is connected to a second device anchored to the seafloor. As the floating unit moves, it drives a piston or a turbine (Figure 7.20).

Chamber systems – oscillating water columns or OWCs – have a chamber containing air. The waves compress the air. The air can escape through an opening to drive a turbine. When the wave recedes, the air also drives a turbine as it returns into the chamber (Figure 7.20).

TapChan is the abbreviation of Tapered Channel. These systems are used in coastal regions. A top basin catches the waves. When the water flows back in the sea, it drives a turbine. The main drawback of such systems is the large amount of space they take up on the coast.

In the past two decades, numerous prototypes of wave power plants have been built, but none of the concepts have been a commercial success. The main problem is the drastically changing conditions at sea. Technical systems have to make do with as little material as possible to reduce costs. At the same time, however, storms with waves several metres tall place great demands on system durability. Quite a few prototypes have been broken in storms. But if robust systems can be made at an attractive price, wave power plants can also become a source of climate-friendly electricity.

8 Geothermal energy

Geothermal resources

During the icy winter it may be hard to imagine, but the earth is a hot planet. Ninety-nine percent of it is hotter than 1,000 °C. Radioactive decay processes within the planet lead to temperatures around 6,500 °C, hotter than the sun's surface. It is, however, not possible for mankind to use this heat from within the earth's core. We are merely made aware of our planet's inner structure when, say, a volcano erupts, thereby bringing parts of the core to the surface in an uncontrolled fashion.

The earth itself has layers a bit like an onion (Figure 8.1). It consists of a core, a mantle, and a crust. The core has a diameter of around 6,900 kilometres. There is an outer fluid core and an inner core consisting of solid iron and nickel at pressures up to 4 Mbar.

Boreholes allow us to reach parts of the earth's crust, which is up to 35 km thick. Relative to a reference temperature of 15 °C, the earth has a total heat content of $1.26 \cdot 10^{31}$ J, $5.4 \cdot 10^{27}$ J of which is estimated to be in the crust. This amount is several times greater than mankind's primary energy demand of around $4 \cdot 10^{20}$ J. The temperature differences between the earth's core and its crust lead to a constant flow of heat of between 0.063 and 0.42 W/m². Such low levels of heat current are by and large not suited to technical exploitation. Usually, deep boreholes are required for the exploitation of geothermal heat.

The average geothermal depth gradient is 1 °C/33 m. At a depth of 3,300 m, the temperature increases by 100 °C on average. However, the temperature does not increase at the same rate everywhere across the planet. There are especially great anomalies in areas where the continental plates meet (Figure 8.2). In areas with good geothermal resources, temperatures reach far above 100 °C at a depth of only a few hundred metres. There are also great differences in Germany. In the Rhine Graben, for instance, temperatures exceed 150 °C at a depth of 3,000 m.

The technical potential of geothermal power generation in Germany is estimated at a total of 300,000 TWh, roughly 600 times as much as the country's annual power supply [Pas03]. Over a period of 1,000 years, geothermal could therefore theoretically meet half of Germany's power demand at around 300 TWh per year. The additional potential of the use of thermal energy in cogeneration is roughly 1.5 times as great as the power generation potential, and it increases to 2.5 times when heat pumps are used. Because geothermal potential exists at relatively great depths in Germany for the required temperatures, Germany is not one of the countries best-placed for the exploitation of geothermal. As a result, the cost of geothermal plants is relatively high compared with other renewable energy systems, particularly because of the great drilling costs. A lot of studies therefore suggest that geothermal will not make up a large share of German power supply until the middle of the twenty-first century.

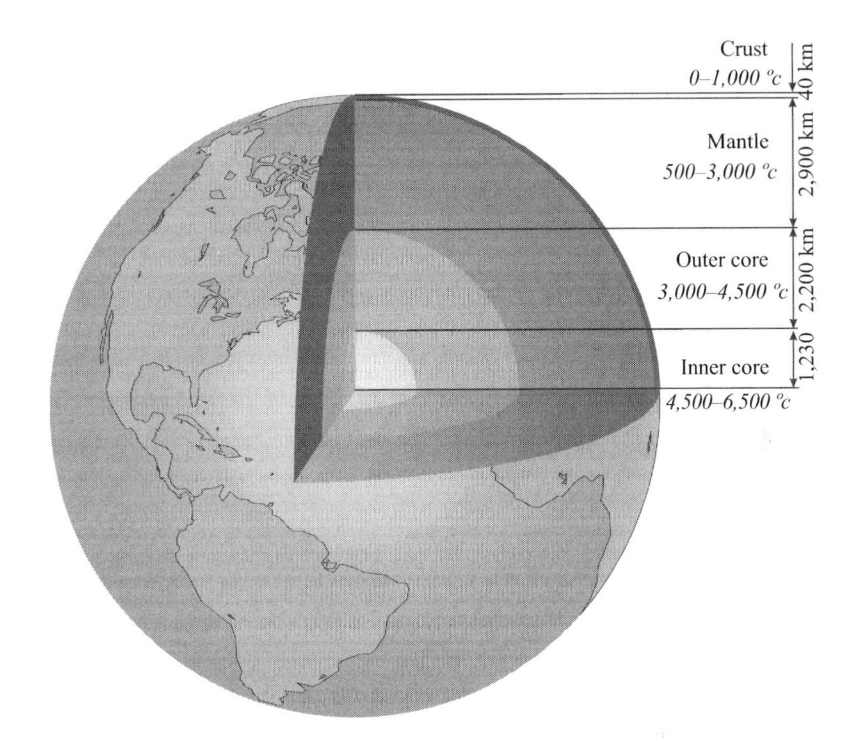

Figure 8.1 Cross-section of the earth [Qua09]

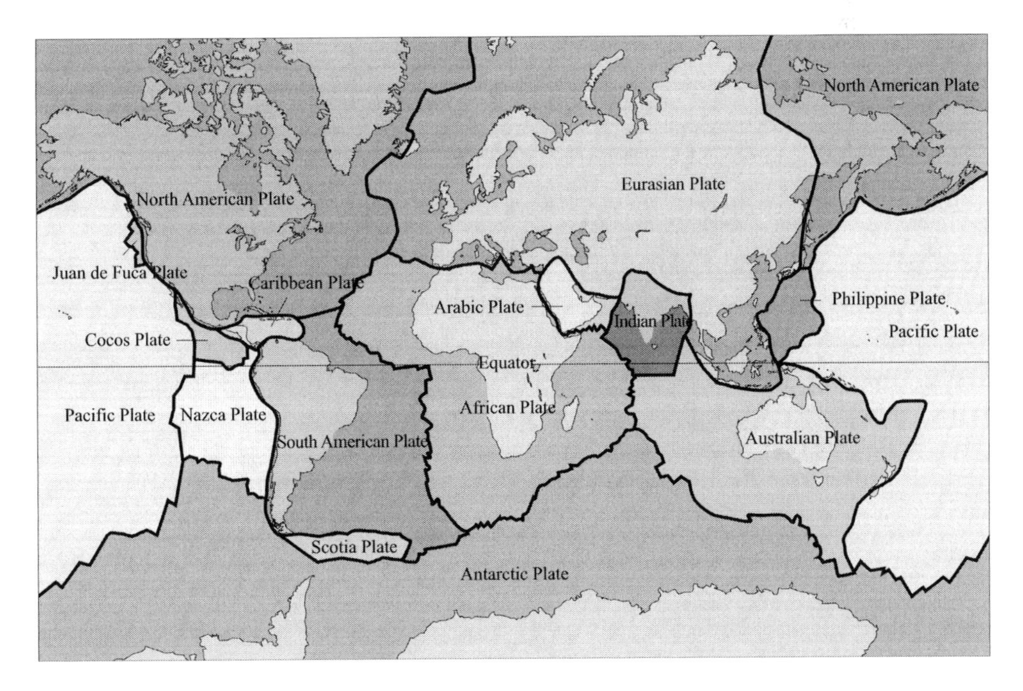

Figure 8.2 Tectonic plates on the earth

(Source: US Geological Survey)

The best conditions in Germany are located in the Rhine Graben (Figure 8.3). Here, temperatures around 150 °C and higher can be reached at depths of 3,000 m. In comparison, such temperatures are reached only a few hundred metres underground in Iceland.

A distinction is made between the following geothermal resources:

- aquifers (hot water)
- underground faultlines
- hot dry rock.

Hot water aquifers are relatively rare. In addition, boreholes drilled in the search for underground sources of hot water can cost €1 million per kilometre, making exploration a great financial risk. Underground hot water can be directly used for technical applications. Often, however, it has a high salt content and low natural radioactive contamination, both of which hamper its use. In practice, two boreholes are needed. Hot water is pumped up through one to the surface. The heat is extracted from it before it is reinjected into the second borehole.

Hot dry rock (HDR) has the greatest potential by far. Here, underground gaps first have to be created for the heat to be extracted. Cold water is first injected into the depths, where it heats up, thereby creating cavities in the hot rock. More cold water is then injected into these cavities, where it heats up before being pumped back up to the surface. HDR is used in depths from 3,000 to 5,000 m. A number of demonstration projects are currently planned or under construction in Germany, France, and Switzerland.

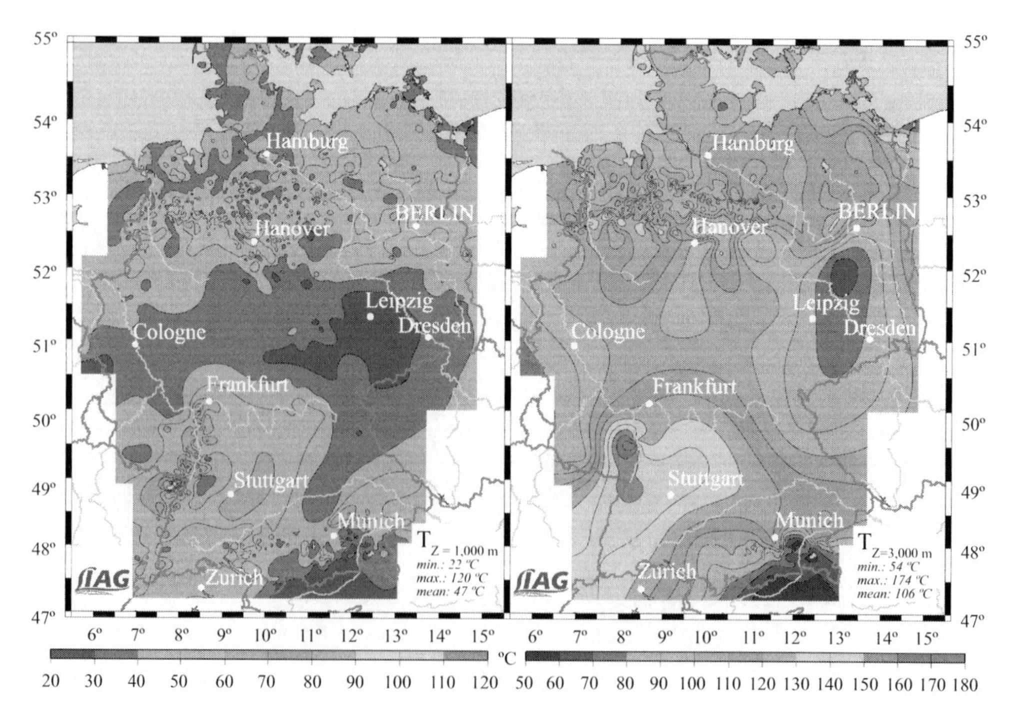

Figure 8.3 Temperatures in Germany at depths of 1,000 and 3,000 m

(Graphics: www.liag-hannover.de [Sch02])

The drilling technology used in the geothermal sector is well known from the oil sector, for instance. In rotary drilling, motors drive a diamond-studded drill bit into the ground. Eventually, the hole is too deep for the drill to be turned at the end of a system of rods; torsion and friction are simply too great. An electric motor or a turbine is then used to directly turn the drill.

On the surface, only the drill tower is visible (Figure 8.4); it holds the drill rods in place. Inside the drill, water is injected into the borehole at pressures up to 300 bar. The process not only allows rock particles that collect around the drill bit to be drawn up, but also keeps the drill cool. It is also possible to drill at an angle to cover a larger underground area from boreholes close to each other on the surface.

Depending on the composition of the underground rock, the walls of the drill hole can be unstable. To prevent it from collapsing, a steel pipe is inserted across large sections and held in place using special cement. The drilling process then continues with a smaller bit. For a long time, the very high salt content of thermal water was a major problem for drills. The salt ate away at the metal, quickly causing corrosion. Now, special coatings are available to prevent this from happening.

The deepest borehole ever was drilled for research purposes on the Russian Peninsula of Kola. It went down 12 km. In Germany, the deepest borehole ever reached 9.1 km. Such depths push the limit of what is currently technically possible. At around 10 km, conditions are extreme, with temperatures exceeding 300 °C and pressures around 3,000 bar. Under such conditions, even rock softens and starts to melt.

Fortunately, geothermal energy can be used at far more shallow depths. The deepest boreholes currently planned for large systems are around 5 km. Nonetheless, the technology required even for such depths is quite complex, and hence expensive.

Figure 8.4 Left: A new and used drill bit. Right: A drill tower

(Photos: Geopower Basel AG)

Once tapped, a borehole can only be used for a certain period of time. As heat is extracted, the underground area slowly cools down. In general, enough space is left between boreholes so that the desired temperatures remain available for around 30 years. After that, the temperatures drop to a level too low for further exploitation. But the areas that cool down are relatively small, so new geothermal systems can generally be set up only a few kilometres away. It takes many decades for disused boreholes to heat up again.

Geothermal heat plants

Once the boreholes have been drilled for the thermal water source, a geothermal heat supply is relatively easy to build. Because the boreholes are quite costly, the heat demand needs

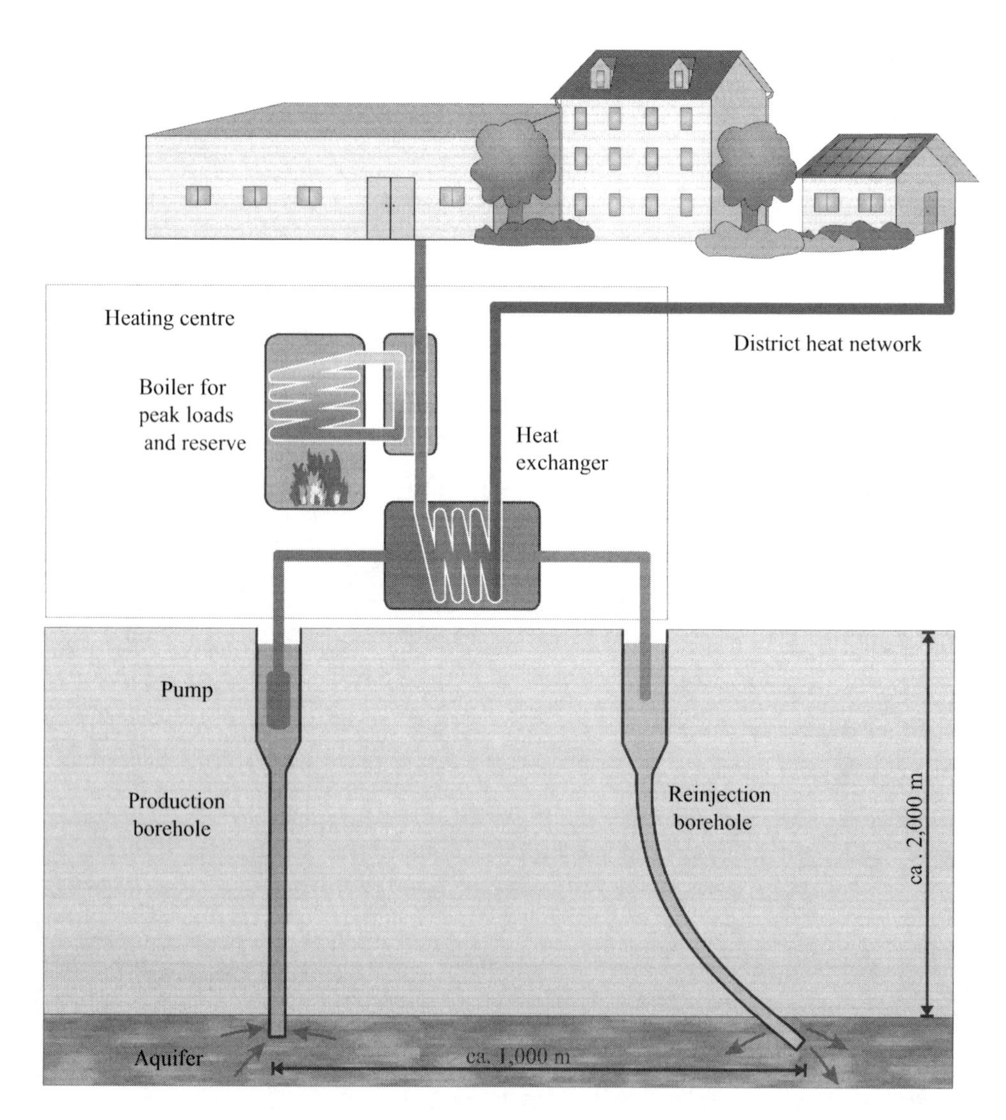

Figure 8.5 How a geothermal heat plant works

to be relatively great for the system to pay for itself. The geothermal heat plant can then be connected to a district heat network.

The plant uses a pump to draw hot thermal water up to the surface through the production hole (Figure 8.5). Because the thermal water often has high salt content and a certain level of natural radiation, it cannot directly be used for heat supply. Instead, a heat exchanger is used to draw off the heat and pass it on to the district heat network. Once the water has cooled down, it is reinjected through a second hole.

For heating purposes, relatively low temperatures below 100 °C suffice. For this purpose, the boreholes do not need to be very deep. In Germany, such systems often need only be around 2,000 metres deep.

The central heating system adjusts the amount extracted to suit heat demand. A peak demand boiler can be used to cover extreme spikes in heat demand. To ensure a reliable heat supply, a reserve boiler is also a good idea in case there are problems, for instance, with the pump or the borehole.

Geothermal power supply

Great store is also set by electricity from geothermal energy for future climate protection. Numerous geothermal plants are already used for power generation in countries with geothermal anomalies, such as Iceland, Italy, and Indonesia. The first geothermal power plant in Germany was completed in 2003 in Neustadt-Glewe with a capacity of 210 kW.

Power plant processes

Geothermal power supply is a bit more complex than heat supply. In particular, the temperatures that can be provided by geothermal energy are relatively low, requiring new power plant concepts, such as:

- the use of direct steam
- flash power plants
- ORC power plants
- Kalina power plants.

If the underground thermal water is already available at temperatures and pressures that produce steam, the steam can be directly used to drive a steam turbine. One drawback is the often very corrosive components of the steam. If not enough steam is available, **flash processes** are used. Here, the deep thermal water is allowed to partially expand under pressure. As the temperature drops, the share of steam increases as a result. A separator and a demister then separate the steam from the rest of the water. The steam can then be used to drive a steam turbine before being reinjected into the underground after condensing. In a **double flash process**, two expansion stages are used in sequence, thereby increasing efficiency slightly.

The **Clausius-Rankine process** (see Chapter 4) used to generate electricity requires temperatures above 150 °C. It is not always possible to attain such temperatures simply by tapping a geothermal heat source. An alternative starting at around 80 °C is the Organic Rankine Cycle or **ORC** (Figure 8.6). Here, a working medium is used in the steam turbine process that has a lower boiling point than water. For instance, isopentane or PF5050 (C_5F_{12}) boils at an ambient temperature of only 30 °C.

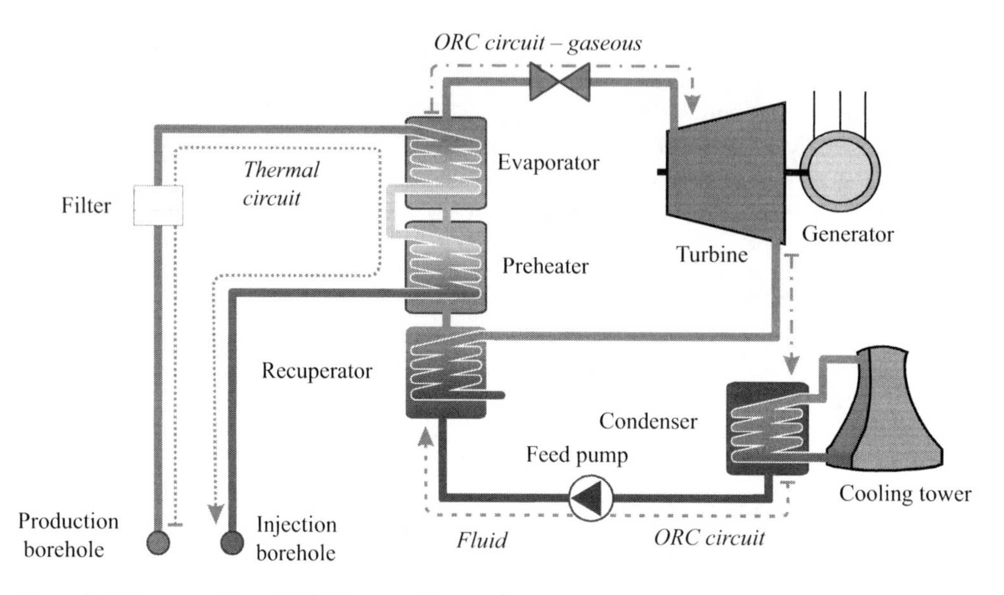

Figure 8.6 How a geothermal ORC power plant works

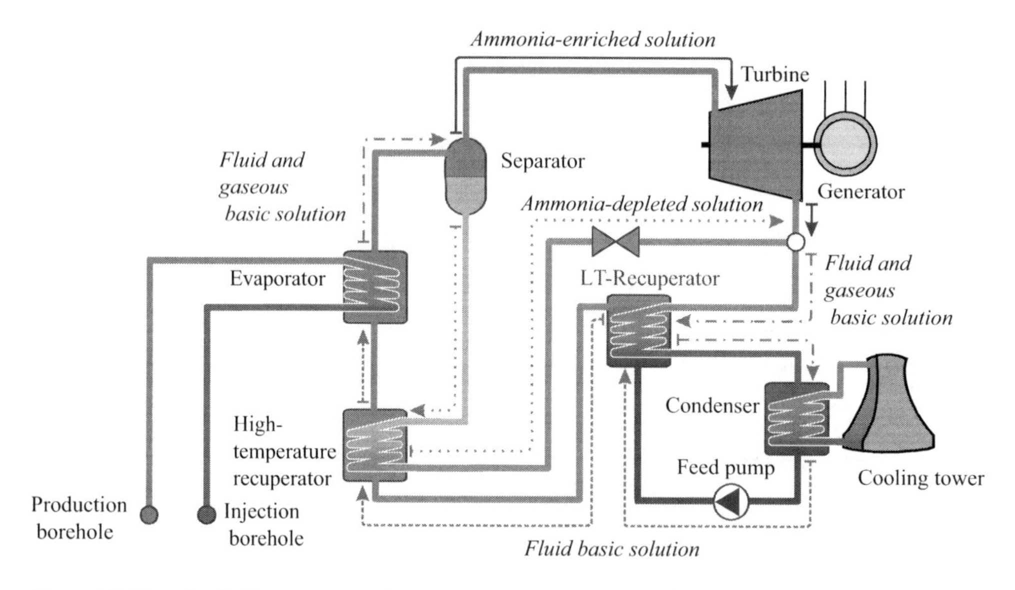

Figure 8.7 How the Kalina process works

The **Kalina process** (Figure 8.7) offers greater efficiency at low temperatures. Here, a mixture of two operating media – such as water and ammonia – is used. When the mixture turns into steam in the boiler, the boiling point does not remain constant. As a result, the thermodynamic process properties are better than in the water vapour and ORC processes. The remaining fluid goes into the separator. The steam with heavy ammonia content drives a turbine. As in the Clausius-Rankine process, the expanded steam returns to a fluid state in the condenser. A feed pump brings this fluid back up to operating pressure. The fluid sent to the separator passes its heat on to the cycle via a recuperator before returning into the cycle just before the condenser.

Despite all of the technical advances, efficiencies are still only moderate at low process temperatures. The Carnot efficiency (see Chapter 4) describes the theoretical limit, which real processes cannot attain. Figure 8.8 shows the typical efficiencies of the low-temperature processes described above in relation to the Carnot efficiency.

Geothermal power plants

Because of the low efficiency, the direct use of heat at temperatures below 150 °C is a better option for geothermal power production both in economic and ecological terms. For instance, if excess heat is available in the summer, geothermal power production makes sense. This principle is applied at Germany's first geothermal heat plant in Neustadt-Glewe.

In Germany, the best conditions for geothermal power production are found at depths up to 5,000 m. However, thermal water sources can rarely be tapped then. Rather, hot dry rock (HDR) is used. To get heat out of the rock, artificial cavities are needed within which the water can heat up. To create these cavities, water is injected into a borehole at high pressure. It expands because of the heat, thereby causing new fractures and expanding existing ones. The result is an underground system of fractures stretching across several cubic kilometres. A second borehole is used to monitor these activities.

When hot water sources are directly tapped, one speaks of **hydrothermal geothermal**, while hot dry rock is also called **petrothermal geothermal**.

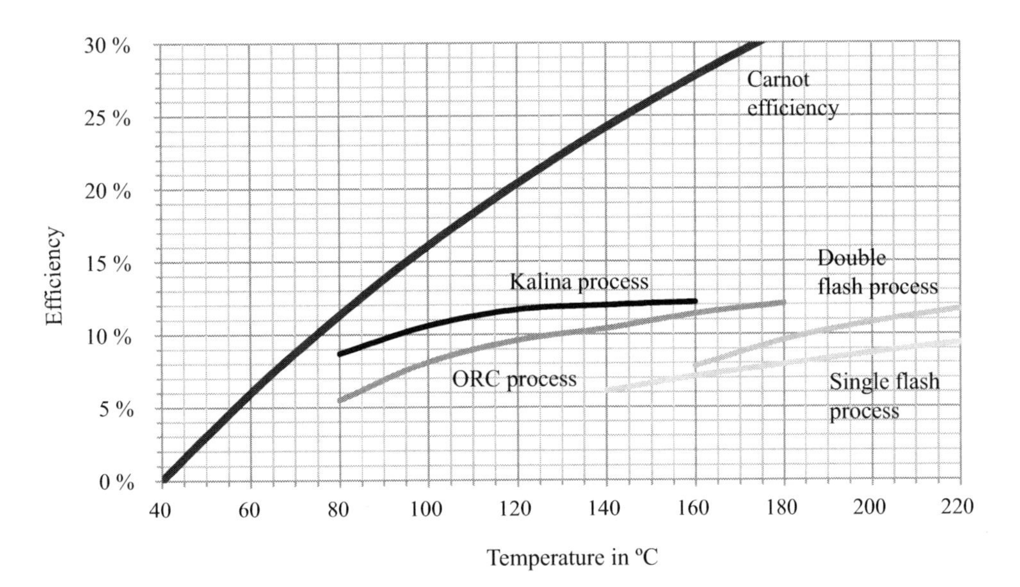

Figure 8.8 Efficiencies of various low-temperature processes compared to the ideal Carnot efficiency (reference 40 °C) relative to the feed temperature

For geothermal power production, a pump is used to inject cold water into the injection borehole, where it travels down fissures and cracks in the crystalline rock, heating up to temperatures around 200 °C in the process. The hot water is drawn back up to the surface through production boreholes, where it passes on heat via a heat exchanger to a power plant process and district heat network.

In the 1970s, the first HDR tests were conducted in Los Alamos, California. In Soultz-sous-Fôrets, France, a European HDR project has been underway since 1987, and a 1.5 MW pilot facility went into operation there in 2008. In 2004, Geopower Basel AG was founded in Switzerland. This firm's goal was to set up the first commercial HDR power plant. The

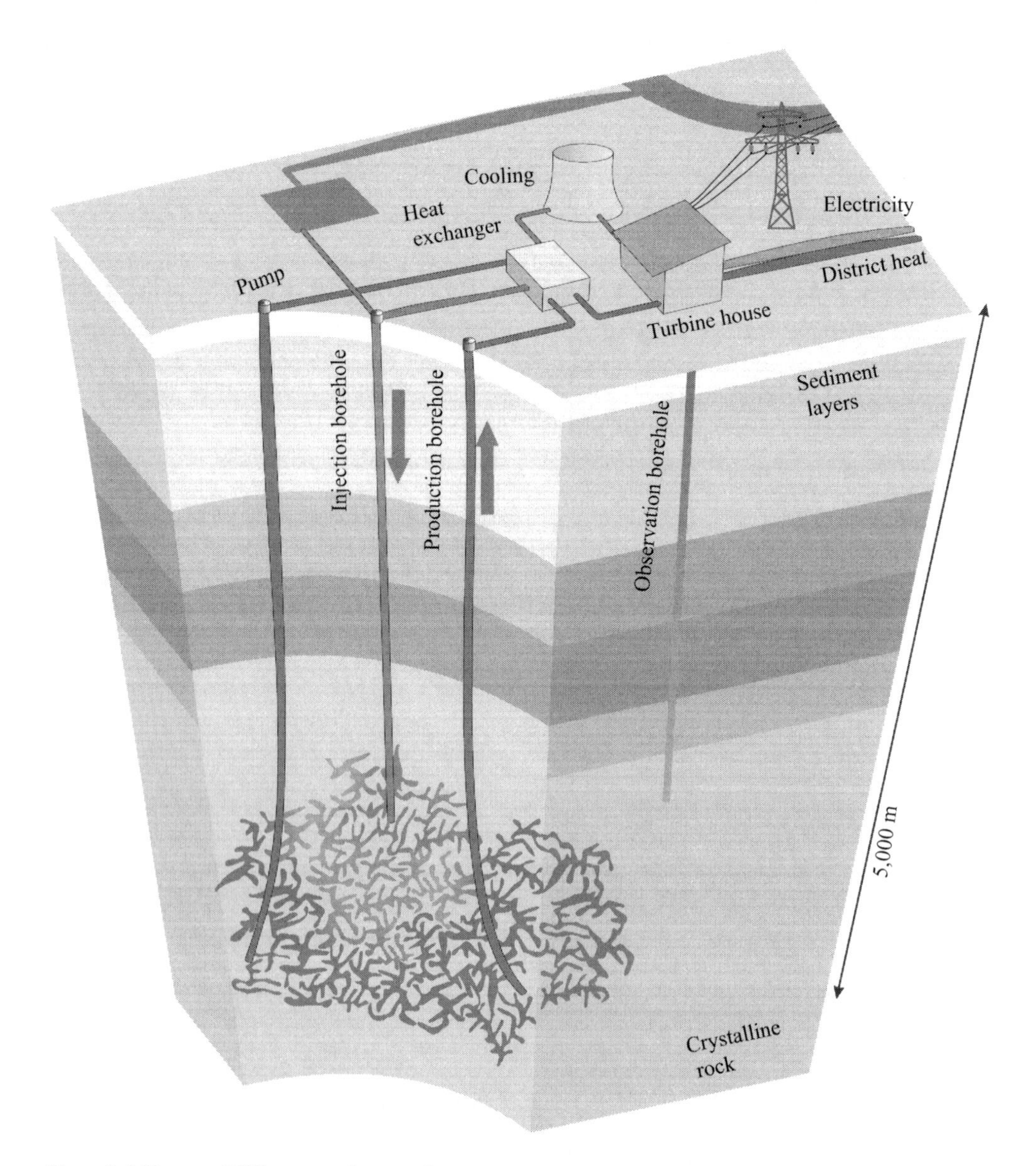

Figure 8.9 How an HDR power plant works

work was discontinued in 2007 when small earthquakes were caused during the process of creating the first underground fractures.

It is currently unclear whether geothermal will also be a market success in regions without considerable geothermal anomalies in the near term. The high cost of deep boreholes will probably not come down much in the foreseeable future, making this option uncompetitive with conventional systems. Nonetheless, geothermal will remain an important component of a renewable power supply in the midterm because geothermal power is always available, unlike photovoltaics and wind power which depend on the weather.

Heat pumps

In general, a heat pump is a machine that uses a pump to generate heat from a low-temperature source. This heat can then be used for space heating, hot water supply, or process heat.

In general, there are three different kinds of heat pumps:

- compression heat pumps
- absorption heat pumps
- adsorption heat pumps.

Heat pumps first became popular during the oil crises because they were a way of providing useful heat without consuming petroleum, the main source of heat at the time. The CFCs used as refrigerants at the time turned out, however, to be problematic ecologically. And as oil prices dropped, so did sales of heat pumps. But in the past few years, sales of heat pumps have picked up again in Germany. If the electrical, mechanical, or thermal energy used to drive the heat pump comes from a renewable source, modern heat pumps can be an effective, carbon-neutral heat supply, though there is still room for improvement in terms of refrigerants. But if the energy to power heat pumps comes from fossil fuels, the economic benefits of heat pumps are hardly greater than modern heating systems fired with natural gas.

Compression heat pumps

Heat pumps based on the compression principle are essentially refrigerators running in reverse. Figure 8.10 shows a compression heat pump used to provide low-temperature heat.

Low-temperature heat is used to boil a refrigerant within a condenser. For heat sources to be tapped at very low temperatures, the refrigerants must have a boiling point below 0 °C. A compressor powered by an electric motor or a motor running on natural gas or gasoline along with external drive energy compresses this gaseous refrigerant to a very high operating pressure. In the process, it heats up considerably. The process is similar to the heat that builds up within a bicycle tire pump when you hold your thumb over the air outlet and pump. This high-temperature heat can then be used as useful heat, such as for space heating or hot water supply. A condenser draws heat away for this purpose, and the gaseous refrigerant turns back into a fluid. The pressurized agent then expands through an expansion valve, cooling off in the process before going back into the reboiler.

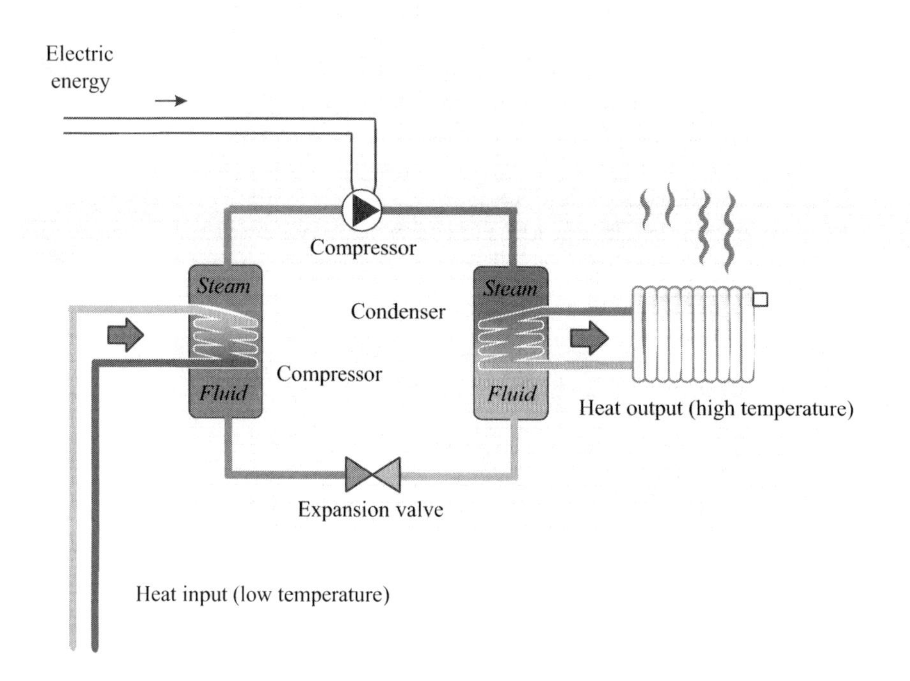

Figure 8.10 How a compression heat pump works [Qua09]

Table 8.1 Physical properties of common refrigerants and their greenhouse gas potential over a period of 100 years

Refrigerant	*Boiling point*	*Condensing temperature at 26 bar*	*Greenhouse gas potential relative to CO_2*
R12 CFC[1)]	−30 °C	86 °C	6,640
R134a HFC	−26 °C	80 °C	1,300
R404a HFC	−47 °C	55 °C	3,260
R407C HFC	−45 °C	58 °C	1,530
R410A HFC	−51 °C	43 °C	1,730
R744 (carbon dioxide)	−57 °C	−11 °C	1
R717 (ammonia)	−33 °C	60 °C	0
R290 (propane)	−42 °C	70 °C	3
R600a (butane)	−12 °C	114 °C	3
R1270 (propene)	−48 °C	61 °C	3

(Data: [Bit04; Fri99])
[1)] Banned since 1995 in new systems because of its impact on the ozone layer

Halocarbons are used as refrigerants in compression heat pumps. While they do not have a detrimental impact on the ozone layer like the CFCs that are now banned, they do have a tremendous greenhouse gas potential (Table 8.1).

Only 1 to 3 kg of refrigerant is needed for heat pumps used in single-family homes, a relatively small amount. However, leaks and losses at the end of life can never be completely ruled out. If 2 kg of R134a escape, the greenhouse gas impact is equivalent to 2,600 kg of CO_2. This amount is also released from the combustion of 13,000 kWh of natural gas. If the electricity needed comes from a conventional fleet of power plants with a large share of coal and nuclear power, the environmental impact of the heat pump worsens even further. If unproblematic refrigerants are used and renewable electricity powers the system, the heat pump can be a major ecological step forward. Because the potential range of applications for heat pumps is very great, it is not unrealistic to assume that installation figures will continue to increase.

Absorption heat pumps

Absorption heat pumps have a thermal compressor instead of a mechanical one as in compression heat pumps. In addition, the chemical sorption process is used. Chemists speak of sorption or absorption when a fluid takes up another gas or fluid. One common example is carbon dioxide in carbonated water. Absorption heat pumps use a refrigerant with a low boiling point, such as ammonia, to be absorbed by the other medium. When water takes up the gaseous ammonia, high temperatures occur.

The absorption heat pump's reboiler brings the ammonia to boil at low temperatures and pressures by means of low-temperature heat, as in a compression heat pump. Water, the solvent, absorbs the gaseous refrigerant in the absorber. The resulting high temperatures can be passed on as useful heat via a heat exchanger. The solvent pump passes the solution on to the compressor. Because the solvent pump does not require the same high pressure as in a compression heat pump, less electricity is needed. The separator then takes advantage of the different boiling points of the refrigerant and solvent – here, ammonia and water – to break down the solution into these two components. In this process, thermal energy

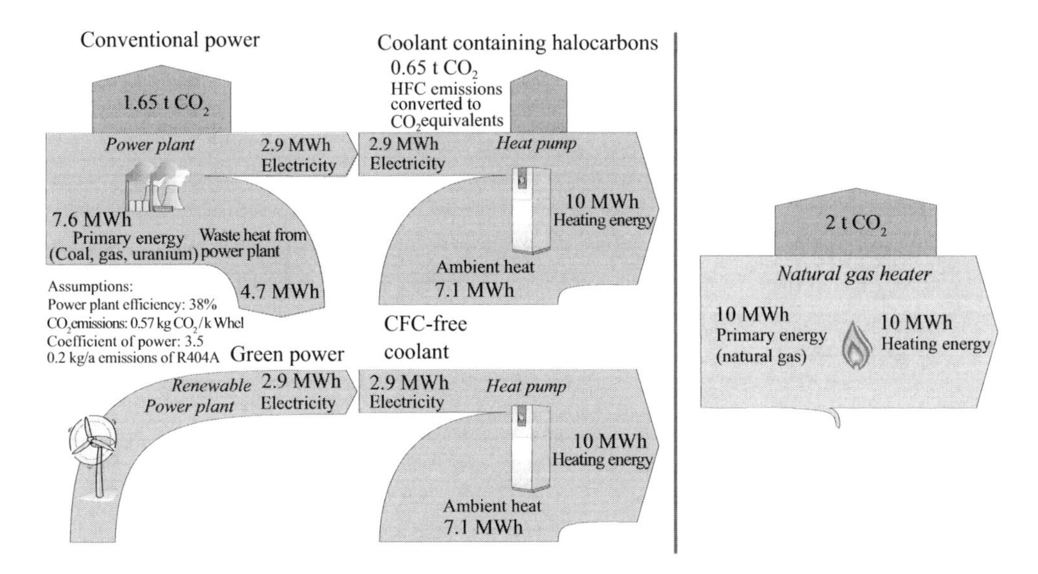

Figure 8.11 The environmental impact of two heating systems with heat pumps compared with a natural gas heater

is consumed. This heat can come from the combustion of natural gas or biogas, but solar thermal and geothermal heat also suffice. Within the condenser, the now separated gaseous refrigerant condenses under high pressure and high temperatures. The condensation heat can also be used as useful heat. The fluid refrigerant passes through an expansion valve to return to the reboiler, and the solvent returns to the absorber, thereby completing the cycle. Figure 8.12 shows how an absorption heat pump works.

Absorption heat pumps are used, for instance, in large units above 50 kilowatts and in camping refrigerators powered by propane. Tests are still being conducted for the use of absorption heat pumps in solar cooling systems. Because the demand for cooling and the availability of solar thermal energy usually coincide, absorption heat pumps represent a good way of using renewable energy in the energy-intensive field of cooling buildings. On the other hand, ammonia is toxic and flammable. Furthermore, ammonia is very corrosive in contact with copper, so that pipes must be made of steel. If there are leaks, ammonia can be soaked up with water, but safety measures have to be taken to prevent injuries and damage.

Adsorption heat pumps

Adsorption heat pumps have solids – activated carbon, silica gel, zeolites, etc. – that adsorb water vapour. During the adsorption process, when a solid takes up water vapour, a strong exothermal reaction takes place at temperatures around 300 °C. Because water is the working medium, a strong vacuum is required so that low-temperature heat can be used. At a pressure of 8.7 mbar, water already turns into steam at 5 °C; at 6.1 mbar, at 0 °C.

An adsorption heat pump works in two phases. A heat exchanger coated with a solid, such as zeolite, is used for this purpose. In the first phase, a gas burner releases the water bound in the solid at high temperatures. The water condenses in a second heat exchanger, which draws from the useful heat at high temperatures. When nearly all of the water has been used up, the second phase begins. Now, the burner is switched off. Low-temperature heat is input via a second heat exchanger. In the process, the water turns back into steam

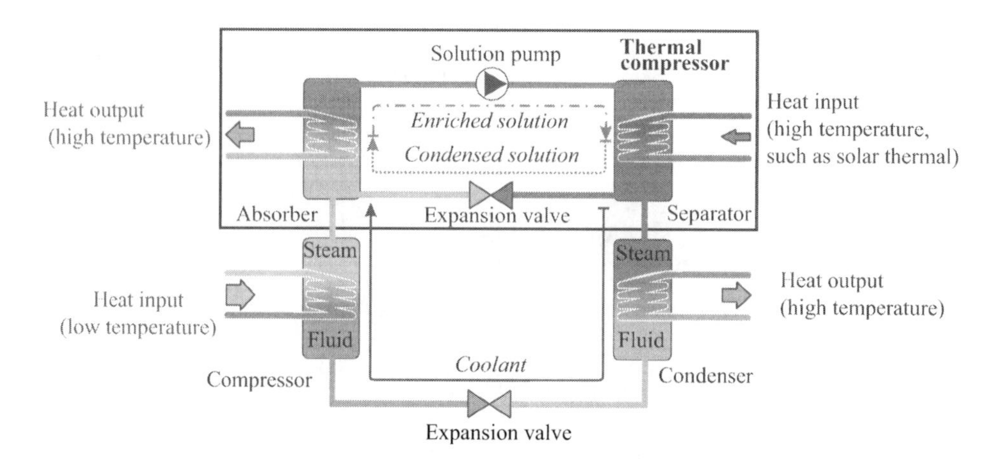

Figure 8.12 How an absorption heat pump works

in the vacuum described above. This heat exchanger coated with zeolite adsorbs this water vapour, producing a high temperature that the heat exchanger can draw off as useful heat.

One benefit of adsorption heat pumps is that no toxic materials are used. The heat needed in the process can come from renewable heat, such as solar thermal, geothermal, or biogas. Most commercial systems, however, currently still run on natural gas.

Applications, planning, and coefficient of performance

A heat pump requires a low-temperature heat source. The greater the temperature of the heat source, the more efficient the heat pump is.

A distinction is made between the following sources of heat:

- ground water and surface water (water/water)
- soil, underground heat exchanger/collector (brine/water)
- soil, boreholes (brine/water)
- ambient air (air/water)
- waste heat, such as from industrial processes (air/water or water/water).

Depending on the heat source and the way heat is used, heat pumps can be divided into air/air, air/water, brine/water, and water/water systems. The first part of each dichotomy is the heat medium input into the process. When ambient air is used, it is air. When constantly frost-free groundwater is used, it is water. When frost is an issue, the lines running underground generally contain a mixture of water and antifreeze, called brine.

The second part of the dichotomy is the heat medium output. In most cases, heat pumps heat up water for heating or service water. In rare cases, heated air is used in air heating systems.

The higher the temperature of the heat source and the lower the temperature required for heating energy, the less electricity is needed to drive the heat pump. A floor heater is preferable to conventional radiators because the larger area makes the use of low heating temperatures more effective.

The heat pump's mechanical power P is lower than the current of heat given off $\dot{Q}_{out}$, which itself is the sum of the mechanical power and the input current of heat $\dot{Q}_{in}$:

$$\dot{Q}_{out} = \dot{Q}_{in} + P \tag{8.1}$$

Depending on the pressure and the working medium used, the heat pump can make use of heat far below 0 °C to produce a much higher temperature. The heat source can be ambient air, boreholes, or groundwater. The ratio of the heat current drawn off $\dot{Q}_{out}$ to the mechanical power input P is called the **coefficient of performance** (COP):

$$\varepsilon = \frac{\dot{Q}_{out}}{P} \tag{8.2}$$

The ideal COP ε_C in accordance with Carnot is determined by the lower temperature T_{in} and the greater temperature T_{out}:

$$\varepsilon_C = \frac{T_{ab}}{T_{ab} - T_{zu}} \tag{8.3}$$

This equation shows that the temperature difference should be low and the temperature of the heat source high for great COPs. In practice, COPs are quite low. The ratio of the real COP to the ideal COP

$$\eta = \frac{\varepsilon}{\varepsilon_C} \tag{8.4}$$

reflects the heat pump's quality. Typical values range from 0.4 to 0.5. Because temperature conditions fluctuate over the year, the COP does as well. The average COP is calculated for the year thus:

$$\varepsilon_a = \frac{\sum_a \dot{Q}_{out}}{\sum_a P} = \frac{Q_{out}}{W} \tag{8.5}$$

It greatly depends on the type of heat source. The average annual COP is essential for the ecological and economical operation of a heat pump. An average annual COP of 4, for instance, means that 2,500 kilowatt-hours of electricity input into a heat pump produces 10,000 kilowatt-hours of heat over the year. If the COP drops to 2, that same amount of electricity only produces 5,000 kilowatts-hours of heat.

The best systems reach a COP of around 4, but in practice the values are often lower. Table 8.2 shows average annual COPs for various heat pump types in a field test conducted in the Black Forest.

Heat pumps that get their heat from the soil perform the best. The annual COP of groundwater heat pumps is only slightly lower. The main reason is that the pump requires greater power to circulate the groundwater in the heat pumps investigated than was the case in a closed brine circuit underground. In addition, dirt accumulated in the groundwater boreholes over time, which further increased the energy required for pumping. Here, closed heat exchangers that do not directly use groundwater are more efficient. Because the ambient air temperatures are lower than the soil/groundwater temperatures in the winter, air heat pumps are the most inefficient.

The operating temperature has a considerable impact on the COP. High yields require the lowest possible feed temperatures, which is generally the case with floor heaters.

In a second phase, the field test investigated innovative heat pumps. A **CO_2 geothermal borehole heat pump** posted an annual COP exceeding 5 [Aue11]. This type of heat pump uses vertical geothermal boreholes. The boreholes contain a closed circuit of fluid carbon dioxide under high pressure, which becomes gaseous due to the heat in the soil; the heat is then passed on to the heat pump's own cooling circuit via a heat exchanger. This

Table 8.2 Typical annual COPs ε_a for electrical heat pumps for space heating [Lok07]

Heat pump	*Heat source*	*ε_a with floor heating*	*ε_a with radiators*
Brine/water	Soil	3.6	3.2
Water/water	Groundwater	3.4	3.0
Air/water	Air	3.0	2.3

principle means that no pump is needed in the CO_2 circuit, which reduces the pumping energy required considerably. Geothermal borehole heat pumps based on CO_2 are also permissible in areas with protected groundwater. They are also much more expensive than conventional heat pump systems.

Table 8.3 shows the possible heat extraction rates of vertical **geothermal boreholes**. At an annual operation of 1,800 h, a heat extraction $\dot{Q}_{in}$ of 60 W per metre of tubing would be possible in a sediment saturated with water. To extract 6 kW of heat, the tubing would have to be 100 m long.

The **geothermal collector** generally consists of plastic tubes arranged as serpentines in a yard. The optimal depth is 1.2 to 1.5 m; the optimal distance between the tubes, around 0.8 m. The length l and area A of the geothermal collector can be calculated from the required heat input $\dot{Q}_{in}$ of the low-temperature heat source and the heat extraction $\dot{q}$ relative to the length and the tube distance d_A:

$$l = \frac{\dot{Q}_{in}}{\dot{q}} \tag{8.6}$$

Table 8.3 Possible specific heat extraction for a double U-shaped geothermal borehole according to [VDI4640] for small systems up to 30 kW and tubing lengths up to 100 m

Ground	*Extraction in W/m at*	
	1,800 h/a	*2,400 h/a*
Unsuitable underground, dry sediment, λ < 1.5 W/(m K)	25	20
Normal solid rock underground and water-saturated sediment λ= 1.5 . . . 3.0 W/(m K)	60	50
Solid rock with high heat conductivity, λ >3.0 W/(m K)	84	70
Gravel, sand, dry	< 25	< 20
Gravel, sand, moist	65 to 80	55 to 65
Gravel, sand, flowing groundwater	80 to 100	80 to 100
Clay, loam, moist	35 to 50	30 to 40
Limestone, solid	55 to 70	45 to 60
Sandstone	65 to 80	55 to 65
Acidic magnetite, such as granite	65 to 85	55 to 70
Basic magnetite, such as basalt	40 to 65	35 to 55
Gneiss	70 to 85	60 to 70

Table 8.4 Possible specific heat extraction levels for geothermal collectors used at 1,800 h/a

Underground	*Extraction in W/m*
Sandy soil, dry	10
Loamy soil, dry	20
Loamy soil, moist	25
Loamy soil, saturated with water	35

and

$$A = l \cdot d_A \tag{8.7}$$

Table 8.4 shows the specific heat extraction levels for geothermal collectors relative to soil composition. For example, the required heat output $\dot{Q}_{out}$ = 10 kW and COP ε = 4 produces a required heat input of $\dot{Q}_{in}$ = 2.5 kW. In this example for dry loamy soil, the tube length is

$$l = \frac{2.5 \text{ kW}}{0.02 \text{ kW/m}} = 125 \text{ m}$$

and at a tube distance of d_A = 0.8 m the geothermal collector surface is

$$A = 125 \text{ m} \cdot 0.8 \text{ m} = 100 \text{ m}^2$$

In cases of doubt, these values can be rounded up.

9 Using biomass

Biomass availability

Biomass is any material from organisms living or dead that grow in nature, including waste products. Fossil energy sources originally started out as biomass but are generally not considered as such.

Solar energy makes plant growth – and hence life on earth – possible. The general formula

$$H_2O + CO_2 + \text{auxiliary materials} + \Delta E \longrightarrow \underbrace{C_kH_mO_n}_{\text{biomass}} + H_2O + O_2 + \text{metabolic products} \quad (9.1)$$

describes how biomass is created. The energy E from the visible spectrum of light splits water H_2O with the assistance of pigments such as chlorophyll. The hydrogen H combines with carbon dioxide CO_2 from air to produce biomass $C_kH_mO_n$. In the process, oxygen O_2 is released. The biomass can then be used in various ways as a source of energy. Usually, CO_2 is then released, but only the amount that the plant took up during growth. Biomass is a carbon-neutral renewable source of energy if the amount consumed does not exceed the amount that can grow back.

The efficiency of various plants was determined for the conversion of solar energy into biomass so that biomass production can be compared with other technical energy conversions. The average efficiency of total biomass production is around 0.14 %.

Table 9.1 shows a number of specific efficiencies for the production of biomass.

To calculate the efficiency, take the calorific value from Table 9.2 for the biomass that grows over a certain area in a certain time frame and divide it by the amount of solar energy incident on that area over that time frame.

Not all biomass can be used as a source of energy. Civilization currently uses around 4 % of annual biomass growth. Two percent is used for food and feed production, while 1 % is devoted to the production of timber, paper, and fibres. Around 1 % of biomass growth is used for energy purposes, usually as firewood, equivalent to around 10 % of global primary energy demand.

C4 plants are the most efficient at converting sunlight into biomass. Because of their fast photosynthesis, they utilize a solar energy quite effectively. The C4 plants include amaranth, millet, corn, sugar cane, and miscanthus. Under optimal conditions, they reach efficiencies of 2 to 5 %.

Table 9.1 Efficiencies in the production of biomass [Kle93]

Oceans	0.07 %	Forest	0.55 %
Freshwater	0.50 %	Corn	3.2 %
Cultivated landscapes	0.30 %	Sugarcane	4.8 %
Grasslands	0.30 %	Sugar beets	5.4 %

Table 9.2 Calorific values of various fuels from biomass [Fac96, FNR08a]

Fuel (dry)	*Calorific value Hu*	*Fuel (dry)*	*Calorific value Hu*
Straw (wheat)	17.3 MJ/kg	Sunflower seed husks	17.9 MJ/kg
Non-woody plant (wheat)	17.5 MJ/kg	Miscanthus	17.4 MJ/kg
Timber without bark	18.5 MJ/kg	Rapeseed oil	37.6 MJ/kg
Bark	19.5 MJ/kg	Ethanol	26.7 MJ/kg
Timber with bark	18.7 MJ/kg	Methanol	19.7 MJ/kg
Olive pits	18.0 MJ/kg	Gasoline (for comparison)	43.9 MJ/kg

When biomass is used, a distinction is made between residue from forest trees and agriculture, on the one hand, and dedicated energy crops, on the other. For Germany, studies have found a total potential of around 1,200 PJ/a (Table 9.3), equivalent to nearly 9 % of the country's primary energy consumption in 2011. Even if comprehensive energy conservation measures are implemented, Germany could only get a fraction of its energy from biomass.

There are many ways to use biomass (Figure 9.1). The greatest potential comes from the use of timber and timber products. Residual material from agriculture and forestry are also important energy sources along with biowaste. In addition to the use of residues, special energy crops can be planted. Because energy crops compete with food crops for land, however, extensive energy crops have raised controversy.

The next step is to process the biomass resources listed above. For instance, they can be dried, pressed, fermented to produce alcohol, converted into biogas, turned into pellets, and converted into fuels in chemical plants. The goal of these processing options is to produce biofuels that are easy to consume.

Table 9.3 Biomass potential in Germany [Kal03]

	Useful amount Mt	*Energy potential PJ/a*
Straw and stalk biomass	10–11	140–150
Timber and timber residue	38–40	590–622
Biogas substrates (waste and residue)	20–22	148–180
Gas from sewage plants and landfills	2	22–24
Energy crop mix	22	298
Sum of biomass potential	92–97	1,198–1,274

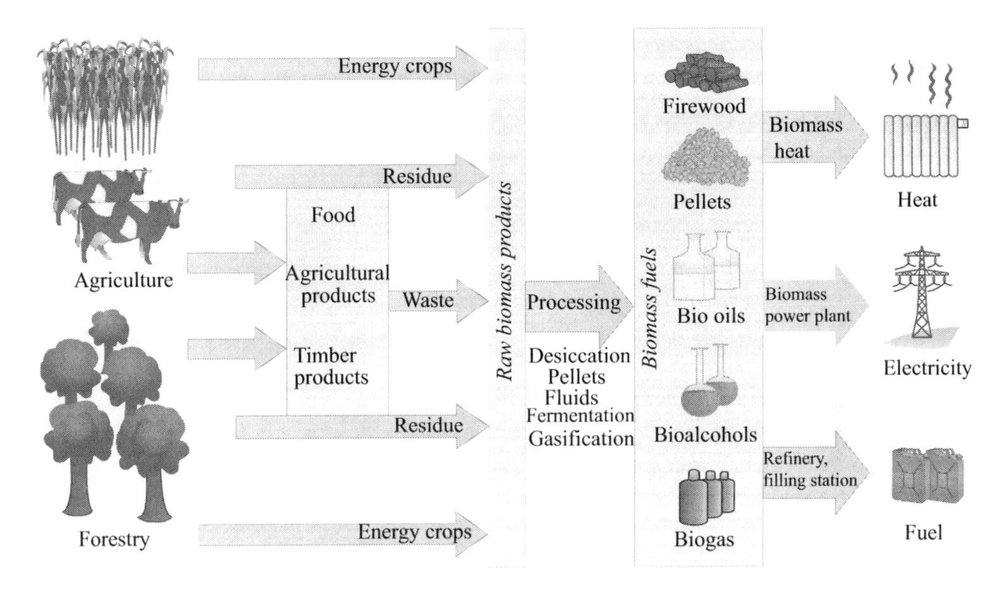

Figure 9.1 Ways of using biomass [Qua09]

The biomass energy sources produced in this way can be used to substitute for coal, petroleum, and natural gas. Biomass power plants can use these fuels to generate electricity, while biomass heat plants provide heat; likewise, biofuels can be used in vehicles.

A distinction is made between the following bioenergy sources:

- solids
- liquids
- gas.

Solid bioenergy

The main examples of solid bioenergy are timber and timber products, solid biowaste, straw, and energy crops. There is a wide range of timber products. In addition to timber boards, bark, and sawmill waste, there are logs, firewood, wood chips, briquettes, and wood pellets.

The calorific value H_i of timber mainly depends on its **water content** w. If the timber contains absolutely no water, $w = 0$. The calorific value $H_{i,dry}$ of waterless timber can be roughly calculated as follows:

$$H_i = H_{i,dry}(1-w) - 2{,}44\ \frac{\text{MJ}}{\text{kg}} \cdot w \tag{9.2}$$

The water content

$$w = \frac{m_{H_2O}}{m_{timber}} \tag{9.3}$$

Figure 9.2 Different ways of processing firewood; from the top left to the bottom right: logs, firewood, briquettes, and pellets

is defined relative to the ratio of the water's mass m_{H_2O} to the entire mass of the timber m_{timber}. The **moisture content of the timber** u is calculated from the water's mass m_{H_2O} within the timber to the mass $m_{timber,dry}$ of absolutely dry timber:

$$u = \frac{m_{H_2O}}{m_{timber,dry}} \tag{9.4}$$

Fresh woody biomass is generally between 40 % and 60 % water, a figure that increases to as much as 80 % for perennial plants. After up to two years of drying in fresh air, the water content drops to 12 to 20 %. An average value can be set at 15 %. This is called air-dried biomass.

The water content of bioenergy products that are dried in special systems, such as wood pellets and briquettes, is below 10 %. When the water content reaches 0 %, we speak of absolutely dry biomass. Such a condition is mainly useful for comparisons, not in practice. Figure 9.3 shows the calorific value of timber relative to its water and moisture content.

The calorific value of different types of timber relative to their mass is generally similar at the same degree of dryness. But because different types of timber have different densities, there are clear distinctions in the calorific value relative to volume. Table 9.4 shows the properties of different types of firewood.

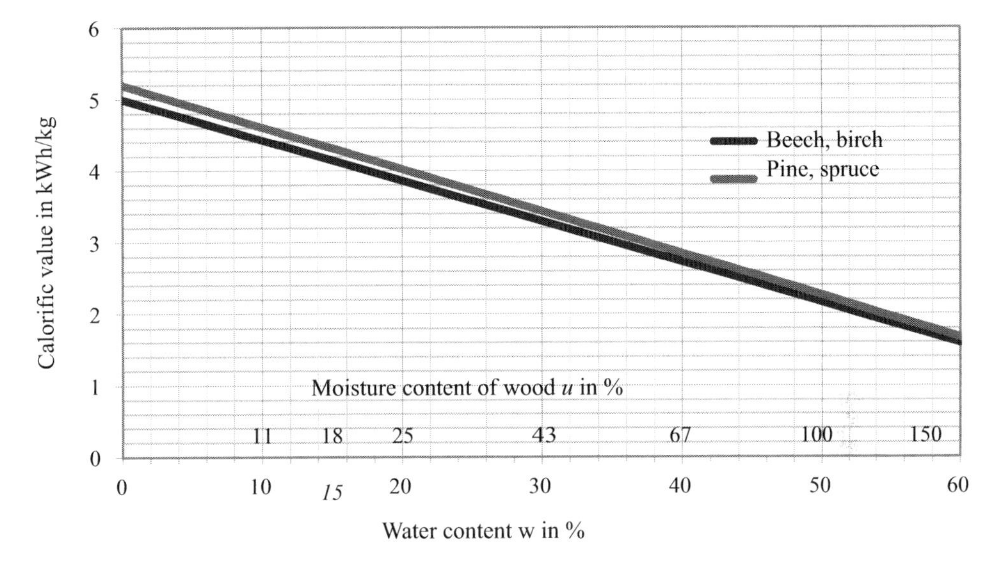

Figure 9.3 Calorific value of timber relative to moisture and water content

Table 9.4 Properties of different types of timber

	Calorific value(w = 0 %)	*Density(bone dry)*	*Calorific value Hi at w = 15 % (air-dried)*		
	Hi0 in kWh/kg	*kg/Fm*	*kWh/kg*	*kWh/Fm*	*kWh/Rm*
Beech	5.0	558	4.15	2,720	1,910
Birch	5.0	526	4.15	2,570	1,800
Pine	5.2	431	4.32	2,190	1,530
Spruce	5.2	379	4.32	1,930	1,350

Table 9.5 Conversion factors for cubic measurements of timber products (approximations) [DGS04]

	fm	*Stere*	*srm*
Solid cubic metre (scm)	1	1.43	2.43
Stere	0.7	1	1.7
Loose cubic metre (lcm)	0.41	0.59	1

The calorific value relative to volume can even differ for a single type of timber depending on the timber product. A distinction is made between a cubic metre of solid wood (scm), a stere equivalent to a cubic metre, and a loose cubic metre (lcm). The solid metre does not include air between logs, whereas the loose cubic metres/stere do; likewise, the loose cubic metre applies for pellets, wood chips, and sawdust. Table 9.5 shows conversion factors for these measurements. They can also vary considerably because timber is a natural product.

Table 9.6 Specifications of wood pellets in accordance with EN 14961–2 / ENplus-A1

Diameter *d*	6 or 8 ±1 mm	Calorific value	16.5–19 MJ/kg
Length *l*	3.15–40 mm	Ash softening point	≥ 1,200 °C
Fine particles < 3.15 mm	≤ 1 %	Mechanical stability	≥ 97.5 %
Loose cubic metre density	≥ 600 kg/m³	Sulphur	≤ 0.05 %
Water content	≤ 10 %	Nitrogen	≤ 0.3 %
Ash content	≤ 0.7 %	Chlorine	≤ 0.02 %

One goal in using biomass as an energy source is to consume biofuels the way fossil fuels are used in fully automated systems. Because wood chips can block up conveyor systems, standardized **wood pellets** are becoming increasingly important. Such pellets are made of sawdust and waste products from sawmills.

Table 9.6 provides an overview of the properties of wood pellets in accordance with EN 14961–2 [DIN11]. The standard specifies quality classes A1, A2, and B. Wood pellets are also certified according to the standard ENplus to ensure a certain quality level. Pellets are produced without any additional adhesive agent. The lignin in wood holds the pellets together after they are compressed. In the process, dry sawdust and wood chips are pressed into a particular shape and length. The energy required for the production and processing of pellets from dry industry wood is relatively low at 2.7 % of energy content [Abs04].

Woodchips vary in size more than wood pellets do because of the production process. Table 9.7 shows classifications based on size in accordance with applicable standards. There are also classes for water and ash content. The letter W or M indicates the classification for water content followed by a number for the maximum. For the storage of suitable wood chips with a water content less than 30 %, class M30 suffices.

Fluid bioenergy

Fluid bioenergy is mainly used as a substitute for fossil fuels, such as heating oil, diesel, and gasoline. The following bioenergy sources are used:

- vegetable oil
- fatty acid methyl ester or biodiesel
- bio alcohols, such as bioethanol
- BtL fuels.

Table 9.7 Specifications for the distribution of particle sizes for wood chips in accordance with ÖNORM M7133 and CEN/TS 14961

ÖNORM M17133 class	*CEN/TS 14961 class*	*Main share > 80 % of mass*	*Fine particles < 5 % of mass*	*Course particles maximum length (max. 1 %)*
G30	P16	3.15 ≤ P ≤ 16 mm	< 1 mm	> 45 mm, all < 85 mm
G50	P45	3.15 ≤ P ≤ 45 mm	< 1 mm	> 63 mm
G100	P63	3.15 ≤ P ≤ 63 mm	< 1 mm	> 100 mm
	P100	3.15 ≤ P ≤ 100 mm	< 1 mm	> 200 mm

Table 9.8 Properties of biofuels compared with conventional fuels [FNR08a]

	Density	*Calorific value*	*Calorific value*	*Calorific value*	*Viscosity at 20°*	*Flash point*
	kg/l	*MJ/kg*	*MJ/l*	*kWh/l*	*mm²/s*	*°C*
Diesel	0.83	43.1	35.9	10.0	5.0	80
Gasoline	0.74	43.9	32.5	9.0	0.6	<21
Rapeseed oil	0.92	37.6	34.6	9.6	74.0	317
RME (biodiesel)	0.88	37.1	32.7	9.1	7.5	120
Bioethanol	0.79	26.7	21.1	5.9	1.5	<21
BtL	0.76	43.9	33.5	9.3	4.0	88

Table 9.8 shows the properties of various biofuels compared to conventional products from petroleum. Clearly, the calorific value of biofuels varies greatly. Pure vegetable oil has the greatest calorific value in terms of volume. Its value is only slightly below the level of conventional diesel.

Vegetable oil

The easiest biofuel to produce is **vegetable oil**. More than 1,000 different plant oils can be used for this purpose. The ones most commonly used are rapeseed oil, soybean oil, and palm oil. Oil mills produce the vegetable oil directly by pressing or extracting. The press cake can be used as animal feed.

But only a few older indirect-injection (prechamber) diesel motors can run properly on vegetable oil without further modifications. Vegetable oil is somewhat more viscous than diesel, and its flashpoint is higher. If upgraded, however, normal diesel engines can run directly on vegetable oil.

Developed by Ludwig Elsbett in the 1970s, the Elsbett engine can run directly on cold-pressed vegetable oils, such as sunflower oil, olive oil, and rapeseed oil. This engine resembles a diesel engine except for a few aspects. For instance, the nozzle is opened and closed during the injection process to prevent it from blocking up.

Biodiesel

Biodiesel has properties much closer to those of conventional diesel than untreated vegetable oils do. But the raw material for biodiesel is vegetable oil or animal fats. Back in 1937, a Belgian engineer named Chavanne applied for a patent for the production of biodiesel. In chemical terms, biodiesel is fatty acid methyl ester (FAME).

In northern Europe, rapeseed is generally used to produce biodiesel. Oil mills are used to make oil from rapeseed. The press cake generally is used as animal feed. A biodiesel production unit turns the rapeseed oil into rapeseed methyl ester (RME).

To produce RME, rapeseed oil and methanol (CH_3OH) are mixed with a catalyst, such as sodium hydroxide and water, at temperatures around 50 to 60 °C in a reaction vat. Rapeseed methyl ester and glycerin ($C_3H_8O_3$) are then produced. The formula for the reaction is as follows:

$$\text{Rapeseed oil} + CH_3OH \xrightarrow{\text{Catalyst}} C_3H_8O_3 + \text{RME (Biodiesel)} \tag{9.5}$$

Biodiesel can be used as a substitute for diesel from petroleum. Numerous filling stations in Germany offer biodiesel. The engine should be approved for the use of biodiesel, however. If it is not, there is a risk that the biodiesel will attack tubes and seals, thereby damaging the motor. Small amounts of biodiesel, however, can be mixed in with conventional diesel without the need for approval. In Germany, biodiesel is blended into all diesel fuels. In 2007, the share of biodiesel in all fuels was 5.6 %.

Because large rapeseed monocultures are required, and these plantations compete with land that could be used for food production, the positive environmental payback of biodiesel is nonetheless controversial.

Bioalcohols

Gasoline engines can also run on biofuels. For this purpose, **bioalcohols** such as ethanol are used. Bioethanol is made from sugar, glucose, or starch in combination with cellulose. These substances can be gained from sugar beets, sugarcane, grain, corn, and potatoes (Table 9.9). Sugar can be directly fermented to produce alcohol. Starch and cellulose, in contrast, first have to be split.

Glucose ($C_6H_{12}O_6$) can be fermented with yeast to produce ethanol (CH_3CH_2OH) in an anaerobic process:

$$C_6H_{12}O_6 \xrightarrow{\text{Fermentation}} 2\ CH_3CH_2OH\ + 2\ CO_2 \tag{9.6}$$

One waste product from this reaction is carbon dioxide (CO_2). Because plants bind carbon dioxide during growth, however, this reaction is practically carbon-neutral. The result of this fermentation is a mash with an ethanol content of around 12 %. This raw alcohol can be distilled to produce a concentration exceeding 90 %. Dehydrogenation using molecular sieves finally produces ethanol with a high degree of purity. The residual materials from the production of ethanol can be used as animal feed. Unfortunately, a lot of energy is needed to produce alcohol. If the energy comes from fossil fuels, the climate balance of bioethanol is not impressive. In extreme cases, it can even be negative.

Bioethanol can easily be blended with gasoline. An E number shows the ratio. E85 means that 85 % of the fuel is bioethanol and 15 % gasoline. In Germany, bioethanol is blended with gasoline in small amounts. Up to a share of 5 % ethanol, blending causes no problem.

Table 9.9 The yield of raw materials used to produce bioethanol [FNR08a]

	Fresh mass	*Required biomass*	*Fuel yield*	*Diesel equivalent*
	t/ha	*kg/l*	*l/ha*	*l/ha*
Corn	9.2	2.6	3,520	2,290
Wheat	7.2	2.6	2,760	1,790
Rye	4.9	2.4	2,030	1,320
Potatoes	43.0	12.1	3,550	2,310
Sugar beets	58.0	9.3	6,240	4,050
Sugarcane	73.8	11.4	6,460	4,200

Normal gasoline engines can generally run on up to 10 % ethanol without modifications. If the share of ethanol is greater, however, the engines have to be modified.

Since the 1970s, bioalcohols from sugarcane have been used on a large scale as vehicle fuels to substitute expensive fossil fuels in Brazil. Today, bioalcohol is available in practically every filling station in Brazil. 'Flexible fuel vehicles' are also widespread in the country. Such cars can run on a wide range of fuel blends with the share of ethanol ranging from 0 to 85 %. In the past few years, a number of systems for the production of bioethanol from rye, corn, and sugar beets have also been built in Germany. Recently, food prices have skyrocketed, worsening the affordability of bioethanol production considerably. Morally, bioethanol is also controversial because it competes with food crops.

Biomass-to-Liquid (BtL) fuels

When vegetable oil, biodiesel, or bioethanol is used, the only plant parts that can be used to produce fuel are those containing oil, sugar, or starch. The second generation of biofuels is to tap a wider range of resources. BtL (Biomass-to-Liquid) is the name for the synthetic production of biofuels. A wide range of raw materials can be used completely for this purpose, such as straw, biowaste, forest waste, and special energy crops. As a result, the potential of biofuels increases tremendously, as does the amount of land that could be used.

It is relatively complex, however, to produce BtL fuels. In the first stage, the biomass raw material is gasified. Oxygen and water vapour are added at high temperatures to produce a synthetic gas consisting of carbon monoxide (CO) and hydrogen (H_2). Across various gas purification stages, the carbon dioxide (CO_2) is separated from other undesirable elements, such as sulphur and nitrogen compounds. The synthetic gas is then converted into fluid hydrocarbons.

The best known process for this purpose was developed in 1925 and is called the Fischer–Tropsch process, which can be described as follows:

$$\mathrm{nCO} + (2\mathrm{n}+1)\mathrm{H}_2 \rightarrow \mathrm{C}_\mathrm{n}\mathrm{H}_{2\mathrm{n}+2} + \mathrm{nH}_2\mathrm{O} \tag{9.7}$$

Named after its developers Franz Fischer and Hans Tropsch, the process takes place at pressures around 30 bar and temperatures above 200 °C with the assistance of catalysts. In the Second World War, Germany lacked oil and began to make fluid fuels from coal. Another process makes methanol out of synthetic gas for further processing to produce fuels.

In the last processing stage, the fluid hydrocarbons are separated and refined into various fuel products. Figure 9.4 shows what the production of BtL fuels looks like.

BtL fuels have not yet reached full serial production. At the moment, a number of firms are building prototype production units for synthetic fuels. The main advantage of BtL fuels is that they can replace conventional fuels without any engine modifications. But the complex production processes for BtL fuels are relatively expensive.

Gaseous bioenergy

A wide range of raw biomass materials is used to produce gaseous bioenergy. These materials can be specially grown for the production of biogas. Corn and grass silage used as animal feed provides an especially high yield (Table 9.10). Silage is produced by storing the plant

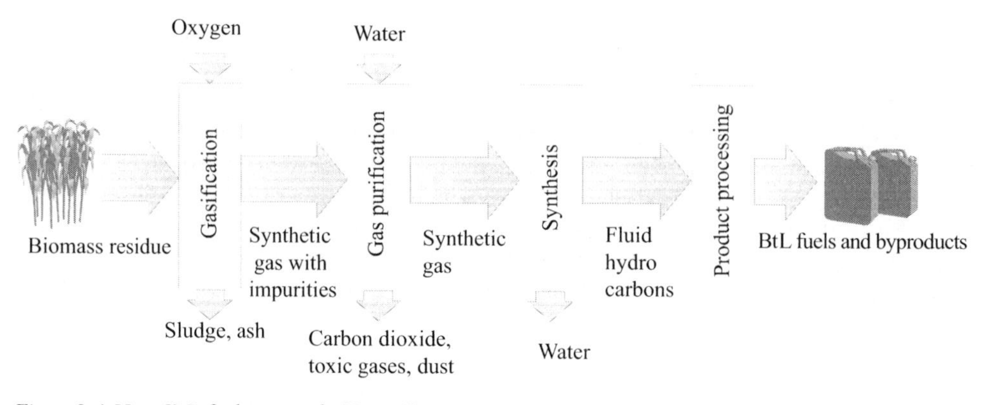

Figure 9.4 How BtL fuels are made [Qua09]

Table 9.10 The biogas yield and methane content of various raw biomass materials [FNR08b]

	Biogas yield	*Methane content*		*Biogas yield*	*Methane content*
	m³/t	*%*		*m³/t*	*%*
Corn silage	202	52	Beet leaves	70	54
Grass silage	172	54	Press cake	67	72
Whole rye silage	163	52	Pig manure	60	60
Beets	111	51	Cow manure	45	60
Sorghum bicolor	108	54	Grain residue	40	61
Biowaste	100	61	Pig slurry	28	65
Chicken manure	80	60	Cow slurry	25	60

material in an airtight, sealed space. But sludge from water purification plants can also be used, as can manure, waste from food and feed production, used fats, and other biowaste.

The raw biomass derails are fermented in an oxygen-free environment by bacteria. The result is flammable **biogas**. It consists largely of methane (CH_4) and carbon dioxide, but also small amounts of hydrogen sulphide (H_2S), ammonia (NH_3) and other trace gases. The methane content ranges from 40 to 75 %. The hydrogen sulphide content makes biogas smell like rotten eggs. It has a density of 1.2 kg/m³ and a flashpoint of 700 °C; its calorific value H_i ranges from 14.4 to 27 MJ/kg depending on the methane content. The calorific value of pure methane is 35.9 MJ/kg.

Biogas can be used to provide heat and electricity. If it is used in cogeneration units, it produces both at the same time. Table 9.11 shows the possible yields of energy crops for biogas production across 200 ha of farmland in Germany.

To reduce transport mileage, it makes sense to build biogas units near the source of the substrate, such as water purification systems and farms. Process heat is produced during biogas production, and it can be used. Biogas production also provides additional benefits, such as the better tolerability of fermented residues and less smell.

There are great hopes for the production of **synthetic gas** from biomass. In an initial pyrolysis process, timber and other dry biomass are turned into biocoke at temperatures

Table 9.11 The potential of various energy crops on 200 ha of farmland [FNR08b]

	Harvest	*Biomass yield*	*Cogen capacity*	*Power yield*
	t_{FM}	*1000 m³*	kW_{el}	MWh_{el}/a
Corn silage	9,000	1,600	360	2,880
Sudangrass	11,000	1,240	300	2,400
Grass silage	7,200	1,090	260	2,080
Whole rye silage	5,200	746	170	1,360

up to 500 °C and then cracked into gas containing tar. These products then move into a second gasification phase in which organic chemical compounds are converted into a synthetic gas consisting of carbon monoxide and hydrogen at temperatures up to 1,600 °C. Synthetic gas is attractive because a wide range of raw materials can be used, from timber to waste. Because a relatively large amount of energy is needed for its production and the conversion of the raw material in the final product is somewhat inefficient, the process has a bigger environmental impact than the direct thermal use of biomass does.

Yield per hectare and environmental balance

As mentioned above, the environmental impact of biomass fuels is controversial. In principle, the use of biomass is carbon-neutral because the biomass only releases the amount of carbon dioxide that it previously bound during growth. Often, however, fossil energy is used in further processing, which releases additional carbon dioxide (Figure 9.5).

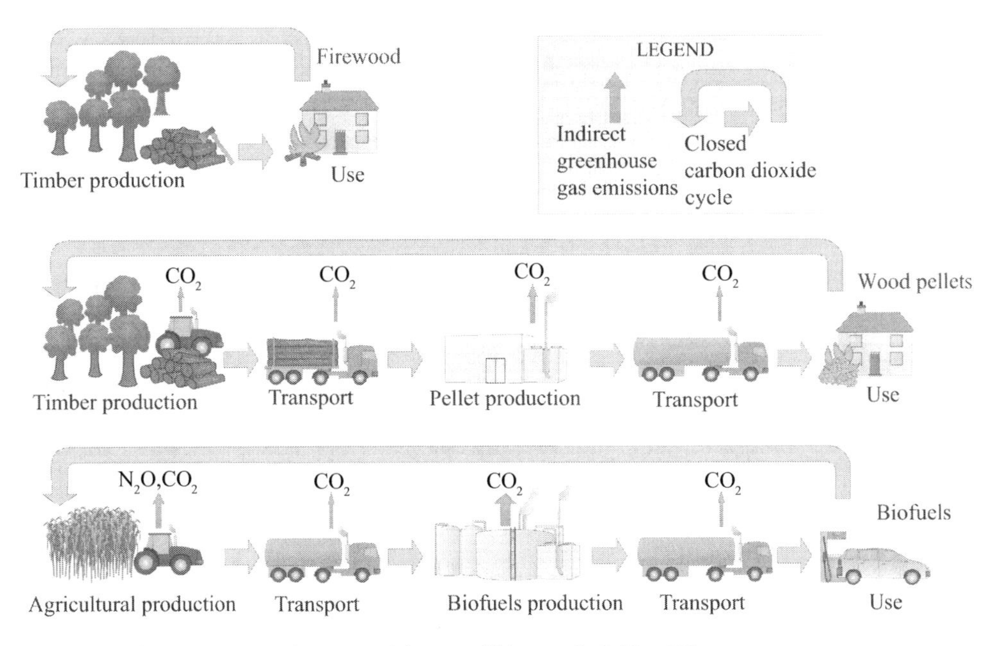

Figure 9.5 The environmental impact of the use of biomass fuels [Qua09]

Table 9.12 Gross biofuel yield on farmland [FNR08a]

Biofuel	*Feedstock*	*Yield*	*Fuel yield*	*Equivalent diesel / gasoline*	*Car range* [1]
		t/ha	*l/ha*	*l/ha*	*km/ha*
Rapeseed oil	Rapeseed	3.4	1,480	1,420	23,300
RME (biodiesel)	Rapeseed	3.4	1,550	1,410	23,300
Bioethanol	Grain	6.6	2,560	1,660	22,400
BtL	Energy crops	20	4,030	3,910	64,000
Biomethanol	Corn silage	45	3,540	4,950	67,600

[1] Car fuel consumption: gasoline 7.4 l/100 km, diesel: 6.1 l/100 km, ha = 1000 m^2 = 2.47 acres

In addition to carbon dioxide, other pollutants are also emitted when biomass is combusted, such as nitrous oxide and particulate matter. But the same kind of filters used to prevent such pollution from fossil energy can also be used with biomass.

Another critical point in the use of biomass as a source of energy is competition with food production and limited land availability. Table 9.12 shows the yields of various biofuels per hectare (ha) of farmland. The energy consumed during fuel production is not included in these numbers. If, for instance, waste products are used to produce BtL fuels or biogas, competition with food crops is practically not an issue.

The following example shows that biofuels are only a limited alternative. Germany has some 12 million hectares of farmland. If all this land was used to grow rapeseed for the production of biodiesel, around 16.9 million litres of diesel (gross) could be produced. If we subtract the energy required during production, we are left with around 12.6 million litres of biodiesel net. In 2012, however, Germany consumed some 35 billion litres of diesel along with 26.1 million litres of gasoline. In other words, all of the farmland in Germany does not nearly suffice to produce enough biofuels for our current consumption.

It is therefore probable that electric vehicles will replace internal combustion engines within the foreseeable future. Electric cars generally consume around 25 kWh/100 km. If 550 billion car-kilometres per year were switched to electric, some 140 TWh of power would be needed.

If photovoltaic systems with proper orientation and an average efficiency of 15 % were used to provide 140 TWh at an average annual irradiation of 1150 kWh/m^2, a mere 0.2 million hectares would be needed if the systems covered a third of the area.

Biofuels will then mainly be used where electric mobility is not a good alternative in the midterm.

Biomass units

Biomass heaters

Biomass has been used as a source of heat for centuries; people burned wood, straw and animal waste. In numerous developing countries, biomass fuels are still among the most important energy sources today. The biomass is used to provide heat from simple open fires, especially for cooking.

For centuries, open fireplaces and stoves have been used for space heating in temperate zones. They were originally fired with wood, but later coal was increasingly added. Today, the renaissance of biomass as a fuel in modern heating systems has begun. The different types of heaters include:

- open fireplaces
- closed fireplaces
- fireplace stoves
- storage furnaces
- pellet heaters
- firewood heaters
- wood chip heaters
- biomass heat plants.

In addition, special bioenergy carriers can also be used in conventional heating systems. Gas heating systems can run on gaseous bioenergy, while oil heaters run on liquids. The fuel can be used with slightly greater efficiency in cogeneration units, which produce both electricity and heat simultaneously.

Modern boilers fired with timber can run automatically or semi-automatically. Figure 9.6 shows a boiler running on firewood. Commercial heating systems running on firewood, wood chips, or pellets are available at various sizes to replace conventional heating

Figure 9.6 Firewood boiler

(Source: © Bosch Thermotechnik GmbH)

systems fired with fossil fuel. **Pellet heaters** are specially made to heat single-family homes and are as easy to use as conventional heating systems are.

A conveyor system or a suction system draws the pellets into the fire. The fire is started and maintained automatically. An electric hot air blower starts the fire. Roughly 70 Wh of electricity are needed. To keep restarts to a minimum and maintain the boiler at its nominal load as much as possible, a buffer storage tank is recommended.

Pellet boilers reach boiler efficiencies exceeding 90 % in nominal operation. Under partial load and when the fire is started, however, the efficiency is much lower. The ash from the pellets consumed is collected in a basin. Because wood pellets have a low ash content, the basin does not need to be emptied out frequently. It is also a good idea to connect a solar thermal system additionally. The wood pellet boiler can then be taken out of operation completely in the summer. Figure 9.7 shows a pellet heating system.

If the pellet storage area is large enough, pellets can be bought for an entire year when the price is low. If the demand for heating energy Q_{heat} and the average boiler efficiency over the year η_{boiler} are known, the calorific value of the pellets H_i = 5 kWh/kg and the loose cubic density ρ_{pellets} = 650 kg/m³ the volume of the storage space

$$V_{\text{storage}} = \frac{Q_{\text{heat}}}{f \cdot \eta_{\text{boiler}} \cdot \rho_{\text{pellets}} \cdot H_i} \tag{9.8}$$

can be calculated. The usage factor f must be taken into account here because only 65 to 75 % of the volume can be used for storage because of the height of the opening and the slanted floor. At a heating demand of Q_{heat} = 10,000 kWh, the factor f = 0.7 and an average boiler efficiency of η_{boiler} = 0.8 would require a storage space of V_{storage} = 5.5 m³.

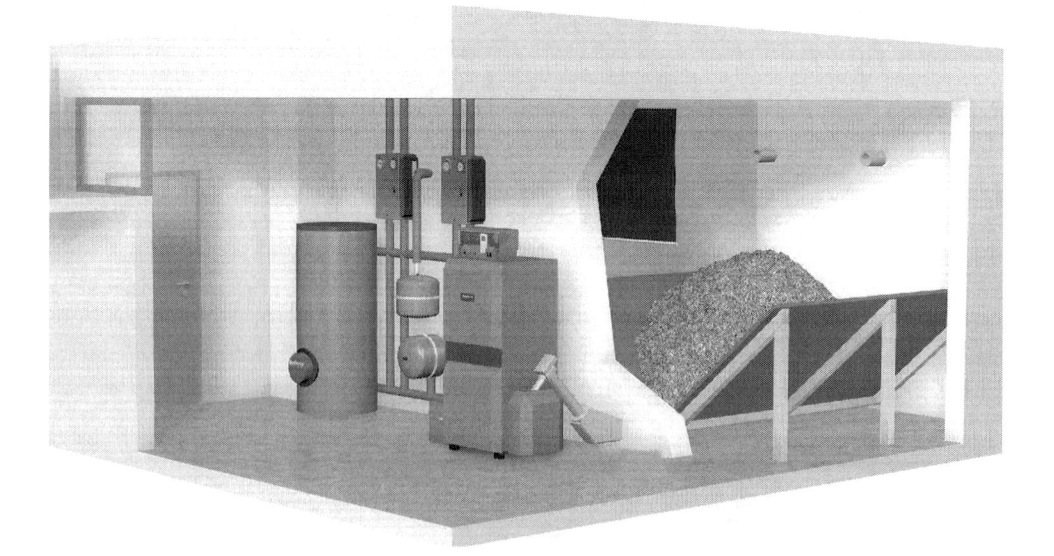

Figure 9.7 Pellet heating system and storage

(Source: © Bosch Thermotechnik GmbH)

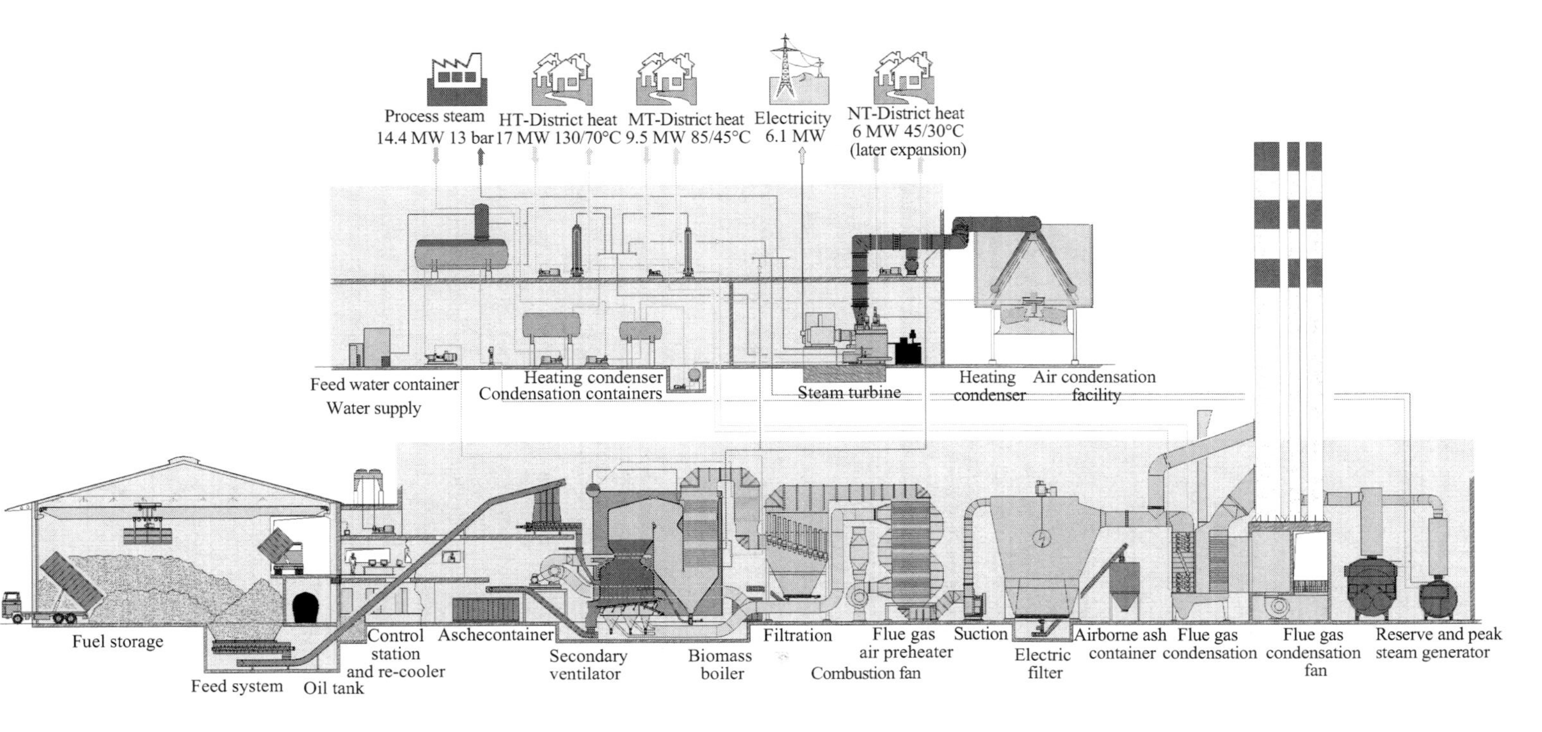

Figure 9.8 The Pfaffenhofen biomass heat and power plant

(Source: Kraftanlagen München GmbH)

But biomass can also be used in large heat plants, not only in domestic units. Heat plants can meet the demand for heat at industrial plants and supply heat to entire urban districts through district heat networks. Ideally, such heat plants would be set up relatively close to where the biomass fuel is produced. If this is done, the supply of raw material will be secure and inexpensive; transport miles are also reduced, thereby minimizing the environmental impact.

Biomass heat plants

In addition to being used as a source of heat, biomass can also be used to produce electricity. Here, the same processes are used as those mentioned before, such as the Clausius–Rankine process (see Chapter 4), for other thermal power plants. Recent policies, such as Germany's **Renewable Energy Act** (EEG), have brought about a wide range of new biomass power plants recently. Figure 9.8 shows the example of the Pfaffenhofen power plant running on biomass based on a design by eta Energieberatung. The power plant went into operation in the summer of 2001. The total investment amount for the plant, the district steam network, and the district heat network was around €45 million. The biomass boiler has an efficiency of 87.2 % without and 95.4 % with flue gas condensation had an output of 23.5 / 26.7 MW. 10,620 kg/h of biomass is consumed. Untreated wood chips and sawmill waste with the water content up to 45 % is used to fire the boiler. The system provides process steam for a nearby industrial plant and produces heat for the district heat network. A condensation turbine runs on steam from the boiler at 450 °C and 60 bar to power a generator. The generator has a maximum output of 6.1 MW. Each year, the biomass heat and power plant produces some 200,000 MWh of heat and 40,000 MWh of electricity for the local grid. More than 60,000 t of CO_2 is offset in the process.

In a fleet of power plants that will run on exclusively renewable sources over the long term, biomass power plants will also be needed to compensate for fluctuations in power from other renewable power plants, such as wind farms and photovoltaic systems. Biomass power plants will therefore play a crucial role in power supply.

10 Hydrogen production, fuel cells, and methanation

Hydrogen and fuel cells have a very positive image among the general public. Large corporations therefore invest tremendous sums of money in the development of fuel cells for the transport and domestic sectors. It is unclear, however, what fuel these fuel cells are to run on. There is practically no renewable hydrogen currently available. And it seems improbable that there will be any at competitive prices in the next few years. The first commercial fuel cells therefore run on natural gas. Unfortunately, such systems do not represent a major ecological advance over other fossil energy systems. Only when hydrogen comes from renewable sources in significant quantities and at competitive prices – and such systems will be marketable in the long term – will the environmental impact be much lower.

Producing and storing hydrogen

Pure hydrogen does not occur in the natural world in significant amounts; rather, it has to be made from other resources. Usually, water (H_2O) or hydrocarbons, such as natural gas, are the starting point. While hydrogen has a relatively high calorific value (Table 10.1), under normal pressure it also has a very low density.

Figure 10.1 shows various ways of producing hydrogen. Industrial processes for the production of hydrogen from hydrocarbons, such as **steam reformation** and **partial oxidization** chemically separate carbon. These processes almost exclusively start with natural gas, petroleum, or coal as the raw material. The carbon then reacts to produce carbon monoxide (CO), which can be used to produce energy. The final product is then carbon dioxide (CO_2). For active climate protection, these methods of producing hydrogen are not a real alternative.

The **Kværner process** also starts with hydrocarbons. But the waste product is pure activated carbon. If this carbon is not further combusted, the direct production of carbon dioxide can be avoided in this method.

Generally, all of these ways of producing hydrogen from fossil energy require high process temperatures. To reach them, they consume large amounts of energy. If this energy comes from fossil sources, further carbon dioxide emissions are the result. In terms of the climate, it is usually a better idea to burn natural gas and petroleum directly than to make hydrogen out of them in a complex process simply to produce an ostensibly 'environmentally friendly' product.

Other processes are needed to produce climate-friendly hydrogen. One option is electrolysis. In **alkaline electrolysis**, electricity is used to split water by means of two electrodes submerged in a watery alkaline electrolyte (Figure 10.2).

Table 10.1 Key energy data for hydrogen under normal conditions (p = 0.101 MPa, T = 273.15 K = 0 °C) [Win89]

Density	*Calorific value*	*Heat of combustion*
0.09 kg/m³ (gaseous)	3.00 kWh/m_n^3	3.55 kWh/m_n^3
70.9 kg/m³ (fluid, –252 °C)	33.33 kWh/kg	39.41 kWh/kg

1 m_n^3 = 1 standard cubic metre, equivalent to 0.09 kg

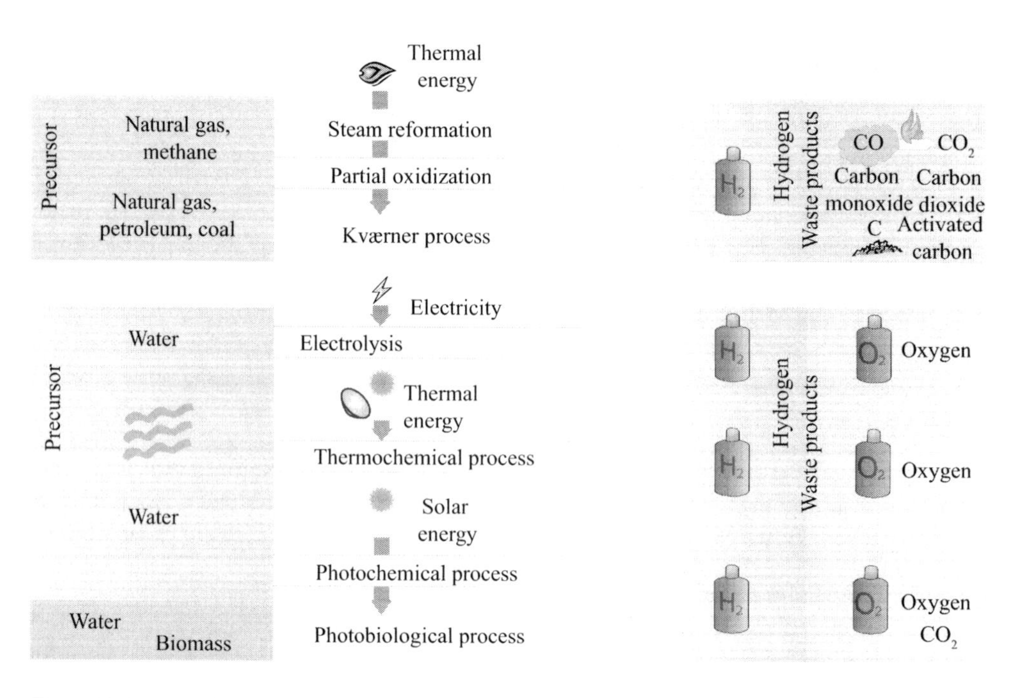

Figure 10.1 Ways of producing hydrogen [Qua09]

In the process, the following reactions take place:

Cathode: $$2H_2O + 2e^- \rightarrow H_2 + 2OH^- \quad (10.1)$$

Anode: $$2OH^- \rightarrow \tfrac{1}{2}O_2 + H_2O + 2e^- \quad (10.2)$$

Efficiencies up to 85 % can be reached today with alkaline electrolysis. In addition to alkaline electrolysis, there are other ways of splitting off hydrogen, such as membrane electrolysis and high-pressure steam electrolysis. Here, a small amount of electricity is needed to maintain the reaction at high temperatures above 700 °C.

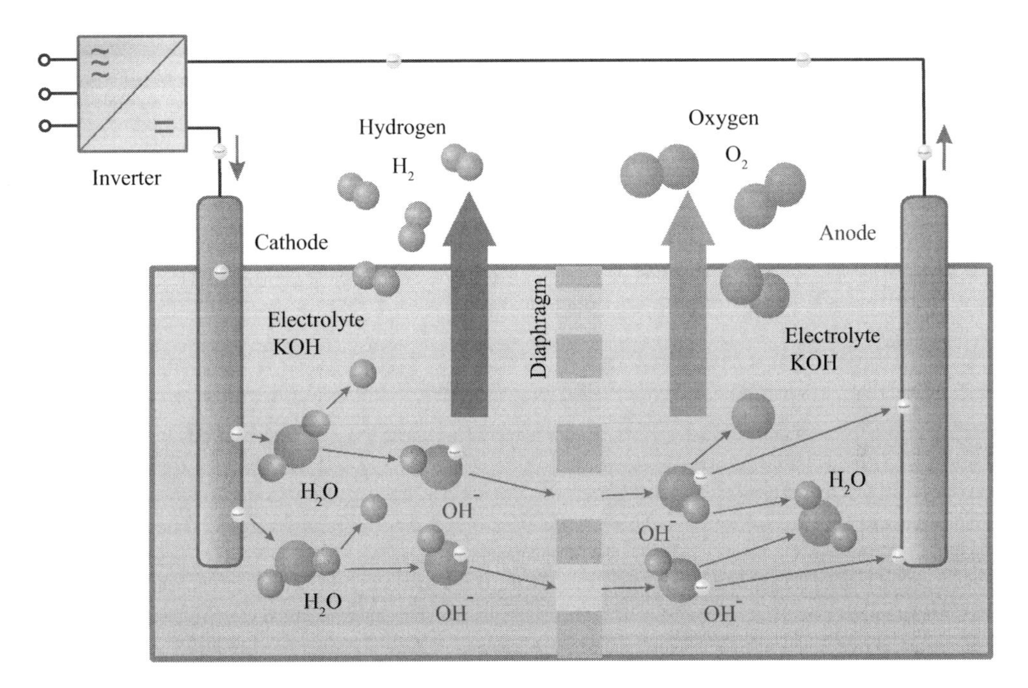

Figure 10.2 Water electrolysis with alkaline electrolyte

Optimally, this electricity for electrolysis would come from renewable power plants so that the hydrogen can be produced without carbon emissions. While electrolysis is already a proven way of producing hydrogen with a small climate impact, other alternatives are under development.

One example is thermochemical processes. At temperatures above 1,700 °C, water directly splits into hydrogen and oxygen. Such temperatures, however, require very heat-resistant materials. Various coupled chemical reactions can reduce the required temperature to below 1,000 °C. Concentrating solar thermal systems (see Chapter 4) can produce such temperatures and have already done so in practice.

Other options include photochemical and photobiological hydrogen production. Here, special semiconductors, algae, and bioreactors are used to split water and hydrocarbons by means of light. These processes are also currently in the research phase. The main challenge is to develop affordable systems with long-term stability.

The various storage technologies available for **hydrogen storage** include:

- pressurized storage
- fluid hydrogen storage
- chemical hydride storage with special alloys
- storage in graphite nanostructures.

In the future, hydrogen will be stored on a large scale, and some of the facilities now used to store natural gas can be exploited for this purpose. For instance, natural gas is currently often stored underground in salt caverns. Underground storage leads to working gas losses of around 3 % per year [Win89]. In addition, energy is required for compression.

Distributed pressurized containers can be used to store smaller amounts of gas. Because of the low density of hydrogen, liquefaction is a good alternative. To this end, hydrogen needs to be cooled down to around –253 °C. One way of doing so is compression and subsequent expansion. But a considerable amount of electricity is required at around 10.5 kWh/kg [Dre01]. And because the molecules are so small, hydrogen atoms diffuse through the storage walls. The losses are around 0.1 % per day for large tanks.

One major advantage of hydrogen is that it is easy to transport. It can be sent through pipelines like those used for natural gas or transported across roads, rails, and waterways as a fluid. A 2,500 km pipeline is expected to have losses ranging from 8 % to 18 % [Dre01].

Hydrogen can be used to **generate electricity** in gas turbines and fuel cells. In gas turbines and combined-cycle turbines, hydrogen can be used much like natural gas. These power plants can also be used to provide heat if the waste heat is recovered (cogeneration). If we start with hydrogen production and move on to storage, transport, and the generation of electricity from hydrogen, the overall efficiency of the process is far below 50 %. It therefore makes economic and ecological sense to transport electricity across high-voltage lines and use it directly. In the future, hydrogen will therefore mainly be used in the transport sector, for emergency power, possibly as a heat source, and perhaps to accommodate fluctuations in power production from a 100 % renewable power plant fleet to a limited extent.

Fuel cells

Introduction

In 1839, the English physicist Sir William Grove discovered the principle of fuel cells. After him, such renowned researchers as Becquerel and Edison further developed the concept. But it was not until the middle of the twentieth century that the technology had become ready for use by NASA, specifically in 1963. Since the 1990s, fuel-cell development has progressed rapidly. Car manufacturers and heating firms have discovered the technology and aim to benefit from its positive image in the midterm.

Fuel cells essentially reverse the electrolysis process by turning chemical energy directly into electricity. A fuel cell always has two electrodes. Depending on the type of fuel cell, pure hydrogen (H_2) or a flammable gas containing hydrogen is fed to the anode, whereas pure oxygen (O_2) or air is fed to the cathode as an oxidation agent. An electrolyte separates the anode and the cathode. As a result, a chemical reaction occurs under controlled conditions. The flow of electrons can be redirected into a power circuit, and ions diffuse through the electrolyte. The only 'waste product' is pure water. Figure 10.3 shows a diagram of a fuel cell running on hydrogen and oxygen with an acidic electrolyte. An individual cell only has around one volt, so in practice multiple cells are stacked in a row to increase voltage.

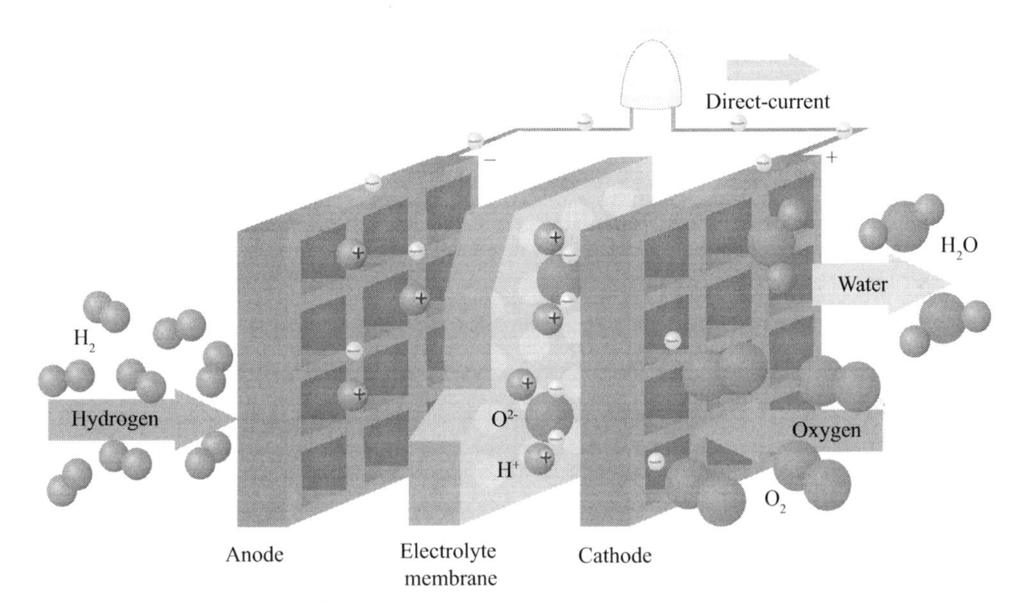

Figure 10.3 A fuel cell with acidic electrolytes

Fuel cell types

There are various types of fuel cells, which mainly differ in the type of electrolytes, the gases that are used as fuel, and operating temperatures. In practice, the following abbreviations are used to describe the different kinds of fuel cells:

- AFC alkaline fuel cell
- PEFC polymer electrolyte fuel cell
- PEMFC proton exchange membrane fuel cell
- DMFC direct methanol fuel cell
- PAFC phosphoric acid fuel cell
- MCFC molten carbonate fuel cell
- SOFC solid oxide fuel cell

Figure 10.4 shows the various fuel gases, oxidation agents, electrolytes, and operating temperature ranges for different types of fuel cells. Table 10.2 shows the chemical reactions that take place at the anode and cathode of each type of fuel cell.

Fully developed in the 1950s, **alkaline fuel cells (AFC)** have a wide operating temperature range of 20 °C to 90 °C and high efficiency. If carbon dioxide reaches the cell, it combines with the alkaline electrolyte to become sodium carbonate, which quickly corrodes the cell. Air therefore cannot be used as the oxidization agent; pure oxygen is required. Because of the cost, this cell is only used in special applications.

The fuel cell used most often is the membrane fuel cell, of which there are two types: **PEFC and PEMFC**. In them, the electrolyte consists of a proton-conducting foil of fluorosurfactant and sulphonated polymers. The poorest gas diffusion electrodes consist of a carbon or metal carrier coated with platinum as a catalyser. The typical operating temperature is around 80 °C. For operation, the cell does not require pure oxygen but can run on normal air. In addition to hydrogen, reformed gas can also be used as a fuel. In the reformation stage, hydrocarbons such as natural gas are chemically split into hydrogen and other components. In general, carbon monoxide (CO) is produced in the process, and it poisons the catalyst. When reformed gas is used in membrane fuel cells, a gas purification stage is therefore also needed to reduce the CO content. The main benefits of membrane fuel cells are that the electrolyte is not corrosive and the energy density can be quite high. Current development focuses partly on extending the service life of membrane and electrode materials.

The **direct methanol fuel cell (DMFC)** has a structure similar to PEMFC. If the membrane is optimized, this cell can run directly on methanol (CH_3OH). The operating temperature is slightly above that of PEMFC.

The **phosphoric acid fuel cell (PAFC)** uses highly concentrated phosphoric acid (H_3PO_4) as the electrolyte. As with PEMFCs, gas diffusion electrodes with platinum are used as catalysts. The electrodes are made of polytetrafluoroethylene (PTFE). The typical operating temperature is around 180 °C. Small amounts of carbon monoxide are converted into carbon dioxide, and only large CO concentrations poison the cell.

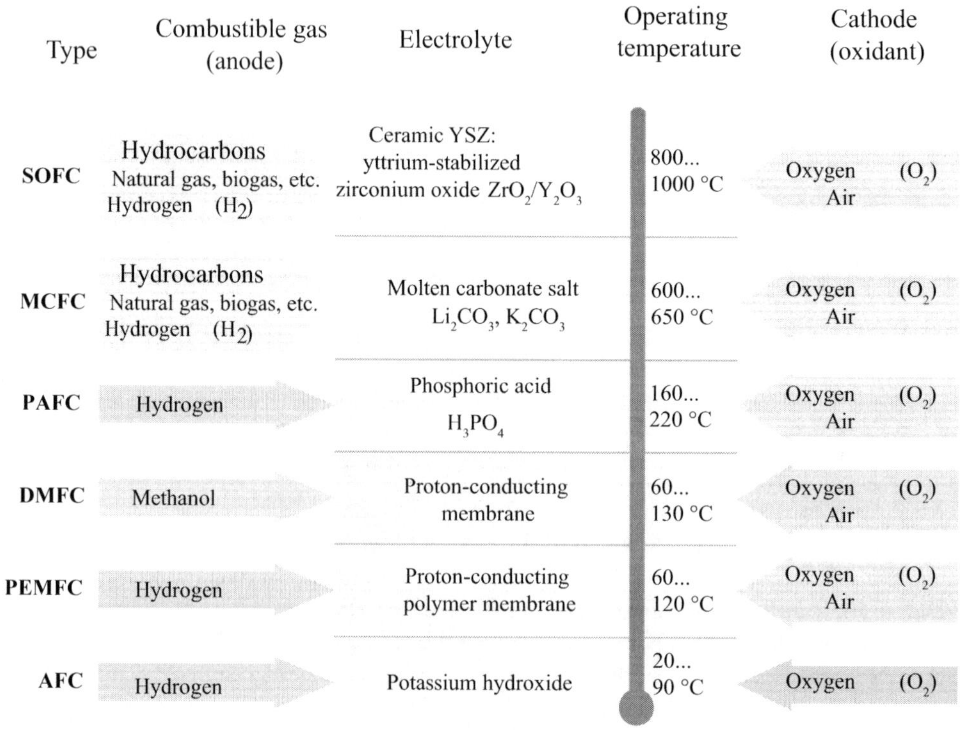

Type	Combustible gas (anode)	Electrolyte	Operating temperature	Cathode (oxidant)
SOFC	Hydrocarbons Natural gas, biogas, etc. Hydrogen (H_2)	Ceramic YSZ: yttrium-stabilized zirconium oxide ZrO_2/Y_2O_3	800... 1000 °C	Oxygen (O_2) Air
MCFC	Hydrocarbons Natural gas, biogas, etc. Hydrogen (H_2)	Molten carbonate salt Li_2CO_3, K_2CO_3	600... 650 °C	Oxygen (O_2) Air
PAFC	Hydrogen	Phosphoric acid H_3PO_4	160... 220 °C	Oxygen (O_2) Air
DMFC	Methanol	Proton-conducting membrane	60... 130 °C	Oxygen (O_2) Air
PEMFC	Hydrogen	Proton-conducting polymer membrane	60... 120 °C	Oxygen (O_2) Air
AFC	Hydrogen	Potassium hydroxide	20... 90 °C	Oxygen (O_2)

Figure 10.4 Fuel gases, electrolytes, operating temperatures, and oxidation agents of various fuel cell types

Figure 10.5 Fuel cell prototype

Molten carbonate fuel cells (MCFC) are high-temperature fuel cells running at around 650 °C. The electrolyte consists of molten alkaline carbonates in a ceramic matrix of $LiAlO_2$. The electrodes are made of nickel. The MCFC integrates carbon dioxide CO_2 in the cell reaction. The anode waste gas containing the CO_2 is mixed into the cathode's feed gas in the process. MCFCs can thus run on hydrocarbon gases like natural gas and biogas. Because of the high working temperature, this technology is generally good for combined heat and power generation (cogeneration).

Solid oxide fuel cells (SOFC) are another high-temperature fuel cell with an operating temperature between 800 °C and 1,000 °C. SOFCs have a ceramic conductor of oxide ions as an electrolyte. The zirconium oxide used only becomes conductive, however, at temperatures around 800 °C for oxygen ions (O^{2-}). Carbon monoxide does not poison SOFCs, making them also useful for the direct use of hydrocarbon gases, like MCFCs.

Table 10.2 Chemical reaction at the anode and cathode of various fuel cell types and ion transport in the electrolyte

Type	*Anode*	*Electrolyte*	*Cathode*
AFC	$H_2 + 2OH^- \rightarrow 2H_2O + 2e^-$	$\leftarrow 2OH^-$	$\frac{1}{2}O_2 + H_2O + 2e^- \rightarrow 2OH^-$
PEMFC	$H_2 \rightarrow 2H^+ + 2e^-$	$2H^+ \rightarrow$	$2H^+ + \frac{1}{2}O_2 + 2e^- \rightarrow H_2O$
DMFC	$CH_3OH + H_2O \rightarrow CO_2 + 6H^+ + 6e^-$	$6H^+ \rightarrow$	$\frac{3}{2}O_2 + 6H^+ + 6e^- \rightarrow 3H_2O$
PAFC	$H_2 \rightarrow 2H^+ + 2e^-$	$2H^+ \rightarrow$	$2H^+ + \frac{1}{2}O_2 + 2e^- \rightarrow H_2O$
MCFC	$H_2 + CO_3^{2-} \rightarrow H_2O + CO_2 + 2e^-$	$\leftarrow CO_3^{2-}$	$CO_2 + \frac{1}{2}O_2 + 2e^- \rightarrow CO_3^{2-}$
SOFC	$H_2 + O^{2-} \rightarrow H_2O + 2e^-$	$\leftarrow O^{2-}$	$\frac{1}{2}O_2 + 2e^- \rightarrow O^{2-}$

Efficiencies and operating response

The efficiency

$$\eta_{FC} = \frac{P}{H_U \cdot \dot{m}} \tag{10.3}$$

of a fuel cell can generally be derived from its electric output P, the calorific value H_U of the anode gas and the mass volume flow $\dot{m}$. In practice, fuel cells reach efficiencies exceeding 60 %. The overall efficiency η_{FC} can also be broken down into other types of thermodynamic efficiencies:

$$\eta_{FC} = \eta_{rev} \cdot \eta_U \cdot \eta_I \cdot \eta_B \cdot \eta_{sys} \tag{10.4}$$

The **thermodynamic efficiency** η_{rev} of an ideal cell is described by Gibbs function

$$\Delta G = \Delta H - T \cdot \Delta S \tag{10.5}$$

which describes the enthalpy change ΔH, absolute temperature T and entropy change ΔS:

$$\eta_{rev} = \frac{\Delta G}{\Delta H} = 1 - T \cdot \frac{\Delta S}{\Delta H} \tag{10.6}$$

The voltage efficiency

$$\eta_U = \frac{U}{U_{rev}} \tag{10.7}$$

is calculated from the real terminal voltage U and reversible cell voltage

$$U_{rev} = -\frac{\Delta G}{z \cdot F} \tag{10.8}$$

derived from Gibbs formula ΔG, the number z of electrons per molecule and the Faraday constant F (F = 96485 As/mol). The thermoneutral or enthalpy cell voltage is calculated in a similar formula:

$$U_{th} = -\frac{\Delta H}{z \cdot F} \tag{10.9}$$

The current efficiency

$$\eta_I = \frac{I}{I_{th}} \tag{10.10}$$

is calculated from the real cell current I and the theoretical maximum cell current

$$I_{th} = \frac{\dot{m}}{M} \cdot z \cdot F \tag{10.11}$$

in accordance with Faraday's Law, with the mass volume flow $\dot{m}$ of the fuel gas and molar mass M (for H_2: M = 2.02 g/mol).

The fuel efficiency η_B describes the excess share of hydrogen not converted into electricity; it indicates how much of the hydrogen put into the cell is actually used. Finally, system efficiency η_{sys} describes losses from the fuel cell system along with the energy required for peripherals, such as pumps, heaters, cooling systems, and compression. Table 10.3 shows some typical values for the parameters described.

The electrical voltage of a real fuel cell is below the reversible cell voltage. Starting from the real open-circuit voltage U_0, cell voltage drops

$$U = U_0 - \Delta U_D - \Delta U_R - \Delta U_{dif} \tag{10.12}$$

as current I increases. Even at small currents, losses ΔU_D occur at the phase contacts between the electrolyte and the electrode as the electrons pass through. Over a wide range, the voltage drop is linear as a result of the voltage reduction ΔU_R from internal resistance. At great currents, the reactants cannot be fed into the system as quickly as the chemical reaction requires. The result is diffusion losses ΔU_{dif} that can quickly lead cell voltage to collapse. Figure 10.6 shows a typical voltage–current curve for a fuel cell.

Table 10.3 Key parameters of hydrogen–oxygen fuel cells under standard conditions (25 °C and 1,013 hPa)

	Two-phase system: fluid water is created	*Gas-phase reaction: steam is created*
Gibbs free energy ΔG	–237.13 kJ/mol	–228.57 kJ/mol
Enthalpy change ΔH	–285.83 kJ/mol	–241.82 kJ/mol
Thermodynamic efficiency η_{rev}	83.0 %	94.5 %
Reversible cell voltage U_{rev}	1.23 V	1.18 V
Thermoneutral voltage U_{th}	1.48 V	1.25 V
Temperature dependence of voltage	–0.85 mV/K	–0.23 mV/K

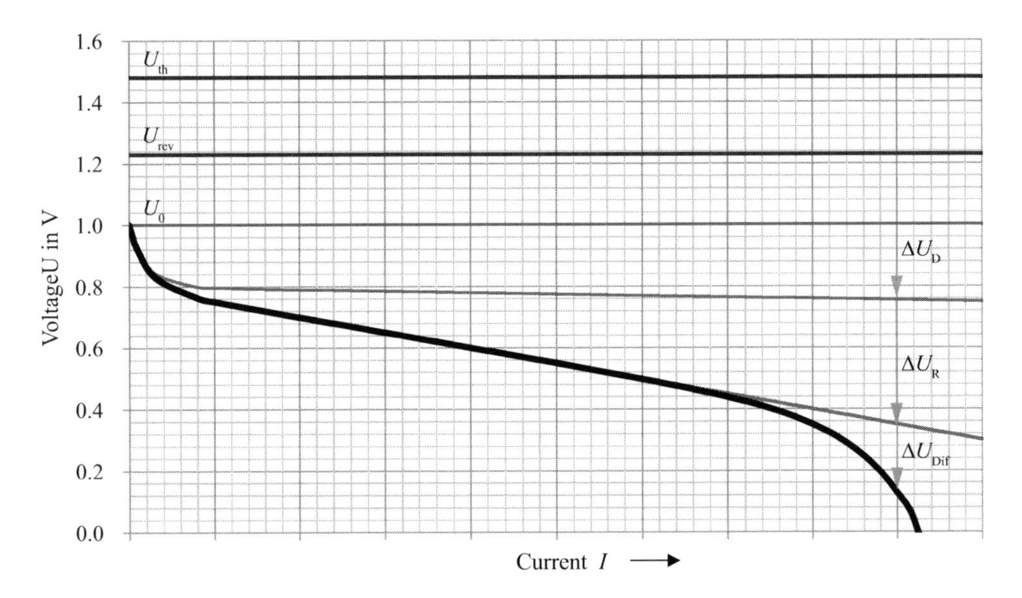

Figure 10.6 A fuel cell's voltage–current curve

Methanation and underground storage

One main drawback of hydrogen production, storage, and use is the lack of hydrogen infrastructure. A completely new distribution and storage system would have to be created. The estimated price tag for the construction of a nationwide network of hydrogen filling stations in Germany alone exceeds €10 billion. The focus is therefore on using the existing natural gas network for storage.

In principle, hydrogen can be blended with natural gas. The presence of hydrogen in gas networks is nothing new. **Coal gas** has a 51 % share of hydrogen, for instance (Table 10.4). In the middle of the twentieth century, coal gas was widespread in Germany but has since been replaced by less expensive natural gas. In West Berlin, the last consumers were not switched over to natural gas until 1996.

When natural gas is used as a fuel, the admissible concentration of hydrogen is currently quite low at 2 %. It is technically possible today to blend 5 % hydrogen into natural gas lines. Some estimates go as high as 20 % [Hüt10]. The problem is the lower calorific value relative to volume as the share of hydrogen increases. For consumers to get the same amount of energy, they need a greater volume flow. While the percentages mentioned above are still tolerable, larger shares of hydrogen will require conversion to the new blend, as was done during the transition from coal gas to natural gas. The amount of work needed would be considerable given the large number of consumers. Another drawback of hydrogen is the greater volume needed for storage.

But an intermediate stage could compensate for this drawback in the switch to hydrogen. In the **Sabatier process**, hydrogen and carbon dioxide are converted into methane:

$$CO_2 + 4H_2 \rightarrow CH_4 + 2H_2O \tag{10.13}$$

At high temperatures and pressures, carbon dioxide and hydrogen become methane and water (Figure 10.7). The process requires a nickel or ruthenium catalyst. The reaction is exothermic, and the waste heat produced can be used. The carbon dioxide required can come from the air by means of an alkaline wash or from downstream combustion processes. The hydrogen required can come from conventional electrolysis utilizing excess renewable electricity.

Pure methane is essentially natural gas H, and the former can substitute directly for the latter. The existing natural gas network and storage facilities can then be used directly without modifications. The first test facilities for the production of methane in this way have already proven successful.

Table 10.4 Characteristic properties of various fuel gases

	Natural gas L	*Natural gas H*	*Coal gas*	*Biogas*	*H_2 gas*	*Methane*
Share of methane in %	80–87	87–99	21	50–75	0	100
Share of hydrogen in %	< 1	< 1	51	< 1	100	0
Calorific value in kWh/kg	11–14	12–15	6–8	4.6–7	39.4	15.5
Calorific value in kWh/m³	9.7–11	11–12	4.5–5.5	5.5–8	3.5	11.1
Density in kg/m³	0.7–0.9	0.7–0.9	0.5–0.8	1.2	0.09	0.71

The figures by volume are given under standard conditions (0 °C and 1,013.25 hPa)

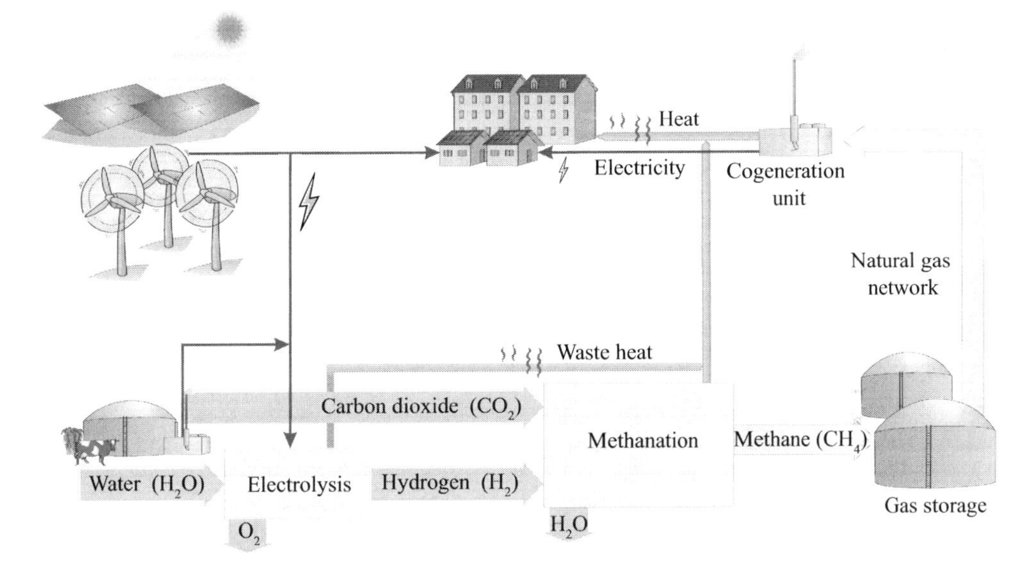

Figure 10.7 Methane production from renewable energy (Power-to-Gas)

Temporary excess electricity will increase as the share of solar and wind power grows. Methanation for natural gas networks is an elegant way of making use of this excess power. If renewable power plants do not provide enough electricity when the wind is still and the sun is not shining, the stored methane can be reused to generate electricity in conventional natural gas combined-cycle plants or fuel cells.

The main drawback of hydrogen and methane production is the relatively low efficiency. Table 10.5 shows the efficiencies for gas production. These numbers drop further when the gas is compressed to 80 bar for transport and 200 bar for storage, for instance. And the losses continue when the stored gas is used to generate electricity again. If we assume an efficiency of 60 % for combined-cycle turbines, the overall efficiency of pressurized hydrogen storage is 43 % at best and only 39 % for methane. If all of the heat from gas production is used and electricity is generated in a cogeneration unit, however, the overall efficiency exceeds 60 %. Still, this level is relatively low compared with the efficiencies of pumped storage and battery storage, which can exceed 80 %. It is therefore very probable that more efficient ways of short-term storage will be used for hourly delivery, with gas storage being employed to cover longer periods, such as days and weeks.

Table 10.5 Efficiencies in the production of hydrogen and methane [Ste11]

Conversion	*Compression*	*Efficiency*
Power → hydrogen	None	64 %–77 %
Power → hydrogen	80 bar	57 %–73 %
Power → hydrogen	200 bar	54 %–72 %
Power → methane	None	51 %–65 %
Power → methane	80 bar	50 %–64 %
Power → methane	200 bar	49 %–64 %

Table 10.6 Capacities for the storage of hydrogen and methane in Germany [Uba10]

Conversion	*Working gas volume in million m^3*	*Storage capacity in TWh for*	
		H_2 gas	*Methane*
Porous storage in operation	13.6	—	136
Cavern storage in operation	7.8	23	78
Cavern storage in operation, under construction, or planned	15.2	46	152
Cavern storage, long-term total potential	36.8	110	368

The main potential of gas storage is that existing natural gas facilities already have extremely large capacity. In Germany alone, underground gas storage has a capacity exceeding 20 billion m^3. The largest natural gas storage facility in Germany is located in in Rehden, Lower Saxony. Here, a former gas field some 2,000 m deep has a total storage volume of 7 billion m^3. For storage facilities, a distinction is made between **porous rock**, which is essentially aquifers or former reservoirs of petroleum and natural gas, and salt caverns, which are especially suitable for hydrogen storage. Additional cavern storage capacity is currently under construction or planned. Further capacity could be created in the long term (Table 10.6).

Methane has a calorific value of around 10 kWh/m^3, so Germany's current gas storage capacity contains more than 200 TWh of thermal energy. Assuming an efficiency of 60 % for power production from this gas, 120 TWh of electricity could be generated, enough to meet Germany's entire power demand for more than two months. Of course, natural gas is also used as a heat source, so not all of this capacity will be available for power storage. Nonetheless, the country's existing gas storage facilities have sufficient capacity to back up a completely renewable power supply.

11 Calculating economic feasibility

Introduction

For renewable energy applications, the question of affordability plays a key role; after all, unaffordability is one of the main arguments that often prevent the use of renewables. During the planning and completion of projects, the focus is usually not on how the project can be optimal in technical or ecological terms, but rather in terms of microeconomics. The result is often a negative macroeconomic impact, and the environmental impact is not sufficiently taken into account. Therefore, the second part of this chapter casts a critical light on conventional microeconomic evaluation criteria.

The systems investigated here all convert energy. The goal of an economic feasibility study is to select the most affordable option – the one that provides the desired form of energy at the lowest price. Generally, different renewable energy systems are compared with each other. Furthermore, the cost of renewable systems is generally compared with conventional energy, usually without consideration of the total macroeconomic benefits.

The product of an economic feasibility study is the **price of an energy unit**. For systems that provide heat, the price is indicated as a kilowatt-hour of heat (€/kWh_{therm}). For systems that produce electricity, the price is a kilowatt-hour of electricity (€/kWh_{el}). Here, all of the costs – from power plant construction to fuel costs, staff costs, and end-of-life costs – are spread across the number of kilowatt-hours produced over the system's service life. When these costs are estimated, errors are often made because the calculations are based on assumptions of future trends that later turn out to be inaccurate. Often, taxpayers and ratepayers – the general public – have to pay for these mistakes, though the costs are often not included in the microeconomic calculation. As a result, a system's actual costs are often not properly reflected. Examples include tremendous underestimations of the costs for decommissioning and storage of nuclear facilities and waste and the rewilding of coal fields.

When costs are compared, the underlying prices sometimes turn out to be from different years. In such cases, **inflation rates** have to be taken into account. It is therefore a good idea to indicate the reference year when listing costs and energy prices (such as $€_{2005}$/kWh). Table 11.1 shows the expenses of private households usually quoted in this context. In addition to general inflation, prices of conventional energy fluctuate greatly, with the oil crises of the 1970s being only one example. Long-term cost calculations of conventional systems thus entail quite a lot of uncertainty.

Table 11.1 Consumer price index in Germany; data before 1991 only refer to the former West Germany; all numbers relative to 2005 (2005 = 100)

Year	*1965*	*1970*	*1972*	*1975*	*1977*	*1980*	*1982*	*1984*	*1986*	*1988*	*1990*	*1992*
Index	31.3	35.3	39.1	47.5	51.3	57.8	64.7	68.5	69.8	70.8	74.7	79.8
Year	*1994*	*1996*	*1998*	*2000*	*2002*	*2004*	*2005*	*2007*	*2008*	*2009*	*2010*	*2011*
Index	85.6	88.3	90.9	92.7	95.9	98.5	**100.0**	103.9	106.6	107.0	108.2	110.7

(Data: [Sta12])

Cost of energy production

Calculations without interest

In classic economic feasibility calculations, investors expect a certain return on their investment, and the amount is largely, though not exclusively, based on the risk. But first, we will consider the parameters aside from capital interest.

Aside from the return on investment, the cost of an energy unit is very easy to calculate. To do so, all of the costs of an energy system over its entire service life are added up and divided by the length of the service life. The result is the total annual cost. These costs are then divided by the number of energy units generated by the facility in a year to produce the cost of an energy unit.

The **total costs** c_{tot} are calculated from the **investment costs** A_0 occurred upfront during construction along with the ongoing costs A_i incurred each operating year i. For instance, operating costs are incurred in the form of property leases, taxes, maintenance, repairs, administration, management, insurance premiums, and – in the case of biomass and fossil fuel systems – fuel costs; they can differ from one year to the other. With a service life of n years, we have the following formula:

$$c_{\text{tot}} = A_0 + \sum_{i=1}^{n} A_i \tag{11.1}$$

Often, the factor f can be used to include the ongoing costs on top of the investment costs:

$$c_{\text{tot}} = A_0 + \sum_{i=1}^{n} f_i \cdot A_0 = A_0 \left(1 + \sum_{i=1}^{n} f_i\right) \tag{11.2}$$

If the ongoing costs A are the same every year, the formula is as follows:

$$c_{\text{tot}} = A_0 + n \cdot A = A_0 \cdot (1 + n \cdot f) \tag{11.3}$$

The total costs c_{tot} are then used to calculate the average total annual costs c_a:

$$c = \frac{c_{\text{tot}}}{n} \tag{11.4}$$

It is assumed that the system will produce an annual amount of energy on average E_a. The **energy production costs** c_E can then be calculated:

$$c_E = \frac{c_a}{E_a} \tag{11.5}$$

Solar thermal systems for hot water supply

Roughly a third of the cost of solar hot water stems from the collector, with another full third being the storage tank and accessories; the remaining third is installation. Depending on the specific design, flat-plate collectors cost anywhere from just under €200/m² to €350/m², while evacuated tube collectors range from €400/m² to €600/m². A 300-litre storage tank costs €700 to €1,100. The installation of a solar thermal system is especially affordable on new buildings or whenever a new hot water storage system is required. There is a wide range of prices for solar thermal. A very inexpensive system with 4 m² of collectors and 300 l of hot water storage costs around €2,000 without installation. The average cost of a system for a family of four without installation ranges from €3,000 to €3,500, with labour for installation bringing the amount up to around €5,000. Public funding can be used in combination with private equity.

In the example below, we calculate this mixed investment for hot water supply. The annual heat demand is 4,000 kWh_{therm}, 50 % of which is to be covered by solar. In other words, 2,000 kWh_{therm} of useful energy has to come from the solar energy system each year. If the conventional system has an efficiency of 85 %, the annual substituted amount of conventional energy (final energy) is 2,353 kWh_{therm}. This amount of energy is assumed to be the amount provided each year in the further calculations below. It is also assumed that the annual maintenance costs are relatively low at €25 on average and that the pump in the collector circuit consumes 60 kWh_{el} of electricity per year at a price of €0.25/kWh_{el} (€15 annually). At an investment cost A_0 of €3,300 and annual operating costs A_i of €40, the calculation is as follows for a service life of n = 20 years:

$$c_{tot} = A_0 + 20 \cdot A_i = 3{,}300\text{ €} + 20 \cdot 40\text{ €} = 4{,}100\text{ €},\ c_a = c_{tot} / n = 205\text{ €}$$
$$c_E = c_a / E_a = 205\text{ €} / 2{,}353\text{ kWh}_{therm} = 0.087\text{ €/kWh}_{therm}$$

Thrifty single-family homes will have much lower energy demand than in the example above. As a result, affordability suffers. A shorter service life also increases the cost. If energy demand is high and investment costs are low (such as with a do-it-yourself system or with public funding), a solar hot water supply can be quite affordable, as shown in Table 11.2.

In all other cases, the heat production costs are usually higher than the cost of a conventional system. Pool heating systems are generally affordable because the specific investment costs are lower. The costs are also far lower than in the example above when the solar thermal system covers multiple buildings as a part of district supply.

Solar thermal power plants

In the past 10 years, the number of newly built solar thermal power plants has been relatively small. Constructed in 2007, the 64 MW Nevada Solar One parabolic trough power plant

Table 11.2 Heat production costs in €/kWh$_{therm}$ for solar thermal hot water without return on investment at an annual operating cost of €40

Substituted amount of energy	*2300 kWh$_{therm}$ annually*		*1150 kWh$_{therm}$ annually*	
	Service life			
Investment costs	*20 years*	*15 years*	*20 years*	*15 years*
€5,000	0.13	0.16	0.25	0.32
€3,500	0.09	0.12	0.19	0.24
€2,500	0.07	0.09	0.14	0.18
€1,800	0.06	0.07	0.11	0.14

(see Chapter 4) entailed around €230 million in investment costs. The plant is designed to generate 129 GWh/a. Completed in 2008, the 50 MW Andasol 1 parabolic trough power plant cost around €300 million, though it has thermal storage to increase the capacity factor. The plant is designed to generate 180 GWh/a. The ongoing costs can be estimated at 3 % of the upfront cost for a solar thermal power plant.

Without the return on investment, the power generation costs for these two power plants are as follows for a service life of n = 30 years:

$$c_{tot} = A_0 \cdot (1 + 30 \cdot 0.03) = 1.9 \cdot 230 \text{ million €} = 437 \text{ million €}$$
$$c_a = c_{tot} / n = 14.6 \text{ million €}$$
$$c_E = c_a / E_a = 14.6 \text{ million €} / 129 \text{ million kWh}_{el} = 0.113 \text{ €/kWh}_{el}$$

or

$$c_{tot} = A_0 \cdot (1 + 30 \cdot 0.03) = 1.9 \cdot 300 \text{ million €} = 570 \text{ million €}$$
$$c_a = c_{tot} / n = 19 \text{ million €}$$
$$c_E = c_a / E_a = 19 \text{ million €} / 180 \text{ million kWh}_{el} = 0.106 \text{ €/kWh}_{el}$$

Considerable cost reductions are expected to come from future power plants, so that the cost of power generation from solar thermal is expected to drop.

Photovoltaic systems

At the beginning of the 1990s, when Germany had its 1,000 Roofs Programme, the cost of a **grid-connected photovoltaic system** still came in far above €10,000/kW$_p$ on average, but the costs are now far lower. Consumer prices for solar modules are now below €1/W$_p$. For complete systems, the costs for inverters, accessories, stands, and labour are added. In the second quarter of 2012, turnkey roof systems with a capacity of up to 100 kW$_p$ came in at an average of €1,702/kW$_p$ in Germany excluding VAT [BSW12]. The specific investment cost decreases as the size of the system increases. As early as 2014, systems in the megawatt range cost less than €1,000/kW$_p$.

Below, we discuss the cost of a kilowatt-hour of electricity from photovoltaics. The figures given here are also used in the additional examples below. A system with an output of 1 kW_p and investment costs of A_0 = €1,500 is to be built. It is assumed that the system will be used for 20 years (n = 20) and produce E_a = 1,000 kWh of electricity per year. In the tenth year of operation, the inverter is to be replaced. That will add a one-off cost of A_{10} = 400 €. In this example, no further costs are incurred. The calculation is thus as follows:

$$c_{tot} = A_0 + A_{10} = 1900\ €,\ c_a = c_{tot}\ /\ n = 95\ €,\ c_E = c_a\ /\ E_a = 0.095\ €/kWh_{el}$$

In the Sahara, the same system would produce 2,000 kWh/a of electricity. Without consideration of return on investment, the power generation cost in this example would then drop from €0.095/kWh_{el} to €0.0475/kWh_{el}. Unlike conventional systems, solar energy systems do not entail any risk of rising fuel prices. While small homeowner systems have very low annual operating costs, utility-scale PV plants generally have operating costs at 2 to 4 % of the upfront investment.

Wind turbines

In the past few years, the overall costs of wind turbines have also dropped, though not as quickly as for photovoltaics.

The upfront investment costs include the cost of the turbine itself along with planning, installation, the foundation, the grid connection, and transport. These ancillary investment costs make up 34.5 % of the purchase price on average [Kle97]. The figure can be much higher, however, for small turbines. The operating costs include land leases, maintenance, repairs, and insurance.

Systems with an output around 1.5 MW have an estimated cost range of €1,000 to €1,600/kW, depending on the location and elevation of the infrastructure that needs to be built. At €1,200/kW, a 1.5 MW turbine would have an installed price tag A_0 of €1,800,000. In addition to these upfront costs, the 1.5 MW turbine would have annual operating costs A_i of ca. €50,000 in addition. The operating costs can be estimated at 2 % to 5 % of the upfront investment cost.

At a location with a wind velocity of nearly 7 m/s at a hub height of 100 m, equivalent to around 4.5 m/s at 10 m, roughly E_a = 3.5 million kWh per year can be generated. When the turbine has a service life n of 20 years, the following calculation applies for a 1.5 MW turbine:

$$c_{tot} = A_0 + 20 \cdot A_i = 1{,}800{,}000\ € + 20 \cdot 50{,}000\ € = 2{,}800{,}000\ €$$
$$c_a = c_{tot}\ /\ n = 140{,}000\ €,\ c_E = c_a\ /\ E_a = 0.04\ €/kWh_{el}$$

As Table 11.3 shows, power generation costs increase considerably at lower wind velocities because the annual yield drops significantly. The wind velocity at hub height – and hence, the amount of energy relative to power – increases with larger towers because wind velocities are faster higher up. In the future, wind turbines will also be optimized to reduce costs further.

The specific costs are much greater for small wind generators. A 70 W wind generator costs around €700 without a tower or labour, equivalent to a specific investment cost of €10,000/kW. And because such small generators are also usually installed on short towers, the available wind velocity is lower, thereby reducing yield. Theoretically, such small generators could be put onto tall towers, but this solution generally is not a good idea practically or economically.

Table 11.3 A wind turbine's energy yield by turbine size and wind velocity v at a given hub height with a Rayleigh distribution of wind velocities

System size		*Average energy yield in MWh_{el}/a*				
Rotor Ø	*Output*	*v = 5 m/s*	*6 m/s*	*7 m/s*	*8 m/s*	*9 m/s*
30 m	200 kW	320	500	670	820	950
40 m	500 kW	610	970	1,360	1,730	2,050
55 m	1,000 kW	1,150	1,840	2,570	3,280	3,920
65 m	1,500 kW	1,520	2,600	3,750	4,860	5,860
80 m	2,500 kW	2,380	4,030	5,830	7,700	9,220
120 m	5,000 kW	5,300	9,000	13,000	17,000	20,000

Hydropower

The cost of hydropower facilities also greatly depends upon the system size. For micro-hydropower, the specific investment costs range from €5,000 to €13,000/kW. When such systems are upgraded or reactivated, the cost range is €2,500 to €5,000/kW. At Germany's largest recent new project, the Rheinfelden run-of-river plant with a capacity of 100 MW, the construction costs are estimated at around €380 million.

Usually, only rough estimations of cost are available for the largest hydropower plant projects worldwide. For instance, it is assumed that Brazil's Itaipú power plant had a price tag of €14 to €15 billion. At a capacity of 14 GW, the specific investment costs are then around €1,000/kW. The cost of the largest hydropower projects are thus as low as those of wind turbines, though the yield of hydropower is 2 to 3 times higher.

Geothermal facilities

Drilling costs make up a large part of the upfront investments in **geothermal power plants**. In Germany, boreholes have to be relatively deep compared with optimal geothermal regions, so geothermal plants in the country have relatively high investment costs. The plants completed in 2007 in Landau had an upfront cost of around €20 million. The plant has a capacity of around 3 MW_{el}. In addition, some 5 MW_{th} of heat is output. The plant thus produces roughly 22 million kWh of electricity and 9.2 million kWh of heat.

For electricity alone, the investment costs come in at €6,667/kW. But because of the cogeneration of heat and power, the investment costs would drop per energy unit quite a bit.

A typical **heat pump** for single-family homes had a price tag of €8,000 to €12,000 in 2010. In addition, there are the costs for tapping the heat source; geothermal collectors cost around €3,000 to €6,000 for such a project.

If a new building is equipped with a heat pump, the cost of a conventional heat system can be subtracted. If a gas heater would have been used, the cost of the gas burner, the gas connection, and the chimney are not incurred. Still, the upfront investment costs of a heat pump system are often greater than those of a conventional gas/oil heating system.

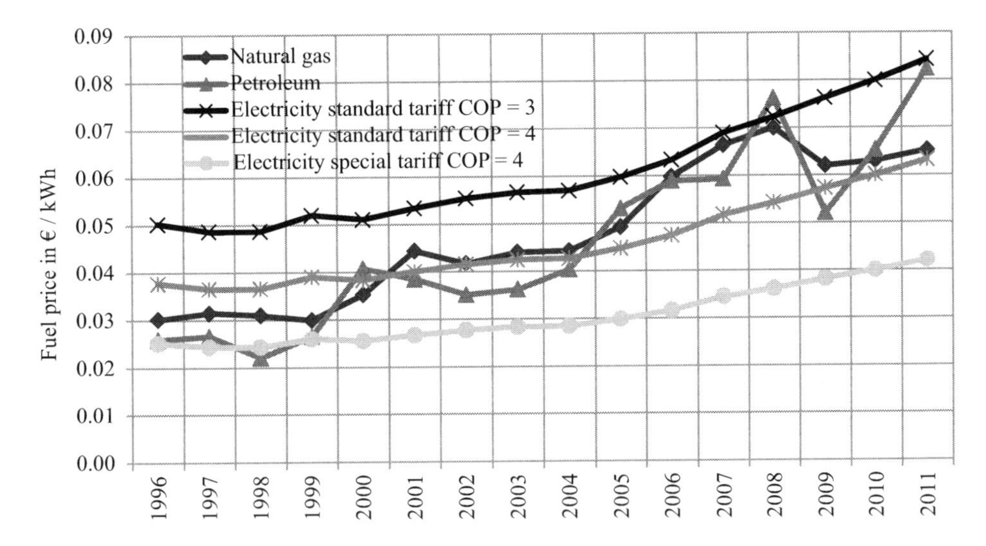

Figure 11.1 Trends in retail prices for natural gas, oil, and electricity for the operation of a heat pump at various coefficients of performance (COPs) in Germany

Heat pumps have low operating costs. Unlike conventional gas heaters, they do not entail expenses for a gas meter, maintenance, and chimney sweeps. Fuel prices – in the case of heat pumps, the electricity consumed – are generally also lower than the cost of oil and gas. The benefits of a heat pump under specific fuel prices depend partly on the pump's COP and the price of electricity. At the usual retail power prices in Germany and a moderate COP of 3, a heat pump's fuel costs can even exceed those of natural gas and petroleum. At an average COP of 4, the heat pump can be slightly less expensive (Figure 11.1).

Wood pellet heating

It is hard to compare the long-term economic trends in biomass fuels relative to fossil fuels, as a look back at the historic trends in wood pellet prices and oil prices reveals (Figure 11.2).

The potential of wood pellets does not merely suffice to cover the current demand on the German market for space heating. If customers increasingly switch over to wood pellets, prices will increase. But because the price of oil will also only go up over the long term, the affordability of wood pellets could continue.

The affordability of a heating systems fired with wood pellets largely depends upon the price of oil and gas. The installation of a wood pellet heat system in a single-family home costs around €15,000, far more than an oil- or gas-fired heating system. In a new building, the cost of a natural gas connection is not incurred, however. The operating costs of a pellet heat system are lower than a heating system running on natural gas or oil because of the lower fuel prices. Depending on how much fuel is consumed and how fuel prices develop, a wood pellet heating system can pay for itself in just a few years. The fuel prices for firewood heating systems are lower than for pellets. They therefore have even lower operating costs.

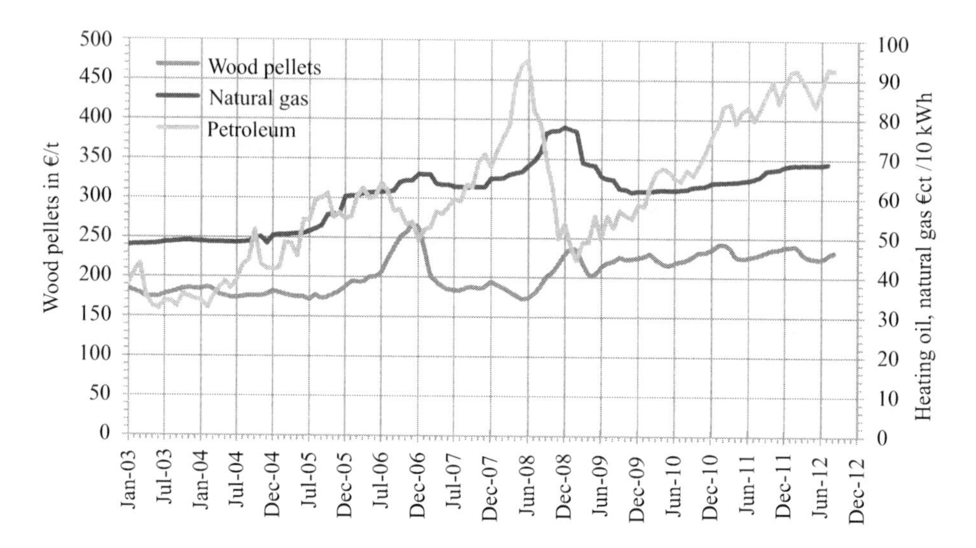

Figure 11.2 A comparison of consumer prices for petroleum, natural gas, and wood pellets (Data: [Sta12b, DEPV12])

For an existing apartment with an annual heat demand of 30,000 kWh_{therm}, a boiler with an efficiency of 0.75 would have an annual fuel consumption of 40,000 $kWh_{pellets}$. If we assume a pellet price of € 250/t (€0.05/$kWh_{pellets}$), the annual fuel costs amount to €1,600. If we assume additional annual operating costs of €200 and the service life of 20 years, then the heat production costs are:

$$c_{tot} = 15\ 000\ € + 20 \cdot (2{,}000\ € + 200\ €) = 59{,}000\ €$$
$$c_a = c_{tot}\ /\ n = 2{,}950\ €$$
$$c_E = c_a\ /\ E_a = 2{,}950\ €\ /\ 30{,}000\ kWh_{therm} = 0.098\ €/kWh_{therm}$$

In a new building with nearly optimal insulation, the annual heat demand can be reduced to 6,000 kWh_{therm}, thereby lowering fuel costs to €400. Although the overall costs are cut in half, the heat production costs increase to €0.225/kWh_{therm}.

Calculating return on investment

In contrast to the calculations above without the return on investment, investors generally want an equivalent return for their investment in line with the usual returns on the capital market for comparable investments. The calculation of capital c_n after a system service life of n years from an initial investment A_0 can be done by using the interest rate p in a formula of compound interest. The formula

$$q = 1 + p \qquad (11.6)$$

produces:

$$c_n = A_0 \cdot (1+p)^n = A_0 \cdot q^n \tag{11.7}$$

If we start with the calculation of the cost of a photovoltaic system, we have an initial investment of A_0 = €1,500 for n = 20 years at an interest rate of p = 6 % = 0.06, producing the following amount of capital including interest after 20 years:

$$c_{20} = 1{,}500\ € \cdot (1 + 0.06)^{20} = 4{,}811\ €$$

If additional payments are made at a later date, interest would also need to be applied, but within a shorter time frame. If an investor pays an additional €400 after 10 years, for instance, that amount would only have 10 years of interest. The capital after 20 years would then be:

$$c_{20} = A_0 \cdot q^{20} + A_{10} \cdot q^{10} = 4{,}811\ € \cdot 1.06^{20} + 400\ € \cdot 1.06^{10} = 5{,}527\ €$$

The following formula can also be used in general:

$$c_n = A_0 \cdot q^n + \sum_{i=1}^{n} A_i \cdot q^{n-i} \tag{11.8}$$

If the payments A are made in different years i in equal amounts, we use the **savings-and-loan formula**:

$$c_n = A_0 \cdot q^n + \sum_{i=1}^{n} A \cdot q^{n-i} = A_0 \cdot q^n + A \cdot \frac{q^n - 1}{q-1} \tag{11.9}$$

For later payments, the question is how much initial capital would be needed if the investment had been made in the beginning at interest rate p to produce the same amount of total capital.

Payment A_i can then be discounted back to year i at the beginning of the investment:

$$A_{i/0} = \frac{A_i}{q^i} = A_i \cdot q^{-i} \tag{11.10}$$

The €400 paid after 10 years in the example above is thus equivalent to the following amount of initial capital:

$$A_{10/0} = 400\ € \cdot 1.06^{-10} = 223\ €$$

In other words, if €223 has a term of 20 years at 6 % interest, it is equivalent to €400 paid 10 years later at 6 % interest.

If multiple payments are **discounted** at various points in time, the following formula can be used:

$$c_0 = A_0 + \sum_{i=1}^{n} \frac{A_i}{q^i} \tag{11.11}$$

If the payments A from different years i are the same, we can use the following formula:

$$c_0 = A_0 + A \cdot \frac{q^n - 1}{(q-1) \cdot q^n} \tag{11.12}$$

The capital after n years can then be calculated using the formula for compound interest:

$$c_n = c_0 \cdot q^n \tag{11.13}$$

No capital remains for investments in the energy sector after a service life of n years. Equipment is probably defective and hence worthless after its service life. The invested capital is paid back from the sale of energy. We now have to calculate what an energy unit has to cost so that investors can get their desired return.

The sales revenue also has to be calculated with interest. If investors receive repayment at the beginning of the first year of operation, they can reinvest it and receive interest across the full service life. If revenue comes in later, the period in which rent is posted is shorter. The invested capital c_0 starting at zero as a point in time is calculated from the investment amount A_0 and the payments A_i discounted back to the beginning as shown above. Repayments Z_i can then be subtracted from that amount. They are also discounted back to the beginning. Payments and repayments that take place within a year are treated as though they were made at the end of the year. For a service life of n years, we then get:

$$K = -A_0 + \sum_{i=1}^{n} \frac{Z_i - A_i}{q^i} = -A_0 - \sum_{i=1}^{n} \frac{A_i}{q^i} + \sum_{i=1}^{n} \frac{Z_i}{q^i} = -k_0 + \sum_{i=1}^{n} \frac{Z_i}{q^i} \tag{11.14}$$

K is called the **net present value**. If the net present value K is greater than or equal to zero, investors do not experience losses.

Below, we assume that repayments Z at the end of each year are made in the same amount. The repayments required each year can be determined by setting the net present value K to zero relative to Z. The formula

$$0 = -c_0 + Z \cdot \sum_{i=1}^{n} \frac{1}{q^i} = -c_0 + Z \cdot \frac{q^n - 1}{(q-1) \cdot q^n} = -c_0 + Z \cdot \frac{1}{a} \text{ leads to}$$

$$Z = c_0 \cdot a \tag{11.15}$$

The factor a is called the **annuity factor** [VDI91]:

$$a = \frac{q^n \cdot (q-1)}{q^n - 1} = \frac{q-1}{1-q^{-n}} \tag{11.16}$$

Table 11.4 provides various service lives n and interest rates p for specific and unity factors a. In literature, the annuity factor is often also just called annuity [Wöh81].

Annuity factor a can now be used to calculate the **energy production costs** c_E for the amount of energy E_a produced in a year:

$$c_E = \frac{Z}{E_a} = \frac{c_0 \cdot a}{E_a} \tag{11.17}$$

Table 11.4 Annuity factors *a* for various service lives *n* and interest rates *p*

n	*Interest rate p = q −1*									
	1 %	*2 %*	*3 %*	*4 %*	*5 %*	*6 %*	*7 %*	*8 %*	*9 %*	*10 %*
10	0.1056	0.1113	0.1172	0.1233	0.1295	0.1359	0.1424	0.1490	0.1558	0.1627
15	0.0721	0.0778	0.0838	0.0899	0.0963	0.1030	0.1098	0.1168	0.1241	0.1315
20	0.0554	0.0612	0.0672	0.0736	0.0802	0.0872	0.0944	0.1019	0.1095	0.1175
25	0.0454	0.0512	0.0574	0.0640	0.0710	0.0782	0.0858	0.0937	0.1018	0.1102
30	0.0387	0.0446	0.0510	0.0578	0.0651	0.0726	0.0806	0.0888	0.0973	0.1061

The interest rate is relative to investment risk. For renewable energy systems, the **risks** partly stem from an overestimation of the solar, hydro, or wind potential in a particular location along with technical uncertainties and policy changes. Because these risks are usually higher than those of a savings account, interest rates are higher. There are practically no formulae or rules to determine the level of an interest rate relative to risk. Rather, the market – and everything that influences moods on it – determines the level, along with the investor's subjective feeling.

A dynamic price approach can also be used in the calculation. In addition to possible price hikes for operations and maintenance, the cost of conventional energy can also rise.

It is difficult, however, to predict how prices will develop over long time frames; in addition to general inflation, the price of energy can also rise as reserves become scarce, and other events – such as wars in areas where oil is produced – can also occur. We therefore do not further investigate dynamic price calculations here. Instead, this chapter ends with a critique of interest rates in general.

Solar thermal systems for hot water supply

For the solar hot water system, we will once again start with the figures used in the calculation above without the return on investment. At an interest rate of 6 %, the annuity factor is 0.1030 and 0.0872 with a service life of 15 and 20 years, respectively. Table 11.5 shows different heat production costs relative to service life and investment cost. Without subsidies, solar hot water is rarely competitive with conventional systems, such as electric water heaters.

Table 11.5 Heat production costs in €/kWh$_{therm}$ from solar thermal hot water systems at an interest rate of 6 % and annual operating costs of €40

Substituted amount of energy	*2300 kWh$_{therm}$ annually*		*1150 kWh$_{therm}$ annually*	
	Service life			
Investment costs	*20 years*	*15 years*	*20 years*	*15 years*
€5,000	0.21	0.24	0.41	0.48
€3,500	0.15	0.17	0.30	0.35
€2,500	0.11	0.13	0.22	0.26
€1,800	0.09	0.10	0.17	0.20

Solar thermal power plants

For an example of a solar thermal power plant, we refer to Nevada Solar One (A_0 = €230 million, A_i = 6.9 Mio. €, n = 30, E_a = 129 million kWh$_{el}$) at an interest rate of p = 8 %:

Capital at the beginning: c_0 = 230 million € + 6.9 million € · 11.26 = 307.7 million €
Annuity factor: $a = (1.08 - 1) / (1 - 1.08^{-30}) = 0.0888$
Required repayments p.a.: Z = 307.7 million € 0.0888 = 27.3 million €
Cost of energy production: c_E = 27.3 million € / 129 million kWh$_{el}$ = 0.212 €/kWh$_{el}$

Photovoltaic systems

For an example of a photovoltaic system (A_0 = 1500 €, A_{10} = 400 €, p = 0.06, q = 1.06, n = 20, E_a = 1000 kWh$_{el}$) used above, we have the following:

Capital at the beginning: c_0 = 1500 € + 400 € · 1.06^{-10} = 1723 €
Annuity factor: $a = (1.06 - 1) / (1 - 1.06^{-20}) = 0.0872$
Required repayments p.a.: Z = 1723 € 0.0872 = 150 €
Cost of energy production: c_E = 150 € / 1,000 kWh$_{el}$ = 0.15 €/kWh$_{el}$

If 2,000 kWh of electricity is produced each year in the Sahara, the energy production costs drop to €0.075/kWh$_{el}$. Nonetheless, the energy production costs are much higher with interest than without (€0.15/kWh$_{el}$ as opposed to €0.095/kWh$_{el}$).

Wind turbines

For an example of a 1,500 kW wind turbine (A_0 = €1,800,000, A_i = €50,000, q = 1.08, n = 20, E_a = 3.5 · 10^6 kWh$_{el}$) and an interest rate of p = 8 %, we have the following:

c_0 = 1,800,000 € + 50,000 € 9.82 = 2,291,000 €; a = 0.1019
c_E = 2,291,000 € 0.1019 / 3.5·10^6 kWh$_{el}$ = 0.067 €/kWh$_{el}$

A large number of wind farms have already been built. Often, these projects only had around 30 % equity for financing. The rest is bank loans, and in Germany low-interest loans are available with rates around 5 %. The providers of equity run the project risks from, for instance, erroneous yield forecasts and fluctuations in wind. They therefore generally require greater returns.

Compensation for renewable energy systems

A number of countries use feed-in tariffs to launch the market for renewable energy because electricity from renewable energy generators is often more expensive than energy from conventional systems in a conventional financial analysis. Feed-in tariffs provide specific compensation levels for particular sources of renewable energy. In Germany, the **Renewable Energy Act (EEG)** formulates the country's feed-in tariffs. The costs incurred are passed on to all power consumers (Figure 11.3).

The feed-in tariffs are reduced annually for new systems. The goal is to make renewable energy completely competitive in just a few years. Already, spot market prices on the power exchange sometimes exceed the feed-in tariffs paid.

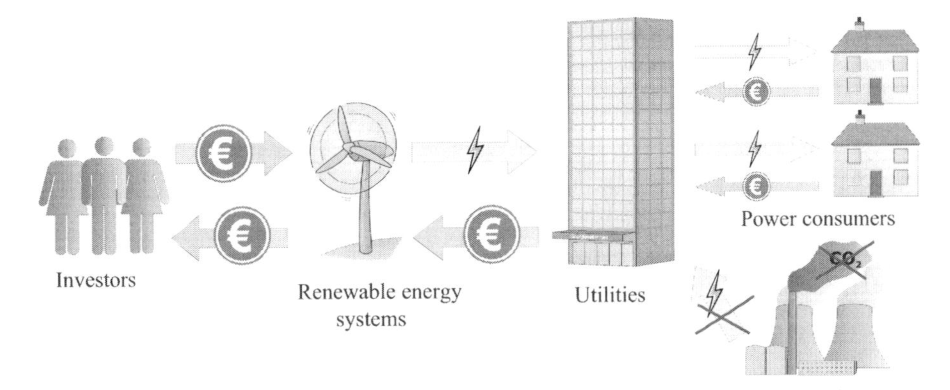

Figure 11.3 How renewable power plants are financed in the German Renewable Energy Act (EEG)

Future development of the cost of renewable energy

The cost of renewable energy systems will continue to drop considerably as it has in the past. In particular, large-scale production will reduce costs through economies of scale and the use of new technologies.

For **wind power**, for instance, the cost of power at a given location will drop when more modern, larger turbines replace old ones. Larger turbines can make more energy because wind velocities are much greater higher up. But because sites with ideal wind conditions are becoming scarce, more wind farms have to be built in areas with lower wind velocities, which detrimentally affects economic feasibility. In the long term, further cost reductions are imaginable offshore. At present, however, offshore wind farms generally produce far more expensive power than onshore wind farms do.

For **photovoltaics**, new materials and improved efficiencies are expected to bring down costs considerably. The cost of thin-films cells, which make do with far less material, is expected to drop in particular. Figure 11.4 shows cost reductions in the past. In just a few years, the cost of electricity from photovoltaic systems is expected to reach that of conventional systems. But the production volume needs to be increased considerably to reach that goal.

The trendline in Figure 11.4 is the **learning curve**. The progress ratio pr can be determined for this cost trendline. This ratio indicates the rate at which costs will drop if the production volume or installed capacity doubles. A value of $pr = 0.8$ means that costs will drop by 20 % to 80 % of the original level when the production volume doubles. The progress ratio pr can be used to determine the learning factor:

$$E = -\frac{\ln pr}{\ln 2} \tag{6.18}$$

The learning value E combined with the cost c_0 of a production volume p_0 can be used to calculate the cost reduction c_1 of an increased production amount p_1:

$$c_1 = c_0 \cdot \left(\frac{p_1}{p_0}\right)^{-E} \tag{6.19}$$

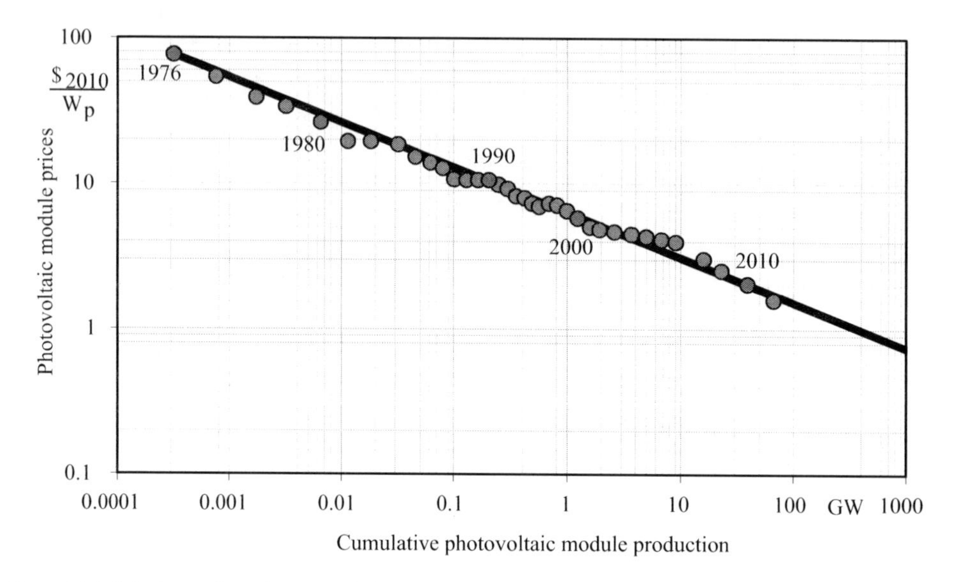

Figure 11.4 Inflation-adjusted prices of photovoltaic modules relative to cumulative photovoltaic module production from 1976 to 2011

A progress ratio of 0.8 is assumed for photovoltaics, compared with 0.88 for solar thermal power plants and 0.92 for wind turbines.

For photovoltaic systems, the learning value is thus $E = 0.322$. With market growth of 25 % per year, the production volume will grow 9.3-fold in 10 years. The cost will then drop to:

$$\frac{c_1}{c_0} = \left(\frac{p_1}{p_0}\right)^{-E} = \left(\frac{1{,}25^{10}\,p_0}{p_0}\right)^{-0{,}322} = 0.49$$

In other words, in 10 years photovoltaics will only cost around 50 % as much. In the past few decades, these reductions have consistently been reached, and even greater growth is likely in the future.

In the next few years, there will be great changes due to cost dynamics, especially for photovoltaics. In 2012, Germany reached **grid parity** for photovoltaics (Figure 11.5). At that point, it is less expensive for households to use their own solar power than to purchase electricity from the grid.

The trick is to consume all of the power from the photovoltaic system directly. If excess power is sold to the grid, a photovoltaic system only pays for itself if that excess is properly paid for – or if it can be stored inexpensively for later consumption. As the cost of photovoltaic electricity drops farther below retail rates, the economic benefits of photovoltaics will increase. Financially, a photovoltaic system will pay for itself in the foreseeable future because so little power needs to be purchased from the grid even if excess solar power is no longer paid for at all.

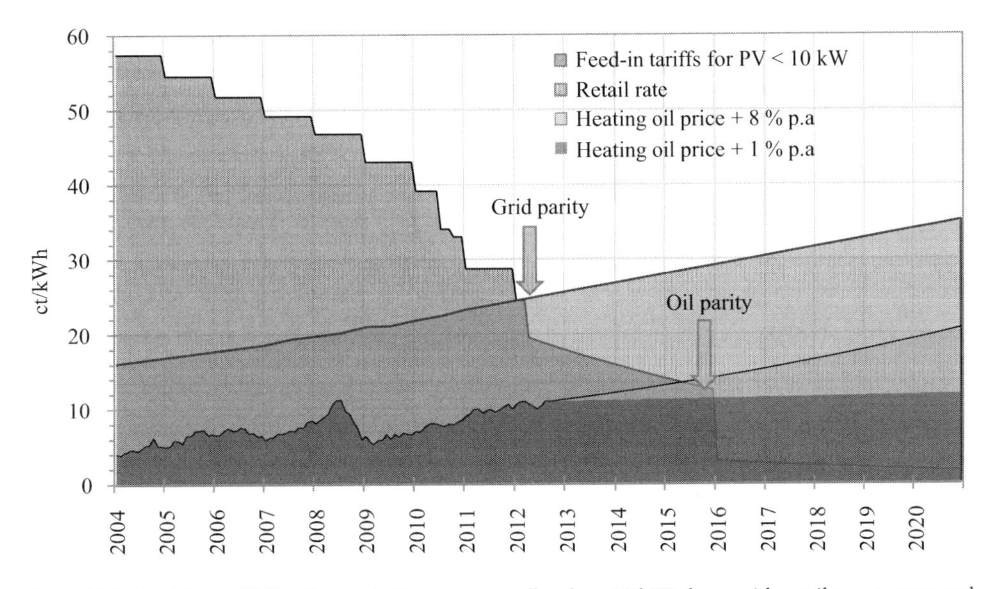

Figure 11.5 Feed-in tariffs for photovoltaic systems smaller than 10 kW along with retail power rates and fuel prices for oil heaters (boiler efficiency 80%) up to 2012 and a projection for trends up to 2020

This excess electricity can also be used to generate heat. Some time around 2020, **oil parity** will also be reached. It will then be less expensive to generate heat directly with photovoltaic electricity than to burn heating oil. Photovoltaics could then also increasingly make solar thermal less attractive. A heating system based purely on photovoltaics will not, however, pay for itself in northern Europe because seasonal storage would be needed, and that option is too expensive.

Renewable power plants, especially photovoltaics, are major new competitors for conventional power plants. Small renewable energy generators can be set up directly where power is consumed. They no longer compete with conventional power plant prices, but with the much higher retail rates, which also contain the cost of distributing electricity along with numerous taxes and levies. There is therefore a public debate about how people who make their own renewable energy can or even should cover these grid costs and taxes.

Cost conventional energy systems

Often, the affordability of renewable energy systems is compared with conventional systems without including external costs, such as environmental impacts. We will come back to these aspects below. The **power generation costs of utility power plants** currently range from €0.05/kWh$_{el}$ to €0.08/kWh$_{el}$ for new coal plants and combined-cycle gas turbines. Older coal plants that have already been written down can produce electricity below €0.05/kWh$_{el}$. As explained above, the affordability threshold is much higher for distributed photovoltaic systems. They compete with retail power rates of around €0.25/kWh$_{el}$ when the power is used directly.

The cost of generating heat from private **oil and gas heaters** is around €0.15 to €0.25/kWh_{therm}. The heat generation costs for hot water are generally higher because conventional systems do not run at full capacity during the summer and are therefore much less efficient. The heat generation costs of electric hot water heaters are above the price of retail electricity at around €0.25/kWhel.

The heat and power generation costs of conventional systems cannot be compared with all renewable energy generators, however. In general, a conventional system is needed in combination with solar hot water systems. In other words, expenses are incurred anyway for the conventional system. The solar thermal unit only saves costs for offset fuel and possibly lower maintenance costs for the conventional system, which may then also have a longer service life. Table 11.6 shows the cost of fuels.

Table 11.6 Average energy prices in Germany 2010/2011 [BAFA12, DEPV12, Sta11b]

Energy source and consumer group	*€/MWh*	*Energy source and consumer group*	*€/MWh*
Hard coal imports (at the border), Q2 2012	11.4	Light heating oil a), 11/2012	95.0
German hard coal, price at mine 2008	21.3	Natural gas b), 1st half 2012	63.7
Petroleum (at the border) 09/2012	56.8	Wood pellets c), 10/2012	48.3
Natural gas (at the border) 06/2012	29.1	Electricity d), 1st half 2012	259.5

a) 3,000 l delivered to a home; b) 30,000 kWh/a; c) 5 t delivered to a home; d) 3,500 kWh; including levies, fees, and taxes

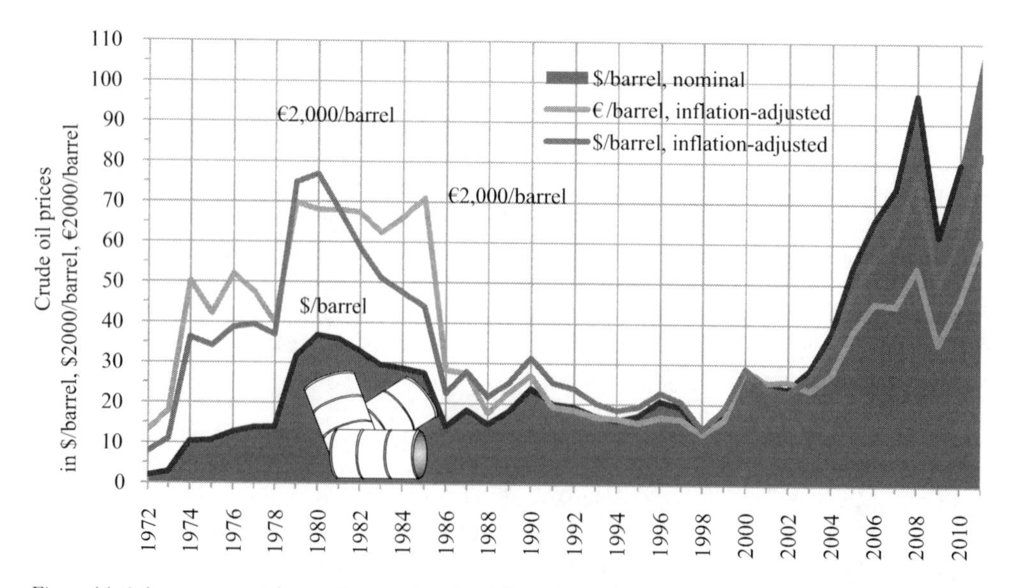

Figure 11.6 Average annual petroleum prices in daily prices adjusted for inflation and exchange rates ($, $2,000 or €2,000, 1 barrel = 158.7579 l)

It is assumed that the cost of fossil fuels will remain the same in the figures for power and heat generation costs. This assumption is unrealistic, however, in terms of the service lives that are common in the power plant sector. Figure 11.6 shows the trend in **petroleum prices**. If we account for inflation and changes in exchange rates, the maximum over the past 30 years has been five times greater than the minimum.

Prices for conventional energy sources can be expected to continue to increase significantly as resources become ever scarcer. Investors currently committed to relatively expensive renewable energy may then suddenly turn out to be the least expensive providers.

External costs of energy consumption

Economic feasibility calculations for conventional energy conversion systems, such as fossil and nuclear power plants, determine energy production costs as described above. In addition to the cost of building and operating power plants, there are also fuel costs – and other external costs that the power plant operators do not cover (completely).

These external costs include state subsidies, state support for research and development, end-of-life costs, and in particular the cost of environmental and health impacts from the use of these energy generators. The price of power and heat generally does not include these indirect, external costs. The result is an uneven playing field between renewable and conventional energy sources because the external costs of renewable energy are usually much lower. It is very difficult, however, to determine all external costs, and the process is often very controversial because of the wide range of assumptions made.

Subsidies on the energy market

It is relatively easy to determine the direct state subsidies in light of the comprehensive statistics available. For many years, Germany subsidized domestic hard coal mining with a surcharge added to the price of a kilowatt-hour of electricity. In addition to this surcharge shown in Table 11.7, other subsidies brought the total amount provided for German hard coal up to €4.5 billion by 1995.

Subsidies continued to be paid through the federal budget after the German Constitutional Court ruled against the surcharge (Table 11.8). In addition, the state of North Rhine-Westphalia had provided more than €400 million by 2010. These coal subsidies expire in 2018. By that time, hard coal mining in Germany will have received around €20 billion in additional subsidies since 2010.

By that time, the total will probably exceed €100 billion, equivalent to more than €140 billion when adjusted for inflation as direct subsidies for the German coal mining

Table 11.7 Subsidies for German hard coal mining (Kohlepfennig) [IZE95]

Year	*1975*	*1980*	*1985*	*1990*	*1995*
As % of power bill	3.24 %	4.5 %	3.5 %	8.25 %	8.5 %
Volume in million €	410	1,020	1,120	2,860	3,120

Table 11.8 State aid from the federal government for Germany's hard coal mining sector [Deu97, Deu99, Bmf12, Bmw06]

Year	*1995*	*1996*	*1997*	*1998*	*1999*	*2000*	*2001*	*2002*	*2003*
State aid in billion €	1,336	5,059	4,637	4,510	4,308	3,972	3,696	3,017	2,695

Year	*2004*	*2005*	*2006*	*2007*	*2008*	*2009*	*2010*	*2011*	*2012*
State aid in billion €	2,224	1,771	1,693	1,907	1,938	1,485	1,425	1,472	1,424

sector. This amount would suffice to build a whopping 80,000 MW of wind turbines, enough to meet a third of German power demand. In addition to direct subsidies, there were other tax incentives. A study conducted on behalf of Greenpeace and the German Wind Energy Association found that €311 $billion_{2012}$ had been provided to the German hard coal sector from 1970 to 2012 when adjusted for inflation, compared to €87 $billion_{2012}$ for German lignite and €213 $billion_{2012}$ for nuclear power. During that same timeframe, renewables received only €67 $billion_{2012}$ [FÖS12].

Federal subsidies are also provided for renewable energy but at a much lower level. The federal **250 MW Wind Programme** was launched to build up the wind power sector between 1989 and 2007. In 1995, €16 million was devoted to this programme, compared with only €1.5 million in 2004. In 2000, the **100,000 Roofs Programme** for photovoltaics began with a budget of €20 to €30 million per year.

In addition to these two programmes, energy audits were provided to show households and companies how to save energy; in 1995, €6 million was devoted to this project, compared with €32 million in 2010. In addition, there were small programmes for the use of renewable energy. In 1995, around €9 million was made available for this purpose, compared with €338 million in 2012.

In 1995, €31.8 million in state aid was made available to renewables and energy conservation/efficiency, less than 1 % of the subsidies for German hard coal mining. In 2012, state aid for renewables and energy conservation/efficiency amounted to 338 million €, roughly 24 % of the expenses for German hard coal mining.

In addition to direct state aid, there are also tax incentives. In 2006, tax incentives for biofuels and biomass for heat amounted to €2.144 billion relative to conventional fuels and heat sources. By 2010, these incentives had been reduced to €80 million. By the end of 2015, they will disappear altogether [Bmf12]. Tax incentives are also found in numerous other areas as well. For instance, there were incentives for overnight heating systems up to 2006 so that nuclear and lignite plants could run at a higher capacity around the clock. Comprehensive tax incentives also generally exist for fuels in the aviation sector. They do not apply for railways, however. Other incentives do not even show up in the German government's report on subsidies. For instance, nuclear power plant operators have to set aside provisions for the disposal of nuclear waste. These provisions amounting to tens of billions can be used, however, as investment capital tax-free.

Germany's **Renewable Energy Act** (EEG) pays for renewable electricity generation. There are no direct subsidies from the federal government in the process. Rather, the additional costs are immediately passed on to all power consumers. In 2013, this EEG surcharge amounted to 5.277 cent/kWh$_{el}$. That year, the differential costs amounted to around €20 billion, an amount that is increasingly used as an argument against the fast growth of renewables.

But the growth of renewables also saves costs elsewhere that more than compensate for these expenditures. For instance, the use of renewable energy for power generation reduced environmental damage by €8.9 billion in Germany in 2011; in addition, €2.9 billion in energy imports was offset. Because renewable power plants mainly push old, expensive conventional units out of the market, the merit order effect led to around €2.8 billion in savings from conventional power generation. Energy providers have not sufficiently passed on the savings to power customers, however. The growth of renewables also leads to municipal added value in the amount of €7.5 billion [Age12]. Companies in the renewables sector generate tax revenue and create new jobs, thereby reducing welfare expenditures.

Spending on research and development

In the past two decades in particular, tens of billions have been spent on research and development for nuclear power as shown in Table 11.9. Up to the mid-1970s, there was practically no support for renewable energy at all. It wasn't until the oil crisis and, in particular, the nuclear disaster in Chernobyl that more financing was made available for renewable energy and conservation.

Table 11.10 shows that the focus in the past few years is still on subsidizing research and development of conventional energy in Germany. Nuclear energy receives almost all of the funding for conventional energy, with around 24 % of the money being devoted to fusion research. For exploration of space, more than €1.05 billion is spent annually along with around €1.02 billion for defence research and technology, neither of which leads to major discussions of economic feasibility.

This unbalanced policy of subsidies leads to an uneven playing field in favour of nuclear energy in particular and to the detriment of renewables. Had such funding been provided for renewables instead of nuclear research, renewable energy would probably already be less expensive today than all conventional sources of energy.

Construction of a prototype fast breeder reactor in Kalkar, Germany, is a good example of mistakes made during research and development. Nearly €4 billion was invested in the facility that, instead of going into operation, became Germany's most expensive unfinished construction project. Today, it is home to an adventure park.

Table 11.9 Federal spending on research and development in the energy sector in millions of euros from 1956–1988 in Germany [Nit90]

Year	*1956–1988*
Coal and other fossil energy	1,907
Nuclear power	18,855
Renewables, conservation, and efficiency	1,577

Table 11.10 Federal spending on research and development in the energy sector in millions of € from 1989–2012 in Germany

Year	*1989*	*1991*	*1993*	*1995*	*1997*	*1999*	*2001*
Coal and other fossil energy	80.4	57.8	36.5	18.2	15.2	21.7	14.4
Nuclear power	522.0	507.9	435.8	435.8	441.8	438.3	411.4
Renewables, conservation, and efficiency	123.2	170.0	177.6	154.6	150.5	148.0	160.3

Year	*2003*	*2005*	*2007*	*2009*	*2010*	*2011*	*2012*
Coal and other fossil energy	8.0	1)	1)	1)	1)	1)	1)
Nuclear power	358.4	450.5	514.3	594.4	558.0	662.3	674.3
Renewables, conservation, and efficiency	202.2	2) 245.1	2) 254.8	2) 369.2	2) 377.2	2) 476.3	2) 506.8

(Data: [BMWi, Bmb12])
1) only reported jointly with renewables, efficiency, and conservation
2) including coal and other fossil energy

The cost of environmental and health impacts

It is especially difficult to determine the cost of environmental and health impacts, which makes the findings all the more controversial.

If a German nuclear plant were to completely melt down, the costs are estimated at around €5 trillion, more than twice the annual gross national income in the country. A nuclear plant's liability, however, is limited to €2.5 billion by law. The rest of the cost would be passed on to society. Insurance coverage for the complete amount would make nuclear power plants unprofitable.

But nuclear power plants in foreign countries can also lead to costs at home. For instance, the meltdown in Chernobyl led to tremendous costs in Germany when radioactively contaminated agricultural products had to be disposed of. And it does not take a reactor meltdown to cause great external costs either. Even when a nuclear plant runs smoothly, there are external costs not covered by the plant owner. The environmental damage from the retired uranium mines run by Wismut AG in Thüringen and Saxony are estimated at least €6.5 billion, while the cost of repairing the Asse nuclear repository is estimated at around €2 billion. When uranium is processed and enriched, used in power plants, transported, and eventually disposed of, the environment is impacted in ways not covered by power plant operators. The operation of nuclear plants can have health impacts such as cancer, and these costs are also passed on to society.

Fossil energy also entails high indirect costs not included in the price of this energy. Some of the most obvious examples are damage to **buildings, acid rain, and health impacts from air pollution**. Air pollution from the use of fossil energy probably causes more than €2 billion in material damage alone each year. The costs of health impacts from fossil energy – such as respiratory diseases, allergies, and cancer – are hard to quantify. In all likelihood, they probably cost several billion euros per year.

The cost of future **damage stemming from the greenhouse effect** is also impossible to determine. International insurance firms say that the number of natural disasters has risen considerably in the past few years (Table 11.11). It cannot be stated with absolute certainty, however, that the increase in natural disasters is the result of the anthropogenic greenhouse effect. Nonetheless, these figures show the costs that the greenhouse effect can cause.

The rise in sea level resulting from the greenhouse effect is also expected to cause major damage. Roughly 2 million km^2 of land will be lost if sea levels rise by just one metre. More than 50 million people would be directly affected along with property worth more than US$ 1,000 billion [Bro06].

Other external costs

When nuclear energy is used, there are especially high costs for the disposal of radioactive waste. In addition to billions for research into and construction of suitable repositories, the government also has to cover additional costs. For instance, public protests in 1997 in Germany led to the country's biggest police presence in the country's history along the transport line for radioactive waste to the intermediate repository at Gorleben. The cost of the roughly 30,000 police officers is estimated at around €50 million.

It is hard to determine other external costs. Among other things, they include administrative costs, such as permits, and the infrastructure needed to operate power plants. Furthermore, there are costs for public radioactivity measurement networks and public disaster relief that would not be needed to the same degree if nuclear plants were not in operation to the current extent.

Internalizing external costs

As the discussions above show, the external costs of renewables are much lower than those of fossil and nuclear energy. To provide a level playing field, a surcharge should be added to conventional energy sources. It could be used to cover the damage caused and help finance the transition to renewable energy, which entails lower external costs. Such considerations are behind discussions about carbon taxation and emissions trading. It is largely up to political decision-makers to regulate markets for the good of the national economy.

It is not easy to quantify external costs for a calculation of such a surcharge. On the one hand, it is hard to attribute specific damage to specific polluters; on the other hand, a lot of the environmental impact is not even known yet, so the external costs cannot be determined. In the past, the estimates of external costs from energy consumption diverged greatly in a number of publications on this topic.

Table 11.11 Major weather disasters (natural disasters excluding earthquakes) and the damage they cause [Mun12]

Time period	*1950–59*	*1960–69*	*1970–79*	*1980–89*	*1990–99*	*2000–09*
Number of major natural disasters	13	16	29	44	74	28
Macroeconomic damage in US$ billion$_{2009}$	53.1	72.4	97.5	155.7	528.0	435.2
Insured damage in US$ billion$_{2009}$	1.6	8.1	15.0	29.0	125.7	193.8

One of the first comprehensive investigations was conducted by Hohmeyer [Hoh89, Hoh91]. In addition to the costs of environmental damage that can be quantified, this study investigated the cost of exploiting resources for fossil and nuclear energy. Because fossil and nuclear fuels will run out within just a few generations, subsequent generations will not have these options. Provisions therefore have to be set aside to compensate for higher energy costs in the future. In addition, there are costs for public goods and services, subsidies, and research and development funding. The psycho-social follow-up costs of illnesses and deaths, indirect environmental impacts, the environmental cost of the nuclear fuel cycle, hidden subsidies, and the cost of the greenhouse effect are not taken into consideration, to mention just a few items.

The external cost of **nuclear power** has been estimated at up to €0.36_{1982}/kWh_{el}. The external costs of fossil energy are somewhat lower. Germany's **current power supply** consists partly of a combination of fossil energy and nuclear power, and the external costs were found in this study to range from €0.026_{1982}/kWh_{el} to €0.133_{1982}/kWh_{el}. Power plant operators should have to cover these costs when they sell electricity in order to compensate for the external costs from the operation of these power plants. If these costs were added to the power price, it would be much higher and even double in extreme cases.

Wind power and **photovoltaics** also involve external costs, though they are much lower than those of fossil and nuclear energy. In return, there are external benefits from the offset external costs of fossil and nuclear energy that is substituted. A transition to renewables would also produce additional macroeconomic benefits when new jobs are created. A kilowatt-hour of wind power was found to produce **external benefits** ranging from €0.026 to €0.133_{1982}/kWh_{el}, meaning that costs within that range could be offset. These costs should therefore be paid to the owner of the wind turbine in addition to revenue from power sales.

Hohmeyer's study finds that our failure to account for the social and external costs of conventional power puts renewables at a tremendous disadvantage. In other words, renewable energy sources are not able to exploit their true competitive potential and will be developed much later than would be the optimal case if the entire cost situation was taken into account, including the social costs and macroeconomic benefits for society.

A critical view on economic feasibility analyses

There are several ways to determine costs. In practice, approaches from economics are used, and though they generally include a calculation of returns, they exclude the external costs explained above. In the section above, the limits of classic economic feasibility analyses is explained; it is hard for them, if not impossible, to take account of external costs. But even aside from external costs, there are still aspects of classic economic feasibility calculations that deserve closer scrutiny.

Infinite capital growth

We can start with an example given in [Goe94]. Let us assume that around the year zero AD one € cent is invested at an interest rate of 4 %. We can now calculate what this investment would be worth 2,000 years later. The formula

$$c_{2000} = 0.01\text{ €}\ (1+0.04)^{2000} = 1.166 \cdot 10^{32}\text{ €}$$

produces the unimaginably large sum of €$1.166 \cdot 10^{32}$. If gold costs €44,000/kg, this amount could be used to purchase an astonishing $2.65 \cdot 10^{27}$ kg of gold. Gold has a density of 19.29 kg/dm³, equivalent to a volume of $1.37 \cdot 10^{14}$ km³ for that amount. In contrast, the earth only has a volume of $1.1 \cdot 10^{12}$ km³, meaning that this amount of gold would have 125 times the volume of this planet.

This example shows that **constant capital growth** over very long time frames is not possible; this compound interest calculation leads into infinity over very long time frames. No one on earth would be able to pay this interest. Over very long time frames, the classic calculation approach therefore makes no sense. Unfortunately, no one can say at what point calculations of compound interest need to be abandoned. At an interest rate of 8 %, the capital invested increases tenfold every 30 years; in 100 years, it is already 2,200 times more valuable. And yet, these are the time frames used by investors in the energy sector. The longer the time frame and the greater the interest rate, the greater the probability that this capital will be lost or depreciated. In the past, wars, currency reforms, and increasingly environmental disasters have led to a total loss of capital. In the long term, current economic thought practically requires such events because the outcome of uninterrupted capital growth would be infinite assets, as the example above shows.

Returns far above economic growth are only possible if capital is redistributed on a large scale. It is not possible to make such returns from mere economic action. The economic crisis of 2009 showed once again that promised returns and capital losses are closely related. The last 50 years of the twentieth century were a time of relative stability, but the first half of the last century was characterized by the First and Second World Wars and the economic Depression of the 1930s, which led to a loss of large amounts of capital within less than 20 years. Today, the rising number of natural disasters caused by our current energy consumption and climate change increasingly make losses more likely.

Often, a capital investment can lead to a **total loss**. One example is German war bonds from the First World War. They were issued to finance the war with a promise of high returns from the state. In the end, investors not only lost all of their capital, but even used this capital to destroy additional assets beyond that investment. Investments in the energy sector are similar. If a nuclear plant meltdown, investors not only lose all of their invested capital, but also much greater assets. Investments in fossil energy speed up the greenhouse effect. The impact of the greenhouse effect, such as more frequent hurricanes, will also lead to great losses. These investments do not necessarily, however, lead to a loss for these particular investors.

For these reasons, it makes sense to pay attention not only to the greatest possible growth of capital, but also to protecting current capital. To do so, capital can be invested in technologies that reduce negative impacts in other areas. An investment in renewable energy systems should also be viewed against this backdrop.

The responsibility of capital

For individual investors, the consequences of an investment are often hard, if not impossible, to assess. The greater the return, the greater the risk, as the saying goes. But the risks in the energy sector not only concern the capital invested, but also assets beyond the investment itself. To be sure, not all investments that promise a return lead to a loss of the initial investment. But every investment entails responsibility for the capital invested, especially in the energy sector. Article 14.2 of the German Basic Law puts it well: 'Property entails obligations. Its use shall also serve the public good.' [Deu94])

At this point, it is worth asking whether there should be returns at all on investments in renewables. When people purchase luxury sports cars, no one asks about economic feasibility. There are far less expensive ways of moving people from one place to another. Nonetheless, tens of thousands and sometimes hundreds of thousands of euros are spent without the question of economics ever being asked. But when a similar amount is to be invested in renewables, suddenly the return becomes an issue. With cars, the argument is greater comfort and quality of life. Do these arguments not apply to renewables? Energy sources that have fewer negative impacts on the environment and people's health definitely increase standards of living, and such benefits are hard to measure in monetary terms.

For these reasons, so many citizens in Germany invested in photovoltaics and solar thermal at a time when profits were unlikely. A number of companies have also built 'unprofitable' systems. Sometimes, the firms aimed to improve their image. Prestige is a value that is hard to measure economically. Perhaps one day an attractive, easily visible photovoltaic system on an attractive building will have the same sex appeal as a luxury sports car or an extravagant holiday.

If all of the aspects listed above are taken into account, renewables will quickly make up a far larger share of energy supply. As a result, the consequences of the greenhouse gas and possible nuclear risks will be minimized, and the earth will be maintained as a habitable place for future generations.

Appendix

Important constants

Electron mass	$m_e = 9.1093897 \cdot 10^{-31}$ kg
Proton mass	$m_p = 1.6726231 \cdot 10^{-27}$ kg
Atomic mass unit	$u = 1.660565 \cdot 10^{-27}$ kg
Elementary charge	$e = 1.60217733 \cdot 10^{-19}$ A s
Magnetic field constant	$\mu_0 = 4\pi \cdot 10^{-7}$ V s /(A m)
Electric field constant	$\varepsilon_0 = 8.85418781762 \cdot 10^{-12}$ A s/(V m)
Boltzmann constant	$k = 1.380658 \cdot 10^{-23}$ J/K
Stefan–Boltzmann constant	$\sigma = 5.67051 \cdot 10^{-8}$ W/(m^2 K^4)
Light speed (in a vacuum)	$c = 2.99792458 \cdot 10^{8}$ m/s
Planck constant	$h = 6.6260755 \cdot 10^{-34}$ J s
Faraday constant	$F = 96{,}485$ As/mol
Solar constant	$E_0 = 1{,}360.8 \pm 0.5$ W/m^2
Surface temperature of the sun	$T_{sun} = 5{,}777$ K
Radiant emittance of the sun	$M_{e,S} = 63.3$ MW/m^2
Divergence of sunlight	$\alpha_D = 0.53° = 0.0093$ rad
Maximum concentration factor	$C_{max} = 46211$
Betz coefficient of power	$c_{P,Betz} = 16/27 = 0.59259$

Major English and American units

Length	1 ft (foot) = 12 in (inch) = 1/3 yd (yard) = 0.3048 m
Length	1 mile = 1,760 yd = 1609.344 m
Mass	1 lb (pound) = 16 oz (ounce) = 0.45359237 kg
Volume	1 gal (gallon) = 231 in^3 = 3.7854345 l
Volume	1 ptr.bbl (petroleum barrel) = 42 ptr.gal (petroleum gallon) = 158.9873 l (crude oil)
Temperature	ϑ_F (in °F, degrees Fahrenheit) = 9/5 · ϑ (in °C, degrees Celsius) + 32
Power	1 h.p. (horsepower) = 745.7 W
Pressure	1 inHg (inch mercury) = 25.4 mmHg = 3,386.3788 Pa
Energy	1 Btu (British thermal unit) = 1.05505585262 kJ = 0.000293071 kWh
Irradiation	1 ly (langley) = 1 cal/cm^2 = 4.1868·10^4 J/m^2 = 0.01163 kWh/m^2

Bibliography

Chapter 1

[AGEB12] AG Energiebilanzen e.V.: *Daten und Infografiken*. Internet: www.ag-energiebilanzen.de, 2012

[atw12] atw Redaktion: Kernenergie Weltreport 2011, atw 57. Jg. (2012) Heft 4, pp. 271–276.

[Bec92] Becker, M.; Meinecke, W.: *Solarthermische Anlagentechnologien im Vergleich*. Berlin: Springer, 1992

[BP12] BP: *BP Statistical Review of World Energy 2012*. London: 2012

[BMU] Bundesministerium für Umwelt (Hrsg.): *Erneuerbare Energien in Zahlen*. Berlin, verschiedene Jahrgänge bis 2012

[BMWi] Bundesministerium für Wirtschaft und Technologie (Hrsg.): *Energiedaten*. Berlin, verschiedene Jahrgänge bis 2011

[BWE11] Bundesverband WindEnergie e.V. (BWE): *Potenzial der Windenergienutzung an Land*. Berlin, 2011

[Dewi12] Deutsches Windenergie-Institut (DEWI): *Statistik, Status 31.12.2011*. Internet: www.dewi.de, 2012

[EEA10] European Environment Agency (EEA): *The European Environment – State and Outlook 2010, Understanding Climate Change*. Kopenhagen 2010

[EIA12] US Energy Information Administration (EIA): *International Energy Statistics*. Internet: www.eia.gov/countries/, 2012

[EnB12] EnBW: *EnBW Geschäftsbericht 2011*. Internet: www.enbw.com, 2012

[Enq95] Enquete-Kommission 'Schutz der Erdatmosphäre' des 12. Deutschen Bundestages (Hrsg.): *Mehr Zukunft für die Erde*. Bonn: Economica Verlag, 1995

[eon12] e.on: *Facts & Figures 2012*. Internet: www.eon.de, 2012

[EWG06] Energy Watch Group (Hrsg.): *Uranium Resources and Nuclear Energy*. Ottobrunn: EWG, 2006

[EST03] European Solar Thermal Industry Federation (ESTIF): *Sun in Action II*. Brüssel: ESTIF, 2003

[EST12] European Solar Thermal Industry Federation (ESTIF): *Solar Thermal Markets in Europe*. Brüssel: ESTIF, verschiedene Jahrgänge bis 2012

[Hil95] Hiller, Karl: Erdöl: Globale Vorräte, Ressourcen, Verfügbarkeiten. In: *Energiewirtschaftliche Tagesfragen* 45. Jg. (1995) Heft 11, pp. 698–708

[Hof95] Hoffmann, Cornelis: Bereitstellungsnutzungsgrade elektrischer Energie. In: *Elektrizitätswirtschaft* Jg. 94 (1995) Heft 11, pp. 626–632

[IEA12] International Energy Agency IEA (Hrsg.): *Key World Energy Statistics 2012*. Paris: 2012

[IEA12b] International Energy Agency IEA-PVPS (Hrsg.): *Trends in Photovoltaic Applications between 1992 and 2011*. Paris: 2012

[IEA12c] International Energy Agency Solar Heating and Cooling Programme IEA-SHC (Hrsg.): *Solar Heat Worldwide*. Paris, verschiedene Jahrgänge bis 2012

[IPC00] Intergovernmental Panel on Climate Change (IPCC): *IPCC Special Report Emissions Scenarios*. Nairobi: IPCC, 2000

[IPC01] Intergovernmental Panel on Climate Change (IPCC): *Third Assessment Report of Working Group I, Summary for Policy Makers*. Shanghai: IPCC, 2001

[IPC07] Intergovernmental Panel on Climate Change (IPCC): *Climate Change 2007, Synthesis Report.* Valencia: IPCC, 2007

[Nat12] Naturfreunde Deutschland e.V.: *Atomausstieg Selber Machen.* Internet: www.atomaustieg-selber-machen.de, 2012

[NOAA12] National Oceanic & Atmospheric Administration (NOAA), Earth System Research Laboratory (ESRL): *The NOAA Annual Greenhouse Gas Index.* Internet: www.esrl.noaa.gov/gmd/aggi, 2012

[Qua00] Quaschning, Volker: *Systemtechnik einer klimaverträglichen Elektrizitätsversorgung in Deutschland für das 21.Jahrhundert.* Düsseldorf: VDI Fortschritt-Berichte Reihe 6 No. 437, 2000

[Qua09] Quaschning, Volker: *Erneuerbare Energien und Klimaschutz.* München: Hanser Verlag, 2009

[Qua12] Quaschning, Volker: *Datenservice Regenerative Energien und Klimaschutz.* Internet: www.volker-quaschning.de/datserv, 2012

[RWE12] RWE: *Strom und Wärmeerzeugung.* Internet: www.rwe.com, 2012

[Sch12] Schaeffer, Michiel, Hare, William, Rahmstorf, Stefan, Vermeer, Martin: Long-term sea-level rise implied by 1.5° C and 2° C warming levels. In: *Nature Climate Change* 6/2012

[Sel90] Selzer, Horst: Windenergie, Studie A.2.2a. In: *Energie und Klima*, Band 3 Erneuerbare Energien. Enquete-Kommission 'Vorsorge zum Schutz der Erd atmosphäre' des Deutschen Bundestages, Economica Verlag, 1990

[Sti94] Stiftung Warentest: *test-Heft* 7/1994, Berlin

[UNF98] United Nations Framework Convention on Climate Change UNFCCC: *Methodological Issues While Processing Second National Communications: Greenhouse Gas Inventories.* Buenos Aires: FCCC/SBSTA, 1998

[UNF12] United Nations Framework Convention on Climate Change UNFCCC: *National Greenhouse Gas Inventory Data for the Period 1990–2010.* Internet: www.unfccc.de, 2012

[Vat12] Vattenfall: *Vattenfall Annual Report 2011.* Internet: www.vattenfall.de, 2012

[WBG08] Wissenschaftlicher Beirat der Bundesregierung Globale Umweltveränderungen WBGU: *Kassensturz für den Klimavertrag – Der Budgetansatz.* Berlin: Sondergutachten, 2009. Internet: www.wbgu.de

[Wei96] v. Weizsäcker, E.U.; Lovins, A.B.; Lovins, L.H.: *Faktor Vier.* München: Droemersche Verlagsanstalt Knaur, 1996

Chapter 2

[DIN4710] Deutsches Institut für Normung e.V. (DIN): *DIN 4710, Meteorologische Daten zur Berechnung des Energie Verbrauchs von Heiz und Raumlufttechnischen Anlagen.* Berlin: Beuth Verlag, 1982

[DIN5031] Deutsches Institut für Normung e.V. (DIN): *DIN 5031, Strahlungsphysik im optischen Bereich und Lichttechnik.* Berlin: Beuth Verlag, 1982

[DIN5034] Deutsches Institut für Normung e.V. (DIN): *DIN 5034 Teil 2, Tageslicht in Innenräumen.* Berlin: Beuth Verlag, 1985

[DIN9488] Deutsches Institut für Normung e.V. (DIN): *DIN EN ISO 9488, Sonnenenergie – Vokabular.* Berlin: Beuth Verlag, 1999

[Die57] Dietze, Gerhard: *Einführung in die Optik der Atmosphäre.* Leipzig: Akademische Verlagsgesellschaft Geest & Portig K.G., 1957

[IEC95] International Electrotechnical Commission (IEC): *IEC 904–9: Photovoltaische Geräte – Teil 9: Leis tungsanforderungen an Sonnensimulatoren.* Genf: IEC, 1995

[JRC10] European Commission Joint Research Centre (JRC): *Photovoltaic Geographical Information System PVGIS.* Internet: http://re.jrc.ec.europa.eu/pvgis. ISPRA, 2010

[Kam90] Kambezidis, H.D.; Papanikolaou, N.S.: Solar position and atmospheric refraction. In: *Solar Energy* Vol. 44 (1990), pp. 143–144

[Klu79] Klucher, T.M.: Evaluation of models to predict insolation on tilted surfaces. In: *Solar Energy* Vol. 23 (1979), pp. 111–114

[Kop11] Kopp, Greg; Lean, Judith L.: A new, lower value of total solar irradiance: Evidence and climate significance. In: *Geophysical Research Letters*, Vol. 38, L01706, 7 pp. 2011

[Pal96] Palz, W.; Greif, J.: *European Solar Radiation Atlas*. Berlin: Springer, 1996

[Per86] Perez, Richard; Stewart, Ronald: Solar irradiance conversion models. In: *Solar Cells* Vol. 18 (1986), pp. 213–222

[Per87] Perez, Richard; Seals, Robert; Ineichen, Pierre; Stewart, Ronald; Menicucci, David: A new simplified version of the Perez diffuse irradiance model for tilted surfaces. In: *Solar Energy* Vol. 39 (1987), pp. 221–231

[Per90] Perez, Richard; Ineichen, Pierre; Seals, Robert; Michalsky, Joseph; Stewart, Ronald: Modeling daylight availability and irradiance components from direct and global irradiance. In: *Solar Energy* Vol. 44 (1990), pp. 271–289

[Qua96] Quaschning, Volker: *Simulation der Abschattungsverluste bei solarelektrischen Systemen*. Berlin: Verlag Dr. Köster, 1996 – ISBN 3–89574–191–4

[Rei89] Reindl, D.T.; Beckman, W.A.; Duffie, J.A.: Diffuse fraction correlations. In: *Proceedings of ISES Solar World Conference 1989*, pp. 2,082–2,086

[Sat87] Sattler, M.A.; Sharples, S.: Field Measurements of the Transmission of Solar Radiation through Trees. In: *Proceedings of ISES Solar World Conference 1987*, pp. 3,846–3,850

[Sch70] Schulze, R.: *Strahlenklima der Erde*. Darmstadt: Steinkoff, 1970

[TÜV84] TÜV-Rheinland: *Atlas über die Sonnenstrahlung in Europa*. TÜV-Verlag, 1984

[Wal78] Walraven, R: Calculating the position of the sun. In: *Solar Energy* Vol. 20 (1978), pp. 393–397

[Wil81] Wilkinson, B.J.: An Improved FORTRAN Program for the Rapid Calculation of the Solar Position. In: *Solar Energy* Vol. 27 (1981), pp. 67–68

Chapter 3

[Bun92] Bundesministerium für Forschung und Technologie (Hrsg.): *Erneuerbare Energien*. Bonn, 1992

[BdE96] Bund der Energieverbraucher (BdE, Hrsg.): *Phönix Solar Projekt. Informationsschrift*. Rheinbrettenbach: BdE, 1996

[DIN06] Deutsches Institut für Normung e. V. (DIN): *DIN EN 12975–2, Thermische Solaranlagen und ihre Bauteile – Kollektoren – Teil 2: Prüfverfahren*. Berlin: Beuth Verlag, 2006

[Fac90] Fachinformationszentrum Karlsruhe (Hrsg.): *Transparente Wärmedämmung (TWD) zur Gebäude heizung mit Sonnenenergie*. BINE Projekt Info-Service No. 2/1990

[Fac93] Fachinformationszentrum Karlsruhe (Hrsg.): *Erfahrungen mit solarbeheizten Schwimm bädern*. BINE Projekt Info-Service No. 8/1993

[Fac95] Fachinformationszentrum Karlsruhe (Hrsg.): *Wärmedämmung für Warmwasser speicher, Heizkessel und Kühlzellen*. BINE Projekt Info-Service No.11/1995

[Gie89] Gieck, K.: *Technische Formelsammlung*. Heilbronn: Gieck Verlag, 1989

[Hah94] Hahne, E.; Kübler, R.: Monitoring and simulation of the thermal performance of solar heated outdoor swimming pools. In: *Solar Energy* Vol. 53 (1994), pp. 9–19

[Hum91] Humm, Othmar: *Niedrig Energiehäuser*. Staufen: Ökobuch Verlag, 1991

[Kha95] Khartchenko, N.: *Thermische Solaranlagen*. Berlin: Springer, 1995

[Kle93] Kleemann, M.; Meliß, M.: *Regenerative Energiequellen*. Berlin: Springer, 1993

[Lad95] Ladener, Heinz: *Solaranlagen*. Staufen: Ökobuch Verlag, 1995

[Smi94] Smith, Charles C.; Löf, George; Jones, Randy: Measurement and analysis of evaporation from an inactive outdoor swimming pool. In: *Solar Energy* Vol. 53 (1994) No. 1, pp. 3–7

[The85] Theunissen, P.-H.; Beckman, W.A.: Solar transmittance characteristics of evacuated tubular collectors with diffuse back reflectors. In: *Solar Energy* Vol. 35 (1985) No. 4, pp. 311–320

[Ung91] Unger, J.: Aufwindkraftwerk contra Photovoltaik. In: *BWK* Vol. 43 (1991) No. 7/7, pp. 375–379

[VDI2067] Verein Deutscher Ingenieure VDI (Hrsg.): VDI 2067 Blatt 4. *Berechnung der Kosten von Wärme versorgungsanlagen; Warmwasserversorgung*. Düsseldorf: VDI Verlag, 1982

[Wag95] Wagner & Co. (Hrsg.): *So baue ich eine Solaranlage, Technik, Planung und Montage*. Cölbe: Wagner & Co. Solartechnik GmbH, 1995

Chapter 4

[Dud94] Dudley, Vernon E.; Kolb, Gregroy J.; Mahoney, A. Roderick; Matthews, Chauncey W.: *Test Results SEGS LS-2 Solar Collector.* Sandia Report SAN94–1884. Sandia National Labaratories. Albuquerque: 1994

[Her12] Hering, E.; Martin, R.; Stohrer, M.: *Physik für Ingenieure.* Berlin: Springer Verlag, 2012

[Hos88] Hosemann, G. (Hrsg.): *Hütte Taschenbücher der Technik, Elektrische Energietechnik, Band 3 Netze.* Berlin: Springer, 29. Auflage 1988

[Kle93] Kleemann, M.; Meliß, M.: *Regenerative Energiequellen.* Berlin: Springer, 1993

[Lip95] Lippke, Frank: *Simulation of the Part-Load Behavior of a 30 MWe SEGS Plant.* Sandia Report SAN95–1293. Sandia National Labaratories. Albuquerque: 1995

[Pil96] Pilkington Solar Internation (Hrsg.): *Statusbericht Solarthermische Kraftwerke.* Köln: 1996

[Qua05] Zukunftsaussichten von Solarstrom. In: *Energiewirtschaftliche Tagesfragen* 55. Jg (2005) Heft 6, pp. 386–388

[Sch02] Schlaich Bergermann und Partner (Hrsg.): *EuroDish-Stirling System Description.* Stuttgart: 2002

[Sti85] Stine, William B.; Harrigan, Raymond W.: *Solar Energy Fundamentals and Design.* New York: John Wiley & Sons, 1985

Chapter 5

[DGS08] Deutsche Gesellschaft für Sonnenenergie (DGS, Hrsg.): *Leitfaden Photovoltaische Anlagen.* Berlin: 2008

[DIN03] Deutsches Institut für Normung e.V. (DIN): DIN EN V 61000 Teil 2–2, VDE 0839 Teil 2–2. Elektromagnetische Verträglichkeit (EMV), *Verträglichkeitspegel für nieder fre quente Störgrößen und Signalübertragung in öffentlichen Niederspannungsnetzen.* Berlin: Beuth Verlag, 2003

[DIN05] Deutsches Institut für Normung e.V. (DIN): DIN EN 61000–3–2, VDE 0838 Teil 2. Elektro magnetische Verträglichkeit (EMV), *Grenzwerte für Oberschwingungsströme* (Geräte-Eingangsstrom 16 A je Leiter). Berlin: Beuth Verlag, 2005

[DIN05b] Deutsches Institut für Normung e.V. (DIN): DIN EN 61215 / IEC 61215 Ed. 2. *Terrestrische Photovoltaik-(PV-)Module mit Silizium-Solarzellen – Bauarteignung und Bauartzulassung.* Berlin: VDE Verlag, 2005

[DIN08] Deutsches Institut für Normung e.V. (DIN): DIN EN 61646 / IEC 61646 Ed. 2. *Terrestrische Dünnschicht-Photovoltaik-(PV-)Module – Bauarteignung und Bauartzulassung.* Berlin: VDE Verlag, 2008

[Goe05] Goetzberger, Adolf; Hoffmann, Volker U.: *Photovoltaic Solar Energy Generation.* Berlin: Springer, 2005

[Gre78] Gretsch, Ralf: *Ein Beitrag zur Gestaltung der elektrischen Anlage in Kraftfahrzeugen.* Nürnberg-Erlangen: Habilitationsschrift, 1978

[Has86] Hasyim, E.S.; Wenham, S. R.; Green, M.A.: Shadow tolerance of modules incorporating integral bypass diode solar cells. In: *Solar Cells* Vol. 19 (1986), pp. 109–122

[Her12] Hering, E.; Martin, R.; Stohrer, M.: *Physik für Ingenieure.* Berlin: Springer Verlag, 2012

[Köt96] Köthe, Hans K.: *Stromversorgung mit Solarzellen.* München: Franzis, 1996

[Las90] Lasnier, F.; Ang, T.G.: *Photovoltaic Engineering Handbook.* Bristol: Hilger, 1990

[Lec92] Lechner, M.D.: *Physikalisch-chemische Daten.* Berlin: Springer, 1992

[Lew01] Lewerenz, H.J.; Jungblut, H.: *Photovoltaik.* Berlin: Springer, 2001

[Mic92] Michel, Manfred: *Leistungselektronik.* Berlin: Springer, 1992

[PRE94] Fachbereich Physik, Arbeitsgruppe regenerative Energiesysteme (PRE), Universität Oldenburg: *Handbuch zu INSEL (Interactive Simulation of Renewable Energy Supply Systems).* Oldenburg, 1994

[Qua96a] Quaschning, Volker: *Simulation der Abschattungsverluste bei solarelektrischen Systemen.* Berlin: Verlag Dr. Köster, 1996 – ISBN 3–89574–191–4

[Qua96b] Quaschning, Volker; Hanitsch, Rolf: Höhere Erträge durch schattentolerante Photovoltaikanlagen. In: *Sonnenenergie & Wärme technik* 4/96, pp. 30–33

[Qua09] Quaschning, Volker: *Erneuerbare Energien und Klimaschutz*. München: Hanser Verlag, 2009

[Qua12] Quaschning, Volker; Weniger, Johannes; Tjarko, Tjaden: Photovoltaik – Der unterschätzte Markt. In: *BWK* Vol. 64 (2012) No. 7/8, pp. 25–28

[Tie02] Tietze, U.; Schenk, Ch.: *Halbleiter-Schaltungstechnik*. Berlin: Springer, 2002

[VDE11] VDE: *VDE-AR-N 4105 Anwendungsregel: Erzeugungsanlagen am Niederspannungsnetz*. Berlin: VDE-Verlag, 2011

[Wag06] Wagner, Andreas: *Photovoltaik Engineering*. Berlin: VDI Springer, 2006.

[Wag07] Wagemann, Hans-Günther; Eschrich, Heinz: *Photovoltaik*. Wiesbaden: Teubner Verlag, 2007

[Wen12] Weniger, Johannes; Tjaden, Tjarko; Quaschning, Volker: Solare Unabhängigkeitserklärung. In: *Photovoltaik* 10/2012, pp. 50–54

[Wol77] Wolf, M.; Noel, G.T.; Stirn, R.J.: Investigation of the double exponential in the current-voltage characteristics of silicon solar cells. In: *IEEE Transactions on Electron Devices* Vol. ED-24 (1977) No. 4, pp. 419–428

Chapter 6

[BDEW08] Bundesverband der Energie und Wasserwirtschaft e.V. BDEW (Hrsg.): *Technische Richtlinie Erzeugungsanlagen am Mittelspannungsnetz*. Berlin: BDEW, 2008

[Bet26] Betz, Albert: *Windenergie und ihre Ausnutzung durch Windmühlen*. Staufen: Ökobuch, Unveränderter Nachdruck aus dem Jahr 1926

[Chr89] Christoffer, Jürgen; Ulbricht-Eissing, Monika: *Die bodennahen Windverhältnisse in der Bundesrepublik Deutschland*. Offenbach: Deutscher Wetterdienst, Ber. No. 147, 1989

[Ene06] Enercon: *Enercon Windenergieanlagen – Produktübersicht*. Aurich: Enercon GmbH, 2006

[Fis06] Fischer, Rolf: *Elektrische Maschinen*. München: Hanser Verlag, 2006

[Gas07] Gasch, Robert; Twele, Jochen (Hrsg.): *Windkraftanlagen*. Stuttgart: Teubner, 2007

[Hau96] Hau, Erich: *Windkraftanlagen*. Berlin: Springer, 1996

[Her12] Hering, E.; Martin, R.; Stohrer, M.: *Physik für Ingenieure*. Berlin: Springer Verlag, 2012

[Kle93] Kleemann, M.; Meliß, M.: *Regenerative Energiequellen*. Berlin: Springer, 1993

[Mol90] Molly, Jens-Peter: *Windenergie*. Karlsruhe: C.F. Müller, 1990

[Mül94] Müller, G.: *Grundlagen elektrischer Maschinen*. Weinheim: VCH Verlagsgesellschaft, 1994

[Ris09] Risø National Laboratory, Wind Energy Division (Hrsg.): *WAsP – Wind Atlas Analysis and Application Program*. Riskilde: Risø Nat. Laboratory, 2009

[Tro89] Troen, Ib; Petersen, Erik L.: *European Wind Atlas*. Roskilde: RisØ National Laboratory, 1989 – ISBN 87–550–1482–8

[VDI94] Verein Deutscher Ingenieure (Hrsg.): Stoffwerte von Luft. In: *VDI-Wärmeatlas*. Düsseldorf: VDI-Verlag, 1994

[Ves97] Vestas: Technische Unterlagen zu den Windkraftanlagen V42 und V44. Husum: Vestas Deutschland GmbH, 1997

Chapter 7

[BAFU12] Schweizerisches Bundesamt für Umwelt BAFU: *Hydrologische Daten*. Internet: www.hydrodaten.admin.ch/de/, 2012

[Bar04] Bard, J.; Caselitz, P.; Giebhardt, J.; Peter, M.: Erste Meeresströmungsturbinen-Pilotanlage vor der englischen Küste. In: Tagungsband *Kassler Symposium Energie-Systemtechnik* 2004

[Böh62] Böhler, Karl: *Pumpspeicherkraftwerk Vianden*. Sonderdruck aus Die Wasserwirtschaft, Heft 12/1961 and Heft 1/1962. Stuttgart: Franckh'sche Verlagshandlung

[Brö00] Brösicke, Wolfgang: *Sonnenenergie*. Berlin: Verlag Technik, 2000

[EIA12] US Energy Information Administration (EIA): *International Energy Statistics*. Internet: www.eia.gov/countries/, 2012

[Gie03] Giesecke, J.; Mosonyi, E.: *Wasserkraftanlagen*. Berlin: Springer, 2003

[Gra01] Graw, Kai-Uwe: *Nutzung der Tidenenergie.* Universität Leipzig, Grundbau und Wasserbau, 2001

[Ita04] Itaipu Binacional (Hrsg.): *Itaipu Binacional Technical Data.* Internet: www.itaipu.gov.br, 2004

[Kön99] König, Wolfgang (Hrsg.): *Propyläen Technikgeschichte.* Berlin: Propyläen Verlag, 1999

[LGRP] Landesamt für Gewässerkunde Rheinland-Pfalz (Hrsg.): *Deutsches Gewässerkundliches Jahrbuch Rheingebiet.* Mainz: Landesamt für Gewässerkunde, verschiedene Jahrgänge

[LUBW] Landesamt für Umweltschutz Baden-Württemberg (Hrsg.): *Deutsches Gewässerkundliches Jahrbuch, Rheingebiet Teil I.* Karlsruhe: Landesamt für Umweltschutz, verschiedene Jahrgänge

[Qua09] Quaschning, Volker: *Erneuerbare Energien und Klimaschutz.* Hanser Verlag München, 2009

[Raa89] Raabe, Joachim: *Hydraulische Maschinen und Anlagen.* Düsseldorf: VDI-Verlag, 1989

[Vat03] Vattenfall Europe (Hrsg.): Wasserkraft Goldisthal – Aus Wasser wird Energie. Berlin: 2003

[Voi08] Voith Siemens Hydro Power Generation (Hrsg.): *Turbinen.* Internet: www.vs-hydro.com, 2008

[VIK98] Verband der Industriellen Energie und Kraftwirtschaft e.V. VIK (Hrsg.): *Statistik der Energiewirtschaft.* Essen, verschiedene Jahrgänge, letzter Jahrgang 1996/97, erschienen 1998

[Wik12] Wikipedia: *Gezeitenkraftwerk Shiwa-ho.* de.wikipedia.org/wiki/Gezeitenkraftwerk_Sihwa-ho, 2012

Chapter 8

[Aue11] Auer, Falk; Schote, Herbert: Ein großer Beitrag zum Klimaschutz. In: *Sonnenenergie* 6–2011, pp. 32–34

[Bit03] Bitzer Kühlmaschinenbau GmbH (Hrsg.): *Kältemittel-Report.* Sindelfingen, 2004

[Fri99] Frischknecht, Rolf: *Umweltrelevanz natürlicher Kältemittel.* Bern: Bundesamt für Energie, 1999

[Lok07] Lokale Agenda-Gruppe 21 Energie/Umwelt in Lahr (Hrsg.): *Leistungsfähigkeit von Elektro wärme pumpen.* Zwischenbericht. Lahr, 2007

[Pas03] Paschen, H.; Oertel, D.; Grünwald, R.: *Möglichkeiten geothermischer Stromerzeugung in Deutschland.* Karlsruhe: Büro für Technikfolgen-Abschätzung TAB, Arbeitsbereicht No. 84, 2003

[Qua09] Quaschning, Volker: *Erneuerbare Energien und Klimaschutz.* Hanser Verlag München, 2009

[Sch02] Schellschmidt, R.; Hurter, S.; Förster, A.; Huenges, E.: Germany. – In: Hurter, S.; Haenel, R. (Hrsg.): *Atlas of Geothermal Resources in Europe.* Office for Official Publications of the EU, Luxemburg, 2002

[VDI4640] Verein Deutscher Ingenieure VDI (Hrsg.): VDI 4640. *Thermische Nutzung des Untergrunds.* Düsseldorf: VDI Verlag, 2008

Chapter 9

[Abs04] Absatzförderungsfonds der deutschen Forst und Holzwirtschaft (Hrsg.): *Pelletsheizungen – Technik und bauliche Anforderungen.* Bonn: Holzabsatzfonds, 2004

[DGS04] Deutsche Gesellschaft für Sonnenenergie, DGS (Hrsg.): *Leitfaden Bioenergieanlagen.* München: DGS, 2004

[DIN11] Deutsches Institut für Normung e.V. (DIN): DIN EN 14961–2. *Feste Biobrennstoffe – Brennstoff spezifikationen und-klassen – Teil 2: Holzpellets für nichtindustrielle Verwendung.* Berlin: Beuth Verlag, 2011

[Fac96] Fachinformationszentrum Karlsruhe (Hrsg.): *Biomasse, Energetische Nutzungs möglich keiten.* BINE Projekt Info-Service No. 9/1996

[FNR08a] Fachagentur Nachwachsende Rohstoffe e.V., FNR (Hrsg.): *Biokraftstoffe Basisdaten Deutschland.* Gülzow: FNR, 2008

[FNR08b] Fachagentur Nachwachsende Rohstoffe e.V., FNR (Hrsg.): *Biogas Basisdaten Deutschland.* Gülzow: FNR, 2008

[Kal03] Kaltschmitt, M.; Merten, D.; Fröhlich, N.; Moritz, N.: *Energiegewinnung aus Biomasse: Externe Expertise für das WBGU-Hauptgutachten 2003.* Internet: www.wbgu.de/wbgu_jg2003_ex04.pdf

[Kle93] Kleemann, M.; Meliß, M.: *Regenerative Energiequellen.* Berlin: Springer, 1993

[Qua09] Quaschning, Volker: *Erneuerbare Energien und Klimaschutz.* München: Hanser Verlag, 2009

Chapter 10

[Dre01] Dreier, T.; Wager, U.: Perspektiven einer Wasserstoff-Energiewirtschaft. In: *BWK* Vol. 53 (2001) No. 3, pp. 47–54.

[Fac90] Fachinformationszentrum Karlsruhe FIZ (Hrsg.): *Wasserstoff – Ein Energieträger und Speicher für die Zukunft.* BINE Projekt Info-Service No. 8/1990

[Hüt10] Hüttenrauch, Jens; Müller-Syring, Gert: Zumischung von Wasserstoff zum Erdgas. In: *Energie Wasser Praxis* 10/2010, pp. 68–71

[Qua09] Quaschning, Volker: *Erneuerbare Energien und Klimaschutz.* Hanser Verlag München, 2009

[Ste11] Sterner, Michael; Jentsch, Mareike; Holzhammer, Uwe: *Energiewirtschaftliche und ökologische Bewertung eines Windgas-Angebotes.* Gutachten des Fraunhofer IWES, Kassel, 2011

[Uba10] Umweltbundesamt (Hrsg.): *Energieziel 2050–100 % Strom aus erneuerbaren Quellen.* Dessau, 2010

[Win89] Winter, C.-J.; Nitsch, J. (Hrsg.): *Wasserstoff als Energieträger.* Berlin: Springer, 1989

Chapter 11

[Age12] Agentur für Erneuerbare Energien: *Bilanz positiv: Nutzen Erneuerbarer Energien überwiegt die Kosten bei weitem.* Internet: www.unendlich-viel-energie.de, 2012

[BAFA12] Bundesamt für Wirtschaft und Ausfuhrkontrolle (BAFA): *EnergieINFO.* Internet: www.bafa.de, 2012

[Bec00] Becker, Gerd; Kiefer, Klaus: Kostenreduzierung bei der Montage von PV-Anlagen. In: Tagungsband *15. Symposium Photovoltaische Solarenergie.* Banz, 2000, pp. 408–412

[Bmb12] Bundesministerium für Bildung und Forschung (BMBF, Hrsg.): *Forschung und Innovation in Deutschland 2006, 2008, 2010 and 2012.* Berlin, 2006 bis 2012

[Bmf12] Bundesministerium für Finanzen (BMF, Hrsg.): 19, 20, 21, 22 and 23. *Subventionsbericht, Bericht der Bundesregierung über die Entwicklung der Finanzhilfen des Bundes und der Steuervergünstigungen für die Jahre 2002 bis 2012.* Berlin, 2003, 2006, 2008, 2010, and 2012

[BMWi] Bundesministerium für Wirtschaft (BMWi, Hrsg.): *Wirtschaft in Zahlen.* Bonn, verschiedene Jahrgänge

[Bmw06] Bundesministerium für Wirtschaft (BMWi, Hrsg.): *Finanzplanung bis 2009.* Berlin, 2006

[Bro06] Brooks, Nick; Nicholls, Robert; Hall, Jim: *Sea Level Rise: Costal Impacts and Responses. Externe Expertise für das WBGU-Sondergutachten 2006.* Berlin: WBGU, 2006. Internet: www.wbgu.de

[BSW12] Bundesverband Solarwirtschaft (BSW): *Infografiken.* Internet: www.bsw-solar.de, 2012

[DEPV12] Deutscher Energie-Pellet-Verband e.V.: *Pellet Preisentwicklung.* Internet: www.depv.de, 2012

[Deu90] Deutscher Bundestag (Hrsg.): *Gesetz über die Einspeisung von Strom aus erneuerbaren Energien in das öffentliche Netz (Stromeinspeisegesetz) vom 7.12.1990.* BGBl. I, p. 2633

[Deu94] Deutscher Bundestag (Hrsg.): *Grundgesetz für die Bundesrepublik Deutschland.* Bonn, 1994

[Deu97] Deutscher Bundestag (Hrsg.): *Bericht der Bundesregierung über die Entwicklung der Finanzhilfen des Bundes und der Steuer ver güns ti gungen für die Jahre 1995 bis 1998.* Berlin, Bundestagsdrucksache 13/8420, 1997

[Deu99] Deutscher Bundestag (Hrsg.): *Bericht der Bundesregierung über die Entwicklung der Finanzhilfen des Bundes und der Steuervergünstigungen für die Jahre 1997 bis 2000.* Berlin, Bundestagsdrucksache 14/1500, 1999

[Deu06] Deutscher Bundestag (Hrsg.): *Bericht der Bundesregierung über die Entwicklung der Finanzhilfen des Bundes und der Steuervergünstigungen für die Jahre 2003 bis 2006.* Berlin, Bundestagsdrucksache 16/1020, 2006

[FÖS12] Forum ökologisch-soziale Marktwirtschaft FÖS (Hrsg.): *Was Strom wirklich kostet.* Berlin, FÖS, 2012

[Goe94] Goetzberger, Adolf: Wirtschaftlichkeit – Ein neuer Blick in Bezug auf Solaranlagen. In: *Sonnenenergie* 4/1994, pp. 3–5

[Hoh89] Hohmeyer, Olav: *Soziale Kosten des Energieverbrauchs.* Berlin: Springer, 1989

[Hoh91] Hohmeyer, Olav; Ottinger, Richard L. (Hrsg.): *External Environmental Costs of Electric Power.* Berlin: Springer 1991

[IZE95] Informationszentrale der Elektrizitätswirtschaft e.V. (IZE): Was kommt nach dem Kohlepfennig. In: *Stromthemen* 2/1995, pp. 1–2

[Kle97] Kleinkauf, W.; Durstewitz, M.; Hoppe-Kilpper, M.: Perspektiven der Windenergie-Technik in Deutschland. In: *Erneuerbare Energie* 4/97, pp. 11–15

[Mun12] Münchener Rück (Hrsg.): *NatCatService, Informationsplattform über Naturkatastrophen.* Internet: www.munichre.com, 2012

[Nit90] Nitsch, J.; Luther, J.: *Energieversorgung der Zukunft.* Berlin: Springer, 1990

[Sta12] Statistisches Bundesamt: *Verbraucherpreisindizes für Deutschland.* Internet: www.destatis.de, 2012

[Sta12b] Statistisches Bundesamt: *Daten zur Energiepreisentwicklung.* Internet: www.destatis.de, 2012

[VDI91] Verein Deutscher Ingenieure: VDI-Richtlinie 2067: *Berechnung der Kosten von Wärme versorgungs anlagen.* Düsseldorf: VDI-Verlag, 1991

[Wöh81] Wöhe, G.: *Einführung in die Allgemeine Betriebswirtschaftslehre.* München: Franz Vahlen, 1981

Index